DEVELOPMENTS IN SEDIMENTOLOGY 25A

DIAGENESIS IN SEDIMENTS AND SEDIMENTARY ROCKS

FURTHER TITLES IN THIS SERIES

1. *L.M.J.U. VAN STRAATEN, Editor*
DELTAIC AND SHALLOW MARINE DEPOSITS

2. *G.C. AMSTUTZ, Editor*
SEDIMENTOLOGY AND ORE GENESIS

3. *A.H. BOUMA* and *A. BROUWER, Editors*
TURBIDITES

4. *F.G. TICKELL*
THE TECHNIQUES OF SEDIMENTARY MINERALOGY

5. *J.C. INGLE Jr.*
THE MOVEMENT OF BEACH SAND

6. *L. VAN DER PLAS*
THE IDENTIFICATION OF DETRITAL FELDSPARS

7. *S. DZULYNSKI* and *E.K. WALTON*
SEDIMENTARY FEATURES OF FLYSCH AND GREYWACKES

8. *G. LARSEN* and *G.V. CHILINGAR, Editors*
DIAGENESIS IN SEDIMENTS

9. *G.V. CHILINGAR, H.J. BISSELL* and *R.W. FAIRBRIDGE, Editors*
CARBONATE ROCKS

10. *P. McL. D. DUFF, A. HALLAM* and *E.K. WALTON*
CYCLIC SEDIMENTATION

11. *C.C. REEVES Jr.*
INTRODUCTION TO PALEOLIMNOLOGY

12. *R.G.C. BATHURST*
CARBONATE SEDIMENTS AND THEIR DIAGENESIS

13. *A.A. MANTEN*
SILURIAN REEFS OF GOTLAND

14. *K.W. GLENNIE*
DESERT SEDIMENTARY ENVIRONMENTS

15. *C.E. WEAVER* and *L.D. POLLARD*
THE CHEMISTRY OF CLAY MINERALS

16. *H.H. RIEKE III* and *G.V. CHILINGARIAN*
COMPACTION OF ARGILLACEOUS SEDIMENTS

17. *M.D. PICARD* and *L.R. HIGH Jr.*
SEDIMENTARY STRUCTURES OF EPHEMERAL STREAMS

18. *G.V. CHILINGARIAN* and *K.H. WOLF*
COMPACTION OF COARSE-GRAINED SEDIMENTS

19. *W. SCHWARZACHER*
SEDIMENTATION MODELS AND QUANTITATIVE STRATIGRAPHY

20. *M.R. WALTER, Editor*
STROMATOLITES

21. *B. VELDE*
CLAYS AND CLAY MINERALS IN NATURAL AND SYNTHETIC SYSTEMS

22. *C.E. WEAVER* and *K.C. BECK*
MIOCENE OF THE SOUTHEASTERN UNITED STATES

23. *B.C. HEEZEN*
INFLUENCE OF ABYSSAL CIRCULATION ON SEDIMENTARY ACCUMULATIONS IN SPACE AND TIME

DEVELOPMENTS IN SEDIMENTOLOGY 25A

DIAGENESIS IN SEDIMENTS AND SEDIMENTARY ROCKS

EDITED BY

GUNNAR LARSEN

Department of Geology, University of Aarhus, Aarhus (Denmark)

AND

GEORGE V. CHILINGAR

University of Southern California, Los Angeles, Calif. (U.S.A.)

ELSEVIER SCIENTIFIC PUBLISHING COMPANY
AMSTERDAM — OXFORD — NEW YORK 1979

ELSEVIER SCIENTIFIC PUBLISHING COMPANY
335 Jan van Galenstraat
P.O. Box 211, 1000 AE Amsterdam, The Netherlands

Distributors for the United States and Canada:

ELSEVIER/NORTH-HOLLAND INC.
52, Vanderbilt Avenue
New York, N.Y. 10017

Library of Congress Cataloging in Publication Data

Larsen, Gunnar, 1928-
Diagenesis in sediments and sedimentary rocks.

(Developments in sedimentology ; 25A)
Edition for 1967 published under title: Diagenesis in sediments.
Bibliography: p.
Includes indexes.
1. Diagenesis. I. Chilingarian, George. II. Title. III. Series.
QE471.L33 1979 552'.5 78-23961
ISBN 0-444-41657-9

ISBN: 0-444-41657-9 (Vol.25A)
ISBN: 0-444-41238-7 (Series)

Printed in The Netherlands

Dedicated to
N.M. STRAKHOV, W.D. KELLER, N.B. VASOEVICH, R. SIEVER,
F.J. PETTIJOHN and W.C. KRUMBEIN
for their important contributions to the field of diagenesis

LIST OF CONTRIBUTORS

H.J. BISSELL
Brigham Young University, Provo, Utah (U.S.A.)

G.V. CHILINGAR
University of Southern California, Los Angles, Calif. (U.S.A.)

E.C. DAPPLES
Department of Geological Sciences, Northwestern University, Evanston, Ill. (U.S.A.)

E.T. DEGENS
Geologisch-Paläontologisches Institut, Universität Hamburg, Hamburg (Germany)

G. LARSEN
Department of Geology, University of Aarhus, Aarhus (Denmark)

K. MOPPER
University of Gothenburg, Gothenburg (Sweden)

M. TEICHMÜLLER
Geologisches Landesamt Nordrhein-Westfalen, Krefeld (Germany)

R. TEICHMÜLLER
Geologisches Landesamt Nordrhein-Westfalen, Krefeld (Germany)

K.H. WOLF
Jiddah (Saudi Arabia)

D.H. ZENGER
Department of Geology, Pomona College, Claremont, Calif. (U.S.A.)

CONTENTS

Chapter 1

INTRODUCTION—DIAGENESIS OF SEDIMENTS AND ROCKS

GUNNAR LARSEN and GEORGE V. CHILINGAR

The book *Diagenesis in Sediments* was published in 1967, almost a hundred years after introduction of the term "diagenesis" in geological literature, i.e., in 1868, in Von Guembel's major work *Geognostische Beschreibung des ostbayerischen Grenzgebirges.* Many decades passed, however, before research into diagenetic processes and products really got underway. It could be stated that up to the 1950's, the topic of diagenesis formed only a very minor part of geological research literature (see for example, Trask, 1951, and Süjkowski, 1958). Rapid development took place during the next fifteen years, as illustrated by the scope and contents of the above-mentioned work *Diagenesis in Sediments.*

The demand for that book was such that the need arose for a new and revised edition. Because of the scope of the subject and the proliferation of literature on the subject, it has been necessary to publish the new edition in two volumes. This reflects the growth which has occurred in the research into diagenetic phenomena since the publication of the first edition.

DEFINITION OF DIAGENESIS

Diagenesis can be defined as the changes which occur in the character and composition of sediments, beginning from the moment of deposition and lasting until the resulting materials (rocks) are either moved into the realm of metamorphism or become exposed to the effects of atmospheric weathering. Other, more restrictive, definitions can be found which, for example, consider that diagenesis ceases when the sediment becomes a sedimentary rock. For practical reasons, the latter definition is strongly recommended by one of the editors (G.V.C.), whose speciality is in the field of carbonate oil and gas reservoir rocks. This definition is used in the USSR and some other countries (see Table 1-I, obtained from the Russian translation by N.B. Vassoevich of the first edition of *Diagenesis in Sediments*), and is briefly discussed in this chapter.

According to the *Glossary of Geology and Related Sciences* (1962) sedimentation can in short be defined as "... that portion of the metamorphic cycle from the separation of the particles from the parent rock, no matter what its origin or constitution, to and including their consolidation into

TABLE 1-I

Classification of various stages in postsedimentation alteration of sedimentary rocks according to different Soviet authors (after Vassoevich, 1971, p. 14.)

Strakhov (1957)	Vassoevich (1957)	Kossovskaya, Logvinenko, Shutov (1957)	Strakhov and Logvinenko (1959)	Vassoevich (1962)	Vassoevich, Vysotskiy, Guseva, Olenin (1967, 1968)	Logvinenko (1968)
* diagenesis	lithogenesis: diagenesis	diagenesis	diagenesis	diagenesis	diagenesis (peat stage)	diagenesis
metagenesis: epigenesis	lithogenesis: catagenesis	epigenesis: initial	catagenesis	catagenesis: protocatagenesis	catagenesis: protocatagenesis (brown coal stage)	catagenesis: early catagenesis (coals B, *D*, *G*)
		epigenesis: deep burial		catagenesis: mesocatagenesis	catagenesis: mesocatagenesis (coalification stages *D*, *G*, *Zh*, *K* and *OS*)	catagenesis: late catagenesis (clinkering coals)
metagenesis: early metamorphism or protometamorphism	? ?	metagenesis: early	metagenesis or initial metamorphism	catagenesis: apocatagenesis (starting with lean coals)	catagenesis: apocatagenesis (lean coal stage *T*, semianthracites and anthracites)	metagenesis: early metagenesis (lean coals anthracites)
		metagenesis: late				
	metagenesis (actually metamorphism)	? regional metamorphism	?	metagenesis (regional metamorphism)	metagenesis: (regional metamorphism)	metagenesis: late metagenesis (ultra-anthracites, graphitized anthracites)
						regional metamorphism

* Lithogenesis

another rock . . .". This portion of the metamorphic cycle comprises a large number of richly varied types of events, controlled by factors such as: (a) physical and chemical processes, and (b) tectonic and morphological conditions in both the field of accumulation and in that of denudation. These events can be grouped into a number of more or less distinct stages, as for instance the following three groups:

(1) The disintegration and decomposition of the parent material, i.e., the effect of the mechanical and chemical weathering.

(2) The separation of the weathering products, i.e., the effect of the processes of erosion, transportation and deposition, leading to the formation of the many different types of sediments.

(3) The consolidation and cementation of the deposited material, i.e., the diagenetic processes.

It must be emphasized, however, that the above clear classification is at the same time artificial, because distinct systems and sharp boundaries are scarce in nature. For example, the development of typical weathering profiles (Goldich, 1938; Wahlstrom, 1948) evidently is the result of not only weathering processes but also of removal of weathered material, i.e., leaching of soluble salts starts during the weathering itself. Thus, the events which give rise to typical weathering profiles are transitional between the above mentioned groups 1 and 2. The boundary between groups 2 and 3, i.e., between deposition and diagenesis, also can hardly be drawn clearly. It can be stated that diagenesis does not start until the moment the material is deposited, for instance on the sea floor. At the same time, however, it must be added that the diagenetic processes at work between the newly formed sediment and the overlying sea water do not have to be fundamentally different from the processes already acting between the sea water and the particles suspended in it. The processes which comprise halmyrolysis can possibly be considered to represent the even transition between the processes of deposition and those of diagenesis. Strakhov (1953, p. 12; 1960) divided the history of sedimentary rock into the following three different and distinct stages:

(1) Sedimentogenesis, namely formation of sediment.

(2) Diagenesis, or transformation of sediment into sedimentary rock.

(3) Catagenesis, which is a long stage of secondary changes in already formed sedimentary rock.

In Strakhov's classification, which is based on recognition of different stages of transformation of sediments into hard rocks, diagenesis is considered as only one of the stages in this development. Thus, one can say that the concept of diagenesis is used by Strakhov in a "restricted sense".

Like many other authors, Williams et al. (1955), on the other hand, are using the concept of diagenesis in a more "broad sense". They include the catagenesis stage of Strakhov in diagenesis under the term of "late diagenesis", which represents a transition to metamorphism. The same point of view is held by Süjkowski (1958), who stated: "Diagenesis, after all, is but the introduction to metamorphism." Some authors go still further: for example, while speaking of the definition of the metamorphic zeolite facies, Coombs (1960) stated that this facies includes the products of not only conventional metamorphism but also those of hydrothermal activity and diagenesis. An attempt at drawing a borderline between diagenesis and metamorphism is found in the article by Fyfe et al. (1958): "Diagenesis of sandstones results in minor changes in the clay matrix and crystallization of cement minerals in the hitherto open pores. When the coarse clastic grains are also extensively involved in reaction so that the rock becomes substantially recrystallized, the process is classed as metamorphic." Thus, it is impossible in a pressure—temperature diagram to fix a boundary of universal validity between diagenesis and metamorphism, because some types of sediments are less stable than others and, consequently, during deep burial will cross the boundary faster than the more stable sediments.

It is important to mention here that the processes of diagenesis in subaerial environment and in shallow stable seas differ from those in subsiding basins, and were termed "exodiagenetic" by Shvetsov (1960). He also pointed out that in addition to dehydration, the coagulation of colloids, rapid growth of crystals (recrystrallization), formation of concretions, and preservation of textural properties of sediment (such as fissures and borings) are characteristic features of "exodiagenesis".

The scale of different stages of alteration of sediments is first developed for clays, because of (1) their most common occurrence (clay deposits constitute not less than half of the volume of the stratisphere), and (2) their sufficiently distinct response to changing conditions, which influence development of lithogenesis. Clays are especially good indicators of pressure. On the other hand, the most sensitive indicators of temperature are coals and coaly organic matter (kerogen), in general, which is widely distributed in sedimentary rocks including clayey deposits. Consequently, the most detailed subdivision of catagenesis is based on properties of organic matter, especially on reflectance and index of refraction of vitrinite.

The role of time (geologic scale) is also of utmost importance. For example, Kossovskaya and Shutov (1963) and Kopeliovich (1965) pointed out a considerable shift of the border between initial and deep-burial epigenesis (catagenesis) in deposits of various ages: (1) in Neogene deposits of the Apsheron peninsula, this border is situated at a depth less than 5 km; (2) in Mesozoic deposits of Siberia, at a depth of 2.5 km; (3) in Paleozoic deposits of the Russian Platform, at a depth of about 1.3—1.7 km; and (4) in Riffean deposits, at a depth of less than 1 km. The role of time on changes which occur in coaly organic matter was quantitatively evaluated during the past few years.

In 1922, A.E. Fersman (in: Vassoevich, 1971, p. 12) introduced two new terms, *hypergenesis* and *catagenesis*, into geochemical literature. The first term rapidly gained popularity in the USSR, whereas the latter became popular in the USSR only in the 1950's as a result of the efforts of N.B. Vassoevich. Fersman recognized: (1) *syngenesis:* formation of sediment, (2) *diagenesis:* changes of sediment, and (3) *catagenesis*: transformations of sedimentary rocks until beginning of deep-burial metamorphism (or *hypergenesis*).

In 1934, M.S. Shvetsov (in: Vassoevich, 1971, p. 12) proposed to distinguish between *initial diagenesis*, or diagenesis of *sediments*, and *late diagenesis*, or diagenesis of *rocks*. In 1940, L.V. Pustovalov (in Vassoevich, 1971, p. 12) recognized five main stages in the history of sediment deposits: (a) destruction of source rocks; (b) transportation of products of disintegration of rocks; (c) sedimentation (accumulation) of sediments; (d) syngenesis, or early diagenesis; and (e) epigenesis, or late diagenesis. Vassoevich (1971, p. 12) objects against using the term epigenesis, because it was used in studies of ore deposits to designate *all* secondary processes and/or their products. Reference is made in the literature to *epigenetic* minerals, textures, and deposits in contrast to primary *syngenetic* ones.

Strakhov (1953) proposed the term *sedimentogenesis* to designate the first stage of formation of sedimentary rocks, including formation of loose sediment at the earth's surface as a result of reworking (weathering), transportation, and deposition of previously existing mineral masses. In 1956, Strakhov described in detail the diagenetic stages involving the following processes: (1) mineral formation in oxidizing environment in unstable components of sediments; (2) mineral formation in a reducing environment in the unstable components of sediments; and (3) redistribution of authigenic minerals and appearance of concretions together with local compaction of sediments.

In 1957, the term *metagenesis* was proposed almost simultaneously in three published works. Unfortunately, however, the definition of this term was not the same in all three cases. Vassoevich (1957), in using it, referred to metamorphism, whereas A.G. Kossovskaya, N.V. Logvinenko and V.D. Shutov (in: Vassoevich, 1971, p. 14) defined metagenesis as early metamorphism, prior to regional metamorphism. Strakhov (1957, in: Vassoevich, 1971, p. 14) agreed with the latter authors as far as the lower boundary of metagenesis is concerned, but included catagenesis in this term.

Strakhov and Logvinenko (1959) distinguished the following stages of rock formation: hypergenesis, sedimentogenesis, diagenesis, catagenesis, metagenesis, etc. According to them, based on tradition, the terms progenesis, syngenesis, and epigenesis can be used loosely to designate the time and stage of successive rock-forming processes, without definite boundaries (sliding scale). (See Table 1-I.)

It has not been the wish of the editors to impose a particular definition on the contributing authors. As in the first edition, the intention has been to describe the processes and conditions which prevail during the period covered by the broad definition of diagenesis, regardless of the terminology used.

No matter how this phenomenon is defined, diagenetic alteration can be characterized as the reaction of a sediment or sedimentary rock to its physicochemical environment. The most important factors involved are the Eh and pH of the environment, the concentration of various anions and cations, pressure, and temperature.

In Chapter 2 of the first edition, Professor R.W. Fairbridge presented a comprehensive model of the sequence of diagenetic events within the broad spectrum of the term diagenesis (Fairbridge, 1967). His elegant approach is briefly presented here.

Taking deposition in a normal marine environment as a starting point, with sufficient water circulation to provide oxidizing conditions on the sea floor, Fairbridge's model describes diagenesis as a cyclic process in which the main events are referred to as *syndiagenesis*, *anadiagenesis*, and *epidiagenesis*.

For a given sedimentary particle, *syndiagenesis* commences at the moment of deposition on the sea floor. The processes which operate on the newly deposited particle are similar to the halmyrolytic processes which affected it during transport and sedimentation in the marine environment. *Halmyrolysis* can thus be considered as the link between the sedimentation and the first stage of diagenesis (syndiagenesis).

Syndiagenesis is subdivided into two stages: the initial stage and the early-burial stage. During the initial stage, oxygen is present in the sediment pore fluids. Oxygen allows organic activity, e.g., that of aerobic bacteria, the essential nutrient of which consists of organic substances present in the sediment. The aerobic activity of bacteria leads to the production of CO_2 with a consequent decrease in the pH of the pore fluids. The environment, therefore, becomes hostile towards $CaCO_3$, which may be present as calcareous shells in the sediment. As a result of the aerobic activity, the oxygen content is eventually used up. Consequently, oxygen disappears at a certain depth below the sediment—water interface. This boundary (Eh = 0) defines the transition from the initial to the early-burial stage. Obviously, in stagnant marine basins, having oxygen-free bottom water, the initial stage is not present.

Reducing conditions prevail during the early-burial stage, with active population of anaerobic bacteria. The pH increases as a result of this activity, and the calcareous material, which survives the initial stage of syndiagenesis, is protected against further dissolution. A characteristic product forming during the early burial stage is FeS_2. The lower limit for the occurrence of anaerobic bacteria marks the boundary between syndiagenesis and anadiagenesis.

The most characteristic feature of the syndiagenetic stage is, thus, the biological activity. The resulting bacterial effects produced on the sediments are largely controlled by the chemical conditions brought about by this activity.

Anadiagenesis comprises the compaction and maturation stages of diagenesis. In contrast to syndiagenesis, which operated in a restricted depth range and over a geologically short time interval, anadiagenesis occurs down to a great depth (e.g., 10 km) and operates over a long period of time (e.g., 10^8 yr). A negative Eh and moderately alkaline to high-pH environments are typical of anadiagenesis. The controlling factors in anadiagenetic processes are the increasing pressure and temperature with depth. The pressure of the overlying sediments results in compaction and simultaneous expulsion of pore fluids. A large-scale migration of pore fluids through sedimentary sequences results from compaction. The migrating solutions may react chemically with the sediments and rocks resulting in either dissolution or precipitation of various minerals. Structural closures resulting from the compaction can be accentuated by cementation processes. Compaction, dehydration, and cementation, which are the major features of anadiagenesis, can all be considered as the effects of pressure increase with depth. As mentioned previously, whereas clay sediments are sensitive indicators of pressure, organic matter, including both coal and other organic materials (e.g., kerogen), which may occur in a dispersed form in, for example, a clay matrix, is particularly sensitive to changes in temperature. The effect of temperature is mainly responsible for the maturation of carbonaceous rocks and hydrocarbons, whereas the migration of hydrocarbons is a result of the pressure effect.

As mentioned above, anadiagenesis is typically of long duration and, theoretically, can last indefinitely. In some instances anadiagenesis may grade into metamorphism. The question of the boundary between diagenesis and metamorphism, as mentioned before, has been discussed by numerous investigators, e.g., Von Engelhardt (1967) stated: "In the diagenetic zone, the sediments contain interstitial fluids as continuous phases which can be moved by normal flow. These fluids (most commonly aqueous solutions) are, therefore, always involved in the diagenetic reactions, the transport of substances occurring through flow and normal diffusion. Because of the intercommunication and mobility of the interstitial fluids, reactions of the open-system type prevail. Some sediments, as, for instance, fine-grained limestones, can already loose all of their porosity at a very shallow depth. Crystallizations and phase transformations in such dense rocks at moderate temperatures and pressures may be called diagenetic, if other closely associated sediments still have an intercommunicating porosity." . . . "The diagenesis stops at a depth where in all sedimentary rocks the intercommunicating pore spaces have been closed up by physical or chemical processes. In this zone the metamorphism begins, its reactions taking place in the solid state or by diffusion at grain boundaries."

Typically, anadiagenesis occurs in basins during subsidence. In cases where subsidence is replaced by uplift, the sediments may become exposed to the influence of the circulating fresh groundwater, which owes its origin to rain water. The environment during the diagenetic stage in the groundwater (fresh) zone, which is termed epidiagenesis by some geologists, is often characterized by the presence of certain amounts of O_2 and CO_2, and, therefore, can be aggressive towards a series of components formed during anadiagenesis. Consequently, traces of the earlier stages of diagenesis can be camouflaged or destroyed during epidiagenesis. Upwards, epidiagenesis is replaced by weathering, with the groundwater table being the practical boundary between them. During the transition stage from epidiagenesis to weathering, a new cycle of sedimentation commences.

SOME FEATURES OF DIAGENESIS

The processes of diagenesis, even in the above mentioned "restricted sense" of Strakhov (1953, 1960), go on in the marine sediments for hundreds of thousands of years after their deposition (Zaytseva, 1954; cf. Chilingar, 1958). The processes of diagenesis include: (1) formation of new minerals; (2) redistribution and recrystallization of substances in sediments; and (3) lithification.

The diagenetic minerals include sulphides and carbonates of Fe^{2+} and Mn^{2+}, namely siderite, ankerite, rhodochrosite, oligonite, pyrite, marcasite, alabandine, etc. Strakhov et al. (1954, p. 577) pointed out that their diagenetic origin is evidenced by the fact that Mn and Fe in these minerals are in a bivalent form, which could have formed in a reducing environment that existed in the sediments. Typical diagenetic minerals also include silicates of iron (and manganese), in which iron is present in a bivalent form; all leptochlorites belong in this category. Dolomite and ankerite also form during diagenesis. Many clay minerals are of diagenetic origin, such as montmorillonite (from volcanic ash), beidellite, mountain leather, and series of zeolites (mordenite, chabazite, phillipsite, and others). Illite can also form during diagenesis. The zonation of new mineral formations in Recent seas and in Cenozoic and Mesozoic basins is presented in Fig. 1-1.

The upper 1—2 m of Recent sediments usually reveal two early stages of

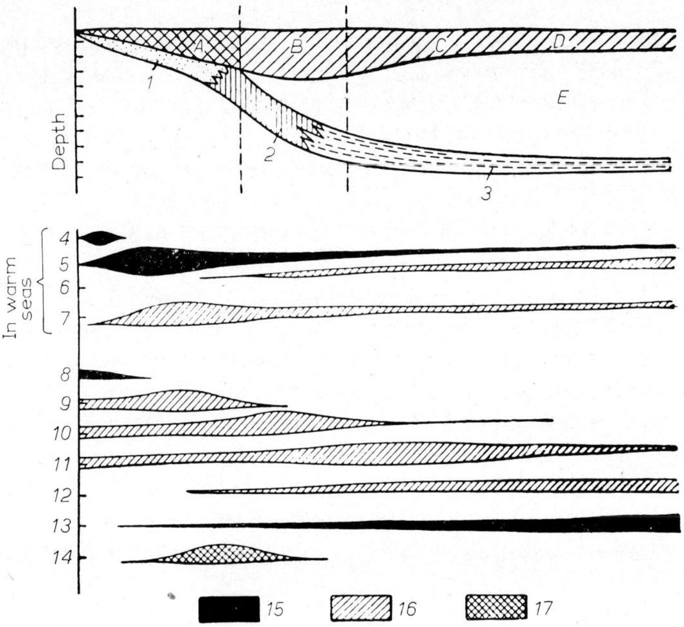

Fig. 1-1. Zonation of new mineral formations in Recent and ancient seas. (After Strakhov, 1954, p. 585.) *A* = turbid zone of fine-grained material, and its carrying out from near-shore zone into more central parts of the basin; *B* = areas of currents usually of circulatory type; *C-D* = surface zone of agitation and wind currents in central parts of basins; *E* = deep, quiet (with very slight movements of water) horizons of pelagic part of basins. *1* = sands; *2* = siltstones; *3* = pelites; *4* = $CaCO_3$, oölites; *5* = biogenous and chemically precipitated $CaCO_3$; *6* = diagenetic $CaCO_3$ (bacterial); *7* = various forms of diagenetic dolomite; *8* = Fe_2O_3, oxides of Mn, Al_2O_3; *9* = leptochlorites; *10* = glauconite; *11* = carbonates of Fe and Mn (in muds without $CaCO_3$ or having very small $CaCO_3$ content); *12* = sulfides of Fe, Mn (Cu, etc.) in muds with high $CaCO_3$ content; *13* = biogenically formed SiO_2; *14* = primary and diagenetic phosphorites; *15* = minerals forming through direct precipitation from water; *16* = diagenetic minerals; *17* = partly primary, partly diagenetic minerals.

diagenesis. The earliest stage occurs when the upper layer of sediment is situated in the oxidizing or neutral environment. In basins with normal oxygen regime the thickness of this layer is around 10—15 cm and can reach 40 cm and higher (Strakhov et al., 1954, p. 578). In basins deficient in oxygen this layer is only a few centimeters (and sometimes a few millimeters) thick, or is completely absent in the central portions of such basins. The duration of this stage also varies from thousands of years to several days. Some iron—manganese concretions and crusts, glauconite grains, phosphorites, and some zeolites form during this stage.

The second stage of early diagenesis is observed in lower parts of cores (up to 10 m) obtained in Recent sediments, and is characterized by the reduction of sulphates, oxides of iron, manganese, etc., and formation of new minerals containing lower(-ous) forms of these elements. The variation in the ratio of Fe^{2+}/Fe^{3+} with different rH_2 and Eh values has been studied in detail by Romm (1950).

Only the first two stages of diagenesis can be studied in detail from cores obtained in Recent sediments; and diagrams such as those prepared by Bruevich and Vinogradova (1947) are of great help in studying diagenesis (see Larsen and Chilingar, 1967, fig. 2). The third stage is apparently characterized by almost complete termination of bacterial activity (and their ferments), due to accumulation of poisonous compounds and, in some cases, disappearance of organic matter. During the third stage of diagenesis there is redistribution of newly formed minerals, formation of concretions and local cementation, and recrystallization of previously formed minerals.

The fourth stage of diagenesis involves transformation of plastic sediment into rigid compact rock (lithification). The squeezing out of interstitial water

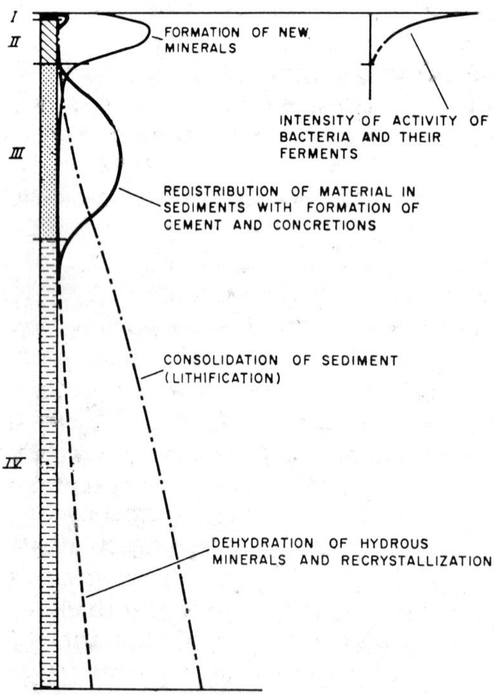

Fig. 1-2. Diagenetic stages in sediments. (After Strakhov et al., 1954, p. 596.)

occurs to a depth of about 300 m (Strakhov et al., 1954, p. 595). Dehydration of minerals occurs during compaction. For example, gypsum changes into anhydrite, hydrogoethite into hydrohematite and hematite, opal into quartz, etc. Fine-grained clays are recrystallized into coarser grains and aggregates, etc., during this stage. The different stages of diagenesis are presented in Fig. 1-2.

One of the earliest diagenetic processes involves consumption of free oxygen by organisms, after which the reduction of hydroxides of Fe^{3+}, Mn^{4+}, V, Cr, etc., and sulphates (SO_4^{2-}) begins. The environment changes from oxidizing to reducing and Eh becomes lower, whereas pH after some initial lowering usually increases (Strakhov, 1960, p. 79). The solid phases present in the sediment (such as SiO_2, $CaCO_3$, $MgCO_3$, $SrCO_3$, etc.) gradually dissolve in interstitial waters. Base exchange occurs between the cations adsorbed on clays and those in the interstitial water. At the same time the organic matter decomposes forming gases (CO_2, H_2S, H_2, N_2, NH_3, etc.) and water-soluble compounds, and some complex compounds which remain as solids in the sediment. As a result of these processes the interstitial water becomes devoid of sulphates, and enriched in Fe^{2+}, Mn^{2+}, SiO_2, organic matter, phosphorous, and minor elements. The O_2 disappears and H_2S, CH_4, CO_2, NH_3, H_2, etc., accumulate instead. The alkalinity becomes high, Eh sharply decreases (—150 to —330) and pH varies from 6.8 to 8.5 (Strakhov, 1960, p. 80). A pronounced exchange of substances occurs between the bottom waters and interstitial waters at this stage. That is why such components as S and Mg are found in higher concentrations in interstitial waters than in originally buried sea water. The eventual saturation of interstitial waters with some components leads to precipitation of diagenetic minerals such as leptochlorites, siderite, rhodochrosite, and sulfides of iron, lead, zinc, etc. The variation in Eh, pH and concentration of various ions in different areas of sediments results in subsequent redistribution (lenses, concretions, etc.) of authigenic minerals.

According to Strakhov (1960, p. 81), the depth at which diagenesis ceases varies from 10—50 m to 200—300 m. It obviously depends on the degree of compaction and the closing off of supercapillary, capillary and subcapillary pores. Consequently the diagenetic processes practically cease on reaching a certain degree of lithification. It is important to note here that lithification of carbonate rocks is accomplished more rapidly than that of other sediments.

Compaction, which is discussed in detail by Rieke and Chilingarian (1974), is very important in the case of clayey sediments. N.B. Vassoevich (1960, in: Klubova, 1965, p. 64) recognized four distinct stages of compaction occurring (1) with ease, (2) with difficulty, (3) with considerable difficulty, and (4) with great difficulty. For clays, the porosity varies from 60 to 85%

during the first stage; and from 35 to 45% during the later stages. The porosity depends not only on the overburden pressure, but also on the mineralogical composition of clays, chemistry of interstitial solutions, etc. Relationship between the remaining moisture content and overburden pressure is presented in Fig. 1-3.

According to Rukhin (1961, p. 307), the average porosity of clayey sediments toward the end of diagenesis (and beginning of epigenesis) is around 40%. This figure, however, can be considerably modified depending on the relative proportions of different clays present (Chilingar and Knight, 1960; Chilingar et al., 1963). Possibly, at a depth of 400 m the sediments can be considered as fully converted to rocks; epigenetic stage sets in at greater depths. Future research work on compaction of sediments will shed more light on this subject (see Rieke and Chilingarian, 1974, and Chilingarian and Wolf, 1975, 1976).

The sedimentary rocks most susceptible to epigenesis and metamorphism are coals. Brown coals are usually encountered at a depth of 1,000 to 1,200 m. They are associated with plastic clays having specific gravity of

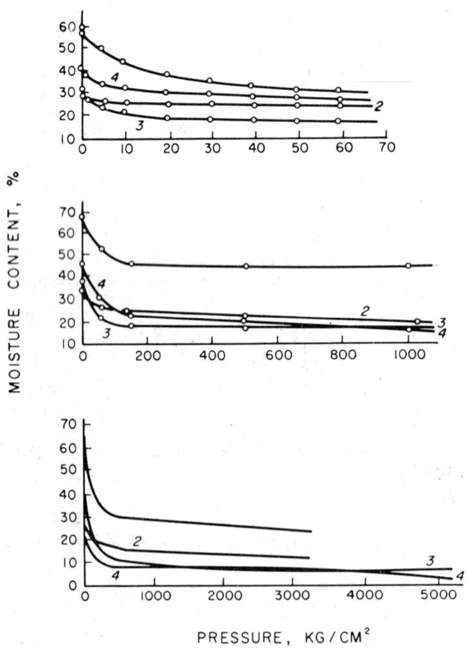

Fig. 1-3. Relationship between moisture content (%) of clays and overburden pressure (kg/cm^2). (After V.D. Lomtadze, 1953; in: Klubova, 1965, p. 56.)

1.4—1.9. The coking coals of the Donetz basin were formed at a depth of 5,000 m, temperature of 160°C and pressure of 1,100 atm (N.F. Balukhovskiy, 1952, in: Rukhin, 1961, p. 305). Anthracites, on the other hand, form at a depth of 8,000 m and a temperature of 240°C. Here again, there is no agreement among geologists as to the boundary between epigenesis (or very late diagenesis) and metamorphism.

Different aspects of diagenesis of various sediments and rocks are being discussed in other chapters of this book, namely: sands and sandstones by E.C. Dapples; silica by E.C. Dapples; organic matter by E. Degens and K. Mopper; coal by M. and R. Teichmüller; carbonate sediments and limestones by G.V. Chilingar, H.J. Bissell and K.H. Wolf; dolomites by G.V. Chilingar, D.H. Zenger, H.J. Bissel and K.H. Wolf.

SANDS AND SANDSTONES

The diagenesis of sands and sandstones is covered by E.C. Dapples in Chapter 2. Three main stages are considered in a progressive development of sandstones, namely, redoxomorphic, locomorphic and phyllomorphic stages.

This classification is comparable to that of R. Fairbridge (syndiagenesis and anadiagenesis) as discussed above. Sandstones can be classified according to texture into grain-supported arenites and matrix-supported wackes, with subwackes as an intermediate type. The diagenetic stage of sands and sandstones is determined by degree and type of crystallization of the interstitial clay, nature of the matrix, and type and amount of cementation. In the uppermost few meters of the deposits, redox processes are particularly active (redoxomorphic stage). The color which the sediment acquires during this stage tends to survive subsequent diagenetic processes. Hardening of the sediment occurs during the next locomorphic stage, as a result of the separation of mineral matter in the pores and the replacement of detrital mineral grains. The final (phyllomorphic) stage is characterized by the increasing crystallinity of illite, by the development of chlorite, and by the alteration of clay minerals to mica. The phyllomorphic stage can grade into metamorphism.

A few comments are added here by the editors concerning another important aspect of the diagenesis of sand deposits, namely, the changes in the composition of the heavy-mineral assemblage which can be made by "interstratal solutions". This topic has been discussed by Pettijohn (1941), Wieseneder (1953), and others. Wieseneder studied the relative stability of the heavy minerals during chemical weathering and diagenetic solution. The results are summarized in Fig. 1-4. Unstable initial material, dominated by hornblende, epidote and garnet, is shown on the left side of the figure. Of these, hornblende is clearly the least stable because it is rapidly removed by both weathering and diagenesis. Garnet is strongly affected by weathering, but is relatively stable during diagenesis, whereas the epidote shows the opposite tendency. The "metamorphic minerals" (staurolite, kyanite, sillimanite) are stable during weathering but not during intense diagenesis. If intense weathering is followed by intense diagenesis, up to 95% of the original heavy-mineral assemblage may be destroyed. There remains a highly

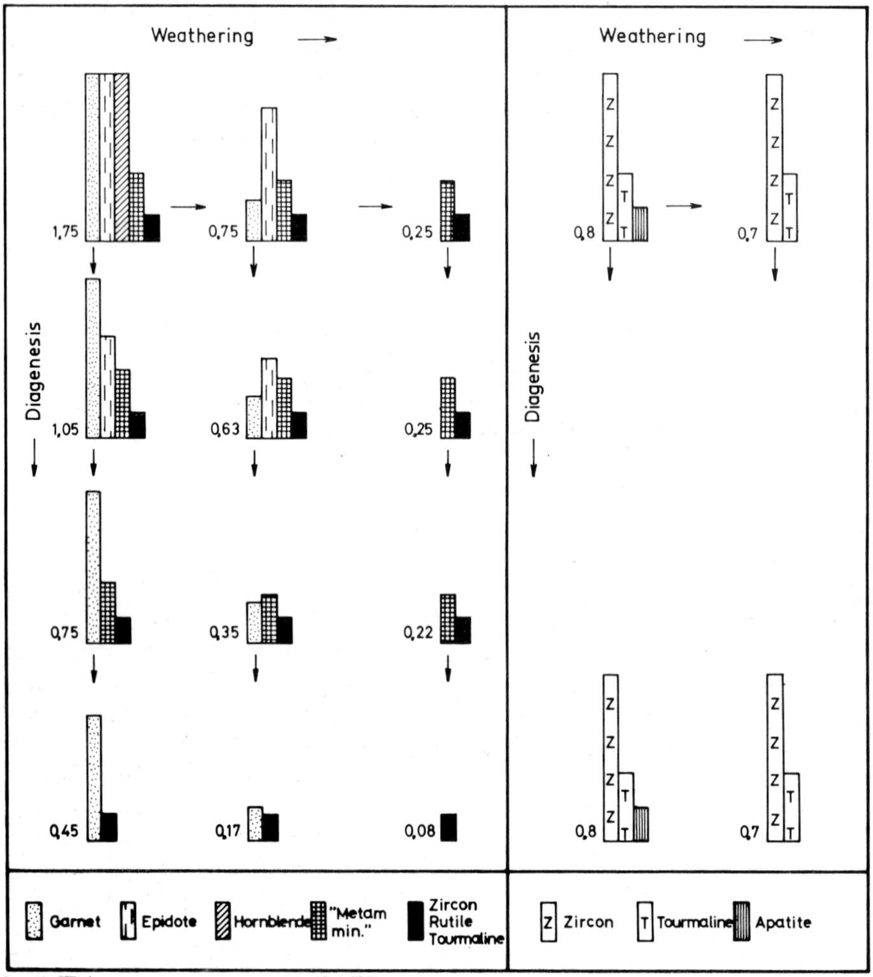

Fig. 1-4. Schematic illustration of the possible effect of weathering and diagenesis on the composition of heavy mineral assemblages. The numbers on the diagram give the percentages of heavy mineral present. (Modified from Wieseneder, 1953, p. 371.)

stable assemblage consisting of zircon, rutile and tourmaline. The occurrence of a stable assemblage does not necessarily mean that considerable solution has taken place, as illustrated to the right of Fig. 1-4. The extent to which a particular stable assemblage is an original association or represents a minor relict after destruction of the less stable minerals, can be investigated by studies of heavy-mineral contents, grain textures, and general petrography of the sediment. Consequently, the significance of diagenesis must be considered when a heavy-mineral analysis is used as a stratigraphic tool. According to Von Engelhardt (1967), both the diagenetic destruction of the heavy minerals and the formation of authigenic minerals in the matrix and cement of sandstones result from reactions between the sediment and the compaction fluids, which pass through the deposits as a result of compaction, particularly in clay-rich sediments. The role of compaction fluids during diagenesis has been discussed in detail by Rieke and Chilingarian (1974) and Chilingarian and Wolf (1975, 1976).

SILICA

In Chapter 3, E.C. Dapples discusses the role of silica in diagenesis. Attention is drawn to the fact that both laboratory experiments and studies of natural occurrences show that the solubility of silica increases with rising temperature. Solubility also increases considerably when pH value is above about 9.

Although the starting materials for the diagenetic silicification are mainly biogenic silica and volcanic glass, unstable silicates can also play an important role. The diagenetic product is typically chert that can occur in a variety of forms, which are clearly dependent on the composition and structure of the host rock. For example, chert in carbonate rocks occurs as concretionary bodies whose form and composition may clearly show that they are the result of a replacement process. Well-documented examples illustrate that the extent of the replacement process can depend on the structure and texture of the host rock (e.g., see Gry and Søndergaard, 1958).

The development of chert can be considered as a three-stage process. In the initial stage, the starting material is mobilized by reaction with the pore fluids which, typically, have high pH. During subsequent migration, possibly as compaction fluids, these solutions can react with other solutions having a lower pH. This may produce oversaturation with respect to silica which may be deposited, possibly contemporaneously with the solution of carbonates. This deposited silica is mainly amorphous, but may partly be ordered as cristobalite. The third stage is a maturation process during which the deposited silica changes to chalcedony or microcrystalline quartz.

This concept of silicification requires migration of the silica solution. The distance of migration is in many cases rather short, e.g., a few meters, but may be of a much larger order of magnitude. There is evidence for the migration of silica solutions from deeply buried geosynclinal sediments into adjacent cratonic deposits. Migration of this type is probably due to compaction of sediments.

ORGANIC MATTER

In Chapter 4, E.T. Degens and K. Mopper discuss the early diagenesis of sugars and amino acids in sediments.

The main part of organic matter in the earth's crust is found in sediments. It is estimated that the total amount of organic matter entrapped in sediments is approximately $3.8 \cdot 10^{15}$ metric tons. Of this, the overwhelming part ($3.5 \cdot 10^{15}$ metric tons) is present in shales as a finely disseminated substance called *kerogen*. Kerogen and other organic substances present in sediments have been developed during diagenetic alteration of different types of originally entrapped organic matter. Amino acids and sugars, which account for about 80—90% of the organic input into sediments, have been selected by Degens and Mopper as indicator elements for biogeochemical alteration processes during early diagenesis.

They concluded that early diagenesis is controlled by: (1) biological degradation at the sediment—water interface; (2) organic—metal ion complexation; (3) organic—mineral interaction; and (4) organic—organic condensations. Independent of environmental settings in terms of marine versus fresh water, or oxic versus anoxic, the pathway of biochemical degradation is essentially the same in all natural habitats; however, kinetics are different.

A reducing environment is a low-energy system, whereas an oxidizing environment is considered a high-energy one. Compared to the severe chemical alteration, which organic matter undergoes during the early stages of diagenesis, the effect of subsequent diagenetic stages on structure and composition of sedimentary organic matter is minor, unless the material is subjected to elevated temperatures (thermal degradation).

Degens and Mopper present an elegant hypothetical model of a structure of organic matter residue in Recent sediments.

COAL

Of the total amount of organic matter in the sediments, the coals are estimated to constitute only 1/500. In spite of this small proportion, the formation of coal is rather well understood, mainly because of its economic importance.

The subject of coalification is discussed by Marlies and Rolf Teichmüller in Chapter 5. Plant matter exhibits a sensitive response to temperature increases, which occur with the burial of peats and coals to greater depths. Criteria indicative of the rank of coal (moisture content, volatile-matter yield, carbon and hydrogen contents, and optical reflectivity) permit a better evaluation of different stages of diageneses than is possible by using other diagenetic parameters. This is particularly true because many sedimentary rocks contain fine coaly inclusions, the rank of which can be determined by microscopic reflectance measurements.

The rank of coal increases more rapidly with depth, the more rapidly the temperature rises during subsidence. Moreover, the duration of heating is of great importance. A strong heating of short duration can have the same effect as low heating over longer periods of time. Thus, the rank of coal is indicative of paleotemperature only if the duration of the heating is known. Although pressure may change the physical properties of coal (e.g., porosity and optical anisotropy), it retards the chemical reactions of the coalification process.

The "geochemical coalification" requires temperatures of more than 50°C. Bituminous coals normally are formed at temperatures of the order of 100—500°C. Anthracitization requires high temperatures, which seldom are attained during subsidence in areas with geothermal gradients of less than 3°C/100 m. Most anthracites and meta-anthracites occur in regions which have received additional heat from large intrusive igneous bodies at depth. Graphitization commonly takes place at temperatures of the order of 300—800°C and is accelerated by strong tectonic shearing stresses. Graphite formation from meta-anthracite corresponds to the beginning of metamorphism of sedimentary rocks other than coal.

The coalification processes, the different rank stages of coal formation, the geological causes of coalification, the relationship between coalification and bituminization (formation of petroleum and natural gas), and the relationship between coalification and diagenesis of sedimentary rocks other than coal, are all discussed in Chapter 5.

CARBONATE SEDIMENTS AND LIMESTONES [1]

As pointed out by G.V. Chilingar, H.J. Bissell, and K.H. Wolf in Chapter 6, entitled "Diagenesis of carbonate sediments and epigenesis (catagenesis) of

[1] The help extended by Dr. K.H. Wolf in writing this section, is gratefully acknowledged.

limestones", diagenesis of carbonate sediments and epigenesis (or catagenesis) of limestones comprise more than thirty different processes which are controlled by both local and regional factors, and which can alter the composition and texture of the sediments. Lithification is either physicochemical or biochemical, and the controversial beach rocks, for example, can be cemented by either process. Destructive processes include corrasion, corrosion, solution, decementation, and disintegration. Textural changes of limestones are often the product of inversion and several types of mechanisms, collectively termed "recrystallization". Under favorable conditions internal sedimentation and chemical infillings can form complex open-space structures (e.g., stromatactis). The sparry calcite is divisible into granular, drusy and fibrous types of which there are various genetic types and some reflect the conditions of formation. Of the processes that form micrite, "grain-diminution" in particular is a relatively new concept. The replacement of carbonates, which may be slight to extensive and is either by other carbonates (e.g., dolomitization, see Chapter 7) or by non-carbonate minerals (e.g., silicification and sulphatization), may or may not follow a predictable sequence.

In general, the paragenesis of carbonate sediments can be grouped into predepositional, syngenetic, and diagenetic processes, products and stages with numerous useful subdivisions related to Recent and ancient deposits.

The sum total of all the syngenetic through epigenetic features can be used as an indicator of environmental conditions, in particular where each stage left some recognizable evidence. Hence, diagenesis and epigenesis are of practical applicability in exploration.

The numerous possible diagenetic and epigenetic aspects discussed indicate that limestone classification schemes should not be used indiscriminately in environmental reconstructions.

The widely accepted division of sedimentary processes into syngenesis, diagenesis, and epigenesis (or catagenesis) may be inadequate for detailed research on carbonate rocks. The stages at which most of the individual processes and products occur cannot be distinctly demarcated. Two or more processes may be active simultaneously. They may overlap, or the termination of one may mark the commencement of another process; and still other alterations may occur independently in both space and time. Many of the interpretations depend on the scale — thin-section, handspecimen, or outcrop — at which observations are made. Hence, "pigeon-holing" of processes without contradictions is difficult, sometimes even impossible. Difficulties in genetic interpretations occur in particular in monomineralic rocks such as limestones. For these and other reasons it is not surprising that no general agreement has been reached on the definition and extent of diagenesis in carbonates (see review by Teodorovich, 1961).

Diagenesis has been restricted to those processes that cause lithification. Such a limited application, however, is arbitrary, artificial and impractical (Newell et al., 1953; Ginsburg, 1957). Not only are there several distinctly different lithification processes which are frequently difficult to recognize and separate, but they are so gradational as to defy precise definition and can occur at any stage during the early history of sediments. It is virtually impossible, therefore, to exclude other early alterations. According to Chilingar et al. in their Chapter 6, it is preferable to apply diagenesis in a wider sense to processes that affect a sediment after deposition and up to, but not beyond, lithification and/or filling of voids. Although these two processes can, and usually do, take place at different times within a sedimentary formation, especially if composed of different facies, the final stage of lithification and/or filling of voids appears to be the most convenient time at which diagenesis can be terminated. [1] Hence, the following rather all-inclusive definition, in general agreement with the concepts of Krumbein (1942) and Ginsburg (1957), has been adopted by Chilingar et al. Diagenesis includes all physicochemical, biochemical and physical processes modifying sediments between deposition and lithification at low temperatures and pressures characteristic of surface and near-surface environments. Postlithification processes grade into epigenesis, and epigenesis passes into metamorphism. [2] Epigenesis near the depositional environments is called juxta-epigenesis ("juxta-" meaning *near*), and epigenesis remote from the surface is named apo-epigenesis [3] ("apo-" meaning *far*, *remote*).

Most limestones have some small voids which have been partly or wholly filled by one or more generations of cement. Hence, it is possible to divide diagenesis into pre-, syn-, and postcementation stages.

In other paragenetic investigations, however, the diagenetic—epigenetic boundary may have to be based on some other criterion to be determined by the individual investigator concerned. No definite rule is possible. As long as the boundary is precisely defined by certain fabric or structural relations,

---

[1] This approach has been found to be of particular use in coarse-grained or open-textured limestones, but may be more difficult to apply in micritic rocks. Nevertheless, the paragenetic model used here is convenient, because after lithification (= infilling of voids by cement) of a rock the intrastratal fluids related to surface conditions cannot penetrate readily the rock framework. In cases where limestones maintain their porosity and permeability for a long period of time, even after being far removed from the original depocenter, the intrastratal fluids occupying the cavities can also be looked upon as either of syngenetic, diagenetic or epigenetic origin.

[2] Strictly speaking, epigenesis (or catagenesis) as defined here, passes into metamorphism only if an increase of pressure and/or temperature occurs.

[3] Weathering not related to the original depositional environment of the sediments, is not included in diagenesis and epigenesis as defined here.

little confusion should occur. A diagenetic—epigenetic boundary established on the basis of a few thin-sections of a local outcrop, however, may have to be revised and shifted up or down the paragenetic scale as soon as the petrologic and petrographic information of the whole formation is available, or it may be found that the termination of diagenesis in one area may be completely unrelated to that of other localities. Not all diagenetic stages are present in limestones. For example, precementation dolomitization may completely alter a limy deposit resulting in an elimination of syn- and post-cementation stages.

The raw material of diagenesis, as Krumbein (1942) called it, consists of organic and inorganic sediment of allochthonous and/or autochthonous origin, interstitial fluids, and other components subsequently formed or introduced into the system. In general, it is possible to subdivide the components that interact during the diagenetic processes into the following (Wolf, 1963b): (1) diagenetic—endogenic; (2) diagenetic—exogenic: (a) supergenic—exogenic; (b) hypogenic—exogenic.

This is merely an expansion of Amstutz' (1959) division: syngenetic—supergenic, syngenetic—hypogenic, epigenetic—supergenic, and epigenetic—hypogenic. In most cases, diagenesis derives its raw material from both endogenic (within the sediments) and exogenic—supergenic (outside source—from above) sources. One or the other may prevail. Under unusual conditions, however, a volcanic, i.e., exogenic—hypogenic, source may supply components for diagenesis without a marked increase in temperature. This would be particularly true for siliceous material introduced into a geosyncline (or other depocenter). During diagenesis of the sediments, quartzose arenites can be converted to orthoquartzites, carbonates can become siliceous, and fossils can be replaced prior to dolomitization.

Diagenesis may express itself in a number of different ways. Krumbein (1942) mentioned that a total of about thirty separate diagenetic processes have been described in the literature. They may result in mineralogical changes, addition and removal of material, and textural and structural modifications and alterations ranging from slight to extensive or complete. Several generations of diagenesis may each leave evidence, or each successive one may obliterate or destroy the products of earlier processes. In many cases, however, diagenesis may appear to be absent if only visually-obtained information is considered. [1]

[1] Note that diagenesis is *not* part of the petrographic (= descriptive) stage but belongs to the subsequent stage of petrology and petrogenesis (= interpretive). Reliable diagenetic reconstructions cannot be made, therefore, on a few local thin-sections, but must be based on as much geochemical, petrographic and stratigraphic information as circumstances permit to be obtained.

To ascribe to certain products merely a genetic term such as "precementation—diagenetic", without relating it to the sediment's history as a whole, invites criticism. It is more accurate to relate all processes and products to a paragenetic sequence. In other words, a paragenetic scheme furnishes less ambiguous information to cases where it seems impossible to define exact syngenetic—diagenetic—epigenetic boundaries. The absolute time of formation may be impossible to determine, but the textural and structural relationships permit the interpretation of relative time of formation. The widely used terms "primary" and "secondary" have very little meaning in diagenetic investigations of carbonates unless precisely defined, although they may be quite useful in a very general colloquial sense.

Paragenetic interpretations are relatively easy and non-controversial if the investigation is made on the scale of one thin-section or handspecimen. Syn-, dia-, and epigenetic processes, however, are not only gradational, and overlap in time and space on a microscopic scale, but especially do so on a regional scale. Regional diagenetic studies may be rather tedious and resemble structural analysis, for example, in that the micro-, meso-, and macroscopically examined features are assembled step by step.

The so-called predepositional (or presyngenetic) processes and products can be deduced from limestone rock fragments (= calclithite fragments of Folk, 1959; extraclasts of Wolf, 1965b), which are derived from other limestones that had undergone diagenesis, e.g., lithification, recrystallization, dolomitization, and silicification, before erosion and transportation.

Diagenetic processes and fabrics in carbonates

Almost 20 years ago, Bathurst (1958) described diagenetic fabrics in some British Dinantian limestones, utilizing 350 thin-sections to complement his petrologic studies of these Lower Carboniferous rocks as exposed mostly at Avon Gorge, North Wales, and Yorkshire. This was a pioneer effort in refining some concepts relating to some postdepositional changes of fabric in limestones. Bathurst (1958, pp. 11—36) pointed out that diagenetic fabrics (textures and structures) that were examined in thin-sections resulted from six processes that were responsible for the change from unconsolidated sediments to limestones. These are: (1) granular cementation and drusy growth; (2) rim cementation (secondary enlargement); (3) pressure solution; (4) grain growth sensu stricto; (5) mechanical deposition in cavities of postdepositional age; and (6) postdepositional formation of cavities by erosion and solution in a carbonate mud. Brief discussion of each of these six processes is warranted in the Introduction Chapter.

Granular cementation and drusy growth. Granular cementation and drusy growth consist of chemical deposition of material from solution onto a free surface causing the outward growth of crystalline material that adheres to the surface. In limestones, the growth of cement in pores of unconsolidated sand and the growth of drusy mosaic on the walls of larger cavities typify these processes. Cement grows between detrital particles, whereas drusy mosaic grows into all other cavities (for example, a shell chamber). Cement, as defined by Bathurst (1958, p. 15), is usually "granular" where the host particle, usually multigranular, lies in a mosaic of cement.

Rim cementation. Bathurst regards rim cementation as a special case of chemical deposition which he treats separately from granular cementation. In a detrital sediment, where each grain is a single crystral (such as a crinoid fragment), carbonate can be chemically deposited on the grain surfaces syntaxially. Intergranular pores are eventually filled and the resultant non-porous fabric is a mosaic of grains each of which has a detrital core partly encased in a rim of syntaxial cement. Accordingly, rim cementation is cementation by secondary enlargement.

Pressure solution. In applying the term pressure solution, Bathurst followed the usage of Pettijohn (1949, p. 481) to describe only the first part of the well-known process involving solution at contact surfaces followed by redeposition; it is essentially solution transfer. In any sediment composed of grains and interstitial solutions, crystal lattices of grains are usually elastically strained in the vicinity of an intergranular boundary and this strain is proportionally increased by the increasing weight of the overburden. Accordingly, greater strain is accompanied by increased solubility. Ions diffuse away from the exposed intergranular boundary to regions of lower concentration. This process is completed before the sediment pores have been closed by cementation. Final development of the junction between grains was the result of cementation after pressure solution had ceased. Bathurst did not discuss stylolites in this section, pointing out that they are the subject of a voluminous literature.

Grain growth. In discussing grain growth, Bathurst (1958, pp. 24—31) noted that many processes, previously and loosely referred to recrystallization, should be treated under this heading. Furthermore, he contends that during grain growth the process is restricted to the monomineralic fabrics of low porosity. Intergranular boundaries migrated causing some grains to grow at the expense of adjacent particles. Bathurst contends that the reaction takes place in the solid state, with ions being transferred from one lattice to another without solution. Larger grains tend to replace the smaller ones and a fine mosaic is gradually replaced by a coarser one. Also, as grain growth continues, many of the enlarged grains are themselves replaced by their more successful neighbors.

It was pointed out that direct evidence for syntaxial enlargement of original grains is shown only by the syntaxial rims. Thus, in examining a coarse mosaic that resulted from grain growth, one should be cognizant of the fact that it replaced a finer mosaic and that large grains must have evolved from smaller ones.

Internal mechanical deposition. Internal mechanical deposition involves deposition in secondary cavities that formed after the deposition of carbonate muds and silts, by bending of laminae, or by internal mechanical erosion or solution. This diagenetic process has been especially characteristic in reef knolls. Accordingly, evidence should point to earlier existence of cavities in the host rock which were formed after deposition of sediments and which were later wholly or partially filled with sediment; any remaining space should be filled with drusy mosaic (calcspar). It is possible that some of the internal sediments are mechanically deposited sediments; an homogeneous and non-fossiliferous nature would suggest such an origin. In calcite-mudstone of reef knolls, cavities are commonly floored with internal sediment that is overlain by drusy mosaic. This mosaic is the "reef-tufa" of some authors; Bathurst contends that the typical drusy fabric of "reef-tufa" is not simply recrystallized calcite-mudstone.

Postdepositional formation of micro-cavities. Bathurst (1958, p. 33) noted that various workers have demonstrated that postdepositional mechanical erosion and solution can produce cavities in fine-grained carbonate sediments. Filling of cavities, he stated, with both sediment and drusy mosaic, seemingly took place while the host rock was mechanically eroded; a certain amount was dissolved and supplied the solute needed for drusy growth. Furthermore, it is known that the movement of interstitial solution in compacting lime muds is commonly restricted to internal channels that may be only 1 or 2 mm in diameter. These channels, like stylolites, are eroded by expelled pore water that may carry fine-textured sediment in suspension.

Lucia (1962, pp. 848—865) studied Devonian crinoidal rocks in the Andrews South Devonian Field, Andrews County, Texas, proving that a clear relationship exists between the original sediment and its diagenetic history. He noted that the most obvious diagenetic process in the limestone is the formation of optically continuous calcite overgrowth on crinoid fragments. Individual plates of modern echinoderm skeletal material are made of optically oriented calcite crystals containing large interstices which become solid single crystals after death. In following the usage of Bathurst (1958), Lucia termed the pore-filling overgrowth *rim cement* and the replacement overgrowth *syntaxial rims.* Rim cementation, Lucia observed, is the dominant process. Rocks with visible porosity were originally particle-supported crinoidal lime sands with a small amount of interparticle lime mud. Lucia (p. 853) stated: "The presence of this small amount of lime mud inhibited the growth of the rim cement, and the lime mud was later selectively leached to form the visible porosity."

Neomorphism. The term "neomorphism" has been given various definitions; Folk (1965) discussed this process at some length. At the outset, one should be aware that neomorphism is not simply the process of pore-space filling, but, contrarily: older crystals must gradually be consumed and their place simultaneously occupied by new crystals of the same mineral or a polymorph. Accordingly, neomorphism involves both polymorphic transformation and recrystallization. Polymorphic transformation is Folk's term "inversion". Wet, in situ, transformation of aragonite is a common example of polymorphic transformation, whereas during the in situ recrystallization the mineralogy remains the same. Folk used the term "aggrading neomorphism" to identify the process, whereby a mosaic of finely crystalline carbonate is replaced by a coarser mosaic that is actually sparry calcite. This process proceeds in a wet state, and includes various processes of polymorphic transformation and recrystallization.

DOLOMITIZATION

The final Chapter 7 by G.V. Chilingar, D.H. Zenger, H.J. Bissell and K.H. Wolf is devoted to dolomitization. The authors of this chapter discussed: (1) "Chemistry of dolomitization", including dolomite synthesis; (2) "Sedimentology and petrology of diagenetic dolomitization", including presentation of some models; (3) "Early versus late diagenetic dolomitization", and (4) "Dedolomitization".

Chilingar et al. attempted to elucidate the "dolomite problem", which according to Zenger (1972) includes: (1) the question of its primary versus secondary origin, (2) the difficulty of its synthesis, and (3) the contrast between relative abundance of dolomite in the Upper Precambrian and Lower Paleozoic rocks and its rarity in the Holocene deposits. In addition, it is not known whether or not high Mg/Ca ratios of dolomitizing solutions are required for dolomitization. Whether high-pH or low-pH environment is conducive to: (a) diagenetic dolomitization, and (b) *precipitation* of primary dolomite, is being continuously debated.

Until 1956, Holocene dolomite was thought to be practically non-existent. This opinion was challenged by Chilingar (1953, 1956). Since 1957, however, there have been numerous discoveries of modern, penecontemporaneous dolomites, primarily in the supratidal zone. There are many analogues in the geologic record. This, unfortunately, resulted in an overgeneralization. In Chapter 7, Chilingar et al. stress the fact that a large number of dolomites lack evidence of being either supratidal accumulations or resulting from supratidal dolomitization. Considerable evidence suggests that dolomite need not be an evaporitic mineral.

The writers of Chapter 7 distinguish between the diagenetic dolomitization of *sediments* and epigenetic (or catagenetic, which is a term preferred by many geologists) dolomitization of *rocks*. The existence of *primary* dolomites, i.e., those precipitated directly from water either above or at the sediment—water interface, have been questioned by many geologists. G.V. Chilingar, however, strongly believes in the existence of such dolomites on the basis of isotope studies and other evidence (see Chilingar et al., 1967a, b). The issue of replacement versus primary dolomites remains significant and, in general, the "dolomite problem" remains unsolved. It should be pointed out here that very early diagenetic dolomites are termed "primary" by some

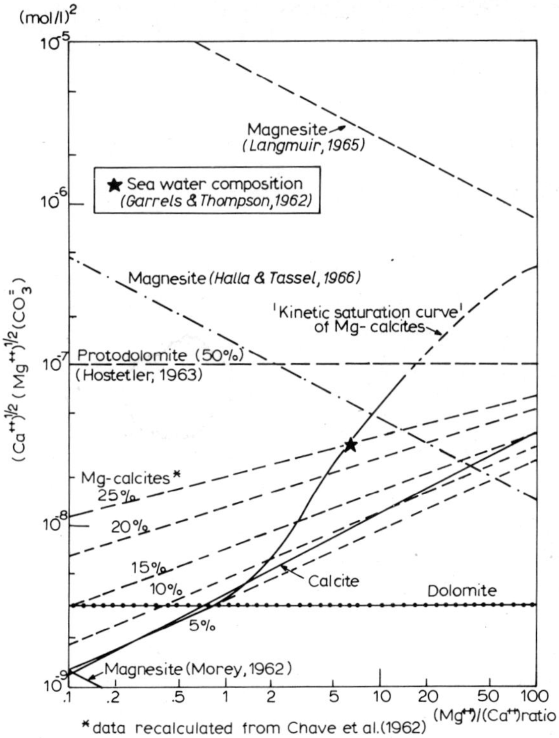

Fig. 1-5. Stability diagram for Mg-Ca carbonates. (After De Boer, 1977, fig. 1, p. 265; reproduced with permission of Pergamon Press.)

investigators. Many geologists have difficulty in distinguishing between the primary precipitates and the early replacement of lime ooze. Consequently, a loosely defined term "primary" exists in the literature for the so-called "primary" dolomites.

Studies on the stabilities of Mg—Ca carbonates have been for a long time a source of tremendous frustrations to geochemists. Recently, De Boer (1977) constructed a diagram in which the stability of Mg—Ca carbonates can be represented by only two parameters (Fig. 1-5). The relative stability of the various carbonates can be easily compared by using this diagram, which can be used as an introduction to the complex study of diagenesis of carbonates.

REFERENCES

Amstutz, G.C., 1959. Syngenese und Epigenese in Petrographie und Lagerstättenkunde. *Schweiz. Mineral. Petrogr. Mitt.*, 39: 2—84.

Amstutz, G.C. and Bubenicek, J., 1967. Diagenesis of sedimentary mineral deposits. In: G. Larsen and G.V. Chilingar (Editors), *Diagenesis in Sediments*. Elsevier, Amsterdam, pp. 417—475.

Bathurst, R.G.C., 1958. Diagenetic fabrics in some British Dinantian limestones. *Liverpool Manchester Geol. J.*, 2: 11—36.

Bruevich, S.V. and Vinogradova, E.G., 1947. Chemical composition of interstitial solutions of the Caspian Sea. *Gidrokhim. Mater.*, 13 (1/2).

Chichua, B.K., 1966. Towards question of determining stages and total volume of catagenetic changes in sedimentary rocks. *Sov. Geol.*, 3.

Chilingar, G.V., 1953. Use of Ca/Mg ratio in limestones as a geologic tool. *Compass*, 30: 29—34.

Chilingar, G.V., 1955. Review of Soviet literature on petroleum source rocks. *Bull. Am. Assoc. Pet. Geol.*, 39 (5): 764—768.

Chilingar, G.V., 1956. Relationship between Ca/Mg ratio and geologic age. *Bull. Am. Assoc. Pet. Geol.*, 40: 2256—2266.

Chilingar, G.V., 1958. Some data on diagenesis obtained from Soviet literature: a summary. *Geochim. Cosmochim. Acta*, 13: 213—217.

Chilingar, G.V. and Knight, L., 1960. Relationship between pressure and moisture content of kaolinite, illite, and montmorillonite clays. *Bull. Am. Assoc. Pet. Geol.*, 44 (1): 101—106.

Chilingar, G.V., Bissell, H.J. and Fairbridge, R.W. (Editors), 1967a. *Carbonate Rocks: Origin, Occurrence and Classification*. Elsevier, Amsterdam, 471 pp.

Chilingar, G.V., Bissell, H.J. and Fairbridge, R.W. (Editors), 1967b. *Carbonate Rocks: Physical and Chemical Aspects*. Elsevier, Amsterdam, 413 pp.

Chilingar, G.V., Rieke III, H.H. and Robertson Jr., J.O., 1963. Relationship between high overburden pressures and moisture content of halloysite and dickite clays. *Bull. Geol. Soc. Am.*, 74 (8): 1041—1048.

Chilingarian, G.V. and Wolf, K.H., 1975. *Compaction of Coarse-Grained Sediments, I.* Elsevier, Amsterdam, 552 pp.

Chilingarian, G.V. and Wolf, K.H., 1976. *Compaction of Coarse-Grained Sediments, II.* Elsevier, Amsterdam, 808 pp.

Coombs, D.S., 1960. Lower grade mineral facies in New Zealand. *Int. Geol. Congr., 21st, Copenhagen, 1960, Rep. Session, Norden,* 13: 339—351

De Boer, R.B., 1977. Stability of Mg—Ca carbonates. *Geochim. Cosmochim. Acta,* 41 (2): 265—270.

Emery, K.O. and Rittenberg, S.C., 1952. Early diagenesis of California Basin sediments in relation to origin of oil. *Bull. Am. Assoc. Pet. Geol.,* 36: 735—806.

Ginsburg, R.N., 1957. Early diagenesis and lithification of shallow-water carbonate sediments in South Florida. *Soc. Econ. Paleontol. Mineral., Spec. Publ.,* 5: 80—100.

Fairbridge, R.W., 1967. Phases of diagenesis and authigenesis. In: G. Larsen and G.V. Chilingar (Editors), *Diagenesis in Sediments.* Elsevier, Amsterdam, pp. 19—89.

Fyfe, W.S., Turner, F.J. and Verhoogen, J., 1958. Metamorphic reactions and metamorphic facies. *Geol. Soc. Am., Mem.,* 73: 259 pp.

Glossary of Geology and Related Sciences, 1962. American Geological Institute, Washington, D.C., 397 pp.

Goldich, S.S., 1938. A study in rock weathering. *J. Geol.,* 46: 17—58.

Gry, H. and Søndergaard, B. 1958. Flint forekomster i Danmark. *Com. Alkali Reactions Concrete, Copenhagen, Progr. Rep. D* 2: 63 pp.

Kaleda, G.A., 1969. About epigenetic changes of Paleozoic deposits of Russian Platform. *Lithol. Miner. Resour.,* 6.

Kashirtseva, N.F., 1970. *Methods of Studying Epigenetic Changes in Loose Sedimentary Rocks.* Nedra, Moscow.

Klubova, T.T., 1965. *Role of Clayey Minerals in the Transformation of Organic Matter and the Formation of Pore Spaces of Reservoirs.* Akad. Nauk SSSR, Moscow, 107 pp.

Kopeliovich, A.V., 1965. Epigenesis of ancient series in the south-western part of the Russian Platform. *Tr. Geol. Inst. Akad. Nauk SSSR,* 1965 (121).

Kossovskaya, A.G. and Shutov, V.D., 1963. Facies of regional epigenesis and metagenesis. *Izv. Acad. Nauk SSSR, Ser. Geol.,* 1963 (7).

Krumbein, W.C., 1942. Physical and chemical changes in sediments after deposition. *J. Sediment. Petrol.,* 12: 111—117.

Kryukov, P.A., 1970. *Rock, Soil and Mud Solutions. Tr. Inst. Neorganich. Khim. Sib. Otdel. Akad. Nauk SSSR,* 220 pp.

Larsen, G. and Chilingar, G.V., 1967. Introduction. In: *Diagenesis in Sediments (Developments in Sedimentology, 8).* Elsevier, Amsterdam, pp. 1—17.

Lisitsyn, A.K., Kondrat'eva, I.A. and Komarova, G.V., 1969. Methods of genetic interpretation of epigenetic changes of sedimentary rocks. *Lithol. Miner. Resour.,* 3.

Logvinenko, N.V., 1968. *Postdiagenetic Changes of Sedimentary Rocks.* Nauka, Leningrad.

Logvinenko, N.V. and Karpova, G.V., 1968. Stages of changes of coal deposits of Big Donbass. In: *Sediment Accumulation and Coal Strata of Carboniferous Age in USSR.* Nauka, Moscow.

Lucia, F.J., 1962. Diagenesis of a crinoidal sediment. *J. Sediment. Petrol.,* 32: 848—865.

Newell, N.D., Rigby, J.K., Fischer, A.G., Whiteman, A.J., Hickox, J.E. and Bradley, J.S., 1953. *The Permian Reef Complex of the Guadalupe Mountains Region, Texas and New Mexico.* Freeman, San Francisco, Calif., 236 pp.

Perozio, G.N., Kazankskiy, Yu. and Lizalek, N.A., 1967. Main factors of postdiagenetic transformations of sedimentary rocks of Siberia. In: *Postsedimentary Transformation of Sedimentary Rocks of Siberia.* Nauka, Moscow.

Petrov, V.P., 1967. *Principles of Studying Ancient Weathering Crusts.* Nauka, Moscow.

Pettijohn, F.J., 1941. Persistence of minerals and geologic age. *J. Geol.,* 43: 610—625.

Pray, L.C. and Murray, R.C. (Editors), 1965. *Dolomitization and Limestone Diagenesis (A Symposium)—Soc. Econ. Paleontol. Mineral. Spec. Publ.,* 13: 180 pp.

Rieke, H.H. and Chilingarian, G.V., 1974. *Compaction of Argillaceous Sediments (Developments in Sedimentology, 16)*. Elsevier, Amsterdam, 424 pp.

Rieke III, H.H., Chilingar, G.V. and Robertson, Jr., J.O., 1964. High-pressure (up to 500,000 p.s.i.) compaction studies on various clays. *Proc. Int. Geol. Congr. New Delhi, Sect. 15, Part 15*, pp. 22—38.

Romm, I.I., 1950. Geochemical characteristics of recent deposits of the Taman peninsula. In: *Recent Analogues of Petroliferous Facies (Symposium)*. Gostoptekhizdat, Moscow.

Rukhin, L.B., 1961. *Principles of Lithology*. Gostoptekhizdat, Leningrad, 779 pp.

Shutov, V.D., 1962. Epigenetic zones in terrigenous deposits of platform crust. *Izv. Akad. Nauk SSSR, Ser. Geol.*, 1962 (3).

Shvetsov, M.S., 1960. Toward the question of diagenesis. *Int. Sediment. Congr., 1960, Rep. Sov. Geol.*, pp. 153—161.

Strakhov, N.M., 1953. Diagenesis of sediments and its significance for sedimentary ore formation. *Izv. Akad. Nauk SSSR, Ser. Geol.*, 1953 (5): 12—49.

Strakhov, N.M., 1956. Towards knowledge of diagenesis. In: *Questions of Mineralogy of Sedimentary Formations*. L'vov Gos. Univ., L'vov.

Strakhov, N.M., 1960. *Principles of the Theory of Lithogenesis, 1. Types of Lithogenesis and Their Distribution on the Earth's Surface*. Akad. Nauk SSSR, Moscow, 212 pp.

Strakhov, N.M., 1962. *Principles of the Theory of Lithogenesis, 2*. L'vov Gos. Univ., L'vov.

Strakhov, N.M., 1970. Evolution of ideas on lithogenesis in Russian geology (from the 70th year of the 19th century to the 70th year of the 20th century). *Lithol. Miner. Resour.*, 2.

Strakhov, N.M. and Logvinenko, N.V., 1959. About stages of sedimentary ore formation and their nomenclature. *Dokl. Akad. Nauk SSSR*, 125 (2).

Strakhov, N.M., Brodskaya, N.G., Knyazeva, L.M., Razzhivina, A.N., Rateev, M.A., Sapozhnikov, D.G. and Shishova, E.S., 1954. *Formation of Sediments in Recent Basins*. Akad. Nauk SSSR, Moscow, 791 pp.

Süjkowski, Zb.L., 1958. Diagenesis. *Bull. Am. Assoc. Pet. Geol.*, 42: 2692—2717.

Teodorovich, G.I., 1946. Minerals of sedimentary formations as indicators of physical-chemical environment. In: *Questions of Mineralogy, Petrography, and Geochemistry. Dedicated to Memory of A.E. Fersman*. Akad. Nauk SSSR, Moscow.

Teodorovich, G.I., 1947. Sedimentary geochemical facies. *Byul. Mosk. Obshch. Ispyt. Prir. Otdel. Geol.*, 22 (1).

Teodorovich, G.I., 1954. Toward the question of studying oil-producing formations (source rocks): *Byul. Mosk. Obshch. Ispyt. Prir. Otdel. Geol.*, 29 (3): 59—66.

Teodorovich, G.I., 1961. *Authigenic Minerals in Sedimentary Rocks*. Gostoptekhizdat, Leningrad, 172 pp.

Trask, P.D., 1951. Dynamics of sedimentation. In: P.D. Trask (Editor), *Applied Sedimentation*. Wiley, New York, N.Y., pp. 3—40.

Vassoevich, N.B., 1957. Terminology used to define stages and sub-stages of lithogenesis. In: *Geology and Geochemistry*. Gostoptekhizdat, Leningrad.

Vassoevich, N.B., 1962. More about terms to designate stages and substages of lithogenesis. In: *Geological Symposium, 7—Tr. VNIGRI*, 1962 (190).

Vassoevich, N.B. (Editor), 1971. *Diagenesis and Catagenesis of Sedimentary Deposits* (Russian translation of: G. Larsen and G.V. Chilingar, Editors, *Diagenesis in Sediments*). Mir, Moscow, 463 pp.

Von Engelhardt, W., 1967. Interstitial solutions and diagenesis in sediments. In: G. Larsen and G.V. Chilingar (Editors), *Diagenesis in Sediments*. Elsevier, Amsterdam, pp. 503—521.

Von Guembel, C.W., 1868. *Geognostische Beschreibung des ostbayerischen Grenzgebirges, I—III,* 700 pp.
Wahlstrom, E.E., 1948. Pre-Fountain and Recent weathering of Flagstaff Mountain near Boulder, Colorado. *Bull. Geol. Soc. Am.*, 59: 1173—1189.
Wieseneder, H., 1953. Über die Veränderungen des Schwermineralbestandes der Sedimenten durch Verwitterung und Diagenese. *Erdöl Kohle*, 6: 369—372.
Williams, H., Turner, F.J. and Gilbert, C.M., 1955. *Petrography*. Freeman, San Francisco, Calif., 406 pp.
Wolf, K.H., 1963a. *Syngenetic to Epigenetic Processes, Paleoecology, and Classification of Limestones; in Particular Reference to Devonian Algal Limestones of Central New South Wales.* Thesis, University of Sydney, Sydney, N.S.W. Unpublished.
Wolf, K.H., 1963b. Limestones. Australian National Univ., Canberra, A.C.T. (unpublished).
Wolf, K.H., 1965a. Petrogenesis and paleoenvironment of Devonian algal limestones of New South Wales. *Sedimentology*, 4: 113—178.
Wolf, K.H., 1965b. Gradational sedimentary products of calcareous algae. *Sedimentology*, 5: 1—37.
Wolf, K.H., 1965c. Littoral environment indicated by open-space structures in algal reefs. *Palaeogeogr. Palaeoclimatol., Palaeoecol.*, 1: 183—223.
Wolf, K.H., 1965d. "Grain-diminution" of algal colonies to micrite. *J. Sediment. Petrol.*, 35: 420—427.
Wolf, K.H. and Conolly, J., 1965. Petrogenesis and paleoenvironment of limestone lenses in Upper Devonian red beds of New South Wales. *Palaeogeogr. Palaeoclimatol., Palaeoecol.*, 1: 69—111.
Wolf, K.H., Chilingar, G.V. and Beales, F.W., 1967a. Elemental composition of carbonate skeletons, minerals, and sediments. In: G.V. Chilingar, H.J. Bissell and R.W. Fairbridge (Editors), *Carbonate Rocks.* Elsevier, Amsterdam, pp. 23—149.
Wolf, K.H., Easton, J.A. and Warne, S., 1967b. Techniques of examining and analyzing carbonate skeletons, minerals, and rocks. In: G.V. Chilingar, H.J. Bissell and R.W. Fairbridge (Editors), *Carbonate Rocks.* Elsevier, Amsterdam, pp. 253—341.
Zaytseva, E.D., 1954. Vertical distribution of biogenous elements in interstitial solutions of Bering Sea. *Dokl. Akad. Nauk SSSR,* 99 (2): 289—291.
Zenger, D.H., 1972. Dolomitization and uniformitarianism. *J. Geol. Educ.*, 20 (3): 107—124.

References added in proof
Folk, R.L., 1959. Practical petrographic classification of limestones. *Bull. Am. Assoc. Pet. Geol.*, 43: 1—38.
Folk, R.L., 1965. Some aspects of recrystallization in ancient limestones. In: L.C. Pray and R.C. Murray (Editors), *Dolomitization and Limestone Diagenesis: a Symposium. Soc. Econ. Paleontol. Mineral., Spec. Publ.*, 13: 14—48.
Pettijohn, F.J., 1949. *Sedimentary Rocks.* Harper and Row, New York, N.Y., 1st. ed., 526 pp.

Chapter 2

DIAGENESIS OF SANDSTONES

E.C. DAPPLES

INTRODUCTION

The coarseness of grain size and the diversity of mineral composition which characterize sandstones make them particularly important in identifying significant diagenetic alterations in the original sediment. Among such changes the principal obvious one involves the cement which has been deposited as a secondary product in the pore space. Some of this cement serves to lithify the sand without any modification in the aspect of the detrital grains. Other cements, however, represent products of complex physicochemical reactions and institute very significant changes in the original sediment by replacement of certain detrital ingredients. Still other cements represent addition of neoformed minerals as marginal overgrowths, or as individual crystals attached to grains of the detrital fraction. Such cements serve to illustrate that their crystallization is directed by the physical conditions of burial and the compositions of the intra-pore solutions.

An important diagenetic alteration which is recognized to be instituted immediately following deposition of a sedimentary layer involves oxidation or reduction reactions. New minerals which are developed, are chiefly clay minerals and related micas, and modification in the state of iron is the principal visual manifestation. In the early stages of the burial history the oxidation—reduction reactions may shift across an equilibrium. As the lithification progresses, the tendency for change in oxidation state diminishes and the condition remains fixed until some drastic dynamic event occurs to institute new reactions.

Although all such modifications are acknowledged to occur, the extent to which such diagenesis has modified the original detritus still remains a subject of considerable diversity in opinion (Pettijohn, 1957, pp. 648—650). Are the modifications so profound that the present-day mineralogy of the sedimentary rocks is only remotely akin to that of the detritus which was deposited? Is it possible that pure-quartz sandstones were never sediments of such composition, but are the products of substratal solution episodes during which episodes all minerals were removed except the most resistant ones (Crook, 1968; Blatt and Sutherland, 1969; Cleary and Conolly, 1972; Heald and Larese, 1973)? If such should prove to be the case, then obviously the existing systems of sandstone classification are not satisfactory devices from which to interpret the genesis of the sediment. It is the purpose of this chapter to clarify some of the uncertainties regarding the importance of diagenetic modification of sand detritus, to demonstrate that stages of diagenesis represent the progress of transformation of the sediment into a rock; and, as lithification is initiated, additional changes in the mineralogy are to be regarded as counterparts to stages of metamorphic grade. This is not precisely a recent concept because its essence was proposed by Harrasowitz in 1927 and reintroduced by Kubler (see Dunoyer de Segonzac, 1968, p. 164). According to these authors, the term *anchimetamorphism* is proposed to represent a zone which transcends the truly unmetamorphosed strata, through changes in mineral content and loss of porosity, to approach the attributes of truly metamorphosed rocks.

Investigators of strata containing graywackes are aware that their common association with features of low-grade metamorphism indicates the possibility that some of the characteristics may be products of partial metamorphism. The common occurrence of authigenic, high-soda zeolites has been reported in thick graywacke sequences, the deep burial and moderate temperature regime of which has helped to establish the zeolite metamorphic grade (Coombs et al., 1959; Zen, 1961; Dickinson et al., 1969; see also Mizutani, 1970). Investigations by Packham and Crook (1960) suggest the existence of additional facies representative of still lower temperature realms which they assigned to epigenesis or a late stage of diagenesis. Other examples of similar strata with a high content of soda are known to contain "fresh" grains of albite—oligoclase of good crystal outline. These have been interpreted as authigenic and the product of low-temperature diagenesis (Middleton, 1972), or that the neoformed minerals were controlled primarily by the initial composition of the sandstone (Brown and Thayer, 1963).

The proposal that deep burial and temperatures exceeding 200° C are not requirements for significant mineral modifications has been supported by the investigations of Raam (1968). He has indicated that a secondary mineral assemblage consisting of albite, chlorite and laumontite has replaced calcic plagioclase as well as volcanic grains. Such alterations have been brought about by diagenesis at a depth of burial not exceeding a kilometer and temperatures which could have been in the order of 35° C.

Strata in which shales preponder illustrate the tendency to undergo transitional changes into the zones of anchimetamorphism (Maxwell and Hower, 1967; Weaver and Beck, 1971). An excellent example has been reported by Dunoyer de Segonzac et al. (1968) who made a detailed study of dark silty shales from a deep boring into the Douala Basin (Cameroun). Individual samples show regular and progressive increases in density from 2000 to 4000 m, culminating in a maximum value of 2.7, and a rock of virtually no porosity. Correspondingly, illite increases in crystallinity, a feature which accompanies degradation of detrital micas needed to supply the potassium. Paralleling the progressive increase in crystallinity of illite are decreases in amounts of montmorillonite and kaolinite until their eventual disappearance, and a growth of chlorite in stages to about equal importance with illite.

Interstitial clays in sandstones do not always behave in such a well-defined progression; however, increase in crystallinity of illite is commonplace and tends to be accompanied by changes in the nature of sand-grain cementation (Bucke and Mankin, 1971). The changes in cementation are observed in cratonic strata which can be shown to have never been buried more than a kilometer or so and, hence, never exposed to temperatures much above those existing at the earth's surface. The precipitation of cements and replacement of grains by later cements show relations indicative of equilibrium reactions. With shift in the equilibrium, a mineral assemblage is no longer stable and is rearranged into a different mineral suite. This has led to a proposal by the author (Dapples, 1962) to establish certain stages in the progress of diagenesis based upon the prominence of certain types of reactions. Among these some clearly are related to the nature of the particle-size distribution, increasing in complexity with an increase in fine-grained matrix content. Other reactions result in simple precipitation of cement. An understanding of the reactions involved concerns the size distribution of the detritus and the various means by which individual grains are bonded into a rock aggregate.

In connection with the variety of the particle-size distribution of the original sediment, the importance of the tectonic control on the composition and texture of the sand debris is now well accepted; and the separation of sediment into shelf, basin, and geosynclinal types is demonstrable. The overall rate of submergence is the dominating influence on the relative fractions of sand, silt, and clay which constitute the ultimate sandstone. Thus, shelf deposits tend to be the most texturally mature and the eugeosynclinal sediments the most immature. The distinction is not drawn so readily with regard to the mineral composition. It is the purpose herein, however, to restrict the consideration to the petrographic modifications which are imposed during diagenesis, and to omit any discussion concerning the origin and the composition of the detritus as it arrived at the depositional site.

GENERAL PROCESSES OF LITHIFICATION

Lithification is the general process of bonding individual particles of sediment together so that an aggregate mass results. Binding of individual grains one to the other can be demonstrated to result from several distinct processes, for example, desiccation, compression and compaction, cementation, crystallization and recrystallization. Each of these crudely understood processes results in some modification in the arrangement, or the constitution, of the original detritus. This modification ranges from minor addition of a mineral precipitated in the available pore spaces to complex interlocking of primary and secondary minerals into a mosaic of very low porosity.

The concept of cementation

Certain authoritative texts have accepted the viewpoint that cementation represents processes involved in the precipitation of mineral matter in voids present in a clastic sediment, the end result of which is induration into a product loosely called rock (e.g., Pettijohn, 1957, p. 650; Krumbein and Sloss, 1963, p. 270). According to views currently prevailing, later cements may replace earlier ones, such relations being observed particularly between quartz and carbonate minerals (Gilbert, 1949; Siever, 1959; Walker, 1962). Also, semi-mechanical processes are responsible for intergranular penetration and consequent lithification (Heald, 1950, 1955, 1956, 1959). There should be no doubt that among the processes of diagenesis in sandstones, cementation and lithification are so closely related that consideration of one by necessity involves the others (Garrison et al., 1969). Moreover, the overlap of separate processes limits the precision of individual definitions so that there is difficulty in assigning a process to cementation or to lithification. Geologists do, however, attribute many diagenetic modifications to processes other than those directly producing lithification (Fairbridge, 1967, pp. 23—27). Among certain sandstones a cement deposited as a pore filling clearly is a lithifying agent. Where clay is a significant constituent of the sand mixture, cohesion between grains involves very different processes, which according to strict definitions stated above, are not true processes of cementation. Attachment of clays to grains of sand sizes involves certain interlattice associations not necessarily the product of precipitation from solution, but the resulting binding of particles of differing sizes is in fact a variety of cementation.

In the USA, a most important investigation of sandstones was reported by Waldschmidt (1941) who concluded that:

(1) Two general groups of cementing materials occur, namely, argillaceous particles and those crystallized from solution.

(2) Certain effects of mechanical compaction occur among sandstones with argillaceous binder.

(3) Sequences occur in the deposition of crystalline cements affecting rock porosity and permeability; and there are variations in the kinds of such cements throughout the geologic column.

(4) Compaction is a product of grain interlock which is the response to

dissolution of silica at the points of grain contact.

A further investigation into the role of grain interpenetration as a mechanism of reduction in porosity was made by Taylor (1950, p. 704). She recognized that the process of simple pore filling by crystallization of a cement involved no change in the shape of the original grains, but when viewed in thin section, showed three types of inter-grain contact: in order of decreasing frequency, tangential, long, and concavo-convex, the latter being rarely observed. A later study by Gaither (1953, p. 181) confirmed her results and demonstrated that such contacts occur in the same frequencies if there is no pore-filling cement. Taylor determined that the proportion of such contacts and a fourth type, sutured, changed with increasing depth of burial. She believed that the observed increases in concavo-convex and sutured types of contacts at greater depth were due to pressure, and that two processes were involved, namely: (1) solid flow and (2) solution and redeposition or removal of the dissolved material. According to Taylor the best evidence of solid flow is provided by rock fragments of heterogeneous mineral composition. Such fragments can be demonstrated to conform to a grain shape along a contact, which preserves the outline of the non-yielding grain.

Current studies have now supported some of the earlier interpretations, particularly that simple crystallization within interstitial pores actually is only one aspect of cementation (Pittman, 1972). Mechanisms which cause discrete sedimentary particles to interpenetrate are to be regarded as varieties of cementation. Still other processes such as chemical interactions between large grains of sand and small particles of silt and clay sizes result in cementing individual particles together. Special reference must be made to the preferential occurrence of different aspects of cementation. Obviously, a narrowly restricted size distribution of sand grains tends to produce a packing with large pores suited to precipitation of cement. Such packing of grains also tends to result in grain penetration and welding of individual grains into an aggregate. A somewhat similar texture is developed as overgrowths are crystallized onto grain boundaries in the open pore regions. In effect a series of separate forms of cementation can lead to a particular type of lithification.

Sands in which the particle sizes include clays tend to show a concentration of small sizes in the pores between sand grains. As the percent of fine particles is increased the large pore openings tend to be obliterated and textures are developed where individual sand grains no longer touch one another. In such sands precipitation of certain cements may be inhibited or it occurs in any available pore space, particularly within clay aggregates. Inter-penetrations or overgrowths on small sand grains are limited to special occurrences. The type of cementation also depends on properties of the clay minerals themselves, such as unsatisfied ionic charges on their surfaces and positions of water molecules. Indeed it may be concluded that the variety of cementation is a response in large measure to the texture, or framework, of the sand as deposited. So important is the influence of the sand framework upon the variety of cementation that a clear understanding of the basic forms of textural framework are reviewed in the following paragraphs.

PRIMARY FRAMEWORKS OF SANDS

Varieties of frameworks

Classification of sandstone frameworks into those which are grain-supported as distinct from those which are matrix-supported has been an attempt to separate two basic types from the continuum extending between the two extremes. Some sands consist of grains with a narrow range of particle size, whereas others include much clay which, in certain examples, may actually dominate the distribution (Krynine, 1948; Folk, 1951; Gilbert, 1954; Klein, 1963a; McBride, 1963; Dott, 1964). One may regard the extremes of such distributions as consisting of one continuous solid phase and one which is discontinuous. In the example of the sand of narrow particle-size range individual grains touch along one or more contacts, constitute the continuous phase, and isolate particles of smaller dimensions in the available pore space. Where the mixture reaches values exceeding 20% (by weight) of clay and fine silt interspersed with grains of medium- and fine-sand sizes, some individual sand grains are isolated by the clay—silt matrix. For such sizes, rather inconclusive measurements indicate that the matrix, as seen in thin section, becomes the continuous phase when it occurs in proportions of approximately 30% or more.

Frameworks consisting of individual sand grains in contact are recognized as *arenites* whereas those in which the sand grains form the discontinuous solid phase are to be herein identified as *wackes* (Nanz, 1954). This definition places the lower limit on the amount of fine sizes (i.e., filling pores or acting as matrix) at somewhat higher values than that established by Gilbert (1954, pp. 289—304). Gilbert recognized a wacke as containing more than 10% argillaceous matrix, whereas all sandstones containing less than that amount are arenites. Accordingly, the nature of the framework tends to differ only in the more extreme values of fine material. A separation between arenites and wackes, therefore, is based upon some observed but nevertheless arbitrary distinctions depending upon the definition selected.

If one uses a difference in grain framework as a device for separating the two major sand groups, a sharply defined boundary does not exist. Instead, there prevails a range of mixtures which within the same hand specimen, or thin section, reveal localized regions of each of the two major frameworks. For such sands and sandstones displaying the two framework types, the writer proposes adoption of the term *subwacke*. Although this term is not to be used as a synonym for subgraywacke, many of the rocks which can be so classified (Pettijohn, 1957, p. 317) illustrate a subwacke framework; it appears that the most common occurrences of subgraywackes are arenites. The three framework types, as recognized best in thin section are to be distinguished on the basis of the relationships between the fine fraction and grains of sand size (ideally 1/2—1/8 mm), and the occurrence, amount, or position of any pore-filling cement must be disregarded.

Arenite framework. Characteristically, sands deposited on the stable craton are arenites and, in the ideal examples, grains of diameter exceeding 1/4 mm commonly are highly rounded and spherical. Such grains touch predominantly with point, or tangential, contacts and average few per grain. For example, in the experimental sand prepared by J. Taylor, she observed an average of 1.6 contacts per grain (Taylor, 1950, p. 707). The study by Gaither (1953) based upon artificially packing St. Peter Sandstone of medium-sand size into a grain-supported framework showed an actual porosity of about 37%. Thin sections of this arenite indicate an average of about 0.85 contacts per grain. Such values are to be compared with those of bulk samples of friable classical arenites such as the Jordan (Cambrian) and St. Peter (Ordovician) sandstones taken by this author undisturbed from outcrop. In these, the number of grain contacts as shown in thin section averages 3—4 per grain, but the numbers increase to continuous grain contact where locally grains are well-cemented by silica. In such examples, local development of overgrowths on quartz grains are present, or granular intergrowth has occurred. Certain grains reveal the original detrital boundary as distinct from that of the later added growth. Other grains, however, show no indication of what is old and what is new unless cathode-ray luminescence is used to distinguish between the two varieties. Sippel (1968) has demonstrated that without the use of luminescent petrography, grain contacts as seen in thin section lead to misleading counts and interpretations. His method presents a far better technique for identifying the contacts of original grains as distinct from those developed, particularly by later overgrowths. Even with the best available techniques there exists uncertainty in distinguishing the existing, secondarily developed grain contacts (number and type), as observed in thin section, from the original ones. For example. Taylor's and Gaither's counts of average number of contacts were based upon original grain packing, whereas the larger number of contacts per grain reported by the present author for the Jordan and St. Peter sandstones clearly indicate the influence of postdepositional modification. Moreover, for such classic arenites the tectonic and environmental conditions of deposition point to the strong possibility that the original porosity locally approached the experimental values reported by Gaither (1953, p. 181). Inasmuch as such space was unoccupied by clay or silt, it was potentially available for various types of cementation (Rittenhouse, 1971a).

Wacke framework. Among wackes the following three attributes are observed commonly:

(1) Various particle-size fractions may be rather uniformly distributed between the extremes of sand and clay; consequently, the observer has difficulty in delimiting the matrix and the enclosed particles. In such examples the number of sand grains in contact with others of sand size is extraordinarily variable ranging to as many as four or five per individual depending upon the local clustering. Because of such variability, average values of numbers of contacts per sand grain are not very reliable as a device for distinction between frameworks, or the amount of inter-grain penetration. Also, the extent to which compaction has occurred influences the amount of open pore space which exists in the vicinity of the clusters of sand grains or pellets of clay (Rittenhouse, 1971b). In the wacke framework, compaction may have caused individual sand grains to be brought into contact, even though following deposition they were separated by small regions of silt or clay. As a result of compaction, aggregates of clay may also be squeezed into positions to isolate grains of sand and likewise modify the original framework (Dickinson, 1970, p. 702). Ideally, the continuous and discontinuous phases exist in three dimensions and should be distinguished by suitable observations in these dimensions. In practice this is not accomplished readily, and observations are based fundamentally upon the thin section. Despite the inability to distinguish in thin section between grains actually surrounded by matrix and those supported in a different plane by other grains, the approximation is regarded as suitable for purposes of classification. Obviously also, where local clusters of sand grains occur, the matrix does not constitute the continuous phase except around the entire cluster. If such clusters are numerous, the framework may be better identified as subwacke.

(2) Quartz grains of sand size tend: (a) to be irregular in cross-sectional outline, (b) to have low sphericity, (c) to show very limited development of overgrowths, if any, and (d) to show corroded margins.

(3) Except for poorly compacted and recently deposited varieties, large pore space unoccupied by some fine-grained material or matrix is exceptionally uncommon. Surfaces of sand grains largely are in contact with some detrital mineral albeit of considerably smaller size. The total porosity of wacke frameworks, however, may be high because of the pore space within the matrix material. A measure of porosity is not regarded as a distinguishing characteristic of a wacke framework.

In examples of wacke where clay and sand constitute separate modal sizes, the framework is readily identified and the clay matrix, having also been squeezed into available openings during compaction, is noted to surround virtually each grain of sand. Bioturbated sands frequently possess ideal wacke frameworks, because the mechanical mixture of the two principal particle sizes has not been the product of settling or current velocites, and sizes segregated into laminae during deposition have been homogenized into abnormal association.

Subwacke framework. Among subwackes, a variability of frameworks exists in the same hand specimen or the same thin section. In microscopic fields where the framework is that of arenites and the sand grains are approximately of the same size, the pore space may consist of large openings. Elsewhere, where the local particles are mixtures of sand, silt, and clay, openings are smaller and of differing shapes. In the latter examples, the number of inter-grain contacts among sand grains tends to be less numerous. Locally, small aggregates of clay, or clay pellets, may be present and they may be squeezed into adjacent openings between sand grains where they serve as a matrix.

Following deposition, each one of the three primary mixtures of sand has the essential elements of its framework; this tends to be emphasized by burial and the accompanying compaction. Closer grain packing is achieved as well as reductions in the dimensions of large pore openings. Also, the details of the framework are subject to modification by additional processes which collectively are grouped within the term diagenesis (Fairbridge, 1967, pp. 31—41). Some diagenetic processes may dominate events which lead to lithification, i.e., the development of particular bonds between adjacent individual mineral grains. Grains of sand may become agglutinated by introduction of organic material, held in position by newly precipitated minerals (Webb, 1974), welded into compound grains or aggregates, or so complexly intergrown with neoformed minerals that individual constituents cannot be separated except by rupture of the intergrowth. All such modifications are varieties of cementation.

PRIMARY BONDS OF SAND MIXTURES

Simple clay bond

Except for exotic cements, such as bituminous matter or ice, bonding of sand grains by clay minerals is one of the first stages of the lithification process in subwacke and wacke frameworks. The behavior is regarded as identical to that among artificially prepared mixtures of sand and clay minerals in molding sands. These are known to have compression strengths which increase with addition of small amounts of water, up to some maximum value of strength, and decrease continuously with additional increments of water. Also, such maximum strengths are clearly greater for some clay minerals than for others. For example, Grim and Cuthbert (1945) showed that the green compression strength of a molding sand increased substantially as increments of sodium montmorillonite were added up to a value of 12% of the mixture. These authors were able to show that a certain amount of water is necessary to develop bonding strength between sand grains and clay minerals. Such water was computed to be the amount needed to develop a sheet 2—4 molecular layers thick on each surface depending upon whether the clay was sodium montmorillonite (i.e., two layers) or calcium montmorillonite (i.e., four layers). The explanation they prefer is, for example, that quartz grains are bonded with a smooth coating of sodium montmorillonite of more or less uniform thickness, except for thicker wedge-shaped blocks at the grain junctions. Where the water film is thin, i.e., 2—3 molecules thick, water molecules occupy fixed positions with respect to the montmorillonite flake units, (oriented as inter-layer water) and, hence, are rigid. As the molecular layers increase in numbers, the inter-atomic connection with the clay-mineral lattice decreases. Such water is not part of the rigid framework, but is liquid, and the bonding strength, therefore, is diminished. Similarly, the water film between quartz and the clay must behave in a like manner, being intimately polarized to connect the two crystal lattices. Grim and Cuthbert (1945) and Davies (1947) suggested that the bonding action which best fits their data is that of a mechanical model of a wedge-shaped block of clay at the position of contact of sand grains. The maximum amount of solidity and rigidity occurs at such a block between quartz grains, which serves to lock the grains into a fixed position. Obviously, with increasing number of such wedges per each sand grain, the bonding strength also increases.

Some clay minerals, principally kaolinitic, tend to remain as aggregates and do not coat quartz grains with a film of uniform thickness. In such mixtures, the wedge areas between sand- and silt-size grains are not uniformly developed, nor do water molecules penetrate the kaolinite lattice as successfully. Consequently, the total bonding effect is weaker than with montmorillonite. In all examples, however, loss of liquid water through air-drying increases the clay—sand bond strength and cementation is initiated.

Films of water coating grains of sand also act as temporary limited cementing agents as illustrated by the well-known distinction between the tendency for moist sand to aggregate in contrast to dry sand. When such films disappear through drying, small quantities of the contained salts, or organic matter, may locally agglutinate grains. Presumably the tensile strength of the water films is the limiting strength value of the temporary bond.

Surfaces of alluvial fans and sedimentary fills in broad stream valleys contain sands which are cemented by the drying process of clays. Similar behavior is also observed among certain so-called "muddy" sandstones. Examples occur among selected beds from the White River and Arikaree formations of western Nebraska, or from the Pennsylvanian Pleasantview Sandstone of Illinois, U.S.A. Locally, these rocks illustrate in part a simple clay bond holding large and small grains into an aggregate when dry, but which tend to disintegrate upon immersion in water. Disaggregation is rarely complete among such sandstones and some aggregates, either large or small, fail to separate into individual grains even with limited gentle stirring. The simple clay bond as described above, therefore, is only partially responsible for the lithification process and some other less reversible mechanism must be involved.

Figs. 2-1 and 2-2 illustrate examples of modifications or progressions of the simple clay bond. Some sand—clay boundaries illustrate bonding achieved by the rigid water and are to be noted by the tendency to separate when wet. Other grain-to-clay boundaries indicate that clay minerals are attached to quartz by some structural intergrowth. Such attachment is clearly illustrated in Fig. 2-3 in which orientation of the clay-mineral plates produces a bond that does not collapse in water. This type of attachment appears best explained in terms of the proposal by DeVore (1956) that hydration structures can be developed on the surfaces of quartz grains. DeVore showed that OH groups in positions very close, relatively, to the arrangement of oxygens in silica tetrahedra could assume orientations to develop a hydrated row, or outer layer. This silica, having composition $H_4Si_4O_{10}$, approximates a sheet structure of phyllosilicates. By this mechanism, incompletely held OH groups on the surfaces or edges of clay plates can orient on the surface of a quartz grain in the form of a lattice intergrowth. In effect, the clay-mineral plate becomes welded to that surface. Relationships shown in Fig. 2-3 are interpreted as illustrating such a weld as shown by authigenic growths of clay mineral on the surface of neoformed quartz. In this example (Pleasantview Sandstone), clusters of delicate crystals of clay mineral have clearly developed from a growth position on the face of a quartz crystal, the relationship being obviously between two groups of neoformed crystals. A similar form of attachment, however, is believed to occur between a detrital clay plate and a detrital quartz grain. Transition from the simple clay bond to the quartz—clay intergrowth is a progressive one in the cementation process (diagenetic modification) and in the final lithification of the sandstone.

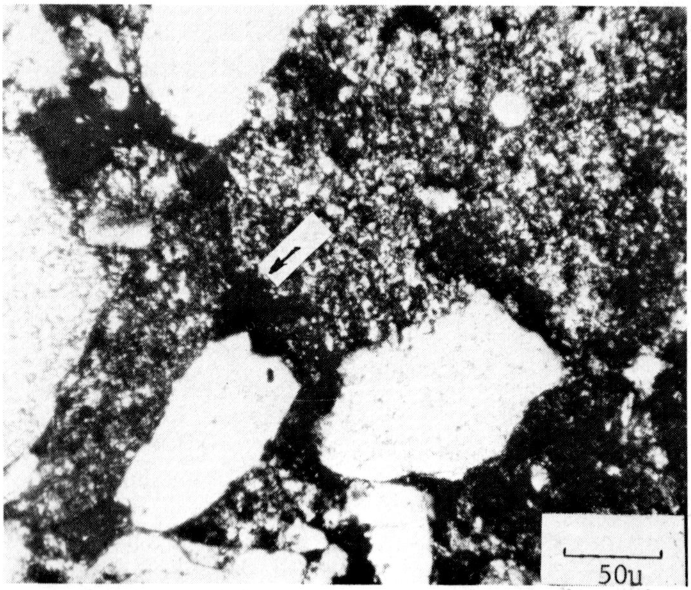

Fig. 2-1. Sandstone of wacke framework illustrating simple clay bond. There is poor adherence of loose aggregate of clay and silt particles (central area), partly squeezed into pore between quartz grains (white) and shale grain (spotted; along left margin). The bond passes through positions of "rigid water" of clay minerals attached to quartz and other grains. Quartz grain near center shows partial intercrystalline intergrowth with clay, interpreted as initial arrangement of OH-ions of clay minerals with unsatisfied bonds of quartz grain to make a surface phyllosilicate layer. This boundary does not collapse with introduction of water as does the simple clay bond. Note large pore (black, below arrow) into which has grown an early overgrowth on the quartz grain.
X-nicols; Tongue River Member of Fort Union Formation, Paleocene; Musselshell County, Mont., U.S.A.

Simple mineral cement: compatible and incompatible varieties

In the typical arenite framework, clay is a very minor constituent and the open pore space between sand grains may be the site for crystallization of mineral cement. Such mineral matter can be regarded as comprising two structural varieties. One type is compatible to individual detrital sand grains, i.e., it may join itself to the crystal lattice of such a grain. Another variety is structurally incompatible and is crystallized in interstitial openings to form a sharp, but crystallographically discordant, boundary with the detrital grains. Boundaries of compatible cements either with a grain, or with other crystals of the cement, are interpreted to be closely related to boundaries between crystals of a single mineral species in metamorphic rocks; i.e., the so-called

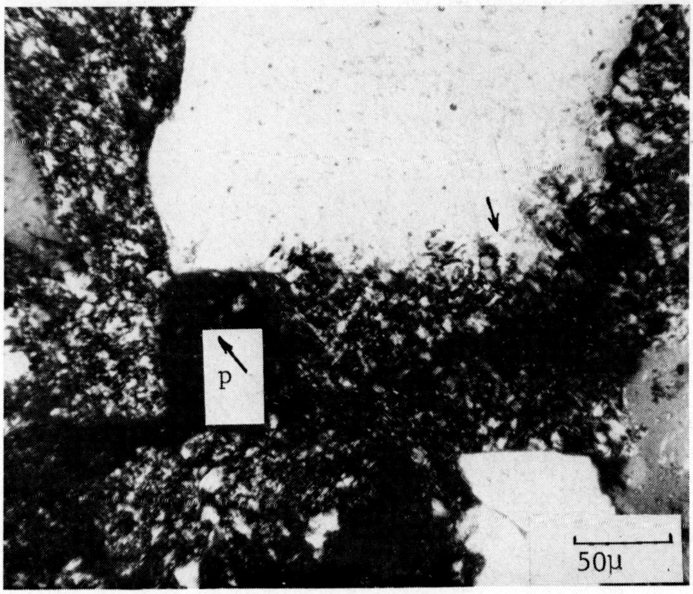

Fig. 2-2. Quartz grain (light gray) illustrating early stage of intercrystalline bond with clay mineral. Note corroded margin of quartz grain (arrow) and tendency for boundary to be transitional from quartz into clay. Contrast this transitional boundary with the sharp separation between clay and quartz along the left margin of quartz grain. This represents attachment of grains by simple bond, i.e., "rigid water". Black area (*p*) is open pore into which has grown epitaxial quartz.
X-nicols; Muddy Sandstone, Upper Cretaceous; near Tensleep, Wyo., U.S.A.

self boundaries (Spry, 1969, p. 20). For example, Voll (1960, p. 509) has applied the term *partly coherent boundary* to denote an intercrystalline junction in which rational lattice planes form the interface of a large part of the lattice sites across the boundary. Such a boundary is to be distinguished from one identified as *coherent*, which indicates an interface that is rational for both lattices in contact. The term *non-coherent boundary* is used where angular misfits occur along the junction of two crystals or grains. Similar lattice intergrowths are regarded as characterizing the compatible mineral bond in sandstone cements, but most intergrain junctions illustrate poor fits between lattices.

The outline of an incompatible boundary is stable and is determined by the shape of the detrital grain, whereas among compatible cements the boundary of grains may be modified by some recrystallization. An incompatible mineral cement acts as a lithifying agent by partially surrounding detrital grains. This cement is recognized readily by the sharp definition of the

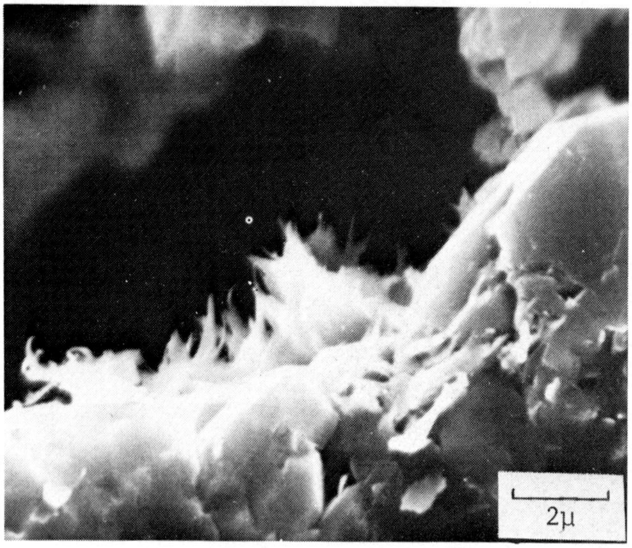

Fig. 2-3. Growth of authigenic clay minerals on surface of neoformed quartz crystals in pore space of subgraywacke sandstone. Delicate crystals (center) of clay mineral are attached to quartz surface and do not separate upon immersion in water.
SEM photo of Pleasantview Sandstone, Pennsylvanian, Rushville, Ill., U.S.A.

boundaries and by a marked difference in mineral composition from that of the detrital grain with which it is in contact. In quartzose arenites, commonplace incompatible simple cements consist of calcite, dolomite and siderite, iron oxide, and anhydrite or gypsum; however, each one of these minerals can behave as a compatible cement wherever it is in contact with a grain of similar composition. Commonly, when in contact with such a grain, the cement displays non-coherent boundaries. At present, attachment of the detrital grains to an incompatible simple cement is regarded as a gluing process. The strength of this bond depends on the extent to which detrital grains are surrounded by the cement, the adhering properties of the cement, and roughness of the detrital grains. Moreover, rupture of the lithified rock usually occurs along the boundary between the incompatible cement and detrital grain.

Ideal examples of the incompatible simple cement occur among arenites of the stable craton where strata tend to have experienced mild tilting only or gentle undulatory folding under shallow depths of burial. Such cements are particularly commonplace where pure-quartz sandstones pass laterally, or vertically, into carbonate strata or evaporites (Fig. 2-4). Frameworks of such rocks show complete transition from those which are grain-

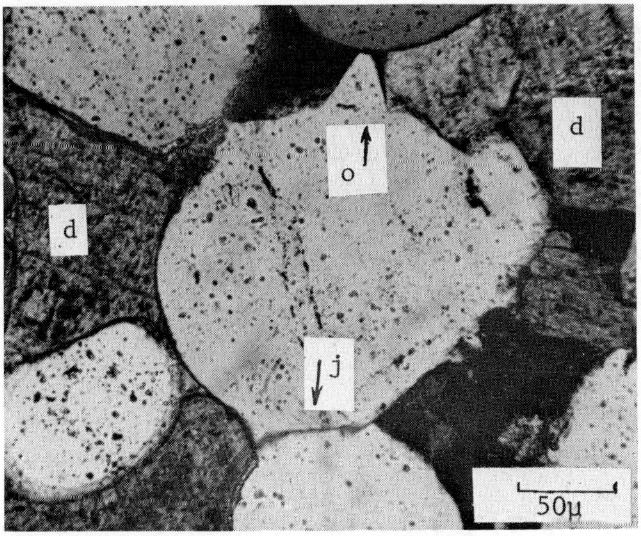

Fig. 2-4. Incompatible cement (dolomite, *d*), later than quartz overgrowths, abuts with sharp boundary against rounded, detrital grains of quartz. Dolomite partly fills pores, does not weld to quartz grains, and assumes outline of detrital grain. Attachment to quartz grain is by surface cohesion. An early stage of a bilateral junction and welding of two quartz grains to form a self-boundary is shown by arrow at position *j*. A coherent self-boundary is shown by overgrowth at position *o*.
X-nicols; Lamotte Sandstone, Cambrian, Flat River, Mo., U.S.A.

supported into those in which individual clastic grains are separated and engulfed by the non-clastic cement. In certain examples, crystallization of the cement is interpreted to have expanded the spaces isolating the detrital grains (Fig. 2-5; Folk, 1965, p. 24; also Dapples, 1971, plates I and II). Morever, grains such as glauconite and micas may be ruptured into fragments by crystallization along cleavage planes or porous zones (Sarin, 1963, p. 339). The process of spar calcite forming room for itself during crystallization of carbonate rocks has been termed *displacive precipitation* by Folk (1965). Expansion of the depositional framework by displacive precipitation is to be noted where cement is crystallized in large amounts, apparently where the fluids moving through the sediments continued to reach a condition of supersaturation of that substance. Crystals grow to large individual size, or a complex mosaic of small crystals is developed in the continuously expanding opening.

Certain incompatible cements, of which iron oxide is an example, occupy interstitial pores, but the adherence of these cements to quartz grains is weak. In such cases, the degree of cementation is limited and introduction of calcite as a second incompatible cement may occur. Where crystallization of a later calcite tends to expand the framework, the earlier-formed iron oxide cement is removed from sites of earlier position and disbursed irregularly throughout the calcite. The second incompatible cement becomes the principal lithifying agent but the earlier-formed iron oxide imparts its color to the rock (Fig. 2-6). Perhaps the most well recognized example of mobilization of an early cement is to be noted where individually large crystals of calcite not only disburse the pre-existing cement but also engulf the detrital grains (Dapples, 1971, plate II). A texture which illustrates detrital grains isolated within a single crystal of cement has been termed *poikilotopic* by Friedman (1965, p. 651).

The compatible cement and the welded boundary

Among arenites, the most commonplace examples of compatible simple cement are illustrated by quartz cement and quartz grains, and grains of carbonate and carbonate cement. Of these two varieties the quartz-to-quartz relationships are most readily identified and can be categorized as demon-

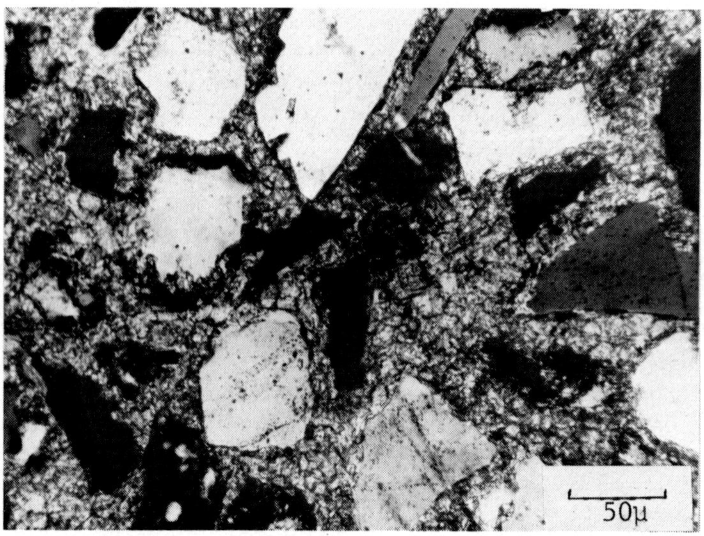

Fig. 2-5. Incompatible cement (calcite) which has expanded arenite framework of quartz (white, gray) and chert (black). This is an example of displacive precipitation by crystallization of a mosaic of calcite crystals, which constitute the continuous phase and act as a matrix surrounding grains touching with point contacts. Irregular boundaries of grains are due to overlap of calcite cement.
X-nicols; sandstone bed in Cody Shale, Upper Cretaceous, Fremont County, Wyo., U.S.A.

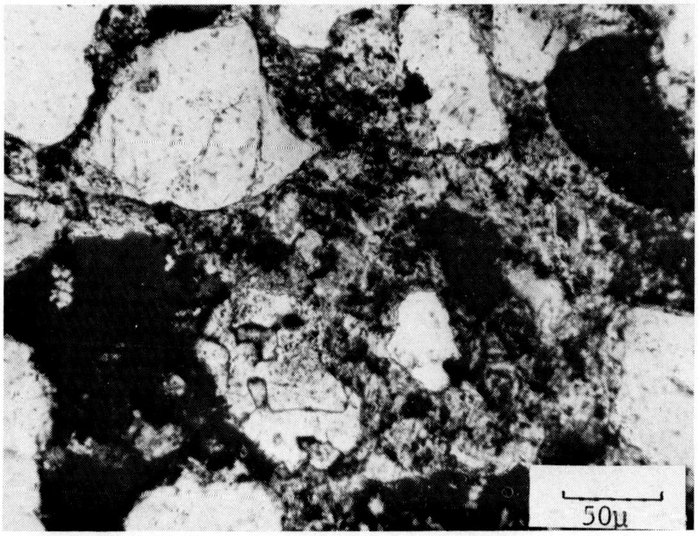

Fig. 2-6. Displacive precipitation of calcite in red-bed sandstone initially cemented by hematite as a primary incompatible cement. The hematite is now dispersed throughout the calcite as inclusions (dark gray and black), primarily in right half of photo. X-nicols; sandstone bed in Pontotoc Formation, Permian, Carter County, Okla., U.S.A.

strating types of partly coherent and non-coherent grain boundaries. Fig. 2-7 serves to illustrate several examples of monocrystalline quartz as compatible cement. One is the welded boundary, as yet incomplete, between two grains of detrital quartz. Here, the junction between grains is only slightly curved but clearly visible in crossed polarized light, indicating that the lattice configurations are quite discordant between the two grains (called *composite grain* by Hubert, 1960, p. 135). The weld, however, is generally strong. A more completely compatible weld is represented by the overgrowths on the central grain which show no visible separation. Such additions have occurred by crystallization into pore space and may involve several generations of monocrystalline quartz (see also Lamar, 1928; Giles, 1932; Heald, 1950, p. 624, 1956, p. 16; Austin 1974).

Ernst and Blatt (1964) were able to produce quartz overgrowths in the laboratory and have demonstrated that they have preferred crystal habits, i.e., coherent and partly-coherent boundaries with the detrital parent. For example, they tend to grow in optical orientation with the parent grain, and may extend outward until they interfere with a neighboring grain or overgrowth. Particular attention is directed to such non-coherent interference boundaries between individual overgrowths extended into the available pore

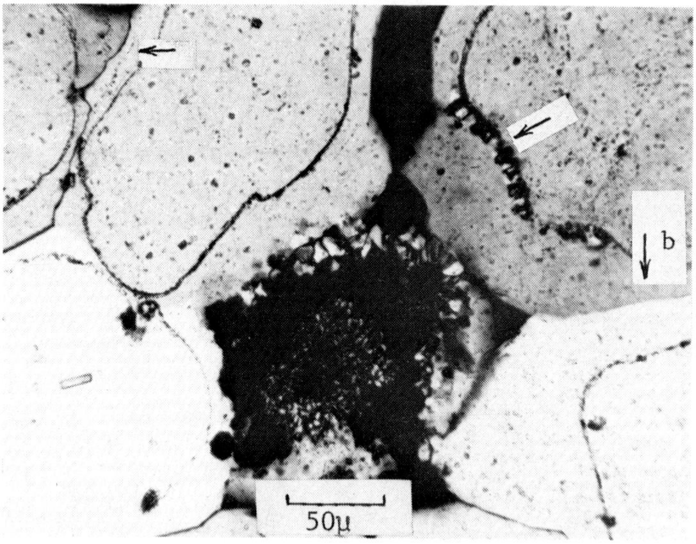

Fig. 2-7. Two or more generations of quartz overgrowths in pure-quartz sandstone. Outline of detrital grain is marked by iron oxide coating. Arrow at position *b* points to noncoherent self-boundary (partially welded), and early stage of compound grain. Repeated epitaxial growths (arrow, upper left) are coherent and show only slight boundary between two generations. The compound grain is more advanced in development. Small crystals along grain boundary (arrow, upper right) show well-developed habits typical of certain initial overgrowths into pore spaces. Area in black (upper center right) represents monocrystalline quartz welded to earlier overgrowths with noncoherent boundary and marks early stage of large compound grain. Crystallization of chert (dark, lower center) was preceded by partial solution of monocrystalline quartz, indicating that polycrystalline quartz was a more stable phase during that crystallization. This illustrates, also, the existence of micro-physiochemical cells which seem to characterize local cementation. X-nicols; Gunter Sandstone, Ordovician, Rock Creek, Mo., U.S.A.

spaces as illustrated by Pittman (1972). With the progress of diagenesis, boundaries illustrative of irregular junctions tend to recrystallize into more stable configurations, i.e., into partly coherent relationships. Such changes also serve to characterize progression in stages of cementation.

A far more uncommon occurrence involves crystallization of polycrystalline quartz on the surfaces of monocrystalline quartz (Fig. 2-7). Some examples show crystallization of the chert directly upon what appears to have been the rounded surfaces of detrital grains (Heald and Larese, 1974, p. 1272), whereas others indicate that chert was crystallized upon earlier overgrowths of monocrystalline quartz. In either occurrence, there appears to be a minutely serrated bounding surface between the two types of quartz, and such a boundary is interpreted to indicate slight solution of the previously deposited monocrystalline quartz before crystallization of the chert. Examination of some of the boundaries, or the surfaces, commonly reveals that they are coated with very small crystals of well-developed habit similar to those illustrated by Waugh (1970, fig. 5). Such clusters of crystals cover the surface of the grain in positions where cementation is in its early stages of development. Where cementation is complete, however, only the chert is identified. Obviously, some control must exist which dictates the form of quartz to be crystallized as the cement, but except for the general agreement that the preferred crystallization must represent the condition of lowest surface free energy, a more precise understanding is not as yet available.

Variations in the simple intergranular weld

In frameworks in which quartz overgrowths had ample opportunity to enlarge, general interlock of crystalline growths, as seen in thin section, is marked by line boundaries between crystals grown from separate grains. Boundaries of this type are recognized as somewhat unstable and subject to recrystallization under conditions of pronounced folding, deep burial, and significant movement of fluids undersaturated in silica. Each of those conditions results in some instability in the initial boundary, particularly where some interstitial clay is present (Pittman and Lumsden, 1968). Irregular migration of the boundary occurs to produce grain inter-penetration of various types (Waldschmidt, 1941; Taylor, 1950; Heald, 1950; Thomson, 1959, p. 108). Boundaries range from simple concavo-convex (Taylor, 1950) to stylolitic (Sloss and Feray, 1948; Heald, 1950; Thomson, 1959, p. 96) and have been generally interpreted as responses to pressure solution.

Renton et al. (1969) have reproduced pitting and deformation of grains of monocrystalline quartz and chert under pressure at temperatures exceeding 272° C. They noted that although solution could occur in distilled water, the most favored condition was that of alkaline solutions which produced pressure solution of the quartz grains and recrystallization of overgrowths. The effect of pressure solution is to reduce porosity and to modify grain shape substantially. As conditions approach those of dynamic metamorphism, annealing of grains occurs first, followed by crushing and destruction of detrital textures (Fellows, 1943, p. 1399).

An appraisal of the degree of complexity of the boundary separating individual quartz grains, indicates only a very approximate relationship between intergranular suturing and the magnitude of folding or depth of burial. Other factors having profound influence, but currently poorly understood, are believed to involve the interstitial fluids and varieties of clay minerals in complex reactions. The effect of such reactions is to irregularly dissolve quartz and lead to the appearance of intergrain boundaries known as micro-stylolitic. Although some gradation between all boundary types appears to exist, the presence of small amounts of clay minerals in the interstices appears to enhance the degree of grain interpenetration. Some uncertainty still prevails regarding the influence of interstitial clay. For example, Whisonant (1970, p. 1023) concluded that there is an inverse relationship between clay matrix and the magnitude of quartz-grain sutures. On the other hand, Renton et al. (1969) found that clay does not have any effect upon pressure-solution phenomena produced in the laboratory. Such apparent contradictions in behavior seem to be explained by the theoretical analysis of Weyl (1959, p. 2018) who stated that "small amounts of clay will enhance pressure solution, large amounts will have the opposite effect". Boundaries which appear to be least associated with so-called pressure solution are either straight lines or slightly curved, particularly in the case of well-developed overgrowths. This configuration is regarded as stable and, in the absence of clay touching the surfaces of grains, is attained by pure quartz sandstones during pronounced folding.

In order for boundaries to achieve the configuration of straight lines,

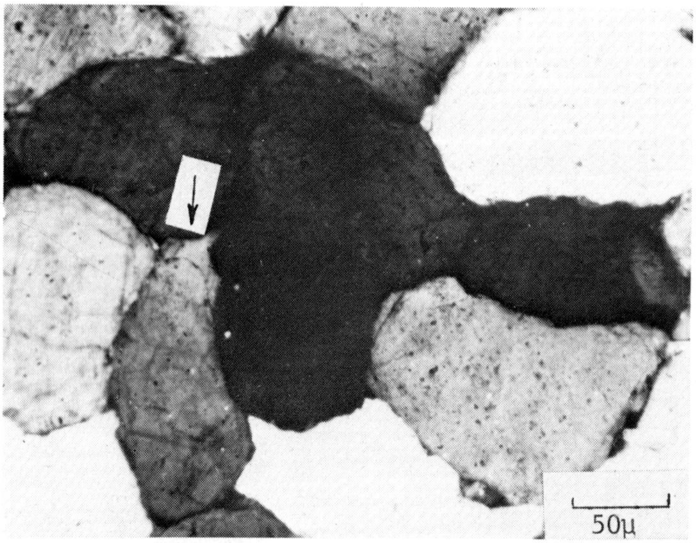

Fig. 2-8. Compound grain (dark gray) developed from four grains in pure-quartz sandstone. Outlines of original grains are poorly visible, but compatible cement welds are generally complete, exhibiting coherent boundaries. Epitaxial growths, visible on some grains (arrow), are not completely recrystallized. Significant reduction has occurred in intergranular porosity.
X-nicols; Oriskany Sandstone, Devonian, Roanoke, Va., U.S.A.

their recrystallization occurs in progressive stages, and grains of angular outline originate through processes generally regarded as responses to Riecke's principle (Renton et al., 1969, fig. 17). The evolution toward straight-line crystal boundaries (i.e., partly coherent) begins with simple compatible welds which may involve self-boundaries between overgrowths or between two detrital grains in close optical orientation. Where the latter occurs through a compatible weld, a *compound grain* is formed. A new grain of irregular outline and shape is produced which differs both in size and shape from the originals and which has a generally uniform optical orientation (Figs. 2-4, 2-7 and 2-8). Compound grains are of commonplace occurrence, and by progressive addition several original grains may constitute a single welded composite or cluster (Aalto, 1972, p. 333). Maximum development is achieved under conditions which facilitate solution and recrystallization of

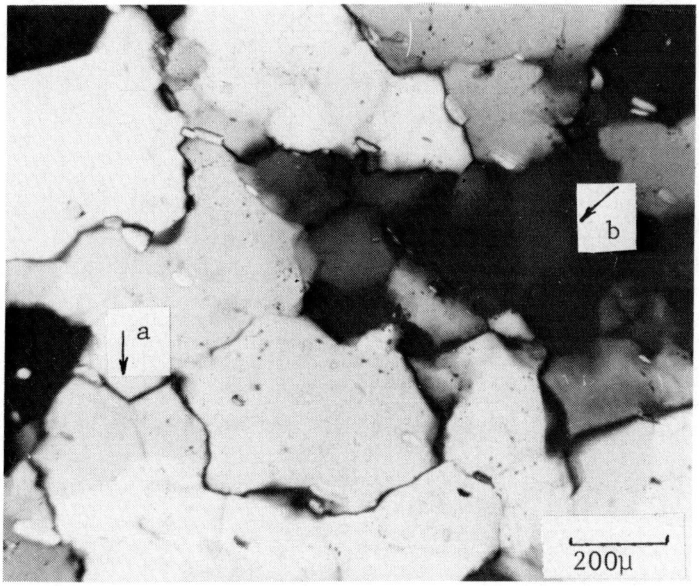

Fig. 2-9. Flexible sandstone (Itacolumite) showing individual compound grains which have reached culminating triple-grain junction texture (arrow, area *a*), but which locally are not welded together. Compound grain growth leads to inter-penetration textures with bilateral and triple-grain junctions. Dark area indicated by arrow (position *b*) illustrates a complex compound grain (in position of extinction). Grains of such shape are products of repeated, partial solution, welding of boundaries, and addition of newly crystallized quartz; this culminates in coherent boundaries. Original grain boundaries no longer are distinguishable.
X-nicols, Itacolumite, Minas Geraes, Brazil.

quartz on separate parts of the same grain. The most stable configuration is attained when the grain interlock achieves a three-dimensional framework of approximately planar boundaries. Such a texture is the counterpart of the triple-grain junctions produced in metamorphic rocks, and similar triple junctions occur in sandstones which illustrate advanced development of compound grains. Certain quartzites and sandstones identified as *ganister* and *flexible sandstone* characteristically display such textures. According to this interpretation, initial intergranular boundaries pass through stages of increased complexity as compound grains are developed. As new crystal growth occurs and compound grains increase in number, however, intergranular boundaries continue to migrate but become increasingly planar to produce bilateral junctions as observed in thin section (Fig. 2-9). Remaining detrital portions of grains become increasingly reduced as the stable arrangements are attained and ideally triple-grain junctions become commonplace (Sippel, 1968; Spry, 1969).

The culminating stage of cementation of pure-quartz sandstones is one of bilateral boundaries and triple-grain junctions which when attained remain

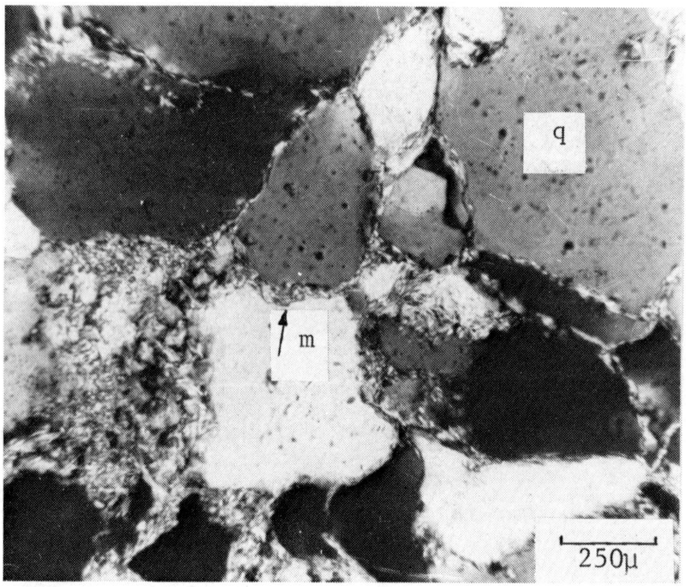

Fig. 2-10. Quartzite showing quartz grains (*q*) with intergranular boundaries characterized by crystallization of secondary muscovite (arrow, area *m*). In this example inter-suturing of quartz grains is virtually non-existent. The fact that the quartzite breaks with conchoidal fracture suggests some inter-suturing and welded boundaries.
X-nicols; Ajibik Quartzite, Precambrian, Marquette, Mich., U.S.A.

stable and irreversible. Only with the onset of strong metamorphism is this configuration modified. Short of this situation, a quartzite of low porosity and permeability persists indefinitely.

Strong contrast is to be noted between the evolution of textures characterized by a culmination of compound grains of bilateral and triple-grain junctions and that of textures of quartz sandstones in which small amounts of clay minerals occur in interstitial space. In these examples progressive intersuturing of grains occurs under the influence of increased pressure and elevations in temperature. The culminating texture is a quartzite of compound grains of highly sutured outline commonly accentuated by recrystallization of the clay minerals into muscovite, biotite or chlorite. These phyllosilicate minerals tend to outline individual and compound grains and serve as indicators of the original granular aspects of the rock (Fig. 2-10).

BONDS DEVELOPED IN THE SUBWACKE FRAMEWORK

Types of simple bonds

Subwacke frameworks tend to be primarily sand-grain supported and only locally are such grains separated by a matrix of clay or silt. Where individual sand grains occur as clusters, local conditions for precipitation of simple cement exist. Elsewhere small amounts of clay promote grain inter-penetration. Much pore space in subwackes contains some detrital clay which tends to bind the sand grains by simple clay bond in the earliest stages of diagenesis. Others show recrystallization of clay minerals (Fig. 2-11) or authigenesis of micas during the stage of diagenesis identified as *phyllomorphic* (Dapples, 1962).

Modifications in the simple cements of subwackes

Crystallization of carbonates. Carbonates, particularly calcite, form simple cement of subwackes and are scattered throughout the rock locally filling pores. Some cement occurs between grains of sand size where there has been little or no intergranular penetration or development of epitaxial quartz. Elsewhere, it is crystallized in pores partially occupied by aggregates of detrital clays.

In local stratigraphic positions, particularly at sites of lateral or vertical proximity to limestones or zones of sideritic concretions, calcite may be crystallized in pore space and as veinlets in extraordinary abundance. Grains formerly in contact are separated and the detrital framework expands by displacive precipitation. Similarly, interstitial clay and/or clay matrix are

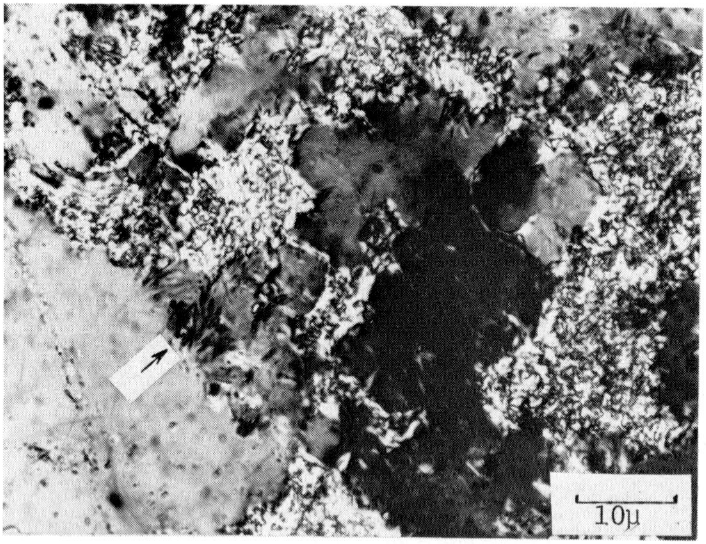

Fig. 2-11. Equilibrium assemblage of chert (gray, center), muscovite (spotted, white), and monocrystalline quartz (gray, lower left) in subgraywacke sandstone. Arrow points to intergrowth of muscovite and chert which has welded to unstable margin of detrital monocrystalline grain. This aspect of cementation illustrates the gradational boundary between former clay matrix and detrital grain of quartz. Presence of large amounts of authigenic muscovite indicates that phyllomorphic stage is well advanced.
X-nicols; sandstone bed in Rome Shale, Cambrian, Kingston, Tenn., U.S.A.

engulfed and generally disbursed within the calcite in an effect which simulates that produced with iron-oxide simple cement previously described. Locally, displacive precipitation tends to form large crystals of poikilotopic texture (Fig. 2-12). Many such crystals have well-developed habits, and in thin section self-boundaries tend to be straight lines, i.e., illustrating stable bilateral junctions.

Corrosion of quartz grains. Replacement of quartz grains by carbonate has been long recognized; currently preferred explanations involve pH control over the solution of quartz and crystallization of calcite as suggested by Walker (1962). Where displacive precipitation by crystallization of calcite cement has occurred, replacement plays a minor role. Corrosion of quartz and replacement by calcite is more frequently noted where introduction of calcite cement is important but is not accompanied by volume expansion. Commonly, replacement cementation appears to be related to carbonate introduced into deeply buried or strongly folded strata, e.g., as may be observed in parts of the Oriskany Sandstone in the region of the middle

Fig. 2-12. Poikilotopic texture developed by growth of large single crystals of calcite (note continuity of twin lamellae). As a result of crystal growth, grains which originally were in arenite and subwacke frameworks have been dispersed. Calcite constitutes an incompatible cement indicated by sharp boundaries between grains and cement. X-nicols; Muddy Sandstone, Upper Cretaceous, Thermopolis, Wyo., U.S.A.

Appalachian Mountains. Seilacher (1968, p. 184) has shown that within parts of the same formation, in the same region, quartz sand grains are strongly replaced by calcite, whereas the associated brachiopods retain much of their fibrous structure. He interpreted this condition as indicating that the solutions from which the carbonate was crystallized affected the quartz grains only and did not modify the carbonate in the already altered brachiopod shell.

Replacement of quartz by calcite in carbonate deposited during the Glacial Epoch is reported by Gavish and Friedman (1969, p. 986) and demonstrates unequivocally that the processes involved are neither pressure- nor temperature-dependent. The association of replacement of detrital quartz by carbonate with strongly folded beds is clearly fortuitous and should not be regarded as a pressure-solution phenomenon.

Crystallization of chert. Among pure quartz sandstones of arenite frameworks, occurrence of chert as a cement is rather uncommon and the typical epitaxial growths are monocrystalline quartz (Fig. 2-7). In subwacke texture, chert may be as abundant as monocrystalline quartz and occurs in two positions: (a) as epitaxial cement on quartz grains which show interpenetration boundaries, and (b) as intergrowths in small masses of interstitial or matrix detrital clay. Abundance of the latter occurrence appears to be directly proportional to the amount of clay present, i.e., as the framework approaches wacke, chert is likely to be the type of quartz cement present. Also, chert which is added to detrital grains as epitaxial quartz, commonly appears to have spread from chert deposited in positions within the concentration of clays (Fig. 2-13). This form of cement is important in lithification of subwackes and wackes. Rock of very low porosity and having exceptional resistance to fracture results in the case of predominance of this cement type.

A satisfactory explanation for the causes of crystallization of chert rather than monocrystalline quartz has not been formulated as yet. Polycrystalline quartz appears to be thermodynamically stable under the diagenetic environment where subwacke frameworks are chiefly produced, namely, in thick clastic deposits associated with basins, on or marginal to the craton. Here,

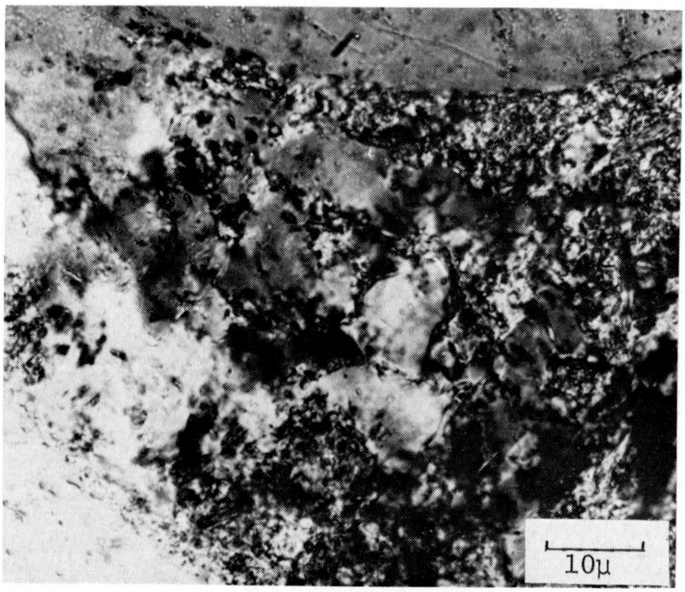

Fig. 2-13. Strained grains of monocrystalline quartz (white, left margin), which have undergone partial solution with epitaxial growth of chert (white with irregular boundaries) to constitute a compatible weld. Chert has crystallized in the interstitial clay and has formed the principal cement of the grains and matrix. Such chert grains approximate a trend toward uniform dimension illustrating coalescent neomorphism. Compare boundary between chert, clay and quartz with that shown in Fig. 2-1, in which grains are held by simple clay bond. Monocrystalline quartz grain (upper margin) is slightly strained, only, and boundary with clay is locally sharp but merges into chert at upper left.
X-nicols; sandstone bed in Rome Shale, Cambrian, Kingston, Tenn., U.S.A.

moderately elevated temperatures and pressures are imposed upon the sediments.

Quartz crystals of small size and having relatively large surface area per unit of bulk volume appear to be the stable forms, whereas grains of sand size tend to interpenetrate and develop compound grains of irregular outline and also relatively large surface area.

The writer regards some of the concepts proposed by Folk (1965, p. 22) for recrystallization of calcite in carbonates to be applicable to the chert—quartz relationship in subwackes. Folk introduces the term *porphyroid neomorphism* to identify a recrystallization process whereby certain crystals throughout a carbonate rock grow more rapidly and attain larger dimension than others. The porphyry-like texture is to be distinguished from that represented by a mosaic of crystals of approximately similar dimensions. The latter is achieved by a process identified as *coalescive neomorphism* by which larger grains are consumed and small grains enlarged until the general grain size of the mosaic approaches uniformity.

Certain individual examples of textures of quartz are regarded as useful in the interpretation of which is the more stable form under specific conditions. In some sandstones, crystals of strained monocrystalline quartz have undergone partial solution, whereas chert was crystallized in the clays of adjacent interstitial pores (Fig. 2-13). Presumably, the chert crystallizing by porphyroid neomorphism would, had the process not been arrested, develop an area of intergrowth crystals having more or less uniform individual sizes. Instability of the strained quartz relative to the chert can be considered an example of strain-induced boundary migration, with the free energy of the chert being less than that of the strained grain of monocrystalline quartz.

As compaction progresses, there occurs corresponding reduction in the size of open pores. Detrital interstitial and matrix clays, as well as clay-aggregate pellets, are squeezed into contact with the predominant quartz sand grains. Where such contact occurs and monocrystalline quartz grains are somewhat strained, crystallization of chert in the clay and solution of the quartz grain develops a welded boundary between the clay and quartz. The chert acts as a welding agent attaching masses of clay to monocrystalline quartz.

Among certain subwacke frameworks there are quartz grains of sand size which do not appear to be strained nor do they show corroded margins. In the adjacent clay matrix, however, chert has been crystallized seemingly as a result of a different mechanism than that described above. For some such examples, the writer considers the chert to be a product of the recrystallization and neoformation of new clay minerals from the detrital mixture. The concept presented is that some initial clays are not stable in the burial environment. Their instability is reflected by reorganization into new clay minerals of somewhat different composition, which are stable under the new conditions of temperature, pressure and ion concentrations of the pore fluids (Powers, 1959). If these clay minerals are stable with quartz, and extra silica was released during the authigenesis, small clusters of quartz intergrowths (chert) would appear scattered throughout the new clay minerals. The individual quartz grain of sand size remains unaffected.

Figs. 2-11 and 2-14 illustrate the products of the authigenesis, as described above, and the cementation which is a product of that alteration. Growth of the chert in a cluster of newly formed individual grains acts as the prime cementing agent within the clay matrix and produces an interlocked weld with the adjacent detrital quartz grain. These relationships are to be observed in many subgraywacke sandstones in which the framework indicates the presence of considerable interstitial and localized matrix clay. Similarly, where quartz grains of silt size are in contact with grains of sand size, they may: (1)

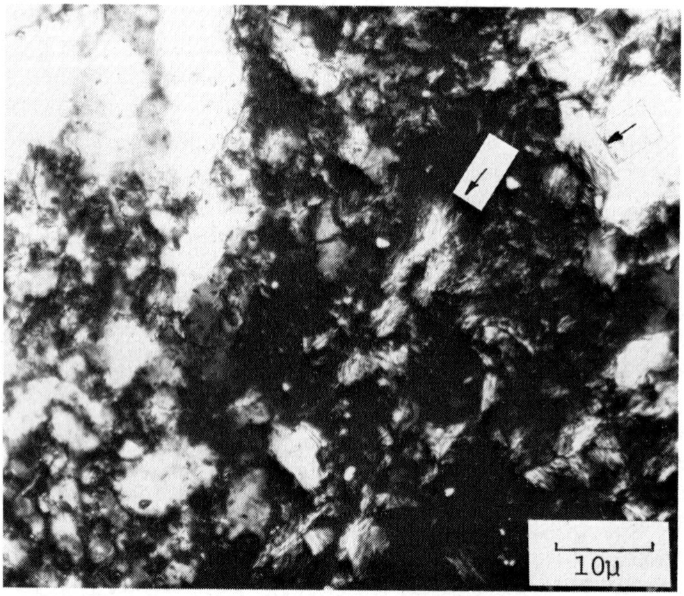

Fig. 2-14. Chert—chlorite mixture in interstitial space of subgraywacke sandstone. Chert (white grains of irregular outline, lower left) has crystallized in clay and is welded by compatible mineral bond to detrital quartz (left margin). Authigenic chlorite (fibrous and white, center area) is associated in an equilibrium mineral assemblage with chert; note intergrowth boundary indicated by arrow. Large sizes of some chert grains illustrate porphyroid neomorphism.
Pocono Sandstone, Mississippian, Dunmore, Pa., U.S.A.

become engulfed in the larger grain as part of a compound grain, or (2) recrystallize as polycrystalline chert, which appears to be somewhat more stable thermodynamically than monocrystalline quartz in the presence of clay minerals. The principal cementing action, however, is carried out by the behavior of the clay-size matrix as a source of crystallizing quartz.

Reorganization of the lattices of detrital clay minerals, which are not stable in the burial environment, into new and more stable configuration is part of the diagenetic stage identified as *locomorphic* (Dapples, 1962). Locally, such reorganization produces small concretion-like masses which in scanning electron microphotographs appear to reveal edge-to-flat surface frameworks typical of electrically stable stacking configurations.

Effects of directed stress. Under conditions of shallow burial and no folding, grains of sand size are held largely by simple clay bonds and simple cements. Where thick accumulations of sediments are found accompanied by rather intense rock deformation, however, a progression of cementation occurs which involves neoformation of minerals. In the extreme cases, this progression may pass into the zeolite grade of metamorphism (Coombs et al., 1959). As these conditions are approached, there are increases in temperature, pressure, and changes in ion concentration which bring about pronounced reorganization of the clay and silt sizes of the detrital mineral fraction. Such aspects of diagenesis dominate the conditions of cementation and lithification.

In those frameworks having only small amounts of matrix, and where many of the grains of sand size touch without any intervening clay, the characterizing feature is the local development of compound grains and well-interlocked margins with other grains. In such local cells the culminating texture is that illustrated by the pure quartz arenites, namely, bilateral and triple-grain junctions. Elsewhere in the same thin section other detrital grains of quartz have boundaries with clay aggregates, along which a common feature is chertification of the clay and welding of chert to the quartz grains. Such interlock of polycrystalline grains is associated with filling of local pore openings and is an important cementing process. The effect is modification of the original boundaries of the detrital grains to produce a gradational aspect between the chert which crystallized in the matrix clay and that which becomes welded to a grain of detrital quartz.

Culminating textures among subwackes show the compound-grain interlock characteristic of arenites, and the gradational boundaries of chert intergrown in the localized clay matrix and the detrital quartz grain. If the localized concentrations of clay are few, the texture will be dominantly that of the quartz arenite. Elsewhere in the same thin sectin, where clay is more abundant, such textures will be supplanted by the gradational boundaries of chert, the indeterminate grain—clay margin, and the interlock cementation produced by the neoformed chert.

Reconstitution of clay minerals into micas. Any local cells of clay concentration are loci of reconstitution of clay minerals and crystallization of micas. Chlorite, biotite and muscovite are the most commonly crystallized minerals, and their behavior as cement differs slightly. Of these micas and mica-like minerals, chlorite is commonly the most abundant and may be intimately interlocked with chert (Fig. 2-14). Also, in the early stages of development of the phyllomorphic stage, a "fringe" of chlorite crystals may form around the border of detrital grains of quartz where pore openings exist (Fig. 2-15). This more or less radial arrangement passes into a zone where some of the newly formed crystals may show a pattern of flow lines developed where clay minerals had been forced during compaction into intergranular space. A localized nonhomogeneous stress is regarded as the agent which caused preferred arrangement of some of the clay plates and neoformed "micas".

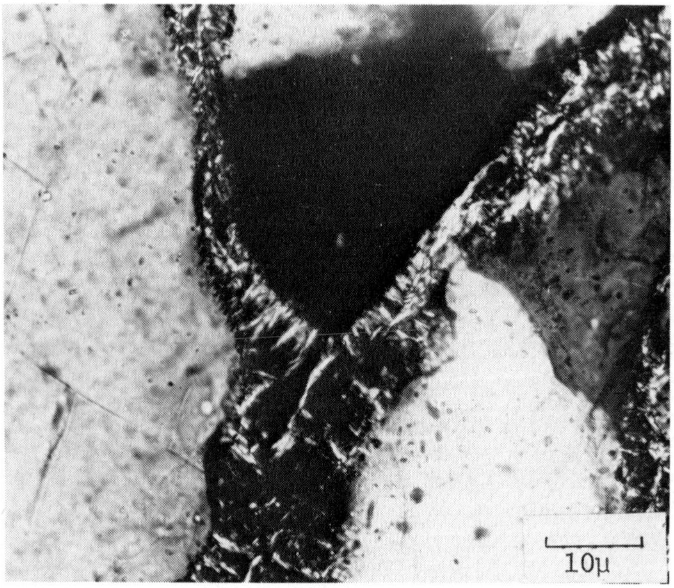

Fig. 2-15. Oriented fringe of authigenic chlorite attached to surface of detrital quartz grain (black) in subgraywacke sandstone, illustrating reconstitution of clay minerals to form authigenic growths in pore spaces. This illustrates locomorphic stage which, where more developed (as in Fig. 2-11), passes into the phyllomorphic stage.
X-nicols; Jacobsville Sandstone, Cambrian, L'Anse, Mich., U.S.A.

Except under uncommon circumstances, biotite and muscovite tend to form within the clay matrix and do not appear to develop fringes around quartz as does chlorite (Fig. 2-11). The effect of the "mica" crystal growth, however, is to produce a cementation interlock within the original clay fabric, and also to form an intergrowth with the margins of detrital quartz grains and certain rock fragments, e.g., shales and schists. In this connection, Frey (1970, p. 270) has documented the development of neoformed micas and their textural aspects in his study of anchimetamorphism in pelitic rocks of the Glarus Alps. For a comprehensive analysis of the behavior of clay minerals in the progression to metamorphic grades, the reader should refer to Dunoyer de Segonzac (1970).

Grains of detrital feldspar tend generally to be unstable in the environment where chlorite is the stable "mica". Such grains frequently are invaded along the twin lamellae, or cleavage lines, by reconstituted "mica". Muscovite and biotite behave similarly, unless where chlorite is present as an important authigenic mineral. In part, this is explained by what appears to be episodes of phyllosilicate mineral crystallization, i.e., the tendency for chlorite to be crystallized earlier than biotite and muscovite. Perhaps this is the result of the requirement of the potassium ion concentration to build up to certain values before biotite or muscovite can crystallize; or, in certain sediments, the Fe^{2+} concentration may be too high to permit muscovite to crystallize.

CEMENTATION OF WACKE

Inhomogeneity of detrital mixture

The processes which have been described as responsible for cementation of grain-supported frameworks tend to be reduced to minor roles in the cementation of wackes. In these, grains of sand size generally are isolated by the fractions of smaller sizes. Self-boundaries between grains of the same mineral occur in isolated clusters only. Where two or more quartz grains of sand size are in contact, however, the original boundary between them is forced to migrate in accordance with the stages described for arenites. Films of clay separating individual grains lead to the development of sutured boundaries, but where such films are absent, bilateral junctions are observed between grains (Fig. 2-8). The latter, however, are rare because compaction of the original mixture tends to force clays into all available original openings.

Wacke fabrics encompass also a wide variety of sand compositions ranging from those of tuffaceous character, in which the most abundant debris is of volcanic origin, to mixtures each rich in rock fragments, or feldspars, or quartz. Mixtures of fragments of splinter, tabular, or elongate shape tend to develop a fabric under the effects of compaction which illustrate the effects of localized movement. Clusters of sand grains having these shapes, are oriented by the small movements, particularly of the clay which shifted into the available openings. The entire effect of such compaction is to produce cells of fragment inhomogeneity, which may range in dimension from microscopic to large structures modifying the original bedding, e.g., load and slide structures. Inhomogeneity of composition and fabric characterize the ideal wackes; however, as the framework approaches subwacke, more orderly arrangement and uniformity in fragment distribution are to be observed. Subwacke framework is accompanied by an increase in quartz sand particles. Correspondingly, localized clusters of such grains will show self-boundaries of various types (Fig. 2-11). The fabric tends to exert a rather important indirect control upon the composition of the local cells men-

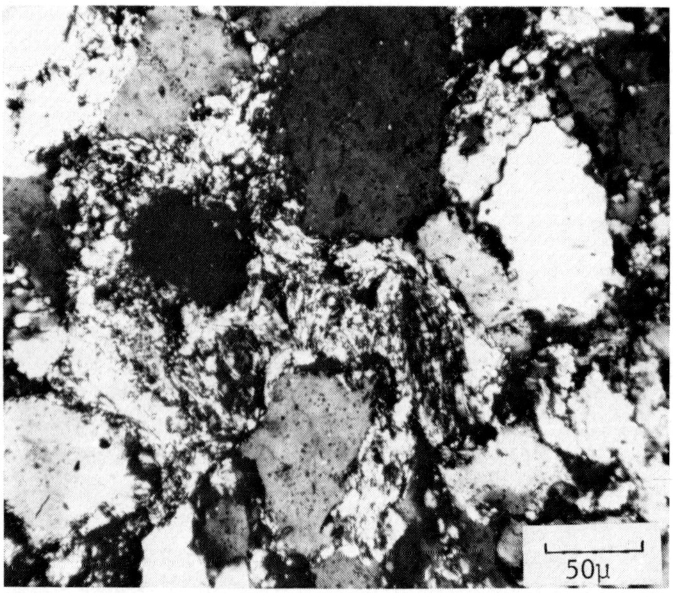

Fig. 2-16. Suggested movement of matrix clay during compaction and diagenesis of a wacke framework. Neoformed micas (areas of white fibre) show arrangement around detrital grains. Note sutured boundaries between quartz grains (upper right) separated by films of clay. These will not culminate in bilateral or triple-grain junctions. Chert crystallized in matrix (white and gray areas, lower right) illustrates coalescent neomorphism and welds to detrital quartz along gradational boundaries.
X-nicols; Mauch Chunk Formation, Mississippian, Rockport, Pa., U.S.A.

tioned above. In the examples of wackes with much clay matrix, local differences in mineral composition are much more abundant than where clay is minor. Because of such localized differences in composition, there are corresponding differences also in the progression of cementation (Fig. 2-16).

STAGES OF DIAGENESIS

Processes in the environment of early burial, that is, during initial compaction, ejection of fluids, and prior to lithification, are dominated by oxidation and reduction reactions. Because such reactions are very important, this stage has been called *redoxomorphic*, to signify the existence of reversible reactions dependent upon oxidation potentials, primarily manifested in the kind of iron-bearing mineral which crystallizes (Dapples, 1962). The redoxomorphic stage tends to establish the bulk final color of the rock. This is most strikingly noted among sediments in which the total iron oxide content exceeds 5%, but red colors can also develop in strata that are lower in total iron content (Picard, 1965; Thompson, 1970, p. 611; Van Houten, 1973, p. 42). Once the reactions have been driven to oxidize or reduce iron, there is much less tendency to reverse the equilibrium during the later history of the rock (Berner, 1969, p. 272). Nevertheless, "bleached areas" and other later color changes are abundant but are presumed to occur on a small scale.

A second and more advanced stage involves significant precipitation of mineral matter in the pore spaces and particularly as replacement of detrital mineral grains. This is the time of primary cementation and the development of induration. Replacement of interstitial clay by chalcedony, calcite by siderite, and quartz by calcite are common examples. In the usual case, the replacement is not pseudomorphic but rather a partial in situ substitution of one mineral for another. This period of important mineral replacement follows in time the principal redoxomorphic reactions, and is identified as the *locomorphic* stage signifying a change of shape in situ. During the locomorphic stage the sand sediment gradually achieves partial or complete lithification and becomes the initial rock. Any additional diagenetic modification which may occur, takes place on a lithified aggregate and for this reason is considered by this author to be associated with its late burial history.

A common characteristic of late burial is increase in crystallinity in illite, development of chlorite, and, in the most advanced stages, alteration of clay minerals into mica. For this reason this late episode of diagenesis is termed the *phyllomorphic* stage. Certain other minerals, particularly feldspar, appear as authigenic growths during crystallization of the phyllosilicates. Although feldspars may crystallize soon after sediment burial, currently they are included as part of the phyllomorphic stage, inasmuch as they are regarded as representing a most advanced stage of diagenesis. In some localities where the phyllomorphic stage is well advanced, there is believed to be gradation into the zeolite and chlorite grade of metamorphism without any currently recognized limits (Packham and Crook, 1960).

Redoxomorphic changes

Oxidation and reduction reactions can be demonstrated to dominate modification of the sediment during and immediately after burial. During this time, compaction is in progress and fluids are being ejected, with concentration gradients toward the depositional interface (Weller, 1959; Von Engelhardt and Gaida, 1963). Principal reactants involved are iron, oxygen, sulfur and carbon. Deposits consisting of a significant fraction of organic matter tend to contain sulfur as well as carbon. Of these, carbon compounds appear to be most rapidly oxidized and may be regarded as contributing electrons to drive the iron into the ferrous state. This results in the fixation of the sulfur as pyrite. As long as the organic fraction remains important, the gray color will prevail and pyrite will be scattered throughout the rock, often in considerable amounts (Love, 1971).

Among red-colored sandstones essentially two somewhat distinct conditions may dominate the redoxomorphic stage:

(1) A situation may exist during which the burial environment is oxygenated by contact with the atmosphere and iron oxides arrive as part of the detritus, or clay minerals with attached iron ions are part of the inorganic suspension load (Carroll, 1958). In the burial environment oxygen gathers electrons principally donated by iron to form hematite and related ferric oxides or hydrates, which along with those having arrived as part of the detritus remain stable (Van Houten, 1973, p. 45). A texture can be observed showing rock fragments, grains of quartz, and other minerals more or less isolated by mixtures of clay minerals and iron oxides as films, matrix, or pore filling. Such a mixture is modified texturally only as a result of differential compaction between the sand and clay sizes. No mineralogic reaction is noted between the iron oxides and the sand grains and the relation is that of inert substances; and grain—matrix boundaries stand out in clearly defined demarcation (Swineford, 1955, p. 155). The detritus composed in part of clay minerals with attached Fe^{3+} ions would tend to be brown or red after deposition (Klein, 1963b). In other circumstances, the iron could arrive in the reduced form and oxidation would take place during deposition; hence, a gray detritus would become red. Inasmuch as the precipitation of iron hydroxide is indirectly controlled by the pH (Garrels, 1960, pp. 116—145), a slightly acid condition of many streams would tend to keep

the iron primarily in the reduced form during transportation, and red color would develop only after deposition in an environment of higher pH, such as a sea or playa lake (McBride, 1974, p. 764).

(2) A more common situation is believed to prevail during which the equilibrium between Fe^{2+} and Fe^{3+} is shifted toward oxidation sometime after burial. In its most obvious form, this can be seen in strata in which the red color transects bedding, and coloration follows fractures or permeable positions related to introduction of water carrying dissolved oxygen. Conversely, bleaching of the red color can be demonstrated for masses of irregular outline or for concentric zones about some center. Paragenetic relations between minerals from such zones show rather persistent association between iron oxides, biotite, chlorite, siderite and calcite (see Greensmith, 1957, p. 410; Thompson, 1970; Horowitz, 1971). The occurrence of biotite, chlorite, siderite and calcite in bleached zones (i.e., where the iron is in the reduced state), and decomposed condition or absence of such minerals where the iron is oxidized, are interpreted to illustrate the following reactions:

(a) Iron-rich silicates such as biotite and hornblende are oxidized to hematite and clay minerals, preferably illite and kaolinite but also montmorillonite (Robb, 1949; Hubert, 1960, pp. 137, 147; Walker et al., 1967; Teisseyre, 1973, p. 459).

(b) In a slightly reducing environment, particularly in the presence of crystallizing calcite, biotite tends to remain stable or alter to chlorite or illite and minor iron oxides (Friend, 1966; Cadigan, 1967, p. 28; Hayes, 1970; Teisseyre, 1973).

In the red-stained parts of the Minturn Formation (Pennsylvanian) of Colorado, immediately surrounding bleached zones, calcite which is present in the bleached portion extends a short distance into the red rock. Here it tends to split detrital grains of strongly oxidized and decomposed biotite along cleavage laminae (Fig. 2-17). In the red-stained rock, biotite is obviously unstable and tends to decompose to clay mineral and ferric oxide, some of which is hematite. Within the bleached zone undecomposed biotite appears stable in the presence of calcite. Biotite grains are not invaded by calcite and the two minerals have sharp and regular boundaries. Similarly, chlorite tends to disappear in the red-stained rock, whereas it is significantly present in the bleached zones. The interpretation currently favored is that in an oxidizing environment detrital biotite is unstable and decomposes first to oxidized biotite and later to clay mineral and ferric oxides. Conversely, in a calcite-precipitating environment ($pH = 8$) biotite is stable and will form as authigenic crystals if a reducing environment prevails. The occurrence of chlorite in greater amounts in the bleached zone in association with calcite and its absence in the oxidized zone, however, is regarded as suggestive that chlorite is more stable than biotite in a calcite-precipitating environment.

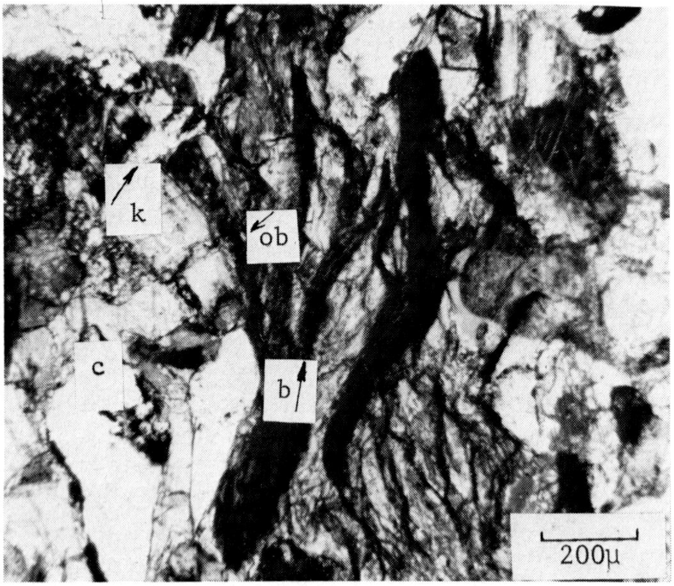

Fig. 2-17. Detrital biotite (arrow, *b*) in red sandstone split along cleavage laminae by calcite cement, illustrating displacive precipitation. Biotite partially altered to iron oxide (*ob*) is locally cut and surrounded by secondary clay mineral (white, *k*). Area *c* represents calcite crystallized by displacive precipitation.
Minturn Formation, Pennsylvanian, near Minturn, Colo., U.S.A.

Occurrence of secondary biotite is common in the reduced zones of many red-bed sandstones, but its distribution is sporadic ranging from local abundance to absence. Crystallization of authigenic biotite within masses of interstitial clay in bleached zones, however, can be established. Miller (1957) reported that in the center of "reduction zones" in Pierce Canyon (Permian, U.S.A.) red beds, biotite is absent whereas secondary calcite is present. Outward from these centers, however, biotite has formed in the presence of precipitated gypsum and calcite. In the same reduced zones, magnetite, which is abundant in the red rock, is absent. This observation had been previously reported by Miller and Folk (1955) in red beds of other ages and localities, and is attributed by them to demonstrate instability of the iron oxide under the conditions which produced the bleached spots. In the Pierce Canyon strata, biotite is stable in an outer shell of the "reduced zone" in association with calcite, and the equilibria indicated above have been driven in favor of crystallization of biotite. The implication to be drawn is that the reactions are more sensitive to relative degrees of oxidation and reduction than to pH, but that chlorite and biotite will crystallize from a solution having a pH = 8. Keller (1953) has shown that green-colored reduced zones in certain red beds contain illite rather than chlorite. Perhaps the presence of illite may be important in determining the variety of mica which crystallizes as a stable phase under certain H-ion concentrations.

Siderite is a common, although not abundant, early precipitated cement in many subgraywacke sandstones (Triplehorn, 1970, p. 842). It is especially observed as concretions and individual crystals in subgraywackes associated with coal beds, and where there is evidence that brackish water prevailed at the depositional site. A necessary item appears to be that enough carbonaceous matter was incorporated to maintain a higher concentration of ferrous than ferric ion. An example is reported by Rusnak (1957, p. 47) in which the order of pore filling in the Pleasantview Sandstone (Pennsylvanian, U.S.A.) shows calcite partially replaced by siderite. Both calcite and siderite were later somewhat replaced by quartz precipitated as overgrowths on detrital grains; hence, crystallization of the siderite is intermediate in time of development. The siderite remains as a stable phase as long as reducing conditions prevail; however, wherever ground waters carrying dissolved oxygen have penetrated the rock, the siderite has at least partially oxidized to limonite.

The presence of calcite and ankerite as small intergrowths in the Bald Eagle and Juniata strata (Ordovician—Silurian, U.S.A.) is suggested by Horowitz (1971) as indicating a diagenetic equilibrium pair. Accordingly, he interpreted their presence as an indication of crystallization from reducing solutions introduced during compaction of underlying marine sediments.

Locomorphic changes

Opal and chalcedony. In many sediments, modifications identified as locomorphic are simple and involve changes in the pore cement only. Such is the case with opal filling in certain sandstones. An example is the Ogallala Formation (Tertiary) in U.S.A. where at certain localities opal has been deposited as a primary simple cement (Fig. 2-18; see also Swineford and Franks, 1959). At least some of this opal is the opal C-T variety (Calvert, 1966, pp. 574—576; Jones and Segnit, 1971), inasmuch as X-ray diffractograms show broad peaks of cristobalite and tridymite. In some specimens opal has partially filled interstitial pores, whereas in others the amount of precipitation has been large, and detrital grains obviously have been separated. Clearly an increase in total volume from that occupied by the unconsolidated sediment has been brought about by opal precipitation.

Other occurrences of precipitation of silica in which opal and cherts are associated are recorded by investigators of silcretes and similar crusts developed on the weathering surfaces (Williamson, 1957; Millot, 1960; Hut-

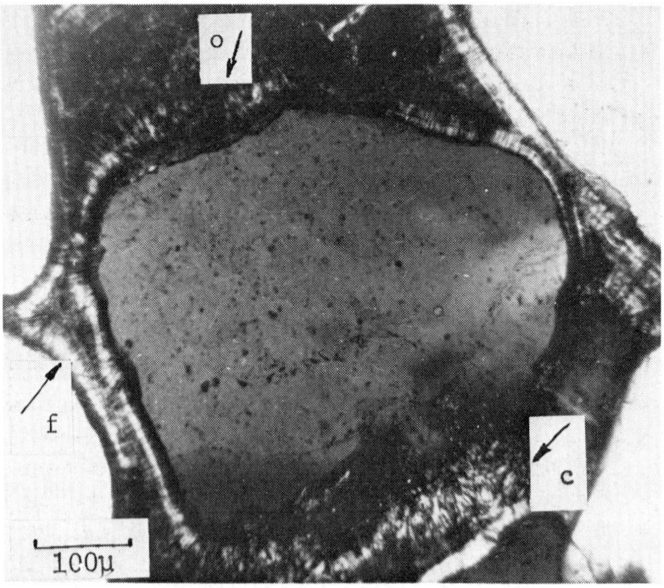

Fig. 2-18. Opal cement (*o*) in interstitial space between quartz grains altered to opal C-T, and fibrous chalcedony (*f*). The latter grades into microcrystalline quartz (*c*) which is regarded as the stable phase.
X-nicols; Ogallala Formation, Paleocene, Smith County, Kans., U.S.A.

ton et al., 1972; Smale, 1973). As currently observed, the deposits are largely in the form of chert associated with some opal and microcrystalline quartz. The occurrences appear as irregular masses and veins, commonly crudely banded, but which locally are intergradational from opal to chalcedony to microcrystalline quartz. Although for the most part the "chert" has been deposited in fractures, much is present also as fillings of original pore spaces. The latter "chert" has produced an expanded intergranular framework described earlier as displacive precipitation. In silcretes, the order of paragenesis has not been established and at present there is no positive evidence that an opal was the initial deposit. Rather, the relationship suggests some form of equilibrium which permitted one variety of silica to be precipitated, later followed by another as the equilibrium trend was shifted.

Although opal as a cement is distributed sporadically in Tertiary or younger sandstones, its occurrence is rare in older beds, a feature which is interpreted to indicate it is not stable in the late burial environment. An example of the replacement relations was reported by Friedman (1954, p. 239) who interpreted a paragenetic sequence as follows: shell calcite is replaced by opal which in turn is transitional into fibrous chalcedony.

Quartz is found only adjacent to the chalcedony, which always separates the quartz from the opal. The replacement was evidently volume for volume (see Matter, 1974, p. 439). Transition from opal to quartz appears to be a product of some aging process, but possibly the ordering in the lattice is inhibited by the presence of some large cations such as K^+ or Ca^{2+} present in the loosely organized opal molecule. During conditions of late burial, such ions could be ejected along with water as the quartz lattice structure develops (Folk and Weaver, 1952). Whatever structural modification is involved, it is unidirectional from opal to quartz and appears irreversible within the physical conditions of stability of sandstone. Moreover, the progressive nature of the change is regarded as some form of opal → cristobalite → chalcedony → microcrystalline quartz (see Mizutani, 1966, p. 77; Calvert, 1971).

Except for the opal generated in the structures of organisms, the inorganic precipitation of opal appears to be an uncommon event but certain physical conditions tend to favor its direct precipitation. One such condition appears to be supersaturation of water with silica, such as develops in hot-spring localities or from surface waters leaching volcanic ash beds (Keller and Reesman, 1963, p. 432). Apparently, large, loosely organized molecules of hydrated silica precipitate as a colloid. The latter becomes organized into a structural lattice of cristobalite during burial, but retains the outward appearance of chert. The other common precipitant of opal is cellulose, or a related carbohydrate, in which by some mechanism substitution of silica and reorganization of the organic molecule occurs as the cellulose is slowly oxidized after burial. Reorganization into chalcedony tends to destroy the details of organic cell structure preserved by the opal.

Much silica appears to be precipitated directly as some form of cristobalite, or, as reported by Hay (1968) and Eugster (1969), through some sodium silicate intermediate compound without going through an opaline phase. According to Greenwood (1973, p. 706), opaline tests of organisms do not alter to cristobalite by recrystallization in the solid state. Rather, the opal dissolves and the silica is reprecipitated as cristobalite or as chalcedony and microcrystalline quartz. Silica precipitated as cristobalite transforms to chalcedony and microcrystalline quartz without the requirement of deep burial, but time is an important factor.

Substitution of chalcedony or microcrystalline-quartz chert for clay matrix and interstitial clay is a locomorphic change of common occurrence (Triplehorn, 1970, p. 843). The replacement is particularly well observed in subgraywacke sandstones, which show some simple clay bonds fastening sand-sized grains (Fig. 2-2). Locally chert can be recognized to have replaced part or all of such interstitial clay. The nature of this substitution is not entirely clear. Limited investigation has not established whether the silica has been precipitated in the clay aggregates or whether it has in fact replaced part of the clay-mineral lattice by substitution in the octahedral layers. The latter mechanism would require migration of aluminum for which there is little evidence at present. Simple precipitation of silica within the openings between individual clay-mineral crystals currently is favored. Inasmuch as the clay-size interstitial material is known to contain small particles of quartz, these could act as nuclei for precipitation of additional silica.

Precipitation of chert in the matrix normally is an early locomorphic process and is considered to follow closely, if not to be contemporaneous with, processes of the redoxomorphic stage. Chert precipitation, however, continues as the locomorphic stage advances and welding of detrital quartz grains gives rise locally to highly silicified sandstones, as, for example, near fault zones. Under such circumstances, the process can be demonstrated to have occurred during the postlithification stage and obviously late in the sandstone's history.

Chert replacing interstitial clay in some subgraywackes appears to grade through zones of coarse microcrystalline quartz to be welded to detrital quartz crystals or may be transitional into quartz overgrowths. Some examples illustrate recrystallization into progressively coarser individual crystals of quartz. This does not appear to be commonplace in those subgraywackes in which calcite is an important cement. Indeed it would appear that presence of calcite tends to inhibit recrystallization of chert to quartz.

The antipathetic association between silica minerals and calcite has been reported so frequently that only brief reference is made herein to this commonplace locomorphic alteration (Walker, 1962; Rapson-McGugan, 1970, p. 401). Suffice it to say that replacement of one by the other is clearly known to be reversible and that the replacement is capable of preservation of the original morphology.

Calcite replacement of clay. A frequent observation among subgraywackes is replacement of clay matrix by carbonates (Rusnak, 1957, p. 41; Laury, 1968, p. 579; Teisseyre, 1973, p. 463). Primarily such replacement is by calcite, but it may also be by dolomite and siderite. The replacement may be so complete as to give the impression that the original particle-size distribution contained virtually no fraction in the clay size (Fig. 2-19). In some sandstones the calcite has actually replaced the clay fraction to such an extent that the clay no longer is present as an insoluble residue in the carbonate. In other examples, at least some of the clay remains as a residue within the calcite (Enos, 1969, p. 22) and can be recovered on solution of the carbonate. The mechanism of such replacement is not understood. It appears, however, that certain clay minerals, known to be primarily illite and kaolinite in the case of some subgraywackes, are flocculated by Ca-ion and occupy less interstitial space, allowing the remainder of pore space to be filled by the precipi-

Fig. 2-19. Calcite (*ca*) replacement of clay matrix in subgraywacke sandstone. Dark and light grains are quartz in various positions of light extinction. The outline of quartz grains suggests little replacement by calcite, whereas the clay matrix appears to be replaced completely.
X-nicols; Fort Union Sandstone, Paleocene, near Wamsutter, Wyo., U.S.A.

tated carbonate. A reaction illustrated below, which was proposed by Eades and Grim (1960), could explain the development of calcium-silicate hydrate or calcium-aluminum hydrate and to account for lattice structure changes upon treating such clay minerals with lime:

kaolinite or illite + Ca^{2+} → calcium-silicate hydrate or calcium-aluminum hydrate

If the new crystals tend to remain very small in size, they could physically move out of the interstitial space with permeating solutions, allowing the precipitating calcite to occupy the former position of the clay. On the basis of what is known of solute precipitation, replacement of clay minerals by calcite is favored by pH above 8 and a high concentration of Ca-ion under which conditions certain clay minerals become unstable.

Another mechanism, which at present is only speculative, considers the calcium-silicate hydrate of the reaction above to be an intermediate product of silica. The reaction proposed is a counterpart of that indicated by Eugster

(1969) by which the sodium-silicate, magadiite, is decomposed by fresh water approaching neutral pH, and silica precipitates as a chert.

Calcite—aragonite replacement. Replacement of aragonite fossil shell material by calcite is an extremely commonplace unidirectional substitution. Precipitation of aragonite appears to be at least partially temperature-sensitive in that aragonitic shells are more abundant among warm-water invertebrates than cold-water forms (Lowenstam, 1954, p. 285). Transition from aragonite to calcite is more rapid under certain conditions than others. Some shells of Mesozoic age still contain aragonite, whereas others of much younger age are completely altered to calcite. Experimental data indicate that transformation of aragonite to calcite is favored by increase in temperature and reduction in pressure (Deer et al., 1962, p. 308). In sandstones, the transformation appears to be independent of the depth of burial, and fossil fragments are composed of calcite even in very young rocks which scarcely have been buried.

Inorganic coprecipitation of aragonite and Mg-calcite as authigenic crystals in small pores of modern warm-water calcareous sands is reported by Alexandersson (1972, fig. 10). Some crystals, which could represent aragonite replacements of Mg-calcite, show in fact incoherent boundaries; the aragonite has crystallized upon previously neoformed crystals of Mg-calcite as an intergrowth.

In a sand off the coast of New Jersey (U.S.A.) crystallization of aragonite cement from cold waters occurs in pore spaces, and other openings, to cement a sand which contains approximately equal parts of ferruginous grains and calcareous shell debris. The aragonite needles do not replace the calcite shell but have crystallized in a complex interlocked overgrowth (Allen et al., 1969, p. 140). Elsewhere, in Recent, non-calcareous sands of the Fraser River delta (British Columbia, Canada) cementation by calcite does occur (Garrison et al., 1969, p. 40). This cementation occurs in pore openings and shows no indication of replacement of aragonite shells.

Apparently in such Recent crystallization of aragonite the calcite—aragonite reaction does not proceed from an equilibrium condition in which calcite and aragonite replace one another. Rather, the precipitation appears to favor crystallization of either phase without partial dissolution of the other. Current understanding of the causes and conditions of calcite—aragonite replacement in siliceous sands is inadequate, and further research is required. Such is not the case with calcareous sediments and the reader is referred to Chapter 6.

Calcite—dolomite replacement. Replacement of calcite by dolomite is present to an important extent in two very distinct groups of sandstones. One is the quartzose group that passes by lateral facies changes into limestone (the quartzite series of Krynine, 1948), which at some later time is subjected to wholesale dolomitization (Sloss, 1963, p. 97; Badiozamani, 1973). In such rocks, quartz grains may be completely engulfed in carbonate, dolomitization may be limited to replacement of calcite, and quartz is unaffected (Glover, 1963, p. 40). Quartz grains are regarded as "floating" within the carbonate and display a well-defined boundary (Dapples, 1971, fig. 1). The interpretation favored at present regarding such a boundary is that no reaction has occured between the carbonate and the quartz, the former having acted as an incompatible cement, locally illustrating displacive precipitation. Among sandstones of this type the calcite characteristically is replaced by dolomite of a somewhat more uniform crystal dimension and euhedral outline. Although replacement of calcite by dolomite represents a commonplace unidirectional reaction, reversals in the order of replacement indicate that authigenic dolomite crystals can become unstable in a calcite-crystallizing environment (Swett, 1965; Zenger, 1973).

A second type is recognized among subgraywacke sandstones in which the primary calcite cement contains isolated rhombs of dolomite and siderite. The association is one which often is accompanied by carbonaceous matter and in which siderite is more common than dolomite (Teisseyre, 1973, p. 473). The presence of Fe^{2+} ions seems to favor precipitation of siderite rather than an iron-rich dolomite, although future investigation may show greater abundance of dolomite than is currently recognized. In rocks of these types, the amount of available magnesium tends to be insufficient to permit the crystallization of large amounts of dolomite; and the common occurrence is as individual rhombs of good crystal outline within masses of calcite. This occurrence suggests that the dolomite may possibly represent some ex-solution phenomenon rather than being the result of introduction of magnesium from some outside source (Goldsmith and Graf, 1955; Goldsmith et al., 1962).

Feldspar—calcite replacement. In certain arkoses, which are known to have attained the phyllomorphic stage, significant replacement of potash feldspar by calcite, precipitated as a cement, is not uncommon. In these arkoses, both quartz and potash feldspar are partially replaced by the secondary calcite (see also Greensmith, 1957, p. 410). The typical alteration represented is interpreted as resulting from a process by which solutions rich in Ca^{2+} and CO_3^{2-} ions are capable of destroying the potash feldspar lattice, possibly by causing the silica tetrahedral units to go into solution under the high pH which must characterize the calcite-precipitating solution. Hay (1957) reported that calcite replaces plagioclase in beds of Eocene age in the Absaroka Range, Wyoming, U.S.A.; hence, the replacement reaction is not restricted to potash feldspar. Replacement, however, does appear to be related to important precipitation of calcite suggesting that a mass action effect is required.

Phyllomorphic changes

Crystallization of micas. Reactions categorized as phyllomorphic are favored by increase in pressure and seemingly also by increase in temperature (Maxwell and Hower, 1967). Sandstones which have been subjected to strong pressures either in folded belts or along fault planes develop micas in interstitial openings and along quartz grain boundaries (Fig. 2-10). There is also an abundance effect in the direct proportion between the quantities of secondary mica which are crystallized and the amount of detrital clay which occurs in the sandstone; the amount of mica present is independent of the intensity of folding. The greatest development of secondary mica, however, is present in the argillaceous sandstones which are well folded or have been deeply buried (Dickinson, 1970, p. 702). In general, crystallization of muscovite appears to be preferred over other micas, either because its lattice is more readily developed from most of the clay minerals or, perhaps, because it is stable under the most common conditions of burial.

Certain pure-quartz sandstones, having limited amounts of interstitial clays, tend to show this relationship clearly (Rex, 1966). Excellent examples of well-terminated crystals of kaolinite, with needles of muscovite approaching the ideal lattice structure having developed upon them, are found in sandstones through which fresh waters have moved. Such crystallization is interpreted as reflecting an equilibrium state between the two minerals. Where quartzose sandstones have been strongly folded, border zones of muscovite crystals bounding the quartz grains and partially penetrating them is a commonly observed feature (Whisonant, 1970, fig. 3B). In others, little penetration is noted and each grain appears isolated. In the third group of folded quartzites, which show strong intersuture and marked granular elongation, quartz is regarded as having been mobile. The latter textures point toward the existence of an equilibrium between solution and precipitation of quartz as reported by Thomson (1959). Growing muscovite crystals could readily become surrounded by quartz during the process of solution and reprecipitation, and thus appear to have penetrated the grain.

In rocks of higher clay content than most quartzose sandstones several types of micas generally are developed. Mention has been made that biotite appears in subgraywackes often associated with carbonaceous fragments, and as occurrences which began crystallization during the redoxomorphic stage. In part this is explained by the frequent presence of iron which may be available either attached to the clay mineral or as Fe^{2+}-ion held by organic compounds. Biotite has not been observed to be as abundant as muscovite except in "reduced zones" in certain red-bed sandstones. If the environment is oxidizing, the iron tends to be stabilized in hematite, and very little clay mineral can alter to neoformed biotite.

Certain sandstones, in which glauconite is distributed as a matrix enveloping quartz grains, indicate a phyllomorphic change involving glauconite, chlorite and various clay minerals, muscovite, biotite and quartz. An example of these alterations is found in the upper portion of the Lamotte Sandstone (Cambrian) near Iron Mountain, Missouri, U.S.A. (see also Ojakangas, 1963). The rock is a quartzose sandstone which in the extreme upper parts contains glauconite matrix. Within the areas of glauconite some poorly-preserved remains of a light gray, non-glauconite clay mineral may be seen in thin section (Fig. 2-20). Such areas also show silt-size grains of quartz highly embayed and replaced by the glauconite (Keller, 1970, p. 795). Current studies on the authigenesis of glauconite do not indicate that dissolution of quartz is involved in the processes (Bailey and Atherton, 1969; Giresse and Odin, 1973). Hein et al. (1974, p. 566), however, noted a crudely inverse

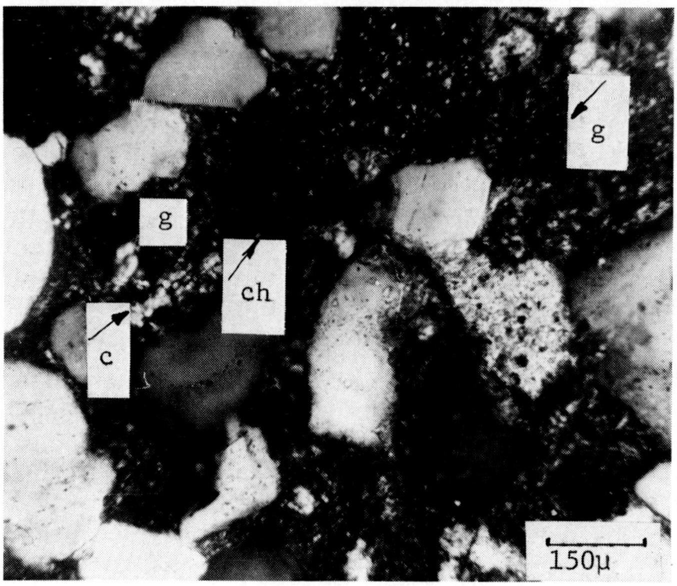

Fig. 2-20. Glauconite (*g*) as matrix in subgraywacke sandstone grades into chlorite (dark area, arrow, *ch*) and non-glauconitic clay mineral (*c*), presumably illitic. Small white irregular patches are quartz silt grains partially replaced by glauconite.
X-nicols; Lamotte Sandstone, Cambrian, Iron Mountain, Mo., U.S.A.

correlation between the amounts of glauconite (Monterey Bay, California, U.S.A.) and of quartz, montmorillonite, kaolinite and chlorite.

Within the mass of glauconite, another equilibrium involving glauconite and chlorite is interpreted as having been established. Irregular masses of chlorite, later in age than the glauconite, indicate that a condition prevailed which resulted in some conversion of the glauconite to chlorite. Presumably, a condition, which caused a second and more advanced phyllomorphic reaction, was superposed upon the glauconite—clay mineral equilibrium (Heller-Kallai et al., 1973, p. 519). Hence, the chlorite—glauconite equilibrium is regarded as marking the most advanced stage attained by these rocks of essentially horizontal attitude and shallow burial.

The equilibrium assemblage suggests also that the K-ion is mobile in the system and may be held in the glauconite, or moves out as a free ion where chlorite is crystallized. In other cratonic quartz—glauconite sandstones, muscovite and biotite rather than chlorite exist in an equilibrium assemblage with glauconite, e.g., in the Tomah Sandstone (Cambrian, Wisconsin, U.S.A.). This may be the result of K-ion remaining in the lattice of the micas, or it may be the response to a different condition than that existing in the case of the chlorite—glauconite equilibrium assemblage.

Local patches within the glauconite can be observed to contain secondary crystals of muscovite or biotite. Presumably the presence of small amounts of Fe^{2+} is the factor deciding which one of the micas would crystallize.

Glauconite does not appear to remain stable under conditions of well-developed folding, although it is present in sandstones which have been rather deeply buried. An arkose layer in the folded Nonesuch Shale (Keweenawan) near Hancock, Mich., U.S.A., contains pelletoidal grains now consisting of a mixture of secondary chlorite and biotite but which on the basis of the residual shape are considered to have been composed formerly of glauconite (Fig. 2-21). The rock is regarded as an example of the above equilibrium having displaced in the direction of well-crystallized micas.

Chlorite as an authigenic product of glauconite is rather common and is more abundant than biotite. Such a preferential development is interpreted as reflecting some general progressive step in the development of a generally lower-temperature mineral phase (chlorite) than the one which is stable under more elevated temperatures (biotite). Muscovite is as common as chlorite and most certainly is developed at equally low temperatures.

A significant association can be demonstrated between the existence of former oxidizing conditions in certain rocks and the crystallization of muscovite from clays. For example, in bright-colored red beds, muscovite is quite stable and can be developed authigenically, whereas biotite clearly is unstable decomposing to iron oxide and clay minerals. In those sandstones where chlorite is the stable phase, the predominance of reducing environ-

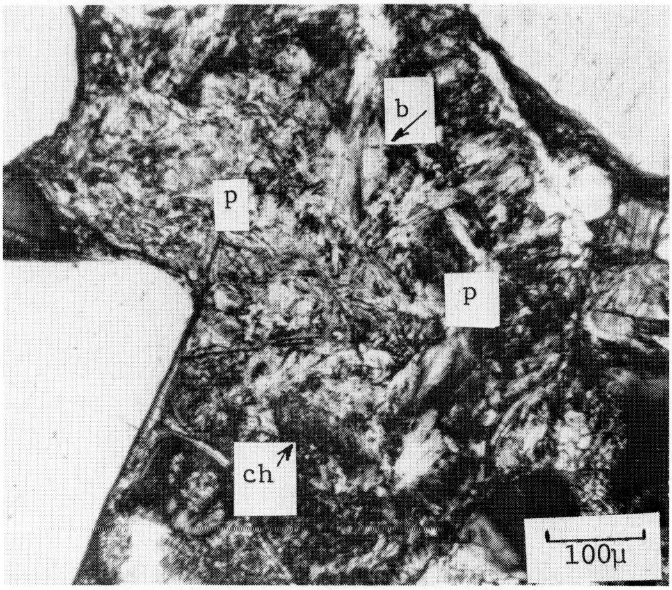

Fig. 2-21. Pellet-shape grain (*p*) and interstitial filling now altered to chlorite (*ch*, dark) and biotite (*b*, light) as an equilibrium mineral assemblage in the phyllomorphic stage. X-nicols; sandstone bed in Nonesuch Shale, Keewenawan, Hancock, Mich., U.S.A.

ment is indicated by the common occurrences of authigenic pyrite, preservation of carbonaceous fragments, and dark (organic) color of the associated shales. Biotite appears to be stable under generally similar conditions, but if both chlorite and biotite occur in the same rock the biotite appears within the chlorite mass as an equilibrium mixture. The presence of K-ion in the bulk composition of the rock as it attains the phyllomorphic grade is not considered to be important in favoring crystallization of biotite over chlorite, inasmuch as the association of glauconite, chlorite, muscovite and authigenic orthoclase is known (Tomah Sandstone, Hudson, Wisc., U.S.A.). For reasons which are not understood, K-ion appears to be rejected during authigenesis of chlorite even though it appears to have been readily available in the glauconite.

Among certain arkoses of green-gray color, biotite has formed from the interstitial clay. An example is found in a few layers of the Stockton Arkose (Triassic), Pa., U.S.A., in which secondary biotite has grown in abundance replacing much detrital clay mineral (Fig. 2-22). This is considered to be an example of authigenesis of a preferred mica in reducing environment in which directed pressure is a minor factor. In the same formation (Stockton

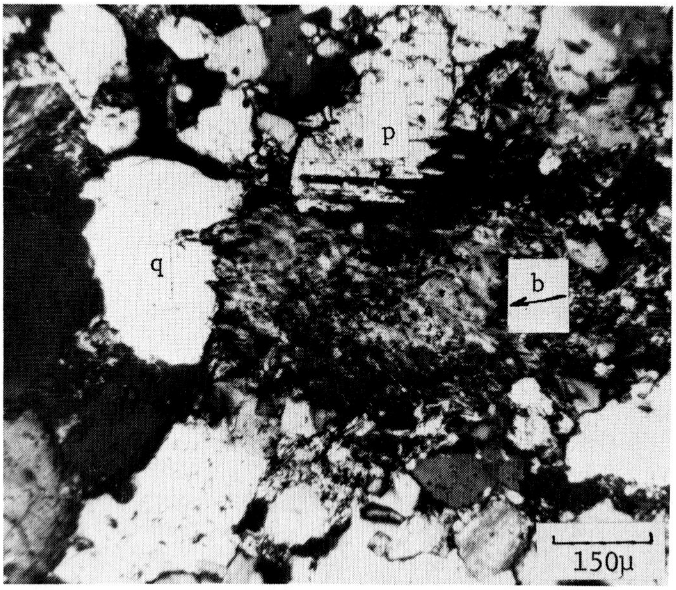

Fig. 2-22. Authigenic biotite (light patches left of arrow, position *b*), which has replaced clay mineral (dark) in matrix of arkose. Corroded margins of detrital quartz (*q*) and plagioclase (*p*) indicate instability of these two minerals.
X-nicols; Stockton Arkose, Triassic, Pennsylvania, U.S.A.

Arkose), but lower stratigraphically and within a portion in which red-colored (hematite) layers are present, the clay matrix has recrystallized into chlorite and to a lesser amount of muscovite (Fig. 2-23) (Glaeser, 1966, p. 74). The interpretation of the association of chlorite—muscovite is that the environment of the phyllomorphic stage was slightly more oxidizing, which favored crystallization of the muscovite rather than biotite as in the preceding example mentioned. Crystallization of the chlorite, however, is interpreted as indicating that the environment probably ranged from very mildy oxidizing, or neutral, to mildly reducing. Hence, chlorite was preferentially formed over biotite, which is considered to crystallize under more strongly reducing conditions.

Crystallization of feldspars. Authigenic feldspar has long been reported (Goldich, 1934; Pettijohn, 1957, p. 644) and is regarded as rather common in sandstones and limestones (Baskin, 1956; Kastner, 1971). Its distribution among sandstones is not as yet well understood because it is locally abundant in certain quartz—glauconite mixtures of the shelf type as well as in strongly folded graywackes. Excellent examples of such authigenic feldspar-

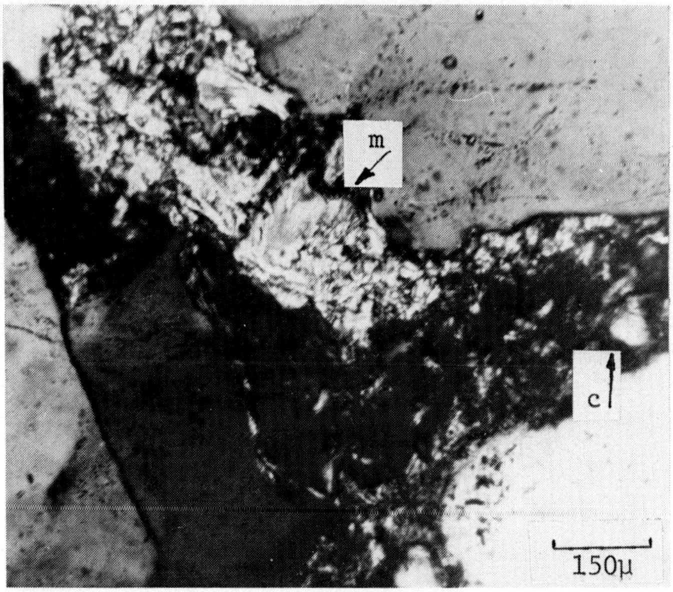

Fig. 2-23. Authigenic muscovite (light patches, arrow, position *m*) and chlorite (medium gray fibrous clusters, arrow, position *c*), which constitute an equilibrium mixture, are derived from altered interstitial clay in red-colored arkose. The authigenic assemblage identifies the phyllomorphic stage.
X-nicols; Stockton Arkose, Triassic, Pennsylvania, U.S.A.

bearing cratonic rocks are the Tomah Sandstone (Cambrian) in Minnesota, U.S.A., and Erwin Quartzite (Cambrian) near Buchanan, Va., U.S.A. Each of these sandstones grades laterally into pure-quartz sandstones containing very little, or no, feldspar. As yet the provenance of the exceptional amounts of potassium which are required has not been clarified, but an internal source from some pre-existing mineral is favored. An example is the postulate by Swett (1968) that dolomitization of illitic limestone could release the required potassium as the illite lattice is destroyed.

Abundant potassic feldspar occurring as veins, cavity fillings and replacement of matrix of graywackes in the Franciscan Formation (Marin County), Ca. (U.S.A.) is reported by Gluskoter (1964, p. 340), and is attributed by him to selective leaching and reprecipitation. A fault system permitted solutions to enter, dissolve detrital potassium feldspar, and transport the necessary ingredients to favorable positions where neoformed potassium feldspar could crystallize. In the example of the graywacke Charny Sandstone (Cambrian, Quebec) studied by Middleton (1972), authigenic albite is interpreted to have replaced detrital potassium feldspar, the soda having been supplied by connate sea waters ejected from the pore spaces during the compaction accompanying deep burial.

The commonplace occurrence of graywackes in strata showing low-grade metamorphism has caused certain authors to regard partial metamorphism of a sandstone essential to their development (Cummins, 1962, p. 65). A matrix primarily composed of chlorite is interpreted as a diagenetic product of detrital grains of volcanic or argillaceous rocks and not an original mud (Dott, 1964; Brenchley, 1969; Whetten and Hawkins, 1970). In this connection, the common presence of albite—oligoclase with sharp crystal outline, as opposed to the extremely embayed outline of other grains, suggests that such plagioclase is not detrital. Crook (1960, p. 543) described graywackes of the Parry Group (Devonian—Carboniferous) of New South Wales having detrital feldspar which was cut by veins of zeolites and albite, and, in some examples, was completely altered to albite. Other examples show similar development of secondary albite—oligoclase and replacement of the anorthite component in plagioclase by laumontite (Raam, 1968, p. 329).

The sequence of stages

In a preceding section, the three subdivisions of diagenesis have been described as though they constituted distinct episodes, each reaching a culmination in development and subsiding in importance before the advent of the next following stage. Obviously this is not the case, because the reactions typical of early stages can occur during the later stages. This is particularly true of reactions of locomorphism of which there may be several generations indicated in the rock. Some replacement reactions occur later than onset of the phyllomorphic stage as indicated by the mineral paragenesis. Certain conditions, as yet unidentified result in telescoping of the diagenetic stages and early onset of phyllomorphism as part of the process of lithification. Among shelf sandstones in the most stable sections of the craton, this stage is infrequently attained, and the sandstone's history has been terminated at the locomorphic stage. In part this is due to the small amount of clay which is present in the interstitial space, but locally where the sandstones intertongue with shales and tend to have matrix clay around the quartz grains, there has been little or no development of well-crystallized mica. Where similar strata are strongly folded, however, the micas are clearly present as a thin border surrounding the quartz grains (Fig. 2-10). Such a relation appears to emphasize the importance of pressure in the development of the phyllomorphic stage. There exist, however, some dramatic exceptions to these occurrences as, for example, the repeated abundance of micas in the more argillaceous subgraywackes, and the occurrence of important amounts of authigenic feldspar in some of the Cambrian sandstones of the classic stable shelf area of Wisconsin, U.S.A. In these examples appeal to moderately elevated temperatures and pressures cannot be made.

Association of very significant quantities of mica, particularly muscovite, in the subgraywackes is connected to thin films of clay. The occurrence of the micas, concentrated on bedding laminae and in the broad troughs of ripple marks, has led investigators to consider such grains as universally detrital, and that they have been deposited in such positions upon settling from the current-transported load. The proof of some secondary mica, however, exists in its development within the interstitial clays. For this reason, much of the unaltered mica occurring in the thin clay films of the bedding laminae is believed to be of similar origin, but as yet no reliable proof of distinction has been available to the author. In the winnowed sand portions, such as stream channel fillings, muscovite is much less abundant. Although this is often explained as the result of removal of the mica during deposition, it may also be the result of the inability to produce secondary mica due to paucity of clay.

Oxidation and reduction reactions primarily involving iron and sulphur may occur late in the rock history. Presence of pyrite as isolated crystals or veinlets can be shown by replacement textures to indicate that it has been among the latest of the authigenic minerals (Condie et al., 1970, p. 2764). The commonplace occurrence, however, is as a product of very early crystallization, particularly as a partial replacement of shell matter or plant cells in organic-rich films of sediment, as reported in Recent deposits by Love (1967). Moreover, if the reducing conditions continue to prevail, small grains of pyrite can be redeposited in finely cross-laminated beds of sands before they become graywackes (Love, 1971).

Occurrences of early and late diagenetic crystallization of pyrite, observed so readily in sandstones associated with high-sulphur coals, suggest that the sulphur previously had been held near the site of the pyrite in some sulphur complex (not necessarily a sulphate) attached to clay. Long existence of strongly reducing conditions may eventually have led to the destruction of the complex and permitted crystallization of pyrite as veinlets, or individual crystals, sometimes much later than lithification of the coal and interbedded sandstone.

From the nature of many ground waters, however, occurrence of dissolved oxygen and oxidizing conditions are restricted largely to near-surface positions; hence, the late crystallized pyrite remains unchanged at depths not reached by aerated waters. Once lithification has been achieved, a slight tendency for a reducing environment seems to prevail, but there is no evidence to point to any extensive shift of the equilibrium in either direction. Rather, the impress of oxidation or reduction during the stage of early burial appears to remain. Hence, red sandstones are known at significant depths, and old red-soil zones developed along unconformities have been preserved to be found now several thousand feet below the earth's surface. Drilling and quarrying have demonstrated that the variety of brown and buff colors of sandstones at the outcrop do not persist at depth and that gray colors ranging from very light to dark are normal. Such colors indicate that iron tends to remain reduced except near the surface, but neither is there evidence that reducing conditions increase in intensity with increasing depth. There do exist local conditions of strong oxidation or reduction at significant depths as evidenced by such features as the "bleached zones" in red beds. Such "bleached zones" and their opposite counterparts, hematite-stained veinlets, can be demonstrated to be responses to a very localized environment or an association of minerals some of which can be oxidized and others reduced. Here, when the zone is well saturated with waters carrying electrolytes, an oxidation—reduction cell can develop and the reaction will be driven preferentially in one direction. The most favorable situation for such conditions, however, exists during the early burial stage. At such a time, organic matter is being oxidized and organisms are extracting oxygen from interstitial water and compounds. Conversely, abundant oxygen may be present in the interstitial water and oxidation may dominate to influence the oxidation state of iron in particular.

As burial progresses and the sediment is compacted, the content of dissolved oxygen diminishes and so also does the organic activity. The oxidation—reduction tendencies are reduced and reactions of this type are less likely to occur. For these reasons, reactions characteristic of the redoxomorphic stage play an increasingly smaller part as diagenesis progresses.

Reaction tendencies of the locomorphic stage

Nature of the reactions. Replacement of one mineral by another can be classified as one of three types, namely: (1) transition from a metastable mineral phase into a more stable form primarily as a function of time; (2) solution of one mineral and precipitation of another as a result of change in solubilities with a change in pH or temperature of permeating waters; and (3) shift in an equilibrium assemblage with change in pH, temperature, oxidation potential, and ionic concentration.

Unidirectional transformation of opal via cristobalite into chalcedony is regarded as an example of the first type of locomorphic replacement, although it may also represent a combination of types 1 and 2. According to Heath and Moberly (1971, p. 1003), marine biogenic opal dissolves and is reprecipitated as a poorly crystallized cristobalite. The latter, however, transforms by gradual increase in crystallinity in the solid state and ends as microcrystalline quartz. Current information indicates that the solid—solid transformation progresses as a function of time.

An example of type-2 locomorphic change is illustrated by replacement of anhydrite by silica during the growth of geodes. Chowns and Elkins (1974) interpreted the progression to begin with initial replacement of anhydrite by quartzine, which preserves the delicate fibrous structure of the anhydrite. Later, as the rate of anhydrite solution exceeds the volume for volume exchange, a secondary porosity develops and microcrystalline quartz is crystallized in the available pore space commonly engulfing some of the anhydrite.

When chert (microcrystalline quartz) is formed in sandstones, this phase appears to have reached a relatively stable condition of crystallinity which, except in the pure-quartz sandstones, remains without perceptible change in crystal dimension despite increases in pressure and temperature resulting from folding or deep burial. Under such conditions, in the very quartzose sandstones, chert tends to be metastable and is replaced by monocrystalline quartz through a reaction of type 2. With the progress of time, this monocrystalline quartz as a result of solid—solid transformation passes through stages of compound grains to attain the culminating texture of planar bilateral and triple-grain junctions as described earlier.

Replacement of interstitial clay by chert can be an important early event of the locomorphic stage and is particularly to be noted among tuffaceous sandstones (Fig. 2-24). In the devitrification process silica is presumed to be released and reprecipitated in an adjacent bed, or within the same specimen, generally in the form of chert. Such chert is accommodated in the interstitial space by crowding or engulfing the detrital clay minerals present. Elsewhere, clay-mineral aggregates appear to have been completely replaced which, if true, indicates destruction of the lattice. Similar chertification occurs to a limited extent where subgraywacke sandstones underlie bentonite beds. In such examples, the occurrence of secondary chert is limited to a zone several centimeters thick, and is attributed to decomposition of glass and release of silica during alteration of the bentonite.

In cases where chert replaces clay matrix in graywackes, much of the durability and tensile strength of the rock is to be attributed to this replacement. Where replacement chert is abundant, there appears to be some association with either unstable volcanic material which has decomposed to release silica, or reprecipitation of silica dissolved from detrital grains of quartz.

The tendency for chert grains to represent an unstable phase of quartz in the pure-quartz sandstones is considered to illustrate the condition of locomorphic replacement type 3, in which some difference in temperature or ionic concentration favors crystallization of silica as overgrowths on the pre-existing monocrystalline quartz grains. In a number of localities where this process is proceeding, the sandstones are aquifers in which the silica content is low, e.g., a few parts per million. Some of these waters indicate pH of 7±

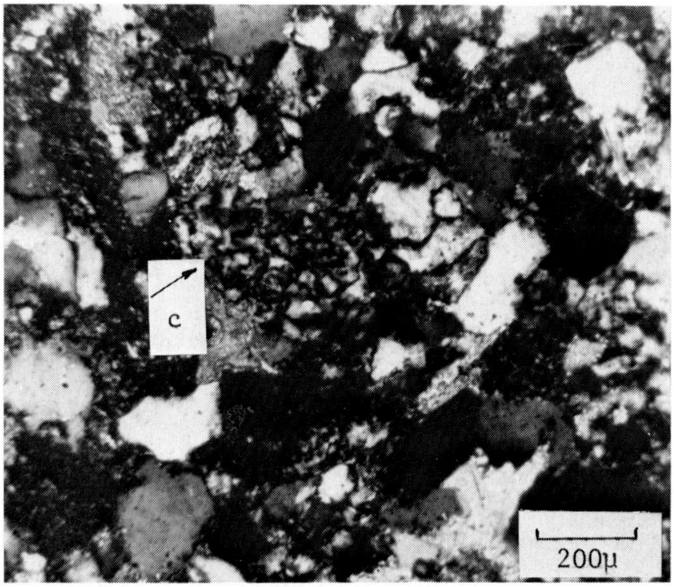

Fig. 2-24. Matrix in tuffaceous sandstone which has been replaced by microcrystalline quartz (spotted light and gray area to the right of arrow, position *c*). The authigenic chert has replaced both matrix and grains in much of the photo and it is exceedingly difficult to identify original material.
X-nicols; Blairmore Formation, Cretaceous, Crowsnest, Alta., Canada.

and general prevalence of oxidizing conditions (Meents et al., 1952, p. 36). The low ionic concentration of silica characterizing some of the prominent aquifers, indicates general saturation with silica at the low temperatures involved (Krauskopf, 1959; Siever, 1962).

Calcite—quartz relations. Calcite is an extraordinarily abundant cement among sandstones, and only an exceptional thin section does not contain some secondary calcite. In some rocks it can be demonstrated to occur as a primary cement (Taylor and Illing, 1969), whereas in other rocks several generations of calcite precipitation are recognized.

Certain examples of precipitation of calcite in large crystals completely engulfing much of the original clastic material is commonplace. Special examples are the so-called "sand crystals" in which siliceous sand grains are cemented into scalenohedrons of calcite habit (Pettijohn, 1957, plate 8). More frequently observed are masses of calcite cement precipitated in optical orientation in siliceous sandstones as concretionary or irregularly shaped masses in which the quartz grains are arranged in a poikilotopic texture (Johnson and Swett, 1974, p. 791).

Precipitation of calcite is directly associated with instability of chert and quartz. In the calcite environment, both show prominent embayments in the grains and well exhibited calcite replacement textures (Walker, 1962). In some rocks solution of silica is illustrated by extreme irregularity of the grain boundary and "ghost" remains of chert or quartz within the mass of calcite. The antipathetic reaction between quartz and calcite was attributed by Correns (1950) to the inverse relation in solubility with change in pH. Calcite decreases in solubility and silica becomes more soluble as the pH is elevated. Although the general relations have not been challenged, the silica solubility curve has been shown by additional laboratory data to remain relatively uniform in concentrations of pH below approximately 9. Important increase in solubility occurs only if pH exeeds this value. Inasmuch as pH values of 9 or higher are reputed to be somewhat exceptional (Baas-Becking et al., 1960), appeal has been made to changes in temperature as having greater control on precipitation of calcite and solution of quartz (Siever, 1962, p. 144). The occurrence of the replacement of quartz by calcite in certain sandstones at the outcrop, particularly in present-day semi-arid climates, suggests to the present author, however, that shift in pH is very important in this locomorphic process despite the very small differences in solubility. Local changes in the partial pressure of CO_2 resulting from temperature differences near the outcrop are considered to be the indirect cause of change in pH values. Similar mechanisms could produce comparable replacement in the subsurface.

Although the most frequently observed relation is replacement of silica by later deposited calcite, there are examples where chert is precipitated and calcite is unstable. Generally, the reaction involves replacement of detrital carbonate grains by silica, but some occurrences of a late silica cement replacing earlier calcite cement can be found. In such examples calcite is unstable and quartz is the stable mineral phase. There is reason to regard, therefore, solution and replacement of quartz and calcite as reversible reactions controlled by changes in pH, partial pressure of CO_2, and temperature (Walker, 1962), and being illustrative of type-3 locomorphic reactions.

Reaction tendencies of the phyllomorphic stage

Nature of the reactions. As defined in the sequence of diagenetic modifications, the phyllomorphic stage is regarded as most advanced and transitional into the lowest grades of metamorphism (Brown and Thayer, 1963; Dunoyer de Segonzac, 1970). The basis for establishment of this rank is primarily two-fold: (1) certain large crystals of authigenic mica transect some of the cement precipitated during the stage of locomorphism, and (2) selected authigenic mineral associations, developed during this stage, are considered to be equilibrium minerals representing limited ranges in bulk compositions. In earlier diagenetic stages, certain reactions can be considered to represent equilibria, in the sense that they are reversible, as illustrated by the oxidized or reduced states of iron. They represent an equilibrium which is displaced in favor of the oxidized or reduced ion depending upon the presence of other oxidizable or reducible compounds in the detrital mixture. Perhaps the most common example of displacement of such an equilibrium is to be observed where there is abundance of readily oxidizable organic matter which reduces iron to the ferrous state.

Reactions characteristic of locomorphism tend to go to completion in one direction; hence, they are not to be considered as representative of typical equilibria. Instead, they can be regarded as reactions displaced strongly in one direction as a result of the primary influence of the ionic concentration of invading solution which is a function of temperature, pH, and solubility products. As the conditions which lead to the precipitation of a mineral are altered, the reaction direction may be reversed and a fraction of the precipitated matter may return to solution, being replaced by whatever else will precipitate under these new conditions. Reactions of this variety can occur during the entire postlithification history of the rock, although they are most importantly recognized as following the stage of prominent oxidation and reduction and preceding the significant phyllomorphic modifications.

A stage of development of authigenic mica assigned to the phyllomorphic stage begins early in the sediment history, probably during the redoxomorphic stage and increases in importance with progress of time. In part this may be attributed to alteration of mixed-layer clay minerals into illites or chlorites of more fixed crystal structure and composition (Rex, 1966). An example of glauconite crystallizing to chlorite and biotite has been presented previously (Fig. 2-21). Also, crystallization of certain micas from clay minerals appears to be influenced by the state of oxidation which prevails and the pH of the permeating waters. Growth of authigenic muscovite, for example, is favored by an oxidizing environment and generally neutral solutions, whereas biotite structure is favored by a reducing condition and a wide range in acidity. With a gradual increase in application of temperature and pressure, clay minerals tend to alter more readily to micas, and the latter increase in crystal size to occupy more of the space formerly filled by clay minerals. Through this process there is an increase in particle size of the matrix material. Consequently, in some graywackes distinction between what is to be called matrix and the enclosed larger particles becomes a purely arbitrary decision.

Equilibrium assemblages. One of the more significant aspects of the phyllomorphic stage is the appearance of what are interpreted to be equilibrium mineral phases. Among the most readily observed are the muscovite—biotite and chlorite—biotite assemblages. These can be shown as intergrown from former masses of clay in various abundance ratios, apparently dictated by conditions not currently recognized. Examples of other assemblages such as albite, quartz—chlorite—prehnite—calcite are reported by Brown and Thayer (1963, p. 414), Raam (1968) and others.

Growth of secondary feldspar, particularly albite, in graywackes proceeds with advance in phyllomorphism. Such development is considered part of an equilibrium assemblage which appears to favor the crystallization of one mineral over others.

Inasmuch as the authigenesis of albite has not as yet been demonstrated to characterize all Precambrian graywackes (Condie et al., 1970, p. 2765), there is some reason to suspect that the bulk composition of the original sediment and not low-grade metamorphism is important in establishing the ultimate minerals of the phyllomorphic stage. For this reason the author prefers to regard certain mineral assemblages as representing attainment of the phyllomorphic stage despite their mineralogical dissimilarity.

The importance of bulk rock composition has been illustrated in the study by Carrigy and Mellon (1964) who have shown that in non-marine fluviatile sandstones of the foothills regions of the Canadian Rockies four distinct authigenic silicate assemblages can be recognized: kaolinite—quartz authigenesis characterize quartzose sandstones, whereas kaolinite—chlorite and illite—chlorite are the equilibrium mixtures in sandstones with small to moderate amounts of volcanic fragments. Chlorite—laumontite assemblage occurs in sandstones with abundant volcanic detritus. These are examples representative of parallel relationships which must exist in sandstones of different bulk compositions as a result of diagenetic modifications that lead to equally significant equilibrium-mineral assemblages (Zen, 1959).

CONCLUSIONS

So frequently has the presence of authigenic silicate minerals been reported to occur in sandstones, that such occurrences can no longer be regarded as accidents of mineral formation. Moreover, the percent occurrence of such minerals in local specimens, or zones, may represent a significant fraction of the total mineral assemblage. Common appearance of secondary feldspars and micas, normally recognized as representative of high-temperature minerals, is suggestive that certain progression has occurred in the development of such minerals. Indeed it can be demonstrated that as conditions suitable to their stability were attained, they were able to form even at low temperatures at the expense of other minerals no longer stable (Mackenzie and Garrels, 1966, p. 1082). Among sandstones, those minerals which appear to be most sensitive to changes in conditions are silica and clay minerals. Silica alters readily into other forms, dissolves, and reprecipitates repeatedly throughout the history of certain sandstones. Clay minerals appear to be exceedingly stable in certain sandstones, whereas in others they can be recognized to have altered to mica or perhaps feldspar. The thesis which has been presented herein has been to regard such authigenesis as not accidental, but rather as part of an orderly sequence of events that lead to the formation of a rock which has crossed the boundary of metamorphism. In this regard the stages of diagenesis have been anamorphic and new minerals take the place of some of those in the original detritus. Ideal representatives of the products of such a process are the true graywackes (those of the original description, Irving and Van Hise, 1892, p. 306; Van Houten, 1958; Huckenholz, 1963). One should not require such rocks to have very narrowly defined limits of composition, but rather to have attained the upper limits of the phyllomorphic stage. One should look for mineral assemblages such as quartz—chlorite—zeolite or biotite—albite as among those principally diagnostic of this stage. Moreover, one should regard such equilibrium minerals as important in designating some of the primary attributes of the petrography of a graywacke.

Diagenesis also moves in an opposite sense, namely, in removal of minerals unstable under the new environment. Such a direction is associated with some aspects of the stage of locomorphism. One mineral is dissolved and another takes its place. In extreme conditions, solution is the only dominating trend and a sandstone of virtually monomineralic composition is the ultimate product of such an environment. Textures such as those indicated in Figs. 2-8 and 2-9 serve to identify such a katamorphic trend of diagenesis and provide the basis for re-interpretation of the origin of certain pure-quartz sandstones.

Progressive stages of diagenesis provide the basis for regarding certain rocks as approaching an ultimate mineralogy in the particular tectonic environment which they have experienced. Sandstones of the typical stable-shelf depositional environment are not to be expected, under normal conditions, to advance far into the phyllomorphic stage, if at all. Rather, they show progressive advancement in the intergranular interlock. Subgraywacke sandstones typical of basins are likely to show significant attributes of the phyllomorphic stage. Those sediments of the eugeosyncline, which contain important amounts of minerals unstable in the postdeposition environments, are the ones which can be expected to show the most clearly defined influence of diagenesis. One is left to regard all such processes as of significant importance and to consider that the sandstone as observed now in the outcrop, or subsurface, is perforce only partially representative of the originally deposited mixture.

GLOSSARY

Anchimetamorphism: processes which produce changes in mineral content of rocks under temperature and pressure conditions prevailing in the regions between the earth's surface and the zone of true metamorphism (Am. Geol. Inst. *Glossary of Geology*).

Arenite: a framework in which individual detrital grains of sand size are in contact.

Bilateral junction: a planar boundary between two detrital grains, which have been welded along the common boundary and tend to be in optical continuity.

Cement, simple clay bond: a cement which forms a hardened aggregate between sand- and clay-sized particles by a process of cohesion between the clay minerals of the matrix of a clastic sediment. The bond is broken in water. In advanced stages, the bond becomes an intercrystalline growth between clay minerals and quartz. This bond is not broken in water.

Cement, simple mineral: a cement which is precipitated in pores between grains of a clastic sediment to cause an aggregated mass. *Compatible* cement involves crystallization against a mineral of the same composition. *Incompatible* cement involves crystallization against a mineral of different composition.

Cementation: any of several processes by which separate mineral or rock fragments are bonded into an aggregate mass. The processes include simple adhesion, crystallization of mineral matter in openings, epitaxial growths, intercrystalline interlock during diagenesis, and replacement of initially present substances by neoformed minerals.

Coalescent neomorphism: development of an intergrowth of authigenic crystals by which small crystals tend to disappear and large crystals tend to assume a uniform dimension.

Coherent boundary: a junction between two crystals of the same mineral along which the lattice of one is continuous into that of the other.

Compound grain: several grains of the same mineral welded along separate boundaries into a single grain of irregular outline but having uniform optical continuity.

Displacive precipitation: crystallization of cement to cause an increase in volume of the original available space.

Intergranular reaction bond: an interlock of grains in a clastic sediment brought about by reaction between adjacent minerals. The process is one of replacement of grains by a newly precipitated mineral.

Intergranular solution bond: an interlock of grains in a clastic sediment resulting from differential solution and irregular inter-penetration of grains. The interlock may be loose, as in itacolumite, or tightly cemented as in some quartzites.

Locomorphic stage: the stage of diagenesis characterized by prominent mineral replacement that is typical of lithification of a clastic sediment. This stage is more advanced than the redoxomorphic stage.

Non-coherent boundary: a junction between two crystals of the same mineral along which there is an angular misfit between the lattices of the two crystals.

Paragenetic relations: the order in which minerals have been precipitated when they occur together in rocks as an association.

Phyllomorphic stage: the most advanced stage of diagenesis in the progression toward metamorphism, characterized principally by authigenesis of micas and also of feldspars. This stage follows the locomorphic stage.

Poikilotopic texture: an interlock consisting of large single crystals of cement within which detrital grains are incorporated. There is displacive precipitation of cement.

Porphyroid neomorphism: development of an intergrowth of authigenic crystals by which certain crystals become preferentially larger than others.

Redoxomorphic stage: the early stage of diagenesis characterized by mineral changes primarily due to oxidation and reduction reactions. This stage is typical of the unlithified sediment and precedes the locomorphic stage.

Subwacke: a framework in which in hand specimen, or thin section, both arenite and wacke frameworks can be observed in local areas.

Triple-grain junction: an intercrystalline intergrowth of three grains of the same mineral along three separate bilateral junctions terminating in a common point.

Wacke: a framework in which individual detrital grains of sand are separated by a continuous phase of matrix having a considerably smaller dimension (ideally clay-size).

REFERENCES

Aalto, K.R., 1972. Diagenesis of orthoquartzites near Bogatá, Colombia. *J. Sediment. Petrol.*, 42: 330—340.

Alexandersson, T., 1972. Intragranular growth of marine aragonite and Mg-calcite: evidence of precipitation from supersaturated seawater. *J. Sediment. Petrol.*, 42: 441—460.

Allen, R.C., Gavish, E., Friedman, G.M. and Sanders, J.E., 1969. Aragonite-cemented sandstone from outer continental shelf off Delaware Bay: submarine lithification mechanism yields product resembling beachrock. *J. Sediment. Petrol.*, 39: 136—149.

Austin, G.S., 1974. Multiple overgrowths on detrital quartz sand grains in the Shakopee Formation (Lower Ordovician) of Minnesota. *J. Sediment. Petrol.*, 44: 358—362.

Baas Becking, L.G.M., Kaplan, I.R. and Moore, D., 1960. Limits of the natural environment in terms of pH and oxidation—reduction potentials. *J. Geol.*, 68: 243—285.

Badiozamani, K., 1973. The dorag dolomitization model—application to the Middle Ordovician of Wisconsin. *J. Sediment. Petrol.*, 43: 965—985.

Bailey, R.J. and Atherton, M.P., 1969. The petrology of a glauconitic sandy chalk. *J. Sediment. Petrol.*, 39: 1420—1431.

Baskin, Y., 1956. A study of authigenic feldspars. *J. Geol.*, 64: 132—155.

Berner, R.A., 1969. Goethite stability and the origin of red beds. *Geochim. Cosmochim. Acta*, 33: 267—273.
Blatt, H. and Sutherland, B., 1969. Intrastratal solution and non-opaque heavy minerals in shales. *J. Sediment. Petrol.*, 39: 591—600.
Brenchley, P.J., 1969. Origin of matrix in Ordovician graywackes, Berwyn Hills, North Wales. *J. Sediment. Petrol.*, 39: 1297—1301.
Brown, S.E. and Thayer, T.P., 1963. Low-grade mineral facies in Upper Triassic and Lower Jurassic rocks of the Aldrich Mountains, Oregon. *J. Sediment. Petrol.*, 33: 411—426.
Bucke, D.P. and Mankin, C.J., 1971. Clay-mineral diagenesis within interlaminated shales and sandstones. *J. Sediment. Petrol.*, 41: 971—981.
Cadigan, R.H., 1967. Petrology of the Morrison Formation in the Colorado Plateau Region. *U.S. Geol. Surv., Prof. Pap.*, 556: 113 pp.
Calvert, S.E., 1966. Accumulation of diatomaceous silica in the sediments of the Gulf of California. *Bull. Geol. Soc. Am.*, 77: 569—596.
Calvert, S.E., 1971. Composition and origin of North Atlantic deep sea cherts. *Contrib. Mineral. Petrol.*, 33: 272—288.
Carrigy, M.A. and Mellon, G.B., 1964. Authigenic clay mineral cements in Cretaceous and Tertiary sandstones of Alberta. *J. Sediment. Petrol.*, 34: 461—472.
Carroll, D., 1958. Role of clay minerals in the transportation of iron. *Geochim. Cosmochim. Acta*, 14: 1—27.
Chowns, T.M. and Elkins, J.E., 1974. The origin of quartz geodes and cauliflower cherts through silicification of anhydrite nodules. *J. Sediment. Petrol.*, 44: 885—903.
Cleary, W.J. and Conolly, J.R., 1972. Embayed quartz grains in soils and their significance. *J. Sediment. Petrol.*, 42: 899—904.
Condie, K.C., Macke, J.E. and Reimer, T.O., 1970. Petrology and geochemistry of Early Precambrian graywackes from the Fig Tree Group, South Africa. *Bull. Geol. Soc. Am.*, 81: 2759—2776.
Coombs, D.S., Ellis, A.J., Frye, W.S. and Taylor, A.M., 1959. The zeolite facies with comments on the interpretation of hydrothermal synthesis. *Geochim. Cosmochim. Acta*, 17: 53—107.
Correns, C.W., 1950. Zur Geochemie der Diagenese. *Geochim. Cosmochim. Acta*, 14: 1—27.
Crook, K.A.W., 1960. Petrology of Parry Group, Upper Devonian—Lower Carboniferous, Temworth-Nundle district, New South Wales. *J. Sediment. Petrol.*, 30: 538—552.
Crook, K.A.W., 1968. Weathering and roundness of quartz sand grains. *Sedimentology*, 11: 171—182.
Cummins, W.A., 1962. The greywacke problem. *Liverpool Manchester Geol. J.*, 3: 51—72.
Dapples, E.C., 1962. Stages of diagenesis in the development of sandstones. *Bull. Geol. Soc. Am.*, 73: 913—934.
Dapples, E.C., 1971. Physical classification of carbonate cement in quartzose sandstones. *J. Sediment. Petrol.*, 41: 196—204.
Davies, W., 1947. The fundamental characteristics of moulding sands. *J. Iron Steel Inst.*, 77: 9—46.
Deer, W.A., Howie, R.A. and Zussman, J., 1962. *Rock Forming Minerals, 5. Non-Silicates.* Wiley, New York, N.Y., 308 pp.
DeVore, G.W., 1956. Surface chemistry as a chemical control on mineral association. *J. Geol.*, 64: 31—55.
Dickinson, W.R., 1970. Interpreting detrital modes of graywacke and arkose. *J. Sediment. Petrol.*, 40: 695—707.

Dickinson, W.R., Ojakangas, R.W. and Stewart, R.J., 1969. Burial metamorphism of the Late Mesozoic Great Valley Sequence, Cache Creek, California. *Bull. Geol. Soc. Am.*, 80: 519–526.

Dott, Jr., R.H., 1964. Wacke, graywacke, and matrix — what approach to immature sandstone classification? *J. Sediment. Petrol.*, 34: 625–632.

Dunoyer de Segonzac, G., 1968. The birth and development of the concept of diagenesis (1866–1966), *Earth-Sci. Rev.*, 4: 153–201.

Dunoyer de Segonzac, G., 1970. The transformation of clay minerals during diagenesis and low-grade metamorphism: a review. *Sedimentology*, 15: 281–346.

Dunoyer de Segonzac, G., Ferrero, J. and Kubler, B., 1968. Sur la cristallinité de l'illite dans la diagenèse et l'anchimetamorphisme. *Sedimentology*, 10: 137–145.

Eades, J.L. and Grim, R.E., 1960. The reaction of hydrated lime with pure clay minerals in soil stabilization. In: R.E. Grim (Editor), *Applied Clay Mineralogy*. McGraw-Hill, New York, N.Y., p. 268.

Enos, P., 1969. Cloridorme Formation, Middle Ordovician flysch, northern Gaspé Peninsula, Quebec. *Geol. Soc. Am., Spec. Pap.*, 117: 66 pp.

Ernst, W.G. and Blatt, H., 1964. Experimental study of quartz overgrowths and synthetic quartzites. *J. Geol.*, 72: 461–470.

Eugster, H.P., 1969. Inorganic bedded cherts from the Magadi Area, Kenya. *Contrib. Mineral. Petrol.*, 29: 1–31.

Fairbridge, R.W., 1967. Phases of diagenesis and authigenesis. In: G. Larsen and G.V. Chilingar (Editors), *Diagenesis in Sediments*. Elsevier, Amsterdam, pp. 19–89.

Fellows, R.E., 1943. Recrystallization and flowage in Appalachian quartzite. *Bull. Geol. Soc. Am.*, 54: 1399–1431.

Folk, R.L., 1951. Stages of textural maturity in sedimentary rocks. *J. Sediment. Petrol.*, 21: 127–130.

Folk, R.L., 1965. Some aspects of recrystallization in ancient limestones. In: L.C. Pray and R.C. Murray (Editors) *Dolomitization and Limestone Diagenesis—Soc. Econ. Paleontol. Mineral., Spec. Publ.*, 13: 14–48.

Folk, R.L. and Weaver, C.E., 1952. A study of the texture and composition of chert. *Am. J. Sci.*, 250: 498–510.

Frey, M., 1970. The step from diagenesis to metamorphism in pelitic rocks during Alpine orogenesis. *Sedimentology*, 15: 261–279.

Friedman, G.M., 1954. Miocene orthoquartzite from New Jersey. *J. Sediment. Petrol.*, 24: 235–242.

Friedman, G.M., 1965. Terminology of crystallization textures and fabrics in sedimentary rocks. *J. Sediment. Petrol.*, 35: 643–655.

Friend, P.F., 1966. Clay fractions and colours of some Devonian red beds in the Catskill Mountains, U.S.A. *Q.J. Geol. Soc. London*, 122: 273–292.

Gaither, A., 1953. A study of porosity and grain relationships in experimental sands. *J. Sediment. Petrol.*, 23: 180–195.

Garrels, R.M., 1960. *Mineral Equilibria*. Harper and Row, New York, N.Y., 254 pp.

Garrison, R.E., Luternauer, J.L., Grill, E.V., MacDonald, R.D. and Murray, J.W., 1969. Early diagenetic cementation of recent sands, Fraser River Delta, British Columbia. *Sedimentology*, 12: 27–46.

Gavish, E. and Friedman, G.M., 1969. Progressive diagenesis in Quaternary to Late Tertiary carbonate sediments: sequence and time scale. *J. Sediment. Petrol.*, 39: 980–1006.

Gilbert, C.M., 1949. Cementation of some California Tertiary reservoir sands. *J. Geol.*, 57: 1–17.

Gilbert, C.M., 1954. In: H. Williams, F.J. Turner and C.M. Gilbert, *Petrography*. W.H. Freeman, San Francisco, Calif., 406 pp.

Giles, A.W., 1932. Textural features of Ordovician sandstones of Arkansas. *J. Geol.*, 40: 97—118.

Giresse, P. and Odin, G.S., 1973. Nature mineralogique et origine des glauconies du plateau continental du Gabon et du Congo. *Sedimentology*, 20: 457—489.

Glaeser, J.D., 1966. Provenance, dispersal, and depositional environments of Triassic sediments in the Newark-Gettysburg Basin. *Bull. Pa. Geol. Surv.*, G-43: 168 pp.

Glover, J.E., 1963. Studies in the diagenesis of some western Australian sedimentary rocks. *J.R. Soc. West. Austr.*, 46: 33—56.

Gluskoter, H.J., 1964. Orthoclase distribution and authigenesis in the Franciscan Formation of a portion of western Marin County, California. *J. Sediment. Petrol.*, 34: 335—343.

Goldich, S.S., 1934. Authigenic feldspar in sandstone of southeastern Minnesota. *J. Sediment. Petrol.*, 4: 89—95

Goldsmith, J.R. and Graf, D.L., 1955. The occurrence of magnesian calcites in nature. *Geochim. Cosmochim. Acta*, 7: 212—230.

Goldsmith, J.R., Graf, D.L. and Witters, J., 1962. Studies in the system $CaCO_3$—$MgCO_3$—$FeCO_3$, 1. Phase relations. *J. Geol.*, 70: 659—687.

Greensmith, J.T., 1957. Lithology with particular reference to cementation of Upper Carboniferous sandstones in northern Derbyshire, England. *J. Sediment. Petrol.*, 27: 405—417.

Greenwood, R., 1973. Cristobalite: its relationship to chert formation in selected samples from the deep-sea drilling project. *J. Sediment. Petrol.*, 43: 700—708.

Grim, R.E. and Cuthbert, F.L., 1945. The bonding action of clays, 1. Clays in green molding sands. *Ill. State Geol. Surv., Rep. Inv.*, 102: 55 pp.

Hay, R.L., 1957. Mineral alteration in rocks of Middle Eocene age, Absaroka Range, Wyoming. *J. Sediment. Petrol.*, 27: 32—41.

Hay, R.L., 1968. Chert and its sodium-silicate precursors in sodium-carbonate lakes of East Africa. *Contrib. Mineral. Petrol.*, 17: 255—274.

Hayes, J.B., 1970. Polytypism of chlorite in sedimentary rocks. *Clays Clay Miner.*, 18: 285—306.

Heald, M.T., 1950. Authigenesis in West Virginia sandstones. *J. Geol.*, 58: 624—633.

Heald, M.T., 1955. Stylolites in sandstones. *J. Geol.*, 63: 101—114.

Heald, M.T., 1956. Cementation of Triassic arkoses in Connecticut and Massachusetts. *Bull. Geol. Soc. Am.*, 67: 1133—1154.

Heald, M.T., 1959. Cementation of Simpson and St. Peter sandstones in parts of Oklahoma, Arkansas and Missouri. *J. Geol.*, 64: 16—30.

Heald, M.T. and Larese, R.E., 1973. The significance of the solution of feldspar in porosity development. *J. Sediment. Petrol.*, 43: 458—460.

Heald, M.T. and Larese, R.E., 1974. Influence of coatings on quartz cementation. *J. Sediment. Petrol.*, 44: 1269—1274.

Heath, G.R. and Moberly Jr., R., 1971. Cherts from the western Pacific, Leg 7, Deep-Sea Drilling Project. In: *Initial Reports of the Deep-Sea Drilling Project*, 7 (2): 991—1007.

Hein, J.R., Allwardt, A.O. and Griggs, G.B., 1974. The occurrence of glauconite in Monterey Bay, California; diversity, origins, and sedimentary environmental significance. *J. Sediment. Petrol.*, 44: 562—571.

Heller-Kallai, L., Nathan, Y. and Zak, I., 1973. Clay mineralogy of Triassic sediments in southern Israel and Sinai. *Sedimentology*, 20: 513—521.

Horowitz, D.H., 1971. Diagenetic significance of color boundary between Juniata and Bald Eagle formations, Pennsylvania. *J. Sediment. Petrol.*, 41: 1134—1144.

Hubert, J.F., 1960. Petrology of the Fountain and Lyons formations, Front Range, Colorado. *Colo. Sch. Mines, Q.*, 55: 1—242.

Huckenholz, H.G., 1963. Mineral composition and texture in graywackes from the Harz Mountains (Germany) and in arkoses from Auvergne (France). *J. Sediment. Petrol.*, 33: 914—919.

Hutton, J.T., Twidale, C.R., Milnes, A.R. and Rosser, H., 1972. Composition and genesis of silcretes and silcrete skins from the Beda valley, Southern Arcoona Plateau, South Africa and Australia. *J. Geol. Soc. Aust.*, 19: 31—39.

Irving, R.D. and Van Hise, C.R., 1892. The Penokee iron-bearing Series of Michigan and Wisconsin. *U.S. Geol. Surv., Monogr.*, 19: 534 pp.

Johnson, D.B. and Swett, K., 1974. Origin and diagenesis of calcitic and hematitic nodules in the Jordan Sandstone of northeast Iowa. *J. Sediment. Petrol.*, 44: 790—794.

Jones, J.B. and Segnit, E.R., 1971. The nature of opal. 1. Nomenclature and constituent phases. *J. Geol. Soc. Aust.*, 18: 57—68.

Kastner, M., 1971. Authigenic feldspars in carbonate rocks. *Am. Mineral.*, 56: 1403—1442.

Keller, W.D., 1953. Illite and montmorillonite in green sedimentary rocks. *J. Sediment. Petrol.*, 23: 3—10.

Keller, W.D., 1970. Environmental aspects of clay minerals. *J. Sediment. Petrol.*, 40: 788—854.

Keller, W.D. and Reesman, A.L., 1963. Dissolved products of artificial silicate minerals and rocks. *J. Sediment. Petrol.*, 33: 426—438.

Klein, G. DeV., 1963a. Analysis and review of sandstone classification in the North American geological literature, 1940—1960. *Bull. Geol. Soc. Am.*, 74: 555—575.

Klein, G. DeV., 1963b. Bay of Fundy intertidal zone sediments. *J. Sediment. Petrol.*, 33: 844—855.

Krauskopf, K.B., 1959. The geochemistry of silica in sedimentary environments. In: *Silica in Sediments — Soc. Econ. Paleontol. Mineral. Spec. Publ.*, 7: 4—20.

Krumbein, W.C. and Sloss, L.L., 1963. *Stratigraphy and Sedimentation.* Freeman, San Francisco, Calif., 2nd ed., 636 pp.

Krynine, P.D., 1948. The megascopic study and field classification of sedimentary rocks. *J. Geol.*, 56: 130—165.

Lamar, J.E., 1928. Geology and economic resources of the St. Peter Sandstone of Illinois. *Ill. State Geol. Surv. Bull.*, 53: 175 pp.

Laury, R.L., 1968. Sedimentology of the Pleasantview Sandstone, southern Iowa and western Illinois. *J. Sediment. Petrol.*, 38: 568—599.

Love, L.G., 1967. Early diagenetic iron sulphide in Recent sediments of the Wash (England). *Sedimentology*, 9: 327—352.

Love, L.G., 1971. Early diagenetic polyframboidal pyrite, primary and redeposited, from the Wenlockian Denbigh Grit Group, Conway, North Wales, U.K. *J. Sediment. Petrol.*, 41: 1038—1044.

Lowenstam, H.A., 1954. Factors affecting the aragonite-calcite ratios in carbonate-secreting marine organisms. *J. Geol.*, 62: 284—323.

Mackenzie, F.T. and Garrels, R.M., 1966. Silica-bicarbonate balance in the ocean and early diagenesis. *J. Sediment. Petrol.*, 36: 1075—1084.

Matter, A., 1974. Burial diagenesis of peletic and carbonate deep-sea sediments from the Arabian Sea. In: *Initial Reports of the Deep-Sea Drilling Project*, 23: 421—469.

Maxwell, D.T. and Hower, J., 1967. High-grade diagenesis and low-grade metamorphism of illite in the Precambrian Belt Series. *Am. Mineral.*, 52: 843—857.

McBride, E.F., 1963. A classification of common sandstones. *J. Sediment. Petrol.*, 33: 664—669.

McBride, E.F., 1974. Significance of color in red, green, purple, olive brown, and gray beds of Difunta Group, Northwestern Mexico. *J. Sediment. Petrol.*, 44: 760—773.

Meents, W.F., Bell, A.H., Rees, O.W. and Tilbury, W.G., 1952. Illinois oil field brines. *Ill. State Geol. Surv., Ill. Petrol.*, 66: 5—38.

Middleton, G.V., 1972. Albite of secondary origin in Charny Sandstones, Quebec. *J. Sediment. Petrol.*, 42: 341—349.

Miller, Jr, D.N., 1957. Authigenic biotite in spheroidal reduction spots, Pierce Canyon red beds, Texas and New Mexico. *J. Sediment. Petrol.*, 27: 177—181.

Miller Jr., D.N. and Folk, R.L., 1955. Occurrence of detrital magnetite and ilmenite in red sediments: new approach to significance of red beds. *Bull. Am. Assoc. Pet. Geol.*, 39: 338—345.

Millot, G., 1960. Silice, silex, silifications et croissance des cristaux. *Bull. Serv. Carte Geol. Alsace Lorraine*, 13 (4): 129—146.

Mizutani, S., 1966. Transformation of silica under hydrothermal conditions. *J. Earth Sci., Nagoya Univ.*, 14: 56—88.

Mizutani, S., 1970. Silica minerals in the early stage of diagenesis. *Sedimentology*, 15: 419—436.

Nanz Jr., R.H., 1954. Genesis of Oligocene sandstone revervoir, Seeligson Field, Jim Wells and Kleberg Counties, Texas. *Bull. Am. Assoc. Pet. Geol.*, 38: 96—117.

Ojakangas, R.W., 1963. Petrology and sedimentation of the Upper Cambrian Lamotte Sandstone in Missouri. *J. Sediment. Petrol.*, 33: 840—874.

Packham, G.H. and Crook, K.A.W., 1960. The principle of diagenetic facies and some of its implications. *J. Geol.*, 68: 392—407.

Pettijohn, F.J., 1957. *Sedimentary Rocks.* Harper and Row, New York, N.Y., 2nd ed., 718 pp.

Picard, M.D., 1965. Iron oxides and fine-grained rocks of Red Peak and Crow Mountain Sandstone members, Chugwater (Triassic) Formation, Wyoming. *J. Sediment. Petrol.*, 35: 464—479.

Pittman, E.D., 1972. Diagenesis of quartz in sandstones as revealed by scanning electron microscopy. *J. Sediment. Petrol.*, 42: 507—519.

Pittman, E.D. and Lumsden, D.N., 1968. Relationship between chlorite coatings on quartz grains and porosity, Spiro Sand, Oklahoma. *J. Sediment. Petrol.*, 38: 668—670.

Powers, M.C., 1959. Adjustment of clays to chemical change and the concept of the equivalence level. *Clays Clay Miner.*, 6: 309—327.

Raam, A., 1968. Petrology and diagenesis of Broughton Sandstone (Permian), Kiama District, New South Wales. *J. Sediment. Petrol.*, 38: 319—331.

Rapson-McGugan, J.E., 1970. The diagenesis and depositional environment of the Permian Ranger Canyon and Mowitch Formations, Ishbel Group, from the southern Canadian Rocky Mountains. *Sedimentology*, 15: 363—417.

Renton, J.J., Heald, M.T. and Cecil, C.B.,1969. Experimental investigation of pressure solution of quartz. *J. Sediment. Petrol.*, 39: 1107—1117.

Rex, R.W., 1966. Authigenic kaolinite and mica as evidence for phase equilibria at low temperatures. *Clays Clay Miner.*, 13: 95—105.

Rittenhouse, G., 1971a. Pore-space reduction by solution and cementation. *Bull. Am. Assoc. Pet. Geol.*, 55: 80—91.

Rittenhouse, G., 1971b. Mechanical compaction of sands containing different percentages of ductile grains: a theoretical approach. *Bull. Am. Assoc. Pet. Geol.*, 55: 92—96.

Robb, G.L., 1949. Red-bed coloration. *J. Sediment. Petrol.*, 19: 19—103.

Rusnak, G.A., 1957. A fabric and petrologic study of the Pleasantview Sandstone. *J. Sediment. Petrol.*, 27: 41—55.

Sarin, D.D., 1963. Petrography and origin of Peguan sandstones exposed at Chauk, Burma. *J. Sediment. Petrol.*, 33: 333—342.
Seilacher, A., 1968. Origin and diagenesis of the Oriskany Sandstone (Lower Devonian, Appalachians) as reflected in its shell fossils. In: G. Müller and G.M. Friedman (Editors), *Recent Developments in Carbonate Sedimentology in Central Europe.* Springer, New York, N.Y., 255 pp.
Siever, R., 1959. Petrology and geochemistry of silica cementation of some Pennsylvanian sandstones. In: *Silica in Sediments—Soc. Econ. Paleontol. Mineral., Spec. Publ.*, 7: 55—79.
Siever, R., 1962. Silica solubility, 0—200°C, and the diagenesis of siliceous sediments. *J. Geol.*, 70: 127—151.
Sippel, R.F., 1968. Sandstone petrography, evidence from luminescence petrography. *J. Sediment. Petrol.*, 38: 530—554.
Sloss, L.L., 1963. Sequences in the cratonic interior of North America. *Bull. Geol. Soc. Am.*, 74: 93—114.
Sloss, L.L. and Feray, D.E., 1948. Microstylolites in sandstone. *J. Sediment. Petrol.*, 18: 3—13.
Smale, D., 1973. Silcretes and associated silica diagenesis in southern Africa and Australia. *J. Sediment. Petrol.*, 43: 1077—1089.
Spry, A., 1969. *Metamorphic textures.* Pergamon, Oxford, 350 pp.
Swett, K., 1965. Dolomitization, silification and calcitization patterns in Cambro-Ordovician oolites from northwest Scotland. *J. Sediment. Petrol.*, 35: 928—938.
Swett, K., 1968. Authigenic feldspars and cherts resulting from dolomitization of illitic limestones: a hypothesis. *J. Sediment. Petrol.*, 38: 128—135.
Swineford, A., 1955. Petrology of Upper Permian rocks in southcentral Kansas. *Kans. State Geol. Surv. Bull.*, 111: 179 pp.
Swineford, A. and Franks, P.C., 1959. Opal in the Ogallala Formation in Kansas. In: *Silica in Sediments—Soc. Econ. Paleontol. Mineral. Spec. Publ.*, 7: 111—121.
Taylor, J.C.M. and Illing, L.V., 1969. Holocene intertidal calcium carbonate cementation, Qutai, Persian Gulf. *Sedimentology*, 12: 69—107.
Taylor, J., 1950. Pore space reduction in sandstones. *Bull. Am. Assoc. Pet. Geol.*, 34: 701—716.
Teisseyre, A.K., 1973. Diagenetic carbonatization due to kaolinization: a hypothesis (with examples from Sudetic Carboniferous sandstones). *Ann. Soc. Géol. Pol.*, 43: 453—482.
Thomson, A., 1959. Pressure solution and porosity. In: *Silica in Sediments—Soc. Econ. Paleontol. Spec. Publ.*, 7: 92—110.
Thompson, A.M., 1970. Geochemistry of color genesis in red-bed sequence, Juniata and Bald Eagle formations, Pennsylvania. *J. Sediment. Petrol.*, 40: 599—615.
Triplehorn, D.M., 1970. Clay-mineral diagenesis in Atoka (Pennsylvanian) sandstones, Crawford County, Arkansas. *J. Sediment. Petrol.*, 40: 838—848.
Van Houten, F.B., 1958. Contribution to the petrology of the Tanner graywacke. *Bull. Geol. Soc. Am.*, 69: 301—314.
Van Houten, F.B., 1973. Origin of red beds — A review 1961—1972. *Annu. Rev. Earth Planet. Sci.*, 1: 39—61.
Voll, G., 1960. New work on petrofabrics. *Liverpool Manchester Geol. J.*, 2: 503—567.
Von Engelhardt, W. and Gaida, K.H., 1963. Concentration changes of pore solutions during the compaction of clay sediments. *J. Sediment. Petrol.*, 33: 919—930.
Waldschmidt, W.A., 1941. Cementing materials in sandstones and their probable influence

on the migration and accumulation of oil and gas. *Bull. Am. Assoc. Pet. Geol.*, 25: 1839—1879.

Walker, T.R., 1962. Reversible nature of chert—carbonate replacement in sedimentary rocks. *Bull. Geol. Soc. Am.*, 73: 237—242.

Walker, T.R., Ribbe, P.H. and Honea, R.M., 1967. Geochemistry of hornblende alteration in Pliocene red beds, Baja California, Mexico. *Bull. Geol. Soc. Am.*, 78: 1055—1060.

Waugh, B., 1970. Formation of quartz overgrowths in the Penrith Sandstone (Lower Permian) of northwest England as revealed by scanning electron microscopy. *Sedimentology*, 14: 309—320.

Weaver, C.E. and Beck, K.C., 1971. Clay water diagenesis during burial: how mud becomes gneiss. *Geol. Soc. Am., Spec. Pap.*, 134: 96 pp.

Webb, J.E., 1974. Relation of oil migration to secondary clay cementation. *Bull. Am. Assoc. Pet. Geol.*, 58: 2245—2250.

Weller, J.M., 1959. Compaction of sediments. *Bull. Am. Assoc. Pet. Geol.*, 43: 273—310.

Weyl, P.K., 1959. Pressure solution and the force of crystallization — A phenomenological theory. *J. Geophys. Res.*, 64: 2001—2025.

Whetten, J.T. and Hawkins, J.W., 1970. Diagenetic origin of graywacke matrix minerals. *Sedimentology*, 15: 347—361.

Whisonant, R.C., 1970. Influence of texture upon the response of detrital quartz to deformation in sandstones. *J. Sediment. Petrol.*, 40: 1018—1025.

Williamson, W.O., 1957. Silicified sedimentary rocks in Australia. *Am. J. Sci.*, 255: 23—42.

Zen, E-an, 1959. Clay mineral-carbonate relations in sedimentary rocks. *Am. J. Sci.*, 257: 29—43.

Zen, E-an, 1961. The zeolite facies: an interpretation. *Am. J. Sci.*, 259: 401—409.

Zenger, D.H., 1973. Syntaxial calcite borders on dolomite crystals, Little Falls Formation (Upper Cambrian), New York. *J. Sediment. Petrol.*, 43: 118—124.

Chapter 3

SILICA AS AN AGENT IN DIAGENESIS

E.C. DAPPLES

INTRODUCTION

Since publication of the first edition of this volume, new information has become available to support the role which silica plays as an agent of diagenesis, particularly in the neoformation of silicates. Much of the information has its source in the many carefully controlled analyses of the minerals involved and their associated interstitial waters. There has been confirmation of the limiting solubilities of amorphous silica and, to a lesser degree, that of quartz. Moreover, occurrences of relatively young cherts associated with Pleistocene brine-lake deposits in Kenya and deep-sea drilling in the Atlantic and Pacific, have provided greater insight into episodes when silica is mobilized and reprecipitated in some form of proto-chert. Although general mechanisms whereby silica is precipitated and crystallized are still somewhat uncertain, the details currently described as controlling the crystallization of chert are based upon reliable physical-chemical data and thermodynamic equations. The distinct difference between the solubilities of amorphous silica and quartz tend to compound the general problems by uncertainty concerning the amount of silica which is in a colloidal, or a non-ionic, state, and the number and constitution of the ions under specific physical-chemical conditions. These are factors which definitely tend to increase the values of "soluble" silica above that in the true solution, and are now regarded as controlling some forms of silica which are crystallized.

Recent studies have strengthened the long-held concept that much of the silica in solution in river waters must be precipitated upon entering the oceans, but as yet no localized concentration of masses of primary gel silica have been discovered (Heath and Moberly, 1971, p. 992). Among geologists there remain some proponents of the concept that primary accumulation of silica in amorphous globules can occur. Nevertheless, there is increasing reason to favor the interpretation that precipitation of silica occurs as very small discrete particles along with clay minerals and carbonates in clay-size dimension. Concentration of silica in the form of concretions is increasingly regarded as an example of accumulation following deposition of the enclosing sediments and the process of concretionary growth is regarded as a migration of "ions" to some center of precipitation. The occurrence of exceedingly large volumes of siliceous sediments associated with former eugeosynclines has been supported by additional observations, but the problems of the source of the silica and the time of its precipitation remain unsolved. Such sediments not only are argillaceous and arenaceous, but also comprise those the composition of which indicates that they are lacking in land-derived aluminum-silicate debris (for example, beds of novaculite).

Carbonates of the stable craton rather commonly are chert-bearing, the chert occurring in the form of nodules, lensing beds, and veinlets. All such occurrences locally are restricted to thin beds separated by intervals of "chert-free" carbonate in which the silica content tends not to exceed 10%. Among such shelf sediments the chert zones are restricted to the carbonates, whereas the shales and sandstones uniformly are free from concretionary silica. Similarly, cherty carbonates commonly constitute some ancient inter-reef deposits, whereas the associated carbonate reefs and banks tend to be characteristically low in chert (Shaver, 1974, p. 946).

A parallel case occurs with evaporites. Bedded anhydrites, salts and associated carbonates are low in silica, whereas commonly carbonates rimming the evaporite basin are cherty. In terms of quantity of silica precipitated, the range clearly is lowest among cratonic sediments, increases in the so-called miogeosynclinal carbonates (but remains low in shales and sandstones), and attains a maximum in the eugeosynclinal strata where sandstones, shales, and carbonates are prominently silicified. Except for the ideal eugeosynclinal sediments, local distribution of silica is sporadic, with highly siliceous strata being interbedded with strata of low silica content despite no apparent change in environment of deposition. The ways and means of the emplacement of the precipitated silica still remain in the realm of uncertainty. Petrographic studies as a whole tend more and more to favor interpretations of a diagenetic origin from primary precipitation of some form of biogenic or inorganic opal. The important processes are those of complex replacement of pre-existent mineral matter and evolution into more stable forms of silica.

The reader must be aware that insofar as demonstrated mechanisms are concerned, there has been little advance in describing the origin of flints and geodes. On the other hand, the origin of bedded cherts is beginning to be well understood. One is, however, in a better position to limit concepts of origin (1) on the basis of restrictions imposed by laboratory studies, and (2) by paragenetic relations between the silica group of minerals. The principal purpose of this chapter is to outline the restrictions on the precipitation of silica, in order to clarify for the reader the currently preferred mechanisms of deposition, and the relationships involved in the development of authigenic forms.

SOLUBILITY OF SILICA

Laboratory studies

Ranges of solubility of amorphous silica have been determined independently by a number of investigators with sufficiently close agreement to permit fixing the limits of silica in solution. There is now no doubt that such silica occurs as tetrahedrally bonded units of SiO_2 together with monomeric H_4SiO_4, which dissociates into $H_3SiO_4^-$ and $H_2SiO_4^{2-}$ at high values of pH, and correspondingly shows a greatly increased solubility under such conditions (Hay, 1968, p. 269). Probably some crudely uniform equilibrium exists between the tetrahedrally bonded units and the ionized forms when the pH is less than 9. Above that value the equilibrium is increasingly displaced toward the ionic species and exceptional values are obtained for the amount of silica in solution. In the laboratory, saturated solutions of silica have been prepared essentially by two methods: (1) allowing supersaturated solutions to come to equilibrium by precipitating SiO_2, and (2) allowing silica to dissolve from suspensions until equilbrium is reached (Krauskopf, 1956; 1959, fig. 2). By either method, ranges of 100—150 p.p.m. silica in solution have been obtained as equilibrium values at temperatures of 22—27°C in about 70 days. Although this appears to be a somewhat sluggish equilibrium with respect to laboratory experimentation, the rate virtually is instantaneous from the viewpoint of geologic rates of deposition. As the temperature is elevated, the solubility rises and at 150°C over 600 p.p.m. of silica are in solution.

Natural hot springs also may contain silica in amounts commensurate with those obtained in the laboratory. In some waters there is evidence of the presence of colloidal silica, or supersaturation, as silica is precipitated from the solution on standing. Silica is also precipitated from such highly concentrated solutions when passing over previously deposited siliceous sinter, whereas algae do not appear to exert any important control on the precipitation (White et al., 1956, p. 39). Certain ground waters rich in silica are saturated approximately at the temperatures of such waters, but many ground waters appear to be close to saturation at the existing temperatures (Table 3-I, Fig. 3-1). There is reason, therefore, to consider that some very special mechanism must be envisaged as responsible for removal of silica from stream and oceanic waters. An exceptional situation is noted at Aqua de Ney, California, where spring water issuing at temperatures between 50—54°C contains 3,400—3,970 p.p.m. SiO_2 (Feth et al., 1961, Table 3-I). The extraordinary high content of silica is regarded by these authors to result from modification of entrapped connate sea water by reaction with calcium sulphate-rich ground waters. According to their interpretation, ion exchange

TABLE 3-I

Partial analyses of natural waters (after Hem, 1959; Feth et al., 1961, 1964)

Variety of water	Issuing from rock at	p.p.m.			pH	Temp. (°C)	Locality
		SiO_2	Fe	Mg			
Silica	flowing well	99	0.04	1.4	9.2	50	Owyhee Co., Idaho
	rhyolite	103	0.0	1.1	6.7	38	Rio San Antonio, New Mexico
	rhyolite, possibly basalt	60	0.01	7.5	—	—	Harney Co., Oregon
	basalt	49	0.01	12	7.8	—	Umatilla Co., Oregon
	volcanic rock	48	0.03	20	—	—	Greenlee Co., Arizona
	conglomerate and sandstone	71	0.0	8.8	7.9	—	Albuquerque, New Mexico
	sand? (Loup River)	62	0.01	4.4	8.1	—	Dunning, Nebraska
	mica schist	29	0.33	1.7	7.1	15	Burke Co., North Carolina
	volcanic rock	3970	0.0	2.6	11.6	54	Aqua de Ney, Siskiyou Co., California
	volcanic rock	3400	0.0	0.9	10.9	50	Aqua de Ney, Siskiyou Co., California
	snowmelt from joints in granite	26	0.0	1.9	7.1	14	Nevada Co., California
	granite soil	5.2	0.01	0.1	6.1	12	Nevada Co., California
	quartz monzonite	22	0.07	1.9	7.6	11	Washoe Co., California
	granite rock	54	0.02	5.5	7.0	11	Calaveras Co., California

Iron	sand and gravel	20	2.3	43	7.6	13	Columbus, Ohio
	sandstone	12	2.9	2.0	7.4	22	Memphis, Tennessee
	dark shale	26	10	8.4	6.4	14	Blount Co., Tennessee
	calcareous sands and clays	41	8.1	17	6.7	—	Elizabeth City, North Carolina
	sandstone (oil well brine)	9.1	32	1,330	7.4	—	Okmulgee Co., Oklahoma
	iron mine (siliceous)	8.1	0.34	17	7.5	15	Ishpeming, Michigan
	stream	21	15	68	3.0	—	Weighscale, Pennsylvania
	sandstone	23	4.8	35	—	—	Ransom Co., North Dakota
	sand	7.9	11	1.5	6.3	17	Fulton, Mississippi
Magnesium	dolomite	8.4	0.24	22	7.4	14	Jefferson City, Tennessee
	dolomite	18	1.4	40	6.7	12	Sidney, Ohio
	olivine tuff breccia	31	—	42	8.2	12	Navajo Indian Res., Arizona
	serpentine	80	5.6	614	—	18	Colusa, California
	dolomitic limestone	11	0.04	0.71	8.3	—	Carlsbad Cavern, New Mexico
	large stream in dolomitic limestone (locally)	11	0.13	8.5	7.0	—	Wisconsin River, Grant Co., Wisconsin
	quartzite	12	0.03	12	7.6	18	Calhoun Co., Alabama
	limestone (primarily)	13	0.01	43	7.4	—	Chaves Co., New Mexico
	dolomite	18	0.39	33	8.2	10	Milwaukee Co., Wisconsin

and sulphate reduction occurs removing H^+ ion to develop the observed values of pH (10.9—11.6). These very high values are regarded as being responsible for the unusual amount of silica in solution.

Quartz is not readily soluble and is acknowledged to be sluggish in reaching the equilibrium-solution value. Following a long-term experiment, however, G.W. Morey and his associates were able to reduce the amount of silica in solution from 80 p.p.m. to 6 ± p.p.m. at 25°C (Morey et al., 1962, p. 1036). They concluded that the latter is the equilibrium value for quartz. Siever's (1962, p. 135) observations caused him to regard the attainment of equilibrium of quartz with its solution at normal near-surface temperatures to be of the order of spans of geologic time. Müller (1961, 1967, p. 150), however, has found authigenic quartz in Recent sediments of the Gulf of

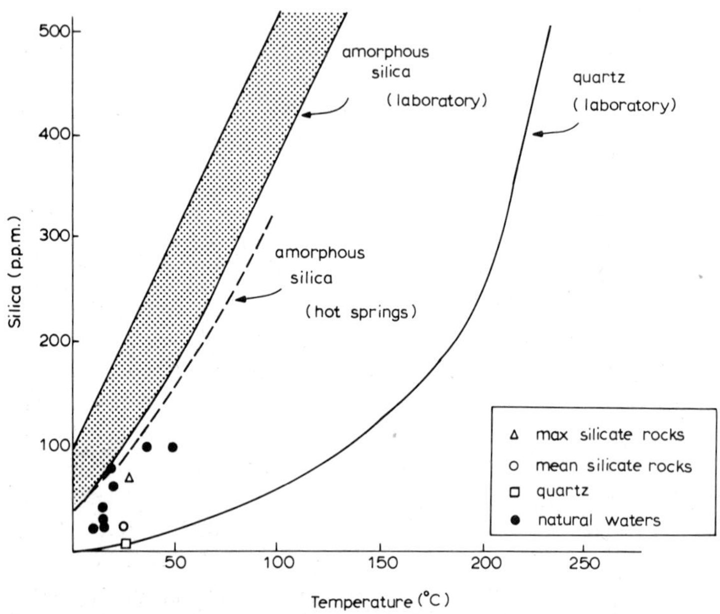

Fig. 3-1. Solubility of silica with change in temperature of solutions. The range of values for amorphous silica are those obtained in the laboratory by Alexander, Okamoto, Krauskopf, and others (Krauskopf, 1959, p. 5). The curve for amorphous silica in hot-spring water is based on data taken from White et al. (1956), whereas the data on solubility of quartz are principally obtained from Siever, Van Lier, and others (Siever, 1962, p. 134). The single points showing solubility values for the maximum and mean for silicate rocks and for quartz are those reported by Keller et al. (1963); these are solubilities obtained by crushing the specimens in distilled water. Scattered values for natural waters are selected analyses of ground waters in silicate rocks (Table 3-I). Such waters are supersaturated with respect to quartz and some tend to approach values of solubility of amorphous silica.

Naples. Authigenic quartz crystals have been observed by the present writer in modern soils developed from siliceous limestone. This observation supports that of Müller (1961), namely that quartz crystallization can be accomplished in a short period, providing the optimum physicochemical conditions prevail. The most recent pertinent information was provided by Mackenzie and Gees (1971), who were able to crystallize authigenic quartz on crushed fragments of quartz in sea water after two years of agitation. The final measured content of the sea water was 4.4± p.p.m. of silica under the experimental conditions of 20° C and pH of 8.1±.

Presently available information tends to support the concept that some natural waters may not be in equilibrium with quartz, whereas they most assuredly are with respect to amorphous silica (e.g., as evidenced by the precipitation of opal in Recent sediments). Sufficiently careful work has been done to establish with small error the range in solubility of quartz with change in temperature. The slope of the curve has the same general trend as that of amorphous silica, but at the same temperature the solubility is much lower (Fig. 3-1). According to Siever (1962, p. 133), the slope of the quartz solubility curve tends to rise somewhat more steeply than that for amorphous silica, and the two are considered to join near the critical temperature of water.

Influence of pH on solubility of silica

Since the classic study of Correns (1941) on the relative solubility of silica with change in pH, additional data have provided points for constructions of a reasonably precise curve showing this relation. The slope of the curve tends to approximate horizontality and to pass through essentially the same values between the pH limits of 2 and 8.5. At higher pH the solubility rises abruptly attaining values approaching 5,000 p.p.m. at pH of 11 (Krauskopf, 1959; Blatt et al., 1972, p. 359). Such extremely basic waters are most exceptional near the earth's surface; hence, only in very localized environments do pH values approach such magnitude. Interstitial waters associated with tidal-flat algal mats and in the sabkha environments of strong evaporation tend to have high pH values; therefore, concentrations of silica, theoretically, can be exceptionally large (Baas Becking et al., 1960). Elsewhere, as with the previously mentioned spring of Aqua de Ney, California (Table 3-I), there are scattered examples of subsurface waters which contain high amounts of dissolved silica when pH of 11 is approached. In certain evaporating lakes, such as Lake Natron, Kenya, described by Hay (1968, p. 260) as a sodium carbonate lake, the amount of dissolved silica reaches values approaching 176 p.p.m. when the pH of the water is 9.7. Hence, the natural conditions clearly parallel the solubility values determined under laboratory conditions.

Observations in the natural state

Data supplied by Keller et al. (1963) on concentrations of silica in solution as a result of grinding of natural silicates show values ranging from 7.5 to 8.6 p.p.m. for quartz to as much as 68.6 p.p.m. for pumice. Their analyses of the silica content of various glacial milks, however, show values which are considerably lower, approximately 0.4—17 p.p.m. (Keller and Reesman, 1963). Presumably, abrasion during erosion is not to be regarded as an important mechanism for releasing silica into solution. Davis (1964, p. 871), however, reported analyses of silica in California stream waters of which 5% contain less than 8 p.p.m., 50% less than 17 p.p.m., and 95% contain less than 36 p.p.m. These concentrations are of the same order of magnitude as those of spring waters draining granite and granodioritic rocks of the Sierra Nevada as reported by Feth et al. (1964). Experiments by Davis (1964, p. 880) simulating natural conditions indicated to him that some waters could acquire almost all of their dissolved silica within a few months or years. For example, in the heavily irrigated eastern side of the Central Valley of California the water brought in for irrigation contains from 8 to 15 p.p.m. of silica, whereas as a result of a short distance of subsurface travel, considered to require about 20 years time, the values of dissolved silica have increased to 30—70 p.p.m. For the sampled streams, the silica content is independent of the volume of discharge. Apparently, storm runoff has gathered its silica content in a few days (Davis, 1964, p. 881). The reader should note that Feth et al. (1964) reported 26 p.p.m. of silica in snowmelt escaping from joints in granite (Table 3-I), values which tend to support Davis' conclusion.

Current opinion based upon information paralleling that presented above indicates that the amount of silica dissolved in ground water is primarily dependent upon the rock type through which the water flows and not the rate of rock abrasion. Largest amounts of dissolved silica are associated with volcanic rocks, whereas intermediate amounts are present in waters draining or flowing through plutonic rocks and sediments containing feldspar and volcanic fragments. In general, the smallest amounts of silica are present in waters associated with carbonate rocks (Table 3-I). According to Davis (1964), the available data fail to reveal any direct correlation between the amount of dissolved silica and salinity, climatic region, surface vegetation, or temperature variation as long as the temperature remains below 35°C.

A representative group of analyses of surface and underground waters in the United States tend to reveal that the silica concentration at the temperatures recorded may be somewhat above that of the solubility curve of quartz, but below that for amorphous silica (Fig. 3-1). For example, Missisippi River water near the delta mouths (temperature of approximately

20° C) averages 4—7.5 p.p.m. of silica (Bien et al., 1959); whereas the average concentration in the surface waters of the Gulf of Mexico is less than 1 p.p.m. (0.11 p.p.m. as reported by Bien et al.). Deep ocean waters tend to contain no more than 6 ± p.p.m. of silica (Garrels et al., 1967). Because these values clearly are lower than the solubility values of amorphous silica at the same temperatures, one can assume that the surface waters of the oceans are undersaturated with respect to silica. It is likely, however, that at the temperatures which prevail in oceanic bottom waters, the latter are approximately saturated with respect to quartz (Siever, 1962). Numerous authors have expressed the opinion that the low values of dissolved silica in the sea waters are due to its extraction by organisms (Siever, 1957). Opaline tests of organisms appear not to return to solution readily as indicated by commonly well-preserved radiolarian and sponge remains. Also, experimental work (D.E. White, as quoted by Krauskopf, 1959, p. 9) showed that opal did not completely saturate a solution of sea water even after 2 years. Many opaline remains of organisms, therefore, are likely to be buried before being dissolved. Carefully controlled studies on the rates of solution of opaline tests of microorganisms, however, indicate that a great amount of solution does occur both in the oceanic waters and in the uppermost part (around one meter) of sediments.

Present-day investigators agree that the quantity of organically supplied silica, which dissolves in the oceanic waters and that present a few meters below the sediment—water interface, is controlled by complex local conditions, some of which remain unknown. According to Moore (1969, p. 2103), Tertiary sediments cropping out in Pacific equatorial regions are, in many cases, richer in biogenous silica than Quaternary ones. Yet, available information indicates that the quantity of radiolaria was sufficient to provide all the required silica for the chert beds known to exist. He concluded that the decrease in amount of silica in the Quaternary beds was due to decrease in weight of the radiolaria tests with the progress of time. For example, Eocene radiolaria have tests four times heavier than those of the Quaternary radiolaria. This feature in itself could account for the decrease in amount of opaline silica without resorting to differential solution following the deaths of the organisms. Lewin (1961) studied the relative influences of pH, temperature, and the presence of certain ions such as Ca^{2+}, Al^{3+}, and Fe^{3+} on the solution of frustules of diatoms following their death. Every treatment which killed the cells led to an increase in the rate of solution of the silica walls. Certain metal ions, particularly Al^{3+}, however, tended to combine with the silica walls and generally reduced their rate of solution. An unexplained observation is that modern material dissolves more rapidly than the fossil counterpart.

An exhaustive study by Calvert (1966) of the Gulf of California reveals that the distribution of opal as weight percent of total sediment is similar to the diatom frustules in the sand fraction as determined by item count. Sediments containing the greatest amount of opal (chiefly biogenous) occur in the deep Guymas Basin and other similar deep basins. Also, above such high concentrations of biogenous silica are the largest standing crops of phytoplankton; hence the source of the opal is indicated directly. The concentration of the silica-producing organisms is related to the current pattern which consists of a dominant surface current (inflow) and a deep-water outflow. During the winter regime, this pattern is modified and the nutrient-rich water flows in at depth and upwells in the central part. Silica concentrations of the outflowing water average 1 p.p.m., whereas at 350 m the inflowing water has about 3 p.p.m. The resulting net gain in silica occurs at a rate of about $5 \cdot 10^{14}$ g/yr. This amount exceeds by two orders of magnitude that supplied via the Colorado and other inflowing rivers, but is close to the rate calculated by Goldberg (1963) to be equivalent to the supply of dissolved silica furnished by the rivers of the world to the oceans ($3.2 \cdot 10^{14}$ g/yr). Calvert's analysis demonstrates that ocean waters are capable of furnishing all the silica known to be accumulating as biogenous opal in certain bodies of water such as Gulf of California. Moreover, for such conditions there is no requirement to seek the source of the silica from decomposition of unstable volcanic glasses in the sediments.

Silica in pore waters

The carefully controlled analyses of Bien et al. (1959) show that Mississippi River water loses silica upon entering the Gulf of Mexico in a quantity greater than that which can be attributed to removal by organisms. According to their interpretation, electrolytes (1) cause silica to be adsorbed on the surface of suspended inorganic particles, or (2) facilitate coprecipitation of silica with such inorganic particles, resulting in a most significant depletion in dissolved silica. In this connection Mackenzie and Garrels (1966a, b) prepared a theoretical analysis demonstrating that lowering of silica content in ocean water with respect to river water can be due, in large measure, to the authigenesis of certain clay minerals from such weathering products as poorly crystalline aluminosilicates and chlorites. The important key to the synthesis of the new clay minerals and the maintenance of the low values of silica content in the oceanic waters, is the partial removal of HCO_3^- ion. This bicarbonate ion is conceived to react with a dissolved silicic acid, e.g., H_4SiO_4, and is converted to CO_2 to be recycled through the atmosphere. Garrels et al. (1967) also demonstrated experimentally that typical clay minerals, in the suspended load of streams, tend to release silica to silica-deficient sea water and, conversely, to remove silica from similar waters enriched in silica. In the latter reactions, abstraction of the silica involves cations in the sea water and is controlled by neoformation of poorly defined aluminosilicates of the clay-mineral type.

In the past few years, important information has been published regarding the composition of interstitial waters of the most recently deposited oceanic sediments. Beneath the depositional interface the tendency for the chemical system to be partially closed is greater than in the supernatant unrestricted water. For example, Siever et al. (1965, p. 63) stated that experimental work has shown that clays must be compacted by differential pressures of at least 300 p.s.i. [1] before they behave as ultrafiltration membranes. Yet, slowly accumulated deep-sea sediments are more compacted than predicted by short-term laboratory experiments. There is reason to believe, therefore, that interstitial waters have more opportunity of reaching equilibrium with the enclosing sediments than identical material at the sediment—water interface. Moreover, the logic of this reasoning has been strengthened by the increasing information pointing to the rapidity with which such an equilibrium can be attained (Garrels et al., 1967).

Years ago, Emery and Rittenberg (1952) published data which showed that interstitial waters in modern sediments in the basins off the coast of southern California contained dissolved silica in excess of the solubility of quartz, but less than that for amorphous silica at the existent temperatures. More recently, Siever et al. (1965, p. 70) reported that dissolved silica in the pore waters of the oceanic sediments, which they analyzed, showed values commonly in the 20—50 p.p.m. range, i.e., greatly exceeding the equilibrium value for quartz but less than that for amorphous silica. They presumed that diatom tests would dissolve until the pore waters would reach equilibrium with amorphous silica, and that such dissolution was the primary source of the dissolved silica. Since publication of their results and those of Bischoff and Ku (1970, p. 969), the general composition of interstitial waters of oceanic sediments has been clarified with respect to silica. The latter is significantly enriched within the uppermost meter of sediment over that in the overlying open water (Hurd, 1973, p. 2260). Although the ranges in values obtained by the various investigators tend to overlap, the mean values differ by significant amounts. For example, the average range in values (10—30 p.p.m.) determined by Bischoff and Ku is lower than that reported by Siever and his co-workers (20—50 p.p.m.). Bischoff and Ku (1971, p. 1011) pointed out that interstitial waters of the sediments on the Mid-Atlantic Ridge average 30 p.p.m. of dissolved silica, i.e., about three times that of the bottom waters, but significantly below the solubility of diatoms (60 p.p.m.).

---

[1] Experimental results obtained by one of the editors (G.V.C.) indicate that this pressure could be as high as 10,000 p.s.i. (1 p.s.i. = 7.031×10^{-2} kg/cm^2.)

Accordingly, they suggested that the various values observed represent the influence of buffering by as yet undetermined silicates. [1]

The restricted investigation by Drever (1971) of the diagenesis of clay minerals accumulating in Banderas Bay, at the mouth of the Rio Ameca in Mexico, demonstrated that silica content of interstitial waters ranges from 27 to 47 p.p.m. In a single core the sample lacked siliceous organisms but the interstitial water contained 27 p.p.m. dissolved silica (pp. 989, 990). Accordingly, Drever interpreted the source of the silica to the presence of some reactive material, perhaps feldspar, some other igneous mineral, or an X-ray amorphous silicate, but not to the equilibrium control of clay minerals alone. Inasmuch as all other samples which he examined contained remains of siliceous organisms, the latter could be the source of dissolved silica.

Hurd's (1973) studies of dissolved silica in interstitial solutions of cores taken in the central equatorial Pacific indicate that values obtained by previous investigators may be considerably in error because of the temperature increase developed by the squeezing effect during removal of the pore water. His results indicate that although biogenic opal is the most soluble silica-containing mineral in the sediments, the interstitial waters remained unsaturated with respect to this mineral. Also, the rate of solution of the biogenic opal is several orders of magnitude slower under natural conditions than in the acid-cleaned material dissolved in the laboratory. For this reason, some biogenic opal can remain in the sediments without being destroyed, albeit this may constitute only 0.05—0.15% of the total opal produced in the surface waters. His findings tend to more accurately define the probability that such ions as Ca^{2+}, Mg^{2+}, Al^{3+}, and Fe^{3+} are strongly attracted to the silica surface under conditions of the buried sediment, and may form coatings on aluminosilicates to inhibit dissolution of the silica.

Despite the ever-increasing precision which characterizes the analytical work, the primary sources of the dissolved silica are not uniquely defined. Moreover, the reasons why the observed values tend to be less than the saturation values of biogenic opal at the in situ temperatures are not uniformly accepted. There is, however, general agreement that biogenic silica is an important contributor to the total dissolved silica. Also, there is increasing verification of the importance of the equilibria resulting in neoformation of clay minerals. In some examples, the amount of dissolved silica is maintained at low values, whereas, under other conditions, excess silica is released eventually to crystallize as a proto-chert. The discovery by Hathaway and Sachs (1965, p. 862) of small veins of authigenic sepiolite as a network cutting a brown laminated clay on the Mid-Atlantic Ridge demonstrated a relationship

[1] In the opinion of the editors, different techniques and pressures used by different investigators in extracting interstitial fluids could account for these differences.

with coccolith ooze and hard white chert. The intergrowth of the minerals was interpreted by them to indicate that silica derived from devitrification of volcanic ash had combined with Mg^{2+} ions to form the neomorphic sepiolite, whereas an excess of silica was crystallized as quartz. Also, distinction between organically and inorganically precipitated opal has been shown by Calvert (1966, p. 573) to be resolved by the X-ray diffractograms in which the biogenic opal appears as a broad area of low intensity lacking any indication of tridymite or cristobalite peaks.

OPAL AS PRIMARY DIAGENETIC MATERIAL

Drilling in the deep-sea basins has proved what had been anticipated by earlier sampling, namely, that the remains of silica-precipitating organisms are abundantly preserved in localized stratigraphic positions as some form of opal in present-day as well as ancient sediments (Goldberg, 1958). Moreover, the relationship opal—chalcedony—macrocrystalline quartz reported long ago from bedded cherts (e.g., Taliaferro, 1934, pp. 206—209) is in fact a diagenetic progression. There has not been a clearly defined indication, however, that long-time aging of cristobalite with an opaline nucleus would result in macrocrystalline quartz. Such a transition does occur with increased temperature as indicated by the unique occurrence of quartz in the cherts enclosed in sediments close to locations of submarine intrusive basalt (Heath and Moberly, 1971, pp. 994, 995). These authors consider that the initial biogenous opal is converted to finely crystalline cristobalite in which form it assumes the aspect of porcelanite. Later this cristobalite converts to quartz. The evidence presented by Heath and Moberly (1971, p. 1003) suggests an initial stage of solution of the organic opal followed by crystallization of cristobalite from solution at the sites of concentration of silica "micelles". A third stage is the progression of the cristobalite into microcrystalline quartz as a solid—solid transformation.

Advances are being made concerning the relationship of opal to quartz, largely from optical and X-ray diffractogram patterns. Jones and Segnit (1971) have separated opal into three distinct groups. Opal *A* is characterized by an X-ray diffractogram pattern of a broad band centered around 4.1 Å. It lacks peaks which can be interpreted as cristobalite or tridymite. Most diatom frustules, sponge spicules, and radiolaria fall into this category, but so also do many precious opals and geyserite. Upon heating to 1000°C, peaks of α-cristobalite dominate the diffractogram pattern (Calvert, 1966, pp. 574—576). Opal *C* shows a pattern of well ordered α-cristobalite with very minor evidence of a tridymite peak. The pattern is not that of a true α-cristobalite, and this type of opal does not invert to β-cristobalite upon heating. Opal *C* is common in cavities of lava flows which probably have undergone re-heating. Opal *C-T* tends toward a fibrous habit and has a broad diffractogram pattern with low but well-defined peaks of cristobalite and tridymite. This is interpreted to indicate that the α-cristobalite shows some stacking disorder as tridymite. Common opal and some tripoli belong to this group. Some porcelanite present in the cores from the North Atlantic has been identified as opal *C-T* by Calvert (1971, p. 282); but others of similar outward appearance are poorly ordered cristobalite and would be more properly catalogued as opal *A*.

Some clarification of the opal to cristobalite transformation has resulted from electron microphotographs, which have shown precious opal to consist of colloidal micelles as fundamental spheres about 1,000 Å in diameter stacked with some arrangement. By dissolution of biotite in acid Jones et al. (1966) produced similar spheres which tended to aggregate into masses of somewhat regular stacking. Individual spheres and aggregates much like those reported in precious opal, but having different size (1,000—8,300 Å) and packing, were discovered by Pollard and Weaver (1973). The spheres are more or less isolated individuals within pores of diatom fragments in a "chert" nodule incorporated in a diatomaceous montmorillonite—attapulgite—sepiolite clay (Hawthorne Formation, Miocene, Georgia, U.S.A.). Because of the absence of stacking, these authors suggest that the opaline spheres appear to have developed freely in open spaces without any strong tendency toward aggregation induced by surface charge distributions. If the spheres in the diatoms are equivalent to the spheres in precious opal, the packing noted in the latter must involve transportation and settling as individuals into a specific arrangement. The spheres in the diatom remains within the "chert" nodule are regarded as having formed where much pure silica existed relatively free from cations such as Fe^{3+}, Al^{3+}, and Mg^{2+}. Where the latter are present, such as in the clay matrix, they tend to cause the silica to combine into plate-like arrangements of authigenic clay minerals or an amorphous opal. Outside the "chert" nodule in the clay, none of the opal spheres was observed in pores of diatom remains.

The significance of the investigations by Jones et al. (1966) and Pollard and Weaver (1973) lies in the implication that some of the silica released to solution from an original combination (i.e., as a biogenic opal, an unstable silicate mineral, or a glass) would tend to develop colloid micelles of spherical form. The minute dimension of such spheres would permit their physical transport through pores in the sediment to some favorable site. Here, they could be aggregated as a result of the presence of suitable cations, or unbalanced surface charges, into a form of opal *C-T*. Elsewhere, particularly in zones of higher temperatures, opal *C* would be the metastable form of arrangement.

OPAL AS A CEMENT

Siliceous crust and silcrete

Opal has been reported as a prominent cement and as a vein material in sandstones which were exposed during important stages of weathering in semiarid regions (Millot et al., 1963). Franks and Swineford (1959) were among the first to report such an opal in the Ogallala Formation (Paleocene), Kansas, U.S.A. Their analysis showed this opal to contain significant amounts of Al_2O_3, Fe_2O_3, CaO, MgO, K_2O, and Na_2O. Moreover, the internal lattice shows spacings which are in accord with those of a disordered form of low cristobalite as reported by Flörke (1955). In thin section, the opal can be observed to be gradational into chalcedony and microcrystalline quartz, indicating that as a cement in sandstones, opal tends to become more crystallographically ordered with aging under near-surface conditions (Dapples, 1967, fig. 2).

Late precipitation of "chert" clearly associated with weathering surfaces has been reported at numerous localities (Williamson, 1957). Primarily, such an occurrence consists of crusts on the weathered surface and of veins filling joints which were enlarged by near-surface weathering. These rocks best known under the term *silcrete*, establish without question that chert can be precipitated along openings intimately related to the present-day weathering surface (Millot, 1960; Hutton et al., 1972). In contrast to other occurrences, however, such late chert has been deposited along the walls of an opening; hence, the problem of making room for the chert in the host rock does not exist. Moreover, the boundaries with the host rock tend to be sharp, following joint planes and weathering surfaces.

In Australia, silcretes tend to occur in peneplaned regions having sporadic rainfall and prolonged effects of chemical weathering. Argillaceous sandstones are found to be the most common host rocks. The allogenic quartz grains may be identified readily where these rocks have been modified to silcrete. Such grains are partially replaced and surrounded by microcrystalline quartz, some fibrous chalcedony, and minor opal to constitute an interlocked quartzitic mass. Smale (1973) has classified their occurrence into the following types:

(1) Terrazzo type, which consists of a framework of quartz grains of varying shape and roundness in a matrix of microcrystalline quartz. Individual grains are corroded and demonstrate unequivocal solution of part of the quartz in the original rock. A later deposition of chalcedony or opal around the corroded grains expanded and cemented the framework.

(2) Conglomerate type, which is made up of pebbles of the "terrazzo" in a red or brown siliceous matrix.

(3) Albertina type, which consists of masses of microcrystalline quartz without a grain framework. Absence of the framework is attributed to complete solution of allogenic grains during the initial solution.

(4) Opaline or massive type, which consists of layers of common opal, chalcedony, or microcrystalline quartz without incorporated detrital fragments.

(5) Quartzitic type, which petrographically resembles an orthoquartzite because overgrowths on the quartz grains have produced a mosaic texture.

A general common origin for the different types described above has been proposed by Smale (1973, p. 1088). He considered them to result from interaction of rising silica-charged solutions with other solutions moving generally downward as interstitial pore water and containing dissolved NaCl, Na_2SO_4, Fe_2O_3, Al_2O_3, and MgO, or perhaps characterized by pH lower than neutral. In order to mobilize the solutions, a semiarid climate is required with evaporation interrupted by rainfall to result in significant fluctuations in the level of ground water.

A somewhat different mechanism of origin of silcretes has been proposed by Hutton et al. (1972, p. 36). They regard them to be of two major varieties. One is produced by intensive leaching capable of attacking the detrital quartz and concentrating Fe, Ti, P, and Zr in suitable stable minerals. A second variety shows a dilution of the resistate minerals by precipitation of large amounts of silica supplied from outside source.

In a broad sense, these concepts of origin parallel the more sophisticated well-documented mechanisms proposed by Eugster (1969) for precipitation of cherts in strata of the former Lake Magadi, Kenya. The important requirements are:

(1) Solutions of high pH must be generated and caused to move through the intergranular pores dissolving quartz.

(2) At some other positions such solutions interact with others of lower pH to produce a situation resulting in supersaturation of amorphous silica.

BEDDED SILICEOUS DEPOSITS

Bedded chert in marine sediments

Certain rather typical strata of eugeosynclinal sites are identified as siliceous shales and bedded cherts. They are distinct from other siliceous accumulations because of the large volumes of silica involved and the strict parallelism of the siliceous strata to bedding. Many of these are exceedingly thinly laminated, suggesting slow deposition in quiet water. Investigators of such strata are uniformly impressed by the characteristics which point to very early precipitation of silica, either as a primary accumulation with the clastic sediment or as a diagenetic product of very early appearance (Taliaferro, 1934, p. 196).

As mentioned above, drilling in the deep oceans has revealed the occurrence of chert interlaminated with sediments possibly no younger than Pliocene and as old as Jurassic. Strata bearing such cherts are widespread throughout the Atlantic and Pacific basins. Although occurring in thin beds, their distribution is believed to be recognizable as seismic reflecting horizons. Based upon such information, individual units reach a few thousand square meters in areal distribution and may be much more extensive (Heath and Moberly, 1971, p. 992). Also associated with the bedded occurrences are nodular forms of chert, ranging from ellipsoidal bodies to very irregular masses including chalk inclusions into which the chert merges. In sediments of the Pacific, nodules tend to be associated with pelagic carbonates, whereas the bedded forms are restricted to siliceous oozes and pelagic clays. In this and other attributes they tend to resemble strata exposed on the continents. Although the nodules have been observed to be as much as 10 cm thick and up to 2 m in greatest diameter, the bedded types usually range between 2.5 cm and 10 cm in thickness and rarely exceed 10 cm (Heath and Moberly, 1971, p. 992).

Cherts taken from the Atlantic Basin appear to be varied in association and occur in foraminiferal silty limestones and red and brown clays (Calvert, 1971, p. 274). Some of the latter contain abundant zeolites (chiefly clinoptilolite), palygorskite (Bonatti and Joensuu, 1968), sepiolite, and some probable radiolarian remains, although it is uncommon for such cherts to be associated with radiolarian-rich sediments. The brown-colored clays are zeolitic and the contained cherts show an intergrowth of zeolite and silica, suggesting an intimate genetic relationship as described earlier.

An important feature concerns the composition of the so-called cherts (Cressman, 1962). Generally, the older ones consist of a mosaic of quartz. Some show the silica to be in the form of cristobalite, and still others are mixtures of cristobalite and quartz. Although the cristobalite is poorly crystallized, it is much more ordered than opal *A*. There is no reason to consider the cristobalite to have formed as overgrowths on aging opal nucleii (Heath and Moberly, 1971, p. 994). The true cherts do not contain opal, but fragments of a porous white siliceous rock (porcelanite) are reported as opaline. Also, Wise et al. (1972) reported that drusy cavities inside the fragments of such porcelanite contain spherulites of cristobalite (opal *C-T*), about 10 μm in diameter.

Cores taken from the Pacific show much of the chert to be associated with radiolarian ooze, whereas in the Atlantic such oozes are absent in Recent sediments but are present in deposits of Eocene or earlier ages. The porous chert in the Pliocene ooze on the Kerguelan Plateau studied by Wise et al. (1972), occurs in a chalk containing, in some pore openings, spherulites of cristobalite mentioned above. The latter crystallized following initial calcite cementation, and marks the first step in silicification of the chalk. Such an occurrence of the cristobalite spherulites indicates that they were not developed by solid—solid transformation, i.e., opal *A* to cristobalite, but rather had to be precipitated from solution with dissolved silica. Presumably, similar solutions could migrate some distance to centers favorable to the nucleation and crystallization of the observed cristobalite, to produce the protochert as nodules.

Certain brown and red clays from the North Atlantic consist of a mosaic of what appears to be clinoptilolite enclosed in an intimate cement of cristobalite. According to Calvert (1971, p. 282), such rocks, which have been described as cherts, should more properly be termed porcelanites, as they are similar to those of the Monterey Formation (Miocene) of California. The X-ray diffractograms show patterns similar to those of opal *C-T* (Flörke, 1955, 1967; Jones and Segnit, 1971), i.e., a cristobalite peak in a broad band and a minor tridymite peak. Calvert (1971, p. 282) has suggested that these porcelanites are intermediate phases in the alteration of opal to quartz (Ernst and Calvert, 1969), particularly because the older (Cretaceous) varieties tend to be composed of mosaic quartz.

At two of the Atlantic drilling sites, Calvert (1971, p. 283) reported the probable replacement by silica of a spicularitic, glauconitic, foraminiferal silt. The neoformed silica is a disordered cristobalite that fills intergranular voids between microfossils and replaces finely-divided biogenic calcite debris. The silica is interpreted as having been originally truly amorphous, later being reorganized into a cristobalite.

As previously mentioned, nodular cherts occur chiefly in chalks. In deposits taken from the western Pacific, Heath and Moberly (1971, p. 996) described the first indication of silica cementation to be represented by "tangential welding" of radiolaria tests, which results in a spongy network of cemented silica. Such a network is not considered to represent a preliminary step in the development of a chert nodule. Rather, its development begins with filling of empty foraminiferal chambers by fibrous chalcedony. A second stage is the replacement of the micritic chalk by finely crystalline cristobalite along a very sharply defined boundary, generally outlining the shape of the nodule and enclosing portions of the carbonate matrix. A third stage is replacement of foraminiferal tests by chalcedony as crystal-to-crystal modification. The process tends to preserve the original radiating crystal arrangement in the walls of the forams. A final, and perhaps much later, stage results in filling of remaining pore space with microcrystalline quartz, but postdates some conversion of early cristobalite to quartz.

In the western Pacific, bedded cherts are associated primarily with laminated clays in which alteration of amorphous volcanic debris and montmorillonitic minerals may have provided the mobile silica. The initial stage of chert development is precipitation of silica in the form of a cristobalite commonly with tridymite, i.e., opal *C-T*. In the second stage this crystallization continues until development of a well-indurated porcelanite very rich in cristobalite, but which is gradational into the soft claystone. This stage is followed by inversion of cristobalite to quartz and elimination of the remaining porosity. Such recrystallization spreads outward from individual centers and creates masses of clear chalcedony, which destroys the tests of siliceous organisms previously preserved in the cristobalite.

Doubtless, the large number of cores containing "cherts" taken during the oceanic traverses has provided the data which have been similarly interpreted by various investigators. A two-stage process appears to characterize the development of the cherts (Heath and Moberly, 1971, p. 1003). Initially, some form of biogenous opal dissolves, or dissolved silica is derived from alteration of unstable silicates. The dissolved silica perhaps migrates as far as several meters to favorable centers where it is reprecipitated as cristobalite and begins the growth of a mass which eventually becomes chert. Why the thermally unstable cristobalite is crystallized is not clear. Upon aging, however, it converts to quartz as an interlocked mosaic of anhedral crystals. Certain observed textures suggest a solid—solid transformation without a solution phase. Using porcelanite from the Monterey Formation, Ernst and Calvert (1969) caused the cristobalite—chalcedony conversion to migrate with an increase in temperature. Hence, they suggested that some of the young deep-sea cherts are the result of locally high temperature gradients. For the most part, however, the passage of time appears to be the primary control of the conversion to fibrous or mosaic quartz, as shown by the increase in the quartz/cristobalite ratio with progressive approach to strata of Jurassic age (Greenwood, 1973).

Bedded chert in nonmarine deposits

Relatively recent investigations have demonstrated the progression of silica diagenesis in strata clearly associated with nonmarine beds and seemingly unrelated to cherts discussed earlier. Examples are to be observed in the Pleistocene beds representing former stages of sodium carbonate-rich lakes of Kenya (Hay, 1968). In deposits of former Lake Magadi, some of the chert occurs as nodules or lenticular beds of undulating outline interbedded with clays that contain sodium-silicate minerals. Single layers of chert are known to range in thickness up to 3 m, but 2—20-cm layers are commonplace and areally may cover many square meters. Eugster (1969) described individual cherts consisting of microcrystalline quartz to be laterally continuous over distances of a few feet and grading into beds of magadiite ($NaSi_7O_{13}(OH)_3 \cdot 3H_2O$). Locally, the surfaces of the chert show large-scale polygonal fracture patterns, which are interpreted as being due to general desiccation, as well as small-scale reticulated (mud-crack) surfaces, crystal casts, and gas escape tubes, also regarded as pre-chert in origin. Soft-sediment deformational folds and fractured beds are common and also predate crystallization of the chert. Replacement of the magadiite by chert, particularly in the near-surface environment and zone of outcrop, indicates some connection between the lake-wide precipitation of magadiite and the subsequent development of the chert.

Lake Magadi present-day sodium-rich brines currently contain as much as 125,000 p.p.m. of Na and 1900 p.p.m. silica in waters locally and seasonally having pH of 10 (Hay, 1968, p. 257; Eugster, 1969, p. 17). According to Eugster, the Na is estimated to have been concentrated by a factor of $3 \cdot 10^3$, whereas silica has been enriched by about 50 times only. The equilibrium silica content is controlled by the solubility of amorphous silica at the high pH and not the initial content of the silica in the waters which were evaporated.

When the seasonal rainfall diluted portions of the lake, and sufficient CO_2 was accumulated from decomposition of organic material in the bottom waters, the pH was decreased, and magadiite would precipitate at pH values slightly in excess of 9. During dry seasons, the waters were very high in pH and silica tended to remain in solution. Such rhythmic precipitation is recorded within the magadiite as fine laminations. Locally, these laminations were contorted into soft-sediment folds due to postdepositional loading or mud-slide phenomena; they also show surfaces which were "mud-cracked" during exposure.

Conversion of the magadiite to chert clearly occurred after the deposition, loading, and drying mentioned above. In fact, it is related to present surface outcrop and positions where rain and fresh ground water are able to penetrate and interact with the magadiite. Eugster (1969, p. 22) has been able to demonstrate that magadiite can transform to chert directly or, perhaps, through an intermediate product, i.e., kenyaite ($NaSi_{11}O_{20.5}(OH)_4 \cdot 3H_2O_{(s)}$). Chert remains after sodium is removed through interaction of fresh waters, a condition which seems to be well supported by the field relationships. Hay (1968, p. 272) found nodules of chert showing presence of chalcedony, which has crystallized as an intergrowth with kenyaite by replacement of the latter. The relationship is complicated, as fibrous kenyaite is deposited upon an earlier formed chert. Subsequently, this kenyaite was replaced by chalcedonic and microcrystalline quartz. Such shifts in precipitation further support the mechanism proposed by Eugster, suggesting equilibria or repititions of transformation from kenyaite to quartz.

Cherts derived from non-aluminous silicates by the processes described above establish the mobility of silica as a product of instability of certain silicate minerals in a low-temperature regime. The reactions involved can be regarded as being driven in the opposite direction from those which develop authigenic zeolites (Hay, 1966, p. 67). In neoformation of zeolites, silica is removed from solution to produce a stable form of aluminosilicate; whereas when non-aluminous silicates such as magadiite are crystallized, they tend to be metastable. In waters of low ionic concentration and, particularly, where they are approximately of neutral pH, such silicates decompose and chalcedony is crystallized.

Bedded cherts and porcelanite

The term *porcelanite* was applied by Taliaferro (1934, p. 196) to a siliceous bedded mass or nodule within shale strata, having the outward porous appearance of unglazed porcelain; it is neither a true chert nor a siliceous shale. As the amount of aluminous impurities increases, the characteristic wax-like luster of chert gives way to a more granular aspect, which because of its increased porosity, assumes a dull porcelain-like appearance. In the more extreme examples of development, chert appears to have recrystallized as elongated quartz crystals, arranged in what is considered to be a random orientation and often having well-developed terminations. Individual crystals are interlocked into a framework which is either porous or dense, depending upon the amount of granular interstitial chert present. The texture appears to be primarily developed by recrystallization of some pre-existent silica material.

Unfortunately, the term was not used uniformly by authors and has been applied to other siliceous deposits including silcretes. Lately, the term *porcelanite* has been restricted once again and is being applied to geologically young cherty rocks having unglazed porcelain-like surface, and of which an X-ray diffractogram shows a poorly ordered pattern of cristobalite (Greenwood, 1973, p. 700). In this respect, it differs from the true chert into which it grades, and which is characterized by chalcedonic and mosaic quartz. No abrupt distinction is observed between cristobalite porcelanite and chalcedonic chert, however, in that not all examples are neatly catalogued; rather commonly there exists an intimate intermingling of the two silica minerals.

Mesozoic and Tertiary bedded cherts tend to be composed of both porcelanite and quartz and, where cryptocrystalline, the latter may contain diatoms and radiolaria (Ernst and Calvert, 1969). Thin sections of the porcelanites show a few frustules of diatoms along with some silt-size grains of quartz and feldspar. In this connection the question arises concerning the differences between the crystallinity and compositional distinctions between porcelanite and chert, and their relationship to the known occurrences of diatom- and radiolaria-rich beds. The presence of opaline silica as some variety of cristobalite serves to distinguish the porcelanite, whereas the varieties of quartz have not been so readily catalogued. In order to establish a basis for greater precision in description of the various forms of quartz occurring in cherts, the following definitions have been proposed by Folk and Pittman (1971; see also Wilson, 1966, p. 1037), who organized inorganically precipitated quartz into:

(1) Megaquartz—generally coarsely crystalline, i.e., exceeding 20 μ in individual crystal width.

(2) Microquartz—crystals less than 20 μ in width: (a) microcrystalline quartz having equal grains generally 1—4 μ in diameter, with small fluid-filled inclusions; (b) chalcedonic quartz consisting of fibrous crusts and spherulites: (i) chalcedonite (optically length-fast); (ii) quartzine (optically length-slow); (iii) lutecite ("c"-axis 30° ± to fiber length).

According to Folk and Pittman (1971, p. 1050), normal length-fast chalcedony occurs generally as a cavity filling. It consists of an orderly arrangement of spherulites each in a radial pattern of delicate fibers of approximately the same length. Quartzine also occurs commonly as a cavity filling, but the fibers tend to be coarse and crudely arranged. Lutecite is found primarily as a replacement of evaporite crystals and nodules, shell material, and spar-calcite cement. Its most commonplace appearance is as symmetrically extinguishing chevron-like grids of coarse fibers crossing at approximately 60°, and grading into normal quartz. Chalcedonite (length-fast) is considered to form at low pH and high concentrations of dissolved silica under which conditions the silica tends to be organized into polymers of $Si(OH)_4$ groups. Under conditions of high pH and high concentrations, the dissolved silica tends to be considerably more ionized and the silica crystallizes as quartzine (length-slow). Replacement of calcite by fibrous chalecedony tends to be controlled by the optical arrangement of the calcite; hence, quartzine and lutecite are typical replacement varieties. Nonfibrous, well-crystallized quartz is regarded as having been crystallized slowly from solutions having low concentrations of dissolved silica.

Studies by Ernst and Calvert (1969) and Greenwood (1973) have helped to clarify the neoformation of cristobalite in porcelanite from biogenic opal. X-ray diffractograms of the latter, believed to be $SiO_2 \cdot n\ Si(OH)_4$, show the characterizing broad and gentle hump in the vicinity of 4 Å, indicating some incipient ordering of SiO_4 tetrahedra (Greenwood, 1973, p. 701). Poorly ordered cristobalite shows distinct peaks corresponding to 4.05 Å and 2.48 Å, and a distinct platform appears to be connected to the principal cristobalite peak. Greenwood interpreted the platform to indicate mixed layering of a tridymite and cristobalite style of arrangement (i.e., opal *C-T*). Similar mixed layering occurs in neoformed cristobalite precipitated in the laboratory in the 150—200°C range, whereas in the form crystallized at higher temperatures (900—1100°C) such mixed layering is not observed. Other experiments indicate that the authigenic, low-temperature cristobalite has an approximate solubility of 27 p.p.m. at 25°C.

Ernst and Calvert (1969) concluded that the conversion of porcelanite to chert at 50°C was a very sluggish transformation and would be accomplished in about 4—5 million years. If circulating fluids undersaturated with silica passed through the beds or if such fluids were alkaline, however, finely crystalline quartz would crystallize in shorter periods of time. Their study of the Monterey Formation shows the alteration of diatomite-to-porcelanite-to-chert to have occurred in selective strata apparently controlled (1) by the alkalinity of the fluids; and (2) in an overall way, by the temperature attained (40—50°C) by the more deeply buried strata.

Porcelanites are gradational also into fine-grained tuffs in which the glass has altered to "chert" for the most part. Such porcelanites have much the same outward appearance as others. In thin section, however, one can note a significant percentage of highly angular fragments of unaltered feldspar. In many such cherty tuffs the percentage of detrital quartz is low; whereas, where the volcanic mixture is diluted by land-derived detritus, gradation into a tuffaceous sandstone occurs. With such a transition, the typical appearance of porcelanite tends to give rise to that of a quartzitic sandstone or siltstone.

Transition of porcelanite into siliceous shale occurs through dilution of the chert by clay minerals and some micas. This transition is accompanied by appearance of well-developed lamination, with corresponding loss of wax-like luster. In thin section, the relation between "chert" and clay minerals is difficult to see. The toughness of the rock and the tendency for small pieces to break with conchoidal fracture, however, suggest that granular interlock exists and that some diagenetic recrystallization may have occurred.

The relationship of porcelanites to carbonate-rich beds is poorly understood. Some porcelanites contain rhombs of siderite or dolomite, which have replaced part of the silica, whereas other examples show that chert has replaced what most certainly must have been carbonate lenses or thin beds (Folk, 1973, p. 717).

Among certain of the shale sequences, there are beds which can be recognized as former limestones, now intensely silicified; locally, these beds have surfaces of porcelanite. Where the strata are as yet calcareous, they may contain as much as 20% silica, which may have been deposited penecontemporaneously with the carbonates. Elsewhere, similar beds can be seen to be for the most part replaced by secondarily deposited chert. Interbedded with such former carbonates are cherts and siliceous shales which by many geologists are considered to represent primary deposits, or to have become silicified very early in their burial history.

Kolodny's (1969, p. 169) studies of phosphatic porcelanites, from upper Cretaceous strata of the Negev (Israel), indicated that early silicification occurred when the associated carbonate beds were still loose sediment. Moreover, such silicification was part of a progression beginning with early partial silicification of some of the detrital carbonate and phosphorite. The initial silicification was followed by: (1) crystallization of some calcite cement; (2) silicification of calcite cement by microquartz, preserving the texture of calcitic-phosphorite; and (3) a final silicification of apatite grains, which in the extreme destroyed the oolite texture.

The Maravillas Formation (Ordovician, Texas, U.S.A.), described by McBride (1970), illustrates the common association of dark bedded chert, dark gray limestones, and dark gray shales in interbedded units. All of the limestones have a fetid odor and consist of calcarenites, micrites, and marlstones which are intergradational. Shales dominated by clay minerals are interbedded with limestones and also brown chert; they may contain up to 40% microquartz. Locally, these shales intergrade into siliceous shales in which microquartz dominates the mineralogy. The cherts occur as distinct beds, fossil replacements, lenses, thin networks, and brecciated masses of black chert cemented by white chalcedony and coarsely crystalline quartz. The siliceous shales appear to have formed through recrystallization of a former variety of silica deposited in localized centers. They differ from the clay shales in that the previously deposited biogenic silica has been mobilized and recrystallized at such centers.

In examples of the lens-like masses of chert, which McBride (1970, p. 1739) has demonstrated to replace portions of the limestone, deposition of the silica is regarded as part of the same postdepositional, but early, wholesale replacement of complete beds of carbonate strata. Similarly, concentration of silica into nodules and lenses is quite early, as small slump folds occur in which the chert has been folded with the rest of the then poorly consolidated sediment. A currently preferred explanation of origin is diagenetic, namely, the high amount of silica which is required to replace the observed volume of replaced carbonate has been supplied by the dissolution of biogenic opal through mechanisms mentioned earlier.

Sufficient information is now available to indicate that decomposition of volcanic ash, etc., in the immediate burial environment, is responsible for local increases in dissolved silica in interstitial water in the upper meter, or so, of present-day oceanic sediments. The concept advocated by Goldstein and Hendricks (1953, p. 441), that the silica in the Bigfork Chert (Ordovician, Ouachita Mts., U.S.A.) was derived from buried volcanic debris because of its chemical instability, has been generally verified. An origin of direct precipitation of inorganic silica from marine waters, later to be mobilized into chert, appears to have little supporting evidence even in localities of active subsea volcanism (Kolodny, 1969, p. 171). Similarly, the indirect supply of silica to organisms by alteration of volcanic debris (or volcanic emanations), which could provide dissolved silica directly to bottom waters, has not been substantiated by the data currently available. Such a mechanism can no longer be regarded as a significant contributor to the development of bedded cherts, porcelanites, and siliceous shales and carbonates.

Radiolarian cherts and siliceous shales

Interdigitation of bedded cherts with siliceous shale is known in many parts of the stratigraphic section, particularly in association with the so-called eugeosynclinal rocks most of which are recognized also as turbidites of deep-water deposition (Grunau, 1965). An example is reported by Oldershaw (1968, p. 258) from Namurian strata of North Wales. He observed three types of bedded cherts:

(1) A thin-bedded shaly variety consisting of beds of brown and gray chert separated by carbonaceous partings. Scattered grains of detrital quartz are enclosed within the cryptocrystalline silica. Such units are interstratified with thin shales and fine-grained sandstones.

(2) Thin-beds of laminated, generally microcrystalline chert with well-developed color banding separated by infrequent occurrences of shales and sandstones.

(3) Massive, laminated chert of various colors ranging from white to black in irregular layers 10—50 cm in thickness. Individual beds are sharply separated and, locally, they may be stylolitic.

In the United States, typical examples have been carefully studied where they constitute part of the Stanley Shale (Mississippian) of the Ouachita Mountains in Arkansas and Oklahoma (Goldstein, 1959; Cline, 1970). Here, the siliceous beds tend to be darker than the enclosing less siliceous beds. They may be: (1) well or poorly laminated, (2) generally silty, (3) micaceous, (4) pyritiferous, and (5) carbonaceous. Where conspicuously laminated, a thin section will reveal crenulated lenses of clay minerals intermingled with cryptocrystalline or isotropic silica. Sponge spicules, radiolaria, spore exines and similar fossils are characteristic; they range from well to rather poorly preserved. Locally, cherts are fractured or brecciated and recemented by silica as a veined mass.

Students of these strata indicate that the cherts are extraordinarily continuous and widespread; this conclusion is based largely upon ease of correlation in the field. This also is true in localities where actual tracing of beds can be accomplished. For example, Fagan (1962, p. 604) stated that individual beds particularly in the Schoonover Formation (Carboniferous, Nev., U.S.A.) can be traced for miles. Finely laminated repetitions of concentrations of radiolaria, carbonaceous matter, flecks of iron oxide and pyrite, and grains of detrital quartz which locally are graded in size, are preserved in the cherts.

Sequences of radiolarian cherts and siliceous and argillaceous shales tend to be commonly, but locally, interbedded with volcanic deposits. In other localities, however, similar beds appear to have no volcanic affinity. An association with ophiolites, generally as pillow lavas, has suggested the source of the silica to be the concurrent volcanism (Trümpy, 1960, p. 866; Grunau, 1965, p. 200). As mentioned above, however, as details of such associations evolve, the volcanic sources for the silica prove to be largely indirect, i.e., the released silica becomes part of interstitial waters as submarine weathering causes decomposition of the volcanics buried at shallow depth.

Novaculite

In the United States, special occurrences of massive to thinly bedded white chert of high purity, dominated by microcrystalline quartz, are identified as *novaculites* when they constitute extensive and generally thick layers. A most important locality is to be found in Arkansas and Oklahoma, and is believed to be continuous in the subsurface into western Texas and Mexico (Flawn et al., 1961). Where such beds crop out in Texas, the formation is known as the Caballos Novaculite (Devonian). Here, it appears as a unit of white chert, locally greenish, tan, or gray brown, generally between 200 and 500 ft in thickness (McBride and Thomson, 1970).

The formation is subdivided into several units of various characteristics, but each is associated with minor siliceous shales, which commonly appear only as partings along the bedding planes. Elsewhere, where the shale is absent, the beds tend to be welded together at some localities by stylolitic interlock. Except for scattered radiolarians and sponge spicules, fossils tend to be extraordinarily rare and poorly preserved. Such paucity in organic remains is attributed to their destruction during recrystallization of the former opal into the dominant microcrystalline quartz. Scarcity of land-derived detritus and absence of volcanic debris associated with the novaculite and chert, along with other attributes, tend to very strongly suggest the original material to have been overwhelmingly composed of biogenic silica. Sometime after burial, this opal was converted to microcrystalline quartz, presumably through the opal—cristobalite—quartz progression described earlier.

Although there appears to be general agreement among later authors that the silica for novaculite originally was of biogenic origin, some controversy exists regarding the depths of water involved. In this connection, Folk

(1973) has presented a very strong argument for shallow-water origin of the Caballos Novaculite. He has reinterpreted a number of sedimentary features as having originated in sabkha supratidal and intertidal environments. Sponges provided the principal source of silica which was accumulated in restricted bays where there could be no entry of water laden with land-derived muds. Elsewhere where evaporites were accumulated, local solution produced collapse breccias. Wherever carbonate beds were deposited, they were replaced after burial by extensive silicification, which spread to marginal areas of the buried and locally dissolved portions of the sabkha sediments. The original sedimentary structures, however, were preserved.

In the light of the studies on the Lake Magadi cherts by Eugster (1969) and Hay (1968), the possibility of interstitial solutions well charged with dissolved silica in the buried biogenic sabkha sediments is a most reasonable expectation. Although the reactions required for silica to replace anhydrite are not established, the possibility cannot be dismissed that development of some intermediate silicate such as magadiite could have occurred under conditions wherein the anhydrite was unstable.

Brecciated chert

The occurrence of beds of brecciated chert has been reported from numerous localities (Rapson, 1962; Folk, 1973, p. 718) and, according to Kolodny (1969, p. 171), there is a strong correlation between the development of brecciated structures and silicification. Also, there is reason to suspect that at certain stages in the development of chert breccia, a non-siliceous matrix was soft, whereas the chert was rigid and part of a continuous band. As this band was ruptured into angular fragments by the tensional stress applied through loading, the soft matrix flowed into position to isolate individual chert fragments. Rapson's (1962, p. 257) study provided rather strong evidence to indicate that where the silica could mobilize and accumulate as individual masses of gel in the unconsolidated carbonate, the surfaces of such gel were brecciated during syneresis. Later, the gel was crystallized and fragments cemented by a microcrystalline chert. A third stage involved selective replacement of the chert by carbonate to isolate splinters of chert. All this occurred before the carbonate rock was lithified, because slight agitation introduced some of the detrital sediment between the chert fragments.

An alternative origin has been proposed by Folk (1973, p. 719) to explain the chert breccias in the Caballos Novaculite (Devonian, Texas, U.S.A.). He attributed the fragmentation to collapse of bedded, lithified, siliceous sediment as solution removed the underlying and interbedded strata, primarily evaporites. A dominantly siliceous residue infiltrated into the rubble to constitute a matrix which eventually became a dark chert, i.e., two "generations" of chertification are to be observed.

SILICA IN CRATONIC CARBONATE ROCKS

Varieties of chert

Chert is commonplace among shelf and intracratonic basin carbonates as nodules, lenses, thin beds, and veins having similar appearance to that observed in so-called miogeosynclinal strata. Some occurrences are lensoid bodies concentrated in the axial positions of small, recumbent, primary folds, which generally parallel the bedding; whereas other chert masses partially replace the carbonate strata (Newell et al., 1953, p. 162). Evidence for replacement of carbonate by chert is abundant, and boundaries between the carbonate and silica are irregular and transitional between the two rock types (Biggs, 1957, pp. 11—16).

Irregular masses of microcrystalline quartz, commonly only 1—3 mm in greatest length, appear as partial replacements of fossils, or are associated with detrital quartz grains. Replacement relations visible on etched polished surfaces demonstrate that such chert nodules are secondary. Inasmuch as their shape is generally elongated parallel to bedding, they are considered to have been gel-like before compaction of the sediments. The source of the silica is not evident and it does not appear to have been derived from solution of associated detrital quartz grains, although commonly their surfaces are pitted and irregular (Lamar, 1950, pp. 30, 32).

Chalcedonite (length-fast fiber), which commonly occurs in cavities of shelf carbonates, owes its origin to crystallization in situ (Pelto, 1956) that appears to follow lithification of the carbonate rock. In some occurrences, isolated carbonate crystals within the chalcedony and along the borders between chalcedony and the host rock are dolomite primarily. Such crystals have different optical properties and composition than calcite of the host rock. The dolomite crystals tend to have well-developed outline and cut primary structures, such as ooids.

West's (1964) study of the Purbeck Beds (Jurassic, Dorset, U.K.) has revealed the stages of diagenesis involved in replacement by chalcedony of once existent anhydrite or gypsum in the strata. As the latter two minerals were undergoing solution, they were replaced by chert and calcite in which ghost crystals of the previously existent anhydrite or gypsum are preserved. Of particular interest are the chalcedonic pseudomorphs of the evaporite crystals, composed of the variety lutecite (extinction inclined to fibers). Also, some crystallization of chalcedony occurred at an early stage, as indicated by deformation of laminae of pelsparite around the nodules of chert. Hence, accumulation of such nodules clearly was an event independent of replacement of evaporite crystals by lutecite.

The puzzling occurrence of lutecite and quartzine (length-slow fibers) as uncommon varieties of chalcedony has recently been explained by Folk and Pittman (1971). Their exhaustive investigation has substantiated the association of quartzine, in particular, with the stratified evaporities locally replaced by chalcedony, or with conditions of high alkalinity in soils as indicated by deposition of caliche. They showed presence of nodules of chert which (1) mimic "flaser structure" in anhydrites, and in which (2) the quartzine crystals are pseudomorphs of a pre-existing evaporite mineral. Since publication of their findings, other investigators have reported similar occurrences elsewhere (Siedlecka, 1972; West, 1973). Jacka (1974), however, reported that skeletal components and ooids in allodapic limestones are replaced by quartzine, such strata being clearly independent of evaporite beds.

Replacement of fossils

Selective replacement of fossils by silica is a common occurrence, but the mechanism of replacement and the localities where it can be expected to occur are known only on an empirical basis. An example is the occurrence in the Capitan Reef (Permian, New Mexico), where virtually no silicification is reported in the reef mass itself (Newell et al., 1953, p. 173). In the flank deposits, however, replacements are abundant and follow a general order (Table 3-II). Carozzi and Soderman (1962, p. 401) reported that in a Mississippian crinoidal limestone in Indiana, calcilutite and spar-calcite matrix between coarse fragments of fossils are replaced locally by silica, whereas the fossils rarely are altered.

As mentioned earlier, Jacka (1974, p. 425) noted that selective silicification of fossil fragments and ooids has occurred in the Getaway Limestone (Cherry Canyon Formation, Permian, Texas, U.S.A.), whereas intraclasts and the micrite matrix remained unsilicified. Moreover, dolomite rhombs occur within the silicified fossils of organisms in which high-magnesium calcite characterized the original test. In shell-matter originally composed of calcite, no such rhombs are present within the chalcedony. Jacka's explanation that exsolution of dolomite and replacement of calcite by silica may have been part of the same event, merits attention. What is clearly demonstrated, however, is that the diagenetic changes involving crystallization of silica are produced in microenvironments influenced only partially by the large-scale physicochemical environment.

Table 3-II lists a few examples where an order of preferential silicification has been reported. If these examples are representative of what would be

TABLE 3-II

Order of silicification of fossils

Increasing tendency to silicification	Permian Capitan Reef (Newell et al., 1953)	Ordovician southeastern Nevada (Hintze, 1953)	Ordovician central Nevada (Hintze, 1953)	Ordovician Fish Haven Dol., Idaho (Gibbs, 1960)	Ordovician Tanners Creek Formation Indiana (Fox, 1962)	Silurian Thornton Reef, Indiana (Ingels, 1963)	Pennsylvanian northeastern Nevada (Dott, 1958)	Ordovician Maravillas Fm., Texas (McBride, 1970)
	bryozoans, corals, punctate brachiopods	trilobites	gastropods	brachiopods	bryozoans	colonial corals	punctate brachiopods, corals	brachiopods
	impunctate brachiopods	brachiopods, bryozoans, and ostracods	brachiopods	solitary corals	brachiopods	stromato-poroids	spirifers, productids, fenestrate bryozoans	bryozoans
	mollusks	gastropods	trilobites	bryozoans, cephalopods	solitary corals		echinoderms	
							forams	corals
	echinoderms	cephalopods		crinoids				
	forams							
	algae and cal-careous sponges							

observed on the basis of future extensive studies, one might be led to conclude that bryozoans, brachiopods, and corals generally are more sensitive to replacement by silica than gastropods, cephalopods, and echinoderms. On the basis of some of the reversals in the orders shown in Table 3-II, however, it may be more reasonable to consider that the preferential replacement is a function more of the distance from centers of silica precipitation and the quantity of silica being precipitated, rather than the crystal habit or composition of the carbonate constituting the fossil shell.

Replacement of calcareous oölites

Among good examples of siliceous oölites, is the occurrence in Cambrian strata near Bellefonte, Pennsylvania. Here, siliceous beds usually only a few centimeters thick are separated by several meters of calcareous oölitic beds, but there are all gradations between siliceous and calcareous ooids. Commonly, where the ooids are embedded in a siliceous matrix, they appear to "float" in the chalcedony, but elsewhere they are in contact with one another. In other examples, individual ooids are fractured and the fissures are sealed by quartz, or carbonate particles of irregular outline are isolated in the silica matrix. Such textures have been interpreted by Choquette (1955) as being illustrative of the effects of a process whereby the silica is a replacement of an original carbonate matrix and not a primary precipitate.

The significant study by Folk and Pittman (1971) regarding the widespread occurrence of quartzine as a replacement of carbonates, originally associated with evaporite strata, has emphasized the need to re-examine the textures of silicified oölites. They found quartzine fibers in a number of oölite occurrences identified as "shrunken". That is, the central parts of an ooid appear to have fallen to the lower outer rind as solution of the interior occurred. Rare occurrences of what are interpreted to be inclusions of anhydrite have suggested to those authors that some causal relationship exists between the presence of siliceous oölites and former evaporite strata. Sufficient evidence has now been assembled to indicate that the processes involved are unequivocally due to replacement of carbonate by quartz, and that no original ooliths of silica ever existed.

Silica in carbonate reefs

Indirect evidence of originally low concentration of silica, precipitated or deposited, in cratonic carbonates appears from analyses of reef-"core" and "flank"-bedded calcarenites. Sufficient analyses are now available to demonstrate that these carbonates tend to be unusually low (±2%) in silica. The accumulations represent the organically precipitated rocky, wave-resistant structures and the wave and current debris eroded therefrom. In essence, such carbonates reflect the absence of original precipitation of silica at such sites. For example, Newell et al. (1953, p. 67) reported that in the carbonate association of the Capitan Reef, fossils in the basin calcarenites have been selectively replaced by silica, whereas those of the inclined reef talus rarely are silicified; the silica content of the coarse talus is low. Important amounts of silica as fossil replacement, nodules, and crusts, however, occur basinward, and micritic basin limestones are cherty. In the inter-reef deposits containing land-derived clastic sediments, the chert content rises, and tongues of such sediments projecting into the reef complex carry nodules and thin beds of chert (Shaver, 1974, p. 946). Currently, the consensus on origin favors original localities of silica-producing organisms, particularly sponges and diatoms, concentrated outside the immediate sites of the reef waters. Later, this opal was transformed to chert by the mechanism involving formation of cristobalite.

In evaporite basins rimmed by reefs, the higher concentration of salts in sea water necessary to produce salt beds has not concomitantly resulted in the precipitation of silica (Dellwig, 1955). Indeed, one is to conclude that the concentration of silica in such waters was so low that virtually no silica could be precipitated even under conditions of high concentrations of electrolytes. Clearly this condition must have represented an entirely different ratio of silica to other salts than that reported in the brine lakes of Kenya (Hay, 1968). Also, unstable silicates, intermediate to crystallization of chalcedony, as reported by Eugster (1969), could not have been precipitated during development of the basin evaporites.

Time-stratigraphic equivalents of such strata, primarily carbonates, contain chert and some interbedded shale or a significant clay-insoluble fraction. The association of chert and clay-size, land-derived detritus appears to be of considerable significance in the diagenesis of silica, as demonstrated in the investigations of the sediments in the Atlantic and Pacific basins, indicated earlier.

Overgrowths on quartz grains in nonsiliceous rocks

A commonplace feature of thin beds of quartzose sandstone and layers of isolated grains of quartz, enclosed in carbonate rocks, is the exceptional sphericity of such grains, suggesting they have been partially dissolved during rolling in transport. At scattered localities, some of the grains show additions of quartz as well-terminated overgrowths replacing carbonate; but rarely are such grains completely surrounded by a a rind of chert (Dietrich et al., 1963, p. 660). Generally, the time of precipitation of such silica is difficult to date, but Dietrich and his associates showed presence of dolomite rhombs, included within the quartz overgrowths. Glover (1963, p. 40) reported an occurrence of quartz overgrowths in the Septimus Limestone (Carboniferous, Western Australia), the composition of which is 6% sparry calcite, 40% quartz, and 53% dolomite. The paragenetic sequence of importance which he reported is: formation of zoned dolomite, precipitation of quartz overgrowths, followed by crystallization of sparry calcite. Precipitation of silica as monocrystalline quartz rather than chert can be substantiated in other carbonate rocks, but only a very crude correlation is noted between precipitation of monocrystalline quartz as overgrowths where detrital quartz is abundant, and precipitation as chalcedony where quartz grains are rare.

Although the order of paragenesis is not always the same, calcite, monocrystalline quartz, chalcedony and dolomite are commonplace associates, which rarely are crystallized in uniform amounts. Rather, the "normal" situation is for one episode of crystallization to dominate and by replacement to destroy partially the evidence of an earlier crystallization. Introduction of any one of the minerals by a secondary crystallization may locally change the texture, e.g., as has been mentioned by Chanda (1963). He reported secondary calcite to have, by replacement of quartz grains, virtually destroyed an original sandstone framework. One of the most perplexing problems, however, is to explain the sources of large amounts of secondary silica which is crystallized as chert intercalated with the normal carbonate strata, rather than as overgrowths on detrital quartz grains. Crystallization of this type of replacement chert shows no indication that it involved the progression of biogenic opal → cristobalite → quartz, well established for porcelanites. Rather, an appeal must be made for gathering the silica from a source outside the beds in question, or to a systematic flow of interstitial waters charged with dissolved silica.

King and Merriam (1969) reported an occurrence of interbedded generally sandy strata and chert in the Morrison Formation (Jurassic, Colo., U.S.A.). Individual small prisms of authigenic quartz, which were crystallized following a primary cementation by calcite, occur in the sandstones. The prismatic quartz appears commonly as overgrowths completely surrounding cores of detrital quartz; it grew in episodes as indicated by pronounced growth lines seen in basal cross-sections. Crystallization of the monocrystalline quartz was terminated and supplanted by deposition of chalcedony in typical beekite ring outlines, but culminated in spheroidal masses. Upon growth, such beekite preferentially enclosed quartz prisms, and also partially replaced them and the calcite cement. In turn, the annular growth of beekite was supplanted by a fibrous chalcedony of a different color. Conditions marking this stage of silicification persisted and widespread, continuous layers of chert were produced largely as a result of replacement of the carbonate cement.

Inasmuch as the Morrison Formation consists of sandstones and shales, the source of the silica is very likely to be intraformational (King and Merriam, 1969, p. 1145). In the example of many cratonic carbonate strata, however, solution of a primary, deposited biogenic opal is not documented with assurance. Moreover, interstitial waters in carbonate sediments tend to be low in silica. Nevertheless, there appears no other choice than to seek the source of the silica in the moving ground waters, from which eventual precipitation of chalcedony occurs at the favorable sites. In this connection, experimental work by Lovering and Patten (1962) suggests that cold, acid solutions supersaturated with silica should, on contact with limestone or dolomite, generate CO_2 and precipitate silica. A cold, neutral solution supersaturated with silica but without Na-ions should be capable of moving long distances through carbonate rocks until making contact with either CO_2 or Na-ions, at which point silica should precipitate. Conceivably interstitial waters in sediment containing very finely divided silica could become saturated with respect to silica. Migration of such waters into positions where Na-ions or higher CO_2 concentrations occur, could result in precipitation according to the mechanism suggested by Lovering and Patten.

Geodes

Occurrence of geodes in shelf carbonates is far more common than in the corresponding basin sediments and is not associated with bedded cherts in the large-scale sense. Geodes characteristically have an exterior rind of chalcedony of subspheroidal outline, which presents a rather sharp boundary with the enclosing carbonate. The relations suggest that beginning with some central nucleus, such as a shell containing an initial opening, chalcedony was deposited on the outer margins and in fissures to expand the dimensions of the opening. Such precipitation was in progress when the enclosing carbonate was still in an unconsolidated condition and, hence, could be pushed aside (Hurst, 1953, p. 228). In other examples, actual replacement of the carbonate by silica was accomplished.

Hayes (1964) has presented evidence for replacement of carbonate concretions by chalcedony to form the rind of the geode. After precipitation of the outer chalcedonic ring, megacrystalline quartz was deposited to complete the internal aspect of the geode. In this connection, there appears to be a relation between the precipitation of varieties of silica and the nature of the surface on which precipitation occurs. Megacrystalline quartz is precipitated in cavities lined with chalcedony and, exceptionally, against the carbonate rock per se. Chalcedony and microcrystalline quartz are preferentially deposited against the carbonate, but rarely upon macrocrystalline quartz.

TIME OF SILICIFICATION

Dolomitization and precipitation of silica

Certain carbonates, which constitute part of the so-called miogeosynclinal strata, contain significant quantities of chert as nodules, thin beds, and veins engulfed within the carbonate mass (Bissell, 1959). Often such strata are extensively dolomitized and the entire formation may be dolomite over distances of many square kilometers. That this dolomitization is a secondary process is well established on the basis of such evidence as the presence of dolomitized fossil remains, which as living organisms precipitated shells of aragonite and calcite. Uncertainty exists as to the time when the dolomitization occurred, but an increasing amount of evidence points to the process as having occurred early in the sediment's history, principally prior to general lithification; it was promoted by slightly positive tectonism in the depositional site (Badiozamani, 1973).

Paragenetic relations between chert and dolomitization are extremely useful in helping to establish the time during which silicification of the carbonate occurred. Individual euhedral crystals of dolomite are found in some cherts. This relation is interpreted to indicate that the dolomitization preceded precipitation of the silica as a postdepositional process.

Silicification, which slightly precedes widespread dolomitization in time, is noted in certain shelf carbonates in which well-preserved fossils occur in the chert, whereas their counterparts in the dolomite show very poor outline. In the same rocks, dolomite is observed to partially replace chert along boundaries of the nodules (Dapples, 1959, p. 49).

Certain outcrop exposures of dolomite reveal broken edges of thin nodules and lenses of chert to overlap in a sort of imbrication. Rupture of the chert obviously has occurred after its solidification, inasmuch as the broken edges of adjacent fragments may be joined into a single piece. Excellent examples of such an occurrence have been described by Dietrich et al. (1963, p. 649) in which the continuous unbroken overlying lenses of chert prove that fracture of the chert is not due to any tectonic forces. They attributed rupture of the chert to syneresis of the gel during a period of exposure to air, drying, and shifting of fragments before deposition of overlying carbonates. The reader's attention is directed to the mechanism and the time of its early development as noted by Rapson (1962).

In similar examples observed by the present writer, it is possible to interpret the fracturing of chert to have occurred after deposition of perhaps several centimeters of carbonate material. Thin laminae of carbonate are continuous, but sag into the gap in the pieces of chert, suggesting that rupture occurred before lithification of the carbonate. The presence of clear dolomite rhombs within iron-oxide-stained chert was regarded by Dietrich et al. (1963, p. 650) to suggest that some dolomitization occurred prior to the development of the chert lenses. This paragenetic sequence appears now to be rather firmly established by repeated observations made throughout the geologic column. Moreover, the common occurrence of pelletoidal texture and gradational boundaries of the chert indicate a replacement of calcium-carbonate sediment. Similar paragenetic sequences can be substantiated in other strata elsewhere where (1) remnants of dolomite exist within the chert, and (2) skeleton outlines of chert nodules have marginally replaced and surrounded masses of the dolomite.

Petrologic studies of carbonate strata indicate that although chertification tends to be an early event, i.e., closely associated with dolomitization, other generations of chert in the form of nodules, lenses, etc., commonly are developed much later in the history of the rock. An excellent example is reported by Banks (1970) to occur in the Leadville Limestone (Mississippian, Colo., U.S.A.). Two varities of early chert are recognized. (1) A black chert was crystallized before final lithification of some of the carbonate as indicated by, among other criteria, abraded fragments of this chert within intraformational limestone conglomerates. Such chert contains anhedral crystals of dolomite which have been embayed by the advancing silica front. (2) A second variety of early chert called "sand chert" occurs in arenaceous dolomite, and does not replace the dolomite crystals. The "sand chert", however, is believed to have been crystallized only where quartz sand grains served as nuclei for radial overgrowths of chalcedony.

A much later chert occurs as nodules, stringers and layers which transgress the bedding. Such chert (1) encloses much carbonate, (2) constitutes as much as 30—40% of the nodules, and (3) when broken, reveals drusy cavities and masses of minute, doubly-terminated quartz crystals. Crystallization of this late chert is associated with an important time of general rock solution during the Early Pennsylvanian resulting in development of a karst topography. Little is known, however, concerning the actual mechanisms involved in the crystallization and, currently, proposed mechanisms are based upon speculation, e.g., similar to the origin of silcrete, or some laboratory-controlled experiment such as the one performed by Lovering and Patten (1962) and described above.

In carbonate rocks in particular, a very local advancing front of silicification can be demonstrated to have occurred. This commonly resulted in mobilization of certain ions, leading to crystallization of certain minerals, commonly dolomite rhombs (Jacka, 1974, p. 425). Swett (1965, p. 933) described the exclusion of iron from an oölite to reprecipitate as hematite in cubes, probably pseudomorphs of pyrite. As noted above, dolomitization and silicification commonly alternate as early events; hence, one may follow the other in more than one generation of crystallization. An example of this has been reported by Swett (1965, p. 933) in the "Durness" Carbonates (Cambro-Ordovician, northwest Scotland) as the following series of events: (1) recrystallization, (2) primary dolomitization, (3) silicification, (4) calcitization, and (5) secondary dolomitization. Swett favored the concept that dolomitization prior to silicification and again following silicification is a commonplace paragenetic sequence regardless of the stratigraphic and geographic occurrence.

CONCLUDING REMARKS

Silica has become recognized as one of the principal agents of diagenesis not only because of the enormous amounts available for recycling, but also because of its tendency to precipitate from solution at low temperatures. Moreover, it shows a strong tendency to combine with common ions such as Al^{3+} to form a variety of minerals generally stable in the near-surface strata. Crystallization of silica in the form of chert is the most abundant occurrence and is now reasonably well demonstrated to take place when the sediments are still unconsolidated (Harris, 1958). There are, however, many examples of crystallization of late chert in lithified strata as crusts, vein fillings, and lens-like masses clearly related to a surface of pronounced solution weathering. In the United States a classic example is the Ozark Dome (Missouri) which illustrates (1) the relation of slight positive tectonism with crystallization of early chert, i.e., before lithification of the sediments, and (2) the pronounced tendency for a late generation of crystallization associated with the present-day weathering surface.

Although proof of the influence of tectonism on the crystallization of chert is not easily obtained, there is clearly an indirect control. Bissell (1959) considered that more chert per unit volume characterizes miogeosynclinal carbonates than their counterparts on the craton. There is some reason, also, to interpret the crystallized silica as the final deposit of a silicification front, which migrates from deeply buried eugeosynclinal or miogeosynclinal sites to its eventual dissipation in cratonic strata. Along cratonic margins of the miogeosynclinal belts much nodular chert is found to interrupt bedding and clearly is of epigenetic origin. In stratigraphic sections composed of thin units of limestone in a dominantly shale matrix, however, the carbonate may be replaced by chert. An example occurs in the Fayetteville Shale (Mississippian) of the south flank of the Ozark Dome passing into the miogeosynclinal strata of the Ouachita Mountains, Ark., U.S.A. (Ogren, 1961, p. 24). Here, silicification of thin limestone beds can be traced progressively. Along the craton margin only a thin irregular zone of chert has replaced upper and lower surfaces of carbonate units, whereas the interior of carbonate beds remains unaltered. Basinward, in the direction of the miogeosyncline, chert replacement of the limestone increases until all the carbonate is silicified, whereas the shale remains unaffected. Where last observed, somewhat deeper into the miogeosyncline, portions of the shale are silicified, chert having been precipitated in the pore space.

The conditions deeper within the miogeosyncline are not known, as the strata do not crop out and drilling has not penetrated such depths. Strata of equivalent age in the eugeosynclinal belt of the Ouachita Mountains, however, are more intensively silicified; hence, there is reason to consider that silicification becomes more pronounced in the shale on approaching the eugeosyncline. In this connection, the aspect of the silica is worthy of note. Whereas chert in cratonic strata of the Ozark Dome occurs as nodules and lenses which transgress bedding, and also in other ways manifests its epigenetic origin, silica in the Fayetteville Shale is of earlier replacement origin. Such replacement carefully follows bedding, first selectively replacing the carbonates, followed by at least partial silicification of the shale. It is important to note that the ultimately produced general aspect is complete replacement of the carbonates by silica. An investigator, who could observe only the final product, would be tempted to conclude that precipitation of the silica was a syngenetic process inasmuch as he would fail to recognize the former presence of the thin, semi-lenticular carbonate beds. Dott (1958, p. 7) has called attention to massive replacement of carbonates early in the burial history of somewhat similar sediments, whereas Bissell (1959, p. 177) favored the possibility of the chert being essentially syngenetic and that it certainly precipitated early in the sediment's history. The recent study by Folk (1973) on the Caballos Novaculite supports generally early crystallization of the chert and its early replacement of associated strata of carbonates and evaporites.

In the ideal eugeosynclinal assemblage, the siliceous shales and bedded cherts show no obvious transgression of bedding. From the studies of chertification in the deep-sea porcelanites an origin of the dissolved silica is now reasonably demonstrated. A primary source from biogenic opal and from unstable volcanic debris is well established, and the mechanism whereby volcanic emanations supply the silica directly to the sea water plays only an insignificant role. In almost all modern sediments where biogenic silica, or volcanic glass, constitute a significant part of the detritus, the interstitial waters in the upper meter of sediments are now known to be significantly enriched in dissolved silica. Under continued burial, these solutions are caused to migrate to a site of supersaturation where opal *C-T* is crystallized around some nucleus and can generate a siliceous, porcelanitic bed. Opal *A* does not progress by solid—solid inversion into cristobalite. Instead the latter is crystallized directly from solution, or perhaps it is developed through structural packing of individual spheres of opal micelles migrating through intergranular pores.

Elsewhere, the ions migrate towards some nucleation center which continues to grow as a concretionary mass, gradually insinuating into the enclosing unlithified sediments. The progression indicated is clearly a response to a diagenetic milieu and the influence of tectonism is important only in establishing the primary conditions necessary for the creation of the environment. On the craton, the sources of silica probably are very much the same as in the oceanic deeps, but the action of poorly organized aluminosilicates to scavenge silica tends to reduce the initial amounts available to crystallize as cristobalite or quartz. Rather, there is reason to believe that only after the clay minerals are well crystallized, the dissolved silica is precipitated in the form of concretions and veins which transgress bedding in lithified or partially lithified sediments.

The influence of the environment on precipitation of silica has now been demonstrated by the studies of Eugster (1969) and Hay (1968) on the sodium-brine lakes. Their investigations have shown the significance of unstable intermediate silicates in temporarily holding the silica in a solid phase until lithification processes have solidified the detrital sediments. As such silicates become unstable in the near-surface ground-water solutions, they decompose into secondary chert which is crystallized in fractures and other megascopic openings. An extension of similar studies into common cratonic carbonate sediments, seeking silicates intermediate to late chert crystallization, appears to be warranted. Also, continued investigation into the neoformation of clay minerals from pre-existent detrital aluminosilicates and the behavior of colloidal micelles of opal is important in this connection. Doubtless new studies will provide the information needed to understand why large amounts of chert are crystallized as well-marked stratigraphic horizons. The mechanisms of accumulation of such widespread lenticular beds appear to be capable of elaboration in the relatively near future.

REFERENCES

Baas Becking, L.G.M., Kaplan, I.R. and Moore, D., 1960. Limits of the natural environment in terms of pH and oxidation—reduction potentials. *J. Geol.*, 68: 243—284.

Badiozamani, K., 1973. The Dorag dolomitization model—application to the Middle Ordovician of Wisconsin. *J. Sediment. Petrol.*, 43: 965—984.

Banks, N.G., 1970. Nature and origin of early and late cherts in the Leadville Limestone, Colorado. *Bull. Geol. Soc. Am.*, 81: 3033—3048.

Bien, G.A., Contois, D.E. and Thomas, W.H., 1959. The removal of soluble silica from fresh water entering the sea. In: *Silica in Sediments—Soc. Econ. Paleontol. Mineral., Spec. Publ.*, 7: 20—35.

Biggs, D.L., 1957. Petrography and origin of Illinois nodular cherts. *Ill. State Geol. Surv., Circ.*, 245: 25 pp.

Bischoff, J.L. and Ku, T.L., 1970. Pore fluids of recent marine sediments, 1. Oxidizing sediments of 20°N, continental rise to Mid-Atlantic Ridge. *J. Sediment. Petrol.*, 40: 960—1072.

Bischoff, J.L. and Ku, T., 1971. Pore fluids of recent marine sediments, 2. Anoxic sediments of 35° to 45°N, Gibraltar to Mid-Atlantic Ridge. *J. Sediment. Petrol.*, 41: 1008—1017.

Bissell, H.J., 1959. Silica in sediments of the Upper Paleozoic of the Cordilleran Area. In: *Silica in Sediments—Soc. Econ. Paleontol. Mineral., Spec. Publ.*, 7: 150—185.

Blatt, H., Middleton, G. and Murray, R., 1972. *Origin of Sedimentary Rocks.* Prentice-Hall, Englewood Cliffs, N.J., 634 pp.

Bonatti, E. and Joensuu, O., 1968. Palygorskite from Atlantic deep-sea sediments. *Am. Mineral.*, 53: 975—983.

Calvert, S.E., 1966. Accumulation of diatomaceous silica in the sediments of the Gulf of California. *Bull. Geol. Soc. Am.*, 77: 569—596.

Calvert, S.E., 1971. Composition and origin of North Atlantic deep sea cherts. *Contrib. Mineral. Petrol.*, 33: 273—288.

Carozzi, A.V. and Soderman, J.G.W., 1962. Petrography of Mississippian (Borden) crinoidal limestones at Stobo, Indiana. *J. Sediment. Petrol.*, 32: 397—415.

Chanda, S.K., 1963. Cementation and diagenesis of the Lameta Beds, Lametaghat, M.P., India. *J. Sediment. Petrol.*, 33: 728—738.

Choquette, P.W., 1955. A petrographic study of the "State College" siliceous oolite. *J. Geol.*, 63: 337—348.

Cline, L.M., 1970. Sedimentary features of Late Paleozoic flysch, Ouachita Mountains, Oklahoma. In: J. Lajoie (Editor), *Flysch Sedimentology in North America—Geol. Assoc. Can., Spec. Pap.*, 7: 85—101.

Correns, C.W., 1941. Über die Löslichkeit von Kiesselsäure in schwachsauren und alkalischen Lösungen. *Chem. Erde*, 13: 92—96.

Cressman, E.R., 1962. Non-detrital siliceous sediments. Data of geochemistry. *U.S. Geol. Surv., Prof. Pap.*, 440-T: 22 pp.

Dapples, E.C., 1959. The behavior of silica in diagenesis. In: *Silica in Sediments—Soc. Econ. Paleontol. Mineral., Spec. Publ.*, 7: 36—54.

Dapples, E.C., 1967. The diagenesis of sandstones. In: G. Larsen and G.V. Chilingar (Editors), *Diagenesis in Sediments.* Elsevier, Amsterdam, pp. 91—125.

Davis, S.H., 1964. Silica in streams and groundwater. *Am. J. Sci.*, 262: 870—891.

Dellwig, L.F., 1955. Origin of salina salt of Michigan. *J. Sediment. Petrol.*, 25: 83—111.

Dietrich, R.V., Hobbs, Jr., C.R.B. and Lowry, W.D., 1963. Dolomitization interrupted by silicification. *J. Sediment. Petrol.*, 33: 646—664.

Dott Jr., R.H., 1958. Cyclical patterns in mechanically deposited Pennsylvanian limestones of northeastern Nevada. *J. Sediment. Petrol.*, 28: 3—15.

Drever, J.I., 1971. Early diagenesis of clay minerals, Rio Ameca Basin, Mexico. *J. Sediment. Petrol.*, 41: 982—995.

Emery, K.O. and Rittenberg, S.C., 1952. Early diagenesis of California basin sediments in relation to origin of oil. *Bull. Am. Assoc. Pet. Geol.*, 36: 735—806.

Ernst, W.G. and Calvert, S.E., 1969. An experimental study of the recrystallization of porcelanite and its bearing on the origin of some bedded cherts. *Am. J. Sci.*, 267-A: 114—133.

Eugster, H.P., 1969. Inorganic bedded cherts from the Magadi Area, Kenya. *Contrib. Mineral. Petrol.*, 29: 1—31.

Fagan, J.J., 1962. Carboniferous cherts, turbidites, and volcanic rocks in northern Independence Range, Nevada. *Bull. Geol. Soc. Am.*, 73: 595—612.

Feth, J.H., Rogers, S.M. and Roberson, C.E., 1961. Aqua de Ney, California: a spring of unique chemical character. *Geochim. Cosmochim. Acta*, 22: 75—86.
Feth, J.H., Roberson, C.E. and Polzer, W.L., 1964. Sources of mineral constituents in water from granitic rocks, Sierra Nevada, California and Nevada. *U.S. Geol. Surv., Water Supply Pap.*, 1535-1: 170 pp.
Flawn, P.T., Goldstein, A., King, P.B. and Weaver, C.E., 1961. The Ouachita System. *Univ. Texas Publ.*, 6120: 401 pp.
Flörke, O.W., 1955. Zur Frage des Hochcristobalit in Opalen, Bentoniten und Glasern. *Neues Jahrb. Mineral., Monatsh.*, 1955 (10): 217—233.
Flörke, O.W., 1967. Die Modificationen von SiO_2. *Fortschr. Mineral.*, 44: 181—230.
Folk, R.L., 1973. Evidence for peritidal deposition of Devonian Caballos Novaculite, Marathon Basin, Texas. *Bull. Am. Assoc. Pet. Geol.*, 57: 702—726.
Folk, R.L. and Pittman, J.S., 1971. Length-slow chalcedony: a new testament for vanished evaporites. *J. Sediment. Petrol.*, 41: 1045—1058.
Fox, W.T., 1962. Stratigraphy and paleoecology of the Richmond Group in southeastern Indiana. *Bull. Geol. Soc. Am.*, 73: 621—642.
Franks, P.C. and Swineford, A., 1959. Character and genesis of massive opal in Kimball Member, Ogallala Formation, Scott County, Kansas. *J. Sediment. Petrol.*, 29: 186—196.
Garrels, R.M., Mackenzie, F.T., Bricker, O.P. and Buckley, F., 1967. Silica in sea water: control by silica minerals. *Science*, 155: 1404—1405.
Gibbs, R.J., 1960. *The Stratigraphy and Paleontology of the Fish Haven Dolomite of South Central Idaho.* Thesis. Northwestern Univ., Evanston, Ill. (unpublished).
Glover, J.E., 1963. Studies in the diagenesis of some western Australian sedimentary rocks. *J. R. Soc. West. Aust.*, 46: 33—56.
Goldberg, E.D., 1958. Determination of opal in marine sediments. *J. Mar. Res.*, 17: 178—182.
Goldberg, E.D., 1963. The oceans as a chemical system. In: R.M. Hill (Editor), *The Sea, II.* Interscience, New York, N.Y., pp. 3—25.
Goldstein Jr., A., 1959. Cherts and novaculites of Ouachita facies. In: *Silica in Sediments—Soc. Econ. Paleontol. Mineral., Spec. Publ.*, 7: 135—149.
Goldstein Jr., A. and Hendricks, T.A., 1953. Siliceous sediments of Ouachita facies in Oklahoma. *Bull Geol. Soc. Am.*, 64: 421—442.
Greenwood, R., 1973. Cristobalite: its relationship to chert formation in selected samples from the deep-sea drilling project. *J. Sediment. Petrol.*, 43: 700—708.
Grunau, H.R., 1965. Radiolarian cherts and associated rocks in space and time. *Eclogae Geol. Helv.*, 58: 157—208.
Harris, L.D., 1958. Syngenetic chert in the Middle Ordovician Hardy Creek Limestone of southwest Virginia. *J. Sediment. Petrol.*, 28: 205—208.
Hathaway, J.C. and Sachs, P.L., 1965. Sepiolite and clinoptilolite from the Mid-Atlantic Ridge. *Am. Mineral.*, 50: 852—867.
Hay, R.L., 1966. Zeolite and zeolitic reactions in sedimentary rocks. *Geol. Soc. Am., Spec. Pap.*, 85: 130 pp.
Hay, R.L., 1968. Chert and its sodium-silicate precursors in sodium-carbonate lakes of East Africa. *Contrib. Mineral. Petrol.*, 17: 255—274.
Hayes, J.B., 1964. Geodes and concretions from the Mississippian Warsaw Formation, Keokuk Region, Iowa, Illinois and Missouri. *J. Sediment. Petrol.*, 34: 123—134.
Heath, G.R. and Moberly Jr., R., 1971. Cherts from the Western Pacific, Leg 7, Deep-Sea Drilling Profect. In: *Initial Reports of the Deep-Sea Drilling Project*, 7 (2): 991—1007.
Hem, J.D., 1959. Study and interpretation of the chemical characteristics of natural

water. *U.S. Geol. Surv., Water Supply Pap.*, 1473: 269 pp.
Hintze, L.F., 1953. Silicification of Ordovician fossils in Utah and Nevada. *Bull. Geol. Soc. Am.*, 64: 1508.
Hurd, D.C., 1973. Interactions of biogenic opal, sediment and seawater in the Central Equatorial Pacific. *Geochim. Cosmochim. Acta*, 37: 2257—2282.
Hurst, V.J., 1953. Chertification in the Ft. Payne Formation, Georgia. *Bull. Ga. Geol. Surv.*, 60: 215—238.
Hutton, J.T., Twidale, C.R., Milnes, A.R. and Rosser, H., 1972. Composition and genesis of silcretes and silcrete skins from the Beda Valley, Southern Arcoona Plateau, South Australia. *J. Geol. Soc. Aust.*, 19: 31—39.
Ingels, J.J.P., 1963. Geometry, paleontology and petrography of Thornton Reef Complex, Silurian of northeastern Illinois. *Bull. Am. Assoc. Pet. Geol.*, 47: 405—440.
Jacka, A.D., 1974. Replacement of fossils by length-slow chalcedony and associated dolomitization. *J. Sediment. Petrol.*, 44: 421—427.
Jones, J.B. and Segnit, E.R., 1971. The nature of opal, 1. Nomenclature and constituent phases. *J. Geol. Soc. Aust.*, 18: 57—68.
Jones, J.B., Biddle, J. and Segnit, S.R., 1966. Opal genesis. *Nature*, 210: 1350—1354.
Keller, W.D. and Reesman, A.L., 1963. Glacial milks and their laboratory-simulated counterparts. *Bull. Geol. Soc. Am.*, 74: 61—76.
Keller, W.D., Balgord, W.D. and Reesman, A.L., 1963. Dissolved products of artificially pulverized silicate minerals and rocks, 1, 2. *J. Sediment. Petrol.*, 33: 191—205; 426—438.
King, R.I. and Merriam, D.F., 1969. Origin of the "welded chert", Morrison Formation (Jurassic), Colorado. *Bull. Geol. Soc. Am.*, 80: 1141—48.
Kolodny, Y., 1969. Petrology of siliceous rocks in the Mishash Formation (Negev, Israel). *J. Sediment. Petrol.*, 39: 166—175.
Krauskopf, K.B., 1956. Dissolution and precipitation of silica at low temperatures. *Geochim. Cosmochim. Acta*, 10: 1—26.
Krauskopf, K.B., 1959. The geochemistry of silica in sedimentary environments. In: *Silica in Sediments—Soc. Econ. Paleontol. Mineral., Spec. Publ.*, 7: 4—20.
Lamar, J.E., 1950. Acid etching in the study of limestones and dolomites. *Ill. State Geol. Surv., Circ.*, 156: 47 pp.
Lewin, J.C., 1961. The dissolution of silica from diatom walls. *Geochim. Cosmochim. Acta*, 21: 182—198.
Lovering, T.G. and Patten, L.E., 1962. The effect of CO_2 at low temperature and pressure on solutions supersaturated with silica in the presence of limestone and dolomite. *Geochim. Cosmochim. Acta*, 74: 787—796.
Mackenzie, F.T. and Garrels, R.M., 1966a. Chemical mass balance between rivers and oceans. *Am. J. Sci.*, 264: 507—525.
Mackenzie, F.T. and Garrels, R.M., 1966b. Silica-bicarbonate balance in the ocean and early diagenesis. *J. Sediment. Petrol.*, 36: 1075—1084.
Mackenzie, F.T. and Gees, R., 1971. Quartz: synthesis at earth-surface conditions. *Science*, 173: 533—535.
McBride, E.F., 1970. Stratigraphy and origin of Maravillas Formation (Upper Ordovician), West Texas. *Bull. Am. Assoc. Pet. Geol.*, 54: 1719—1745.
McBride, E.F. and Thomson, A., 1970. The Caballos Novaculite, Marathon Region, Texas. *Geol. Soc. Am., Spec. Pap.*, 122: 129 pp.
Millot, G., 1960. Silice, silex, silicifications et croissance des cristaux. *Bull. Serv. Carte Geol. Alsace-Lorraine*, 13 (4): 129—146.
Millot, G., Lucas, J. and Wey, R., 1963. Research on evolution of clay minerals and argil-

laceous and siliceous neoformation. *Clays Clay Miner., 10th Nat. Conf.*, pp. 399—412.
Moore, T.C., 1969. Radiolaria: change in skeletal weight and resistance to solution. *Bull. Geol. Soc. Am.*, 80: 2103—2107.
Morey, G.W., Fournier, R.O. and Rowe, J.J., 1962. The solubility of quartz in water in the temperature interval from 25° to 300°C. *Geochim. Cosmochim. Acta*, 26: 1029—1043.
Müller, G., 1961. Die rezenten Sedimente im Golf von Neapel, 2. Mineral Neu-und Umbildungen in den rezenten Sedimenten des Golfes von Neapel. Ein Beitrag zur Unwandlung vulkanischer Glaser durch Halmyrolse. *Beitr. Mineral. Petrogr.*, 8: 1—20.
Müller, G., 1967. Diagenesis in argillaceous sediments. In: G. Larsen and G.V. Chilingar (Editors), *Diagenesis in Sediments*. Elsevier, Amsterdam, pp. 127—177.
Newell, N.D., Rigby, J.K., Fischer, A.G., Whiteman, A.J., Hickox, J.R. and Bradley, J.S., 1953. *The Permian Reef Complex of the Guadalupe Mountain Region, Texas and New Mexico*. Freeman, San Francisco, Calif., pp. 160—182.
Ogren, D.E., 1961. *Stratigraphy of the Upper Mississippian Rocks of Northern Arkansas*. Thesis, Northwestern Univ., Evanston, Ill. (unpublished).
Oldershaw, A.E., 1968. Electron-microscope examination of Namurian bedded cherts, North Wales (Great Britain). *Sedimentology*, 10: 255—272.
Pelto, C.R., 1956. Chalcedony. *Am. J. Sci.*, 254: 32—50.
Pollard, C.O. and Weaver, C.E., 1973. Opaline spheres: loosely packed aggregates from silica nodule in diatomaceous Miocene fullers earth. *J. Sediment. Petrol.*, 43: 1072—1076.
Rapson, J.G., 1962. The petrography of Pennsylvanian chert breccias and conglomerates: Rocky Mountain group, Banff, Alberta. *J. Sediment. Petrol.*, 32: 249—262.
Shaver, R.H., 1974. Silurian reefs of northern Indiana: reef and inter-reef macrofaunas. *Bull. Am. Assoc. Pet. Geol.*, 58: 934—957.
Siedlecka, A., 1972. Length-slow chalcedony and relicts of sulphates—Evidences of evaporitic environments in the Upper Carboniferous and Permian beds of Bear Island, Svalbard. *J. Sediment. Petrol.*, 42: 812—816.
Siever, R., 1957. The silica budget in the sedimentary cycle. *Am. Mineral.*, 42: 821—841.
Siever, R., 1962. Silica solubility, 0—200°C, and the diagenesis of siliceous sediments. *J. Geol.*, 70: 127—150.
Siever, R., Beck, K.C. and Berner, R.A., 1965. Composition of interstitial waters of modern sediments. *J. Geol.*, 73: 39—73.
Smale, D., 1973. Silcretes and associated silica diagenesis in southern Africa and Australia. *J. Sediment. Petrol.*, 43: 1077—1089.
Swett, K., 1965. Dolomitization, silicification and calcitization in patterns in Cambro-Ordovician oolites from northwest Scotland. *J. Sediment. Petrol.*, 35: 928—938.
Taliaferro, N.L., 1934. Contraction phenomena in cherts. *Bull. Geol. Soc. Am.*, 45: 189—232.
Trümpy, R., 1960. Paleotectonic evolution of the central and western Alps. *Bull. Geol. Soc. Am.*, 71: 843—908.
West, I.M., 1964. Evaporite diagenesis in the lower Purbeck Beds of Dorset. *Proc. Geol. Soc. Yorkshire*, 34: 315—330.
White, D.E., Brannock, W.W. and Murata, K.J., 1956. Silica in hot-spring waters. *Geochim. Cosmochim. Acta*, 10: 27—59.
Williamson, W.O., 1957. Silicified sedimentary rocks in Australia. *Am. J. Sci.*, 255: 23—42.
Wilson, R.C.L., 1966. Silica diagenesis in Upper Jurassic limestones. *J. Sediment. Petrol.*, 36: 1036—1049.
Wise, S.W., Buie, B.F. and Weaver, F.M., 1972. Chemically precipitated sedimentary cristobalite and the origin of chert. *Eclogae Geol. Helv.*, 65: 157—163.

Chapter 4

EARLY DIAGENESIS OF SUGARS AND AMINO ACIDS IN SEDIMENTS

EGON T. DEGENS and KENNETH MOPPER

INTRODUCTION

The bulk of organic carbon in the crust of the earth is present in sedimentary rocks as a highly dispersed inert material called kerogen. Since most of this carbon has been cycled through the biosphere, biomolecules such as sugars, amino acids, lipids and phenols are the starting materials from which kerogen is formed.

In the early organic-geochemical studies which began chiefly with the classical works of Treibs (1934, 1936) the presence of sugars, amino and fatty acids, or the bases of purines and pyrimidines in hydrolysis liquors of ancient rocks were used as indication that the original intact polysaccharides, polypeptides, nucleic acids, and other biogenic polymers could survive diagenesis. The present consensus, however, is that during diagenesis new polymers or condensates are synthesized from breakdown products of former biochemical macromolecules.

During early diagenesis in sediment, decay and metabolic degradation of these biomolecules result in randomization of the original well-defined structural order of living matter. Dayhoff (1971) demonstrated that biomolecules are far removed from thermodynamic equilibrium. Diagenesis of organic matter in sediments, therefore, is expected to proceed in a direction of increasing thermodynamic stability, i.e.:

$$\text{biomolecules} \xrightarrow[\text{fast}]{\text{early diagenesis}} \text{metastable products} \xrightarrow[\text{slow}]{\text{late diagenesis}}$$

$$C, CO_2, H_2O, CH_4, CO, \text{asphalts, etc.}$$

The approach to thermodynamic stability, however, is extremely slow and for most organic matter it is never reached (Blumer, 1967). The transformation of biomolecules appears to be kinetically inhibited by the formation of metastable associations during the early stages of diagenesis. In fact, fragile biomolecules such as sugars (Swain, 1969) and amino acids (Degens and Bajor 1960; Hare, 1969) have been extracted from rocks of Paleozoic and Precambrian ages. The usefulness of organic geochemistry for studying biochemical evolution, for understanding the origin of fossil fuels, kerogen and humus, and for discerning paleoenvironments has been limited by a lack of knowledge of the factors affecting the transformation of biomolecules during early diagenesis in sediments. Some pertinent questions concerning early diagenesis are:

(1) What major factors control the cycling of organic matter in oceans and lakes?

(2) What factors determine the quality and quantity of organic matter incorporated into sediment?

(3) What is the organic composition of sediment in different depositional environments (e.g., deep-sea oxic; shallow-sea oxic; deep-sea anoxic; freshwater anoxic; brackish-water anoxic; etc.)?

(4) How does the environment at the sediment—water interface affect the composition and structure of the organic input?

(5) How do metal—organic interactions and biological degradation affect the diagenesis of organic matter in recent sediments?

The emphasis of this chapter is not placed on compiling an inventory of all organic compounds which have been isolated and identified from sediments and rocks. Such an approach is unfruitful in clarifying major biogeochemical processes. Insights into the above questions can be gained by the study even of only a few major organic compounds. Inasmuch as proteins and carbohydrates collectively comprise about 40—80% of the organic matter of most organisms, they represent a large fraction of the organic input of sediments. Therefore, the following discussion is centered upon these compounds. From this discussion, general conclusions about the diagenesis of organic matter in Recent sediment have been drawn.

GEOCHEMICAL BALANCE

Most of the organic debris that has come to rest on earth is found in sediments. Hunt (1962) determined the total organic matter in more than thousand rock specimens collected from two hundred formations in sixty major sedimentary basins; shales averaged 2.1%, carbonates 0.29% (Gehman, 1962), and sandstones 0.05%. Based on these data and the figures reported by Weeks (1958) on the total volume of sediments and the relative proportions of shales, carbonates, and sandstones, the organic matter entrapped in sediments is around $3.8 \cdot 10^{15}$ metric tons. The overwhelming part of it, i.e., $3.6 \cdot 10^{15}$ metric tons, is present in shales as a relatively inert finely disseminated substance called kerogen.

As can be seen in Fig. 4-1, total organic matter is about the same for Recent shell carbonates, limestones, and clay muds. All three samples are representative of a great number of studied specimens from different marine environments. Whereas a mean value of 2% organic matter is approximately

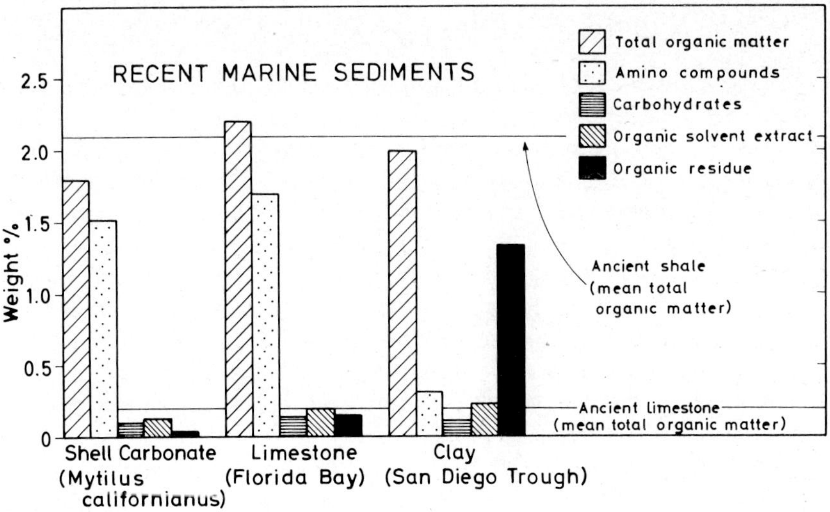

Fig. 4-1. Distribution of major organic components in Recent marine carbonates and clay muds.

maintained in shales of all geologic ages, ancient limestones have on an average ten times less organic matter than their recent counterparts. This is probably due to the chemical differences in the original organic matter.

For comparison, the coal deposits of the world have been estimated to total about $6 \cdot 10^{12}$ metric tons. This is 1/500th of the disseminated organic matter. Estimates on the ultimate primary petroleum reserves run at $0.2 \cdot 10^{12}$ metric tons. This represents only 1/16000th of the total organic matter incorporated in sediments.

The amount of dissolved organic substances in the sea is only a few milligrams per liter (Goldberg, 1961). Fresh waters and some interstitial waters may have considerably higher concentrations, but when compared to the total mass of organic matter in the lithosphere and the ocean, their contribution is small. The flux of organic matter in sea water and sediments has been presented by Degens and Mopper (1975).

In summary, one is reasonably well informed about the range in total organic matter present in various sediments and natural waters. More significant from a geochemical point of view, is knowledge of the chemical nature and distribution pattern of individual organic species. Also, an understanding of the processes and mechanisms responsible for the preservation, alteration, and destruction of organic matter throughout geological history has greater scientific merit than a mere study of the total organic carbon content.

CLASSIFICATION OF ORGANIC MATTER

In sediments the various organic compounds present during early stages of diagenesis not only interact among themselves but also with metal ions and mineral surfaces. As a consequence, the complexity of reaction schemes and resultant organic products becomes immense. Blumer (1974), in a study on fossil porphyrins, estimates that the number of structural permutations for these compounds alone exceeds $2 \cdot 10^6$. If such a variety of compounds arises from just two biochemical precursors, namely chlorophyll and hemoglobin, the number of molecular species that are generated from the main building blocks of life in the course of diagenesis must be almost infinite. So far, over one thousand different organic species have been identified in sediments. The number of compounds isolated from crude oil is on the order of 10,000 (Rossini, 1960). The advent of sophisticated instrumentation (i.e., high-resolution gas-chromatography—mass-spectrometry) has facilitated the compilation of an inventory of molecular species present in sediments and oil, (Simoneit and Burlingame, personal communication).

In order to reasonably handle all these individual organic molecules, they must be grouped into distinct classes of compounds. It is of interest to observe that most organic compounds extracted from geological materials bear some structural resemblance to former living matter. Even constituents synthesized in meteorites and probably also on the prebiotic earth (Bernal, 1965) are both chemically and structurally similar to the monomeric building blocks that constitute today's living things. A geochemical classification of organic substances, therefore, is somewhat similar to a biochemical one. But, whereas the proteins, carbohydrates, and lipids are quantitatively more important in the plant and animal kingdom, the phenolic "heteropolycondensates", hydrocarbons, and asphalts are of greater significance in geological materials. A classification scheme, which includes the major biogeochemical compounds, is proposed here: (1) amino acids and related substances; (2) carbohydrates and derivatives; (3) lipids, isoprenoids, and steroids; (4) heterocyclic compounds; (5) phenols, quinones, and related substances; (6) hydrocarbons; and (7) asphalts and allied substances.

Information on the structural composition and the general geochemistry of organic constituents has been presented in some detail by Degens (1965). For information on the biochemistry and structure of these compounds one should consult a biochemistry text such as that by White et al. (1959).

Fossil organic matter may be classified into compounds that are (1) survivors of diagenesis and (2) products of diagenesis. The first group includes all those organic molecules which are chemically similar or identical to living matter. Compounds such as branched hydrocarbons, that no longer resemble biochemical building blocks, but which were formed from pre-existing organic debris during diagenesis, are representatives of the second group. Metabolic waste matter may be regarded as intermediate, inasmuch as certain metabolites, such as urea, may also arise via thermal degradation in the course of diagenesis.

GENERAL CONCEPTS

Aspects of the biogeochemical cycle in the ocean are shown in Fig. 4-2. Two major organic inputs to sediment are emphasized: (1) biologically labile

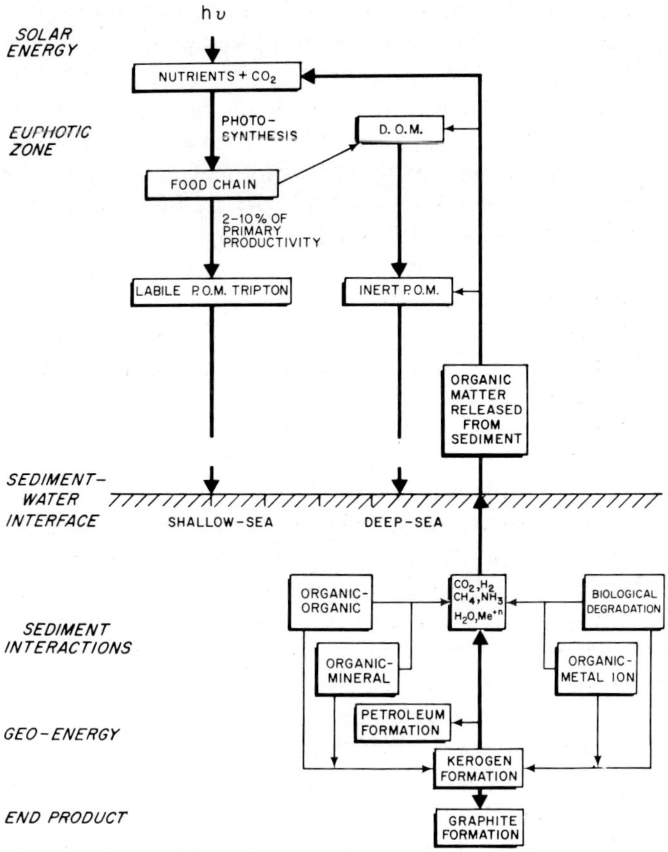

Fig. 4-2. Hypothetical schematic of the biogeochemical cycle in the ocean. The organic input of shallow-sea sediments is dominantly labile seston derived from the local biomass, whereas the organic input of deep-sea sediments is dominantly refractory particulate organic matter derived from both the degradation of seston and the condensation of dissolved organic matter.

particulate organic matter (P.O.M.) or seston, and (2) inert P.O.M.

The organic input of shallow-water sediments tends to be dominated by labile P.O.M. due to the high primary productivity and the low degree of degradation in the short water column. In contrast, organic input of deep-sea sediments is probably dominated by a relatively highly condensed, biologically resistent P.O.M., because the organic input from labile seston is less significant due to extensive degradation in the long water column and also to lower surface water productivity (Mopper and Degens, 1972; and Degens and Mopper, 1974).

Fig. 4-2 also shows the various diagenetic interactions involving sedimented organic matter. Biological degradation is concentrated at and near the sediment surface (~0—20 cm below the sediment—water interface for shallow-water sediments and 0—5 cm for deep-water sediments). In sediments overlain by oxygen-containing water (positive redox potential), the organic input is utilized to varying degrees by burrowing organisms and bacteria. The degree of utilization is dependent on factors such as degree of biological activity, sedimentation rate, and "palatability" effects. Thus, the organic input may be converted partially or wholly to a metabolic waste product bearing little resemblance in structure and composition to the original organic matter. Organic decomposition is further aided by the presence of molecular oxygen and the resultant oxidation processes.

The difference in palatability of the organic input of shallow- and deep-water oxic sediments, as shown in Fig. 4-2, is reflected in animal densities and biomass data (Sanders et al., 1965; Sanders and Hessler, 1969; Rowe, 1971). For example, the benthic metazoan biomass density off New England decreases by a factor of about 500 from the shallow coastal zone (water depth of 10—200 m) to the deep-sea zone (water depth of 4000—5000 m), with the greatest change occurring at the edge of the continental shelf (water depth of 100—200 m). Primary productivity of surface water decreases by only a factor of 2—3 in this interval. Pressure changes in this interval may affect diversity but not the actual biomass. In addition to palatability effects, a deficiency of certain nutrients or growth hormones in the organic food detritus may also limit biomass in the deep-sea zone. Palatability effects and biomass distribution are also reflected in in situ measurements of biological activity; for example, biological oxidation rates range from 1 to 4 g-C/m^2/yr for deep-sea sediments (~2000 m) and 50—150 g-C/m^2/yr for shallow coastal regions (<100 m) (Smith and Teal, 1973). In addition to palatability and possible pressure effects, a deficiency of certain essential nutrients or growth hormones in the organic food detritus may also limit biomass in the deep sea.

The opinion has frequently been expressed that microorganisms alone rapidly degrade sedimentary organic matter even in deep-sea surface environments. In fact, microbial counts may yield up to several million bacteria per gram of deep-sea sediment (e.g., Rittenberg et al., 1963). Based on the work of Jannasch et al. (1971) and Jannasch and Wirsen (1973), who examined in situ microbial degradation of organic substrates at oxidizing sediment—water interfaces, this viewpoint no longer persists. Food stored at a water depth of 1500 m for several months showed negligible bacterial degradation. Organic substrates (e.g., agar, polysaccharides) inoculated at depth, sealed, and embedded for several months in deep-sea sediment were utilized at rates up to several hundred times lower than control samples kept under refrigeration. In contrast, organic substrates placed in shallow-water sediments were utilized at comparable or slightly lower rates than the controls. When clay was added to the substrates, decomposition rates did not increase even though the number of bacteria increased by a factor of 100—1000 relative to clay-free samples.

This decline in microbial activity with water depth may be related to the increase in pressure; however, an alternative suggestion is that microorganisms like burrowing animals, are limited in activity by the increased refractory nature of the organic input in deep-sea sediments. The fact that common microbial substrates, which were inoculated and sealed in the deep-sea sediment, remained virtually intact, suggests that physiological effects exert a greater control on the activity of bacteria than palatability effects.

It is significant to note that samples which were not sealed and thus were open to consumption by both bacterial and metazoan populations were dramatically utilized in comparison to the sealed samples. This observation indicates that it is metazoan activity, not bacterial activity as commonly assumed, which is mainly responsible for degradation of organic matter in oxic surface sediments.

In contrast, sediments which are overlain by waters containing no dissolved oxygen (negative redox potential, a condition which is referred to as either anoxic or reducing) contain no burrowing organisms; the microbial population is limited to organisms such as sulfate and nitrate reducers, methane producers, and fermenters. As a result of the lowered biological activity, organic matter in reducing sediments is not as thoroughly consumed as in oxidizing sediments. This is reflected in the generally higher total organic carbon and nitrogen contents of reducing sediments (Richards, 1965; Mopper and Degens, 1972).

In addition to differences in gross chemical parameters, such as total organic carbon content, one might expect to observe differences in the degree of humification of organic matter in surface sediments deposited under oxic and anoxic conditions. For example, in oxic sediment a higher degree of intermolecular condensation and metal ion—organic matter interaction as a result of the significantly higher biological activity and reworking is expected. Such correlations between biological activity and degree of humification do indeed exist is indicated by aminoacid and carbohydrate studies, which are described in later sections.

There is no doubt that the number of microbes in sediments sharply decreases with depth of burial. For example, the bacterial population in surface sediments of marine origin may amount to 10 million bacteria per gram and more. This would represent approximately 10^{-4} g of organic matter or 0.1 mg/g of sediment. Sediments at a depth of 50—100 cm have considerably lower population densities, i.e., a few thousand bacteria per gram of sediment.

There have been periodic claims in geologic literature on the isolation of viable bacteria from geologically ancient materials — coal, fossil dung, petroleum, connate waters, sulfur domes, salt deposits, sedimentary rocks, igneous rocks, and even meteorites. In many such instances the investigators have stated or implied that the bacteria had survived over the geologic ages involved. In reference to these claims, it is sufficient to point out that the problem of procuring and sampling such materials aseptically, is a very difficult one and, in most instances, a critical appraisal of the published work reveals gross inadequacies in the experiments. The pertinent publications include: Lipman (1928), Waksman et al. (1933), Rittenberg (1940), Porter (1946), ZoBell (1946a, b, 1952, 1963), Emery and Rittenberg (1952), Neher and Rohrer (1959), Baas Becking et al. (1960), Lindblom and Lupton (1961), Oppenheimer (1961), Sisler (1961), and Rittenberg et al. (1963).

Thus the question is still open as to whether or not bacteria play a significant role in organic matter diagenesis below the surface layer. The work of Jannasch et al. (1971) is most relevant in regard to this question. These investigators demonstrated that bacterial activity even at the sediment surface in the deep-sea zone is low, which implies that at greater depths in the sediment where food and hydrogen acceptors (such as O_2) are comparatively scarce, bacterial activity must be vanishingly low. Further in situ experiments are needed to test this hypothesis.

Assuming that microbial activities have virtually ceased at some shallow depth, further diagenesis is dominated by relatively slow abiotic maturation processes (Fig. 4-2). Various functional groups are present on the organic molecules within the heterogeneous framework of sedimentary humic substances (see Fig. 4-19). These groups include: alipathic amino, aromatic amino, carboxylate, enolate, alkoxide, phenoxide, mercaptide, phosphate, phosphonate, carboxyl, ether, ester, amide, thioether, and hydroxyalkyl. A major driving force for polymerization and condensation between functional groups is the dehydration of sediment during compaction.

Some degree of structural order within humic material is also possible by coordination of functional groups to metal ions, resulting in the formation of metal ion coordination polyhedra (Matheja and Degens, 1971). Functional groups of organic molecules may coordinate to surfaces and edges of minerals as well as to metal ions. Furthermore, interlattice positions of certain clay minerals may accommodate and, hence, protect organic molecules from degradation (e.g., Weiss, 1969; Sørensen, 1972).

In addition to coordination and interlayering effects, fragmentation, synthesis, and polymerization of organic molecules may be catalytically promoted by minerals (e.g., Degens and Matheja, 1968; Weiss, 1969; Burton and Neuman, 1971). For example, Harvey et al. (1971) demonstrated that nitrogen heterocycles are synthesized from such simple starting materials as CO_2 and NH_3 in the presence of an aqueous slurry of kaolinite. Further, in the presence of kaolinite saccharides are synthesized from aqueous solutions of paraformaldehyde and fatty acids are esterified to glycerides (Harvey et al., 1972). Combinations of various silicate minerals and lime have been shown to be extremely effective catalysts for the synthesis of simple sugars from paraformaldehyde, and amino acids from paraformaldehyde and urea under aqueous conditions. Yields of up to 15% were not uncommon (Degens, 1974).

Mineral catalyzed synthesis and polymerization undoubtedly occur in Recent sediments; however, their importance appears to be negligible in comparison to the effects of biological degradation and metal complexation (Mopper and Degens, 1972). Mineral catalyzed reactions probably become important as temperature and pressure increase during later stages of diagenesis (e.g., Whitehead and Breger, 1950; Mulik and Erdman, 1963; and Welte, 1965). Synthesis reactions were probably significant during the prebiotic synthesis of organic matter on the primordial earth.

In the context of organic matter—mineral interactions, it is significant to note that organic matter locked inside biologically precipitated minerals is not subject to biological degradation either in the water column or at the sediment—water interface and, consequently, probably represents the only unaltered organic input of sediment. In unrecrystallized sediments, mineralized tissues may follow different diagenetic pathways than the organic matter in the exterior environment (e.g., Mitterer, 1972).

The rate of abiotic diagenesis of organic matter is principally controlled by temperature. Thus in areas of high heat flow, such as along rift zones, one would expect to find a more advanced degree of diagenesis than in areas of normal or low heat flow. Knowledge of thermal stability ranges of organic compounds may reveal information regarding the thermal history of the host strata. For example, the use of thermally sensitive epimeric pairs, such as isoleucine—alloisoleucine (King and Hare, 1972), may be suitable as a geothermometer. In general, however, thermal experiments with pure organic isolates have been unsuccessful in predicting their geological stability. This lack of success is probably attributable to the formation of metastable organic matter—metal-ion and organic matter—mineral associations during early diagenesis. These associations enhance the stability of even relatively fragile organic compounds such as serine.

SUGAR AND AMINO ACID COMPOSITION OF PLANKTON

With the exception of terrigenous organic matter deposited in sediments close to land, especially at river outlets, planktonic debris from the euphotic zone represents the main source of organic input of marine sediments (Degens and Mopper, 1974). Thus a discussion of sediment organic matter logically begins with the composition of plankton.

Sugars

The carbohydrate compositions of plankton from various regions are shown in Table 4-I (Mopper, 1973). The total concentrations range from <0.5% to 5% on a dry-weight basis (100% would correspond to about 6000 μmoles/g). In contrast to the sugar composition of land plants, in which glucose from cellulose and starch is dominant, the sugar composition of marine plankton is not overwhelmingly dominated by any one sugar, although a

TABLE 4-I

Carbohydrate composition of marine plankton (mole%) *

Sugars	Iceland	Oyster Pond, Woods Hole	Cariaco Trench	Bermuda	Peru-Callao	Galapagos Island I	Galapagos Island 2
Rhamnose	2.7	5.4	7.4	4.6	4.0	5.1	5.9
Fucose	6.2	6.3	7.8	6.4	3.0	7.1	9.7
Ribose	37.4	22.1	4.9	5.6	14.8	7.1	5.6
Arabinose	2.5	0.8	2.1	1.1	3.0	6.0	2.9
Xylose	3.2	6.0	5.7	5.5	3.7	19.4	9.0
Mannose	16.0	18.0	15.0	11.9	13.1	13.2	16.0
Galactose	16.8	23.8	39.7	37.3	18.0	22.5	35.9
Glucose	15.3	17.7	17.3	27.6	40.5	19.4	14.3
Total sugars (μ moles/g)	13.1	113.2	144.9	86.9	40.5	44.8	44.3

* In this and the following tables the sugar and amino acid data are generally presented in terms of mole% and residues/1000, respectively. For data in terms of μg/g (or mg/g) the reader is referred to Mopper and Degens (1972) and Mopper (1973).

general order of abundance appears to exist within the HCl extracts [1]: Ga > Gl > M > Ri > X > Fu $\gtrless$ Rh > A (hexoses > pentoses) [2].

These results correlate well with those of Handa and Yanagi (1969). They found that glucose, galactose, mannose, and xylose are the dominant sugars in the water-insoluble carbohydrate fraction of particulate organic matter (P.O.M.) at or near the water—sediment interface in the ocean (0—50 m). These sugars are probably present in plankton as long-chained structural polysaccharides (hemicellulose) and are, hence, more resistant to degradation upon cell lysis. The high abundance of ribose in some plankton samples (Table 4-I) and its low abundance within the water-insoluble carbohydrate fraction within P.O.M. (Handa and Yanagi, 1969) suggests that ribose is dominantly associated with labile compounds such as organo-phosphates. These compounds hydrolyzed rapidly after cell lysis (Harvey, 1960) and are, thus, preferentially lost from P.O.M.

Experiments are presently being undertaken to determine the degree of metal-ion association of the carbohydrates in plankton. By addition of EDTA, a strong chelating agent, to pH-buffered plankton samples, metal ions (especially polyvalent ions) are removed from in situ complexes having a weaker complexing strength; metal-free organic compounds, if soluble, are then released into aqueous solution. EDTA extractions of sedimentary metal-bound carbohydrates and proteins have been successfully achieved and are described in the following sections. Related work on plankton is in progress.

Amino acids

In contrast to the sugar compositions of plankton, which are variable, the amino-acid compositions are remarkably uniform especially considering the geographical distances involved (Table 4-II). The total extractable amino acids range from 14 to 42% on a dry-weight basis. The dominant amino acids detected in the HCl extract are glycine, alanine, glutamic acid and aspartic acid. The amino acid/hexosamine (AA/HA) ratios fall within a narrow range of 13—25.

The degree of metal association of the proteins in a plankton sample is shown in Table 4-III. A substantial part of the acid-extractable amino acids, e.g., 60% of the aspartic acid and 20% of the tyrosine and phenylalanine

---

[1] HCl extract refers to sugars released by acid hydrolysis and is considered to reflect the total sugars or amino acids; this extract is sometimes referred to as total.

[2] The notation used is: Ga = galactose, Gl = glucose, M = mannose, Ri = ribose, X = xylose, Fu = fucose, Rh = rhamnose, and A = arabinose.

TABLE 4-II

Distribution of amino acids and hexosamines in marine plankton (residues/1000)

Amino acids	Iceland	Oyster Pond, Woods Hole	Cariaco * Trench	Walvis Bay	Peru-Callao **	Galapagos Island
Aspartic acid	98	90	102	104	115	109
Threonine	51	49	58	69	57	55
Serine	55	66	62	71	61	60
Glutamic acid	120	106	127	79	138	124
Proline	55	58	69	64	48	54
Glycine	133	82	155	112	159	132
Alanine	106	97	113	118	105	104
Cystine	9	1	2	1	11	13
Valine	63	52	59	76	53	58
Methionine	27	20	22	19	29	23
Isoleucine	44	37	41	54	35	38
Leucine	76	70	76	85	76	76
Tyrosine	8	38	18	40	26	33
Phenylalanine	31	31	26	42	35	32
Lysine	58	93	55	33	25	45
Histidine	6	24	4	6	13	9
Arginine	61	86	11	29	13	34
Total amino acids (mg/g dry weight)	313	419	266	140	312	357
Hexosamines (mg/g)	18.7	36.0	14.3	9.8	7.9	55.6
Ammonia (mg/g)	2.3	3.0	3.2	1.4	0.97	10.8
AA/HA ratio	16.7	11.6	18.6	14.3	35.5	6.4
Organic C(%)	54.4	54.53	43.90	nd.	nd.	48.46
Organic N(%)	5.97	7.28	7.34	nd.	nd.	7.34
Organic H(%)	7.74	7.45	6.24	nd.	nd.	6.98
C/N ratio	9.1	7.5	6.0	nd.	nd.	6.6

* ~50% mineral residue (mainly diatomaceous).
** 79% $CaCO_3$ (coccoliths), removed prior to analysis.

(Table 4-III) are present in the unhydrolyzed EDTA extract [1]. The apparent high degree of metal association of aspartic acid may be interpreted in light of recent research in calcification; aspartic acid has been shown to provide

[1] EDTA exhibits negligible hydrolysis effects; therefore, sugars and amino acids detected after extraction with only an aqueous slurry of EDTA (noted in the Tables as "EDTA") represent an approximation of the in situ metal-bound monomers. Acid hydrolysis of this extract, which will be referred to as the hydrolyzed EDTA extract (noted in the Tables as "EDTA + HCl") represents an estimate of the total metal-bound sugars and amino acids (monomers and polymers). For details of the extraction procedures see papers by Mopper (1973) and Degens and Mopper (1975).

structural sites for the coordination of calcium and magnesium ions (Ghiselin et al., 1967; King and Hare, 1972). The hydrolyzed EDTA extract [1] contains about half of the total acid-extractable amino acids. Hexosamines, which constitute 1—6% of plankton (dry weight), are dominantly EDTA-inextractable and are, therefore, probably present as a long-chained insoluble structural material such as chitin.

EFFECTS OF MICROBIAL DEGRADATION OF SEWAGE SLUDGE

Inasmuch as organic matter generally represents a small fraction of the total sediment (Bordovskiy, 1965), it is difficult to distinguish alterations that arise from biological degradation from those which may arise from non-biological diagenetic processes. In order to clearly discern between these mechanisms, therefore, a system in which biological degradation is principally involved was studied, i.e., sewage sludge treatment. Some of the relevant data for sugars and amino acids are presented in Tables 4-III, 4-IV and 4-VI. Details of these related results are presented by Mopper and Degens (1972), Mopper (1973) and Degens and Mopper, (1975).

Examination of the hydrolyzed EDTA extracts of primary and digested sludges (Tables 4-IV and 4-VI) clearly reveals that microbial degradation induces a large increase in the degree of metal association of the carbohydrates. This increase strongly suggests that metal-bound carbohydrates are biologically inaccessible, even though they include a large fraction of easily degradable monosaccharides (up to 15% of the "total"). Such inaccessibility could be simply due to the low solution levels of free sugars and amino acids, as shown by the unhydrolyzed water extracts (Table 4-IV), or could be due to inhibitory effects of the metal content of the bound carbohydrates.

A similar increase in metal association for the amino acid fraction is not observed (Table 4-III) suggesting that carbohydrates are preferentially consumed over proteins. In fact an actual increase in relative amino acid and organic nitrogen contents is observed after microbial digestion of sewage (Mopper and Degens, 1972). These trends can be understood in light of the fact that carbohydrates (especially cellulose) constitute the bulk of raw sewage sludge. Within sediments, biological consumption takes place over periods of tens to thousands of years instead of a few weeks as with sewage digestion. Thus, in sediment a significantly higher degree of protein consumption, with a concomitant increase in metal association, is expected.

Table 4-V compares heavy-metal analyses of EDTA and nitric acid extracts of sewage sludge. Metals constitute up to 10% of the total (dry weight); most of the metals appear to be nearly totally extracted by EDTA treatment. Thus, despite the high concentrations of potentially toxic heavy

TABLE 4-III

Percentages of amino acids and hexosamines extractable by EDTA-treatment of sediments and organic waste products

		ASP	THR	SER	GLU	PRO	GLY	ALA	VAL
Cariaco Trench	<10 m								
	a *	5	0	0	0	0	9	0	0
	b	13	13	9	7	3	14	4	6
	c	82	87	91	93	97	78	96	94
	>10 m								
	a	12	3	6	0	0	58	0	0
	b	52	8	23	12	7	76	8	20
	c	36	89	71	88	93	—	92	80
Black Sea	15 cm								
(marine)	a,b	31	12	14	7	2	13	7	5
	c	69	88	86	93	98	87	93	95
(interm.)	65 cm								
	a	14	0	4	0	nd.	3	0	1
	b	1	25	24	9	nd.	10	8	4
	c	85	75	72	91	nd.	87	92	95
(fresh)	125 cm								
	a,b	10	22	25	5	nd.	83	2	1
	c	90	78	75	95	nd.	17	98	99
Walvis Bay sediment surface	a	14	1	1	0	0	2	0	0
	b	16	17	14	14	16	15	13	16
	c	70	82	85	86	84	83	87	84
Lake Kivu	130 cm								
	a	17	0	0	0	0	2	0	0
	b	5	4	10	5	4	4	4	4
	c	78	96	90	95	96	94	96	96
	240 cm								
	a,b	6	17	28	10	3	35	6	7
	c	94	83	72	90	97	65	94	93
Argentine Basin									
	0–1 m								
	a	65	4	20	2	5	41	2	1
	b	1	25	26	30		88	11	10
	c	34	71	54	68	95	—	87	89
	3–5 m								
	a	71	5	12	1	5	105	29	11
	b	19	38	28	30		387		
	c	10	57	60	69	95	—		89

MET	ISO	LEU	TYR	PHE	LYS	HIS	ARG	TOTAL	HEX
0	0	0	0	0	3	0	0	1	0
2	3	2	2	2	8	0	0	6	2
98	97	98	98	98	89	100	100	93	98
0	0	0	0	0	4	0	0	6	0
2	2	1	0	0	82	0	0	26	0
98	98	99	100	100	14	100	100	68	100
0	5	5	2	4	14	0	0	10	10
100	95	95	98	96	86	100	100	90	90
0	0	0	0	nd.	0	0	0	2	0
6	4	2	14	nd.	13	0	15	10	10
94	96	98	86	nd.	87	100	95	88	90
0	0	0	0	0	234	0	0	21	2
100	100	100	100	100	—	100	100	79	98
0	1	0	0	0	11	70	0	4	2
0	6	4	2	3	1	0	16	11	6
100	93	96	98	97	88	30	84	85	92
0	0	0	0	0	0	0	0	1	0
1	3	2	6	5	8	14	2	4	8
99	97	98	94	95	92	86	98	95	92
0	4	4	1	1	16	8	5	12	3
100	96	96	99	99	84	92	95	88	97
0	0	0	nd.	nd.	192	nd.	nd.	18	0
15	9	6	nd.	nd.		nd.	nd.	36	8
85	91	94	nd.	nd.	—	nd.	nd.	46	92
33	12	12	nd.	nd.	12	nd.	nd.	21	0
			nd.	nd.	102	nd.	nd.	57	8
67	88	88	nd.	nd.	—	nd.	nd.	22	92

TABLE 4-III (continued)

		ASP	THR	SER	GLU	PRO	GLY	ALA	VAL
Oyster Pond plankton	a	61	13	12	3	14	13	14	9
	b	12	41	32	54	36	47	38	22
	c	27	46	56	43	50	40	48	69
Cow Manure	a	13	1	1	1	0	2	2	2
	b	11	16	16	13	9	15	15	12
	c	76	83	83	86	91	83	83	86
Deer Island sludge, primary	a	9	5	3	1	3	5	5	6
	b	20	14	22	23	22	26	21	20
	c	71	81	75	76	75	69	74	74
digested	a	11	4	4	2	2	4	7	8
	b	14	13	17	18	11	18	14	13
	c	75	83	79	80	87	78	79	79
Cranston sludge, digested	a	11	1	1	0	0	2	2	2
	b	8	18	12	9	13	19	14	21
	c	81	81	87	91	87	79	84	77

* a = (EDTA/total) × 100%; b = [(EDTA + HCl)/total] × 100% — a; c = 100% — (a + b) [non-EDTA-extractable residue]

metals, microbial degradation does not appear to be inhibited. This suggests that these metals are strongly bound to organic matter which in turn greatly reduces their chemical activity and hence their toxicity. Ian et al. (personal communication), for example, measured a stability constant of 6.8 (at pH 7) for complexes of zinc with a sewage sludge organic extract.

It is significant to note that along with an increase in metal association with biological degradation there is a large decrease in carbohydrate carbon relative to total organic carbon (see Fig. 4-18), which suggests that carbohydrates are utilized preferentially to other organic compounds, such as humic and fulvic substances. Alternatively, carbohydrates may become part of humic materials and are thus rendered non-extractable.

EARLY DIAGENESIS OF SUGARS AND AMINO ACIDS IN VARIOUS SEDIMENTS

In order to clarify diagenetic trends in different sedimentary environments, three "case histories" will be presented here in some detail. This is

MET	ISO	LEU	TYR	PHE	LYS	HIS	ARG	TOTAL	HEX
26	16	15	20	21	17	6	17	18	0
22	42	37	28	28	34	32	10	33	5
52	42	48	52	51	48	62	73	49	95
0	1	1	1	1	3	0	0	2	0
3	11	10	15	8	36	7	15	16	34
97	88	89	85	91	71	93	85	82	66
2	5	3	6	2	5	0	2	4	0
18	19	17	24	22	34	12	14	21	3
80	76	80	70	76	61	88	84	75	97
4	6	4	8	4	8	1	1	5	0
0	11	10	5	9	32	19	3	14	6
96	83	86	87	87	60	80	96	81	94
0	1	0	0	0	1	0	1	2	0
2	18	12	14	16	31	19	21	15	34
98	81	88	86	84	68	81	78	83	66

followed by a discussion of general trends which can be deduced from these and other studies of sediments (Mopper and Degens, 1972; Mopper, 1973).

Argentine Basin: oxic, deep-ocean basin

General description. Within the sediment cores examined, two distinct sedimentary units exist, the geological ages of which were determined by carbon-14 dating. The upper unit, approximately 15 cm in thickness, consists of a light red-brown, non-calcareous, unstratified clay. The top of this unit approximately represents the present sediment—water interface, because the gravity coring technique used does not severely disturb the top sediments. The organic carbon content of this unit is about 0.35%, which is the lowest value in the core as shown in Fig. 4-3. The base of the unit was dated at 2000—3000 years B.P.

The lower unit (15—500 cm in thickness) is, in appearance, a grey-green homogeneous non-calcareous clay. Occasional layers (1—2 mm thick) of sand and gravel are present, which probably owe their origin to either ice-rafting

TABLE 4-IV

Carbohydrate composition of organic waste materials (mole%)

	Sugars								total (μmoles/g)
	Rh	Fu	Ri	A	X	M	Ga	Gl	
Deer Island, Ma., sewage sludge, *primary*, HCl	3.1	1.0	0.4	6.9	26.5	25.3	13.6	23.3	276.7
Conc. H_2SO_4	—	—	—	0.7	7.7	8.6	0.8	82.2	695.8
EDTA + HCl	5.3	2.0	1.1	16.3	15.4	11.3	22.3	26.5	117.3
EDTA	4.0	3.8	3.0	48.3	15.5	3.3	12.3	10.0	40.0
H_2O	—	—	6.6	6.6	—	—	—	87.3	1.0
Deer Island, Ma., sewage sludge, *digested*, HCl	5.1	1.4	0.4	9.1	21.7	20.3	15.9	26.1	152.7
Conc. H_2SO_4	3.2	0.5	—	3.3	4.8	11.9	6.0	70.2	167.7
EDTA + HCl	9.2	3.1	1.2	16.9	16.6	13.2	22.9	16.9	128.5
EDTA	6.1	5.1	4.3	48.3	13.4	3.0	11.3	8.2	29.2
H_2O	—	—	25.5	52.9	—	21.6	—	—	0.5
N.Y. Bight sewage sludge, *disposal area*, HCl	9.1	8.1	1.6	8.6	12.9	15.1	21.5	23.1	18.6
EDTA + HCl	10.8	10.8	2.5	10.8	12.5	12.5	22.5	17.5	12.0
EDTA	6.7	14.4	9.2	21.5	21.5	3.5	17.8	5.4	4.04
N.Y. Bight, sewage sludge, Hudson Canyon, *shallow*, HCl	9.5	8.3	4.1	7.3	10.3	16.7	22.7	21.2	6.6
N.Y. Bight, Hudson Canyon, *deep*, HCl	10.0	8.0	2.5	7.3	10.8	15.1	23.7	22.6	9.3
EDTA + HCl	12.4	12.9	6.1	8.3	12.2	10.4	22.6	15.1	5.9

TABLE 4-V

Metal analyses of nitric acid digested and EDTA-extracted sewage sludges (μg/g dry wt.)

Metal			Nitric acid digestion	EDTA * extracted	Extractable with EDTA (%)
Lead	Cranston	(s) ***	365	345	95
	Deer Is.	(p)	868	975	112
	Deer Is.	(s)	722	750	104
Cadmium	Cranston	(s)	12	5	42
	Deer Is.	(p)	36	28	78
	Deer Is.	(s)	55	55	100
Nickel	Cranston	(s)	1490	480	32
	Deer Is.	(p)	153	115	75
	Deer Is.	(s)	197	170	86
Chromium	Cranston	(s)	280	103	37
	Deer Is.	(p)	895	900	101
	Deer Is.	(s)	1376	1675	123
Cobalt	Cranston	(s)	19	—	0
	Deer Is.	(p)	11	—	0
	Deer Is.	(s)	12	—	0
Zinc	Cranston	(s)	6704	4700	70
	Deer Is.	(p)	1747	1950	112
	Deer Is.	(s)	2537	3200	126
Copper	Cranston	(s)	2860	70	2
	Deer Is.	(p)	870	15	2
	Deer Is.	(s)	1200	15 **	1

* Analysis of the EDTA reagent revealed only trace quantities of heavy metals. A water extraction (8 h, 100°C) of the Deer Is. (s) sample released negligible amounts of heavy metals relative to the nitric acid digestion, e.g., 0.25% Cu, 0.70% Cd, 0.34% Zn, and 22% Ni.

** Extraction of the Deer Is. (s) sample with cysteic acid (a sulfur-containing amino acid) released 88 μg/g Cu or 7% of the nitric acid digestion, which is about seven times that released by EDTA. The low Cu yield of the EDTA extract is not clearly understood.

*** s = secondary treated sludge; p = primary treated sludge.

or turbidity currents. The organic carbon content is variable, but generally clusters between 0.6—0.7%. The age at the base of the core is approximately 90,000 years B.P.

In comparison to "normal" open-ocean sediments, which generally contain 0.2% organic carbon (Bordovskiy, 1965), Argentine Basin sediments average between 0.5 and 1.0%. This relatively high organic content is undoubtedly related to the high primary productivity in the surface water which, in turn, is related to upwelling in the area of the subtropical convergence near the edge of the continent (Ryther, 1963; Deacon, 1963).

The distinct temporal variations in organic carbon content as shown in

TABLE 4-VI

Percentages of sugars extractable by EDTA-treatment of sediments and organic waste products (relative to the HCl extract) *

		Sugars:								
		Rh	Fu	Ri	A	X	M	Ga	Gl	total
Cariaco Trench										
<10 m	a *	9	25	93	30	19	2	10	3	14
(6)	b	34	34	— ***	21	23	20	28	28	28
	c	37	41	7	49	58	78	62	69	58
Cariaco Trench										
>10 m	a	12	25	49	19	20	5	10	5	12
(3)	b	48	34	—	19	25	24	29	30	28
	c	40	41	51	62	55	71	61	64	60
Cariaco Trench, oxic sediments	a,b	71	80	74	57	68	46	52	59	60
	c	29	20	26	43	32	54	48	41	40
Black Sea (marine) 15 cm	a	9	23	43	26	15	1	11	1	12
	b	91	43	0	36	10	17	42	32	28
	c	0	34	57	38	75	82	47	67	60
Black Sea (transition) 65 cm	a	20	30	79	22	31	3	18	2	20
	b	50	35	—	20	30	45	29	48	35
	c	30	35	21	58	39	52	53	50	45
Black Sea (fresh) 125 cm	a	13	25	76	32	15	3	11	3	14
	b	71	103	89	41	61	56	44	45	54
	c	16	—	—	27	24	41	45	52	32
Walvis Bay, sediment surface	a	8	15	74	22	19	5	8	1	11
	b	33	18	—	12	28	48	27	29	28
	c	59	67	26	66	53	47	65	70	61
Lake Kivu 130 cm	a	19	38	53	51	32	2	23	1	14
	b	52	31	—	26	36	48	16	20	23
	c	29	31	47	23	31	50	61	79	63
Argentine Basin 5 m (2)	a	30	43	94	39	38	9	20	13	28
	b	52	37	—	14	30	44	50	39	36
	c	19	20	6	47	32	47	30	48	36

		Sugars:								
		Rh	Fu	Ri	A	X	M	Ga	Gl	total
N.Y. Bight sediment surface	a	16	39	123	54	36	5	18	5	22
	b	60	48	—	27	27	49	50	44	43
	c	24	13	—	19	37	46	32	51	35
Hudson Canyon, sediment surface	a,b	78	103	157	72	72	44	60	47	63
	c	22	—	—	28	28	56	40	58	37
Wood in Cariaco Trench sediment, 0.64 m	a	16	63	0	88	23	5	19	1	15
	b	26	7	100	—	—	6	9	0	1
	c **	58	30	0	12	77	89	72	99	84
Oyster Pond, sediment surface	a	10	24	63	34	15	2	8	2	11
	b	44	43	37	31	34	29	39	24	33
	c	46	33	0	35	51	69	53	74	56
Cow Manure	a	5	33	—	44	2	3	9	1	8
	b	53	—	—	25	26	11	57	27	28
	c	42	67	—	31	72	86	34	72	64
Dear Island sewage sludge, primary	a	19	52	120	102	8	2	13	1	6
	b	54	27	—	—	17	17	56	4	11
	c **	27	21	—	—	75	81	31	95	83
Dear Island sewage sludge, digested	a	23	71	194	101	12	3	14	2	17
	b	128	119	45	55	52	52	107	16	60
	c **	—	—	—	—	36	45	—	82	23
Cranston sewage sludge, primary	a	6	19	45	61	2	1	4	2	6
	b	70	51	36	22	6	22	67	46	41
	c	24	30	19	17	92	77	29	52	53
Cranston sewage sludge, digested	a	11	22	150	45	4	1	5	1	8
	b	97	87	0	67	40	52	65	53	60
	c	—	—	—	—	56	47	30	46	32

* a = (EDTA/total) × 100% = EDTA-extracted monomers as % of total; b = (EDTA + HCl)/total × 100% — a = EDTA-extracted polymers as % of total; c = 100% — (a + b) = EDTA-nonextracted residue as % of total.

** Treated with cold concentrated H_2SO_4 followed by dilution to 1.3 *M* for further hydrolysis.

*** = concentration is less than other extracts. ( ) = no. of samples represented.

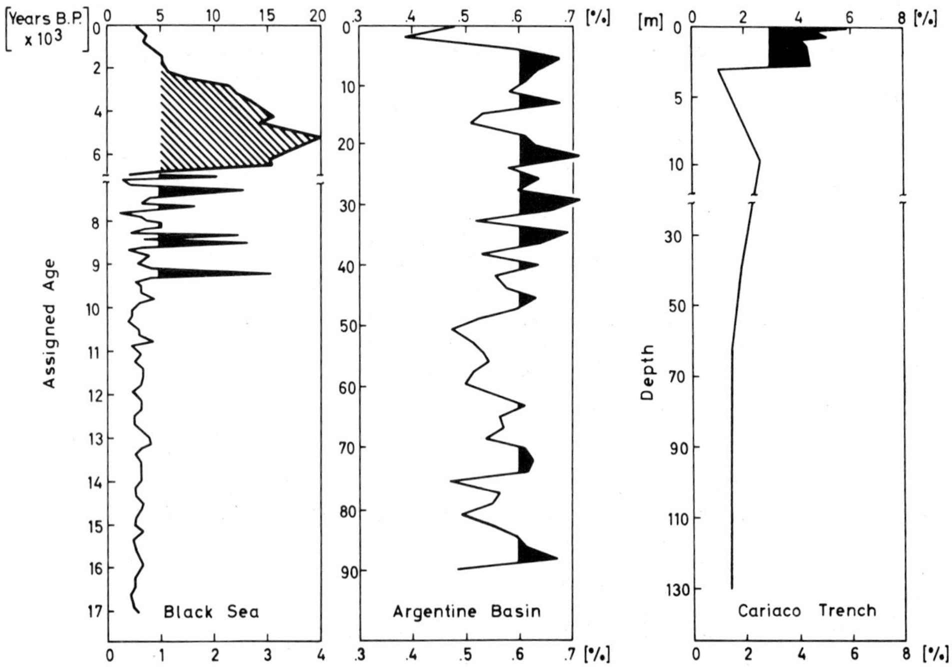

Fig. 4-3. Fluctuations of organic carbon in cores from three different sedimentary environments: (1) Black Sea—inland sea; (2) Argentine Basin—open ocean; and (3) Cariaco Trench—oceanic anoxic basin. The lower portion of the Black Sea curve (>6000 years old) is plotted on an expanded scale in order to enhance the fluctuations.

Fig. 4-3 for the Argentine Basin sediments do not appear exclusively correlative with sea-level fluctuations or glacial periods described by Stevenson and Cheng (1972) for this basin. The present writers attribute the organic carbon fluctuations to: (1) variations in surface productivity as a result of shifts in the subtropical convergence, and (2) variations in land run-off due to eustatic changes in sea level during glacial and interglacial periods.

Sugars in Argentine Basin sediments. The carbohydrate contents of seven sediment samples from the Argentine Basin ranging in age from 0 to 90,000 years B.P. are shown in Tables 4-VI and 4-VII. The compositions of various extracts are remarkably uniform for the time span sampled (Table 4-VII) although a diagenetic relationship between arabinose and xylose in the HCl extract may exist: at the top of the core arabinose > xylose, in the middle arabinose ~ xylose, and at the bottom arabinose < xylose. Furthermore, a striking inverse relationship between glycose and ribose exists as shown in

TABLE 4-VII

Carbohydrate composition of sediment from Argentine Basin (mole%)

Depth (cm)	Rh	Fu	Ri	A	X	M	Ga	Gl	Total (μmoles/g)
	HCl extract								
0	8.8	8.1	6.2	10.2	9.5	13.7	23.7	19.9	4.2
5	8.1	10.1	4.9	11.5	10.1	12.2	23.6	19.6	1.5
15	8.5	9.1	3.2	9.1	9.8	13.9	24.9	21.4	3.2
60	9.0	9.7	9.7	10.3	11.2	13.7	24.3	17.1	3.2
100	8.7	9.3	4.3	10.6	10.3	14.1	23.9	18.8	3.7
300	9.0	10.2	5.6	9.8	10.9	17.5	23.3	16.5	2.6
500	6.7	10.5	5.0	9.6	12.0	14.9	25.4	15.8	3.4
	Hydrolyzed EDTA extract								
0	10.1	9.5	8.9	8.9	13.1	12.5	24.3	12.8	3.4
5	8.3	11.9	6.7	10.1	11.0	11.9	23.8	16.5	1.1
15	10.7	12.0	5.4	8.7	10.7	12.0	25.2	15.3	2.4
60	11.6	12.6	6.3	8.7	12.6	11.6	23.2	13.5	2.1
100	10.4	11.6	2.7	8.4	11.6	11.6	26.9	16.5	2.5
300	10.2	12.9	3.9	8.1	9.7	14.5	25.3	16.7	1.9
500	10.2	13.2	6.3	8.3	12.2	12.2	26.8	10.7	2.1
	Unhydrolyzed EDTA extract								
100	7.9	15.0	15.0	15.0	16.0	4.4	18.0	8.3	1.0
500	9.5	16.7	15.6	13.5	14.6	5.2	16.7	8.6	1.0

Fig. 4-4. This relationship probably resulted from changes in terrigenous and marine organic inputs.

The relative abundance of sugars in the HCl extract of Argentine Basin sediments is: Ga > Gl > M > X > A > Rh > Fu > Ri; hexoses > pentoses (Table 4-VII). In the unhydrolyzed EDTA extract (EDTA-extracted monomers; ~30% of the "total") the reverse trend is observed: pentoses > hexoses. Fucose, ribose, arabinose, and xylose are significantly present as metal-bound monosaccharides (Table 4-VI, line a). In the hydrolyzed EDTA extract the order of abundance is Ga > Gl > M > X > Fu > Rh > A > Ri. No one sugar in any of the extracts is present in overwhelming abundance. EDTA extracts average 64% of the "total" which implies that the bulk of the sugars is present in the sediments as metal-bound monomers and polymers.

When the data on Argentine Basin sediments are compared to that of shallow-sea oxic sediments of the New York Bight and Hudson Canyon (Table 4-IV and 4-VI), many similarities are revealed. For example, the relative abundance of sugars in the extracts are almost identical. The degree of metal complexation is 60—70%; about 25% of the "total" is present as

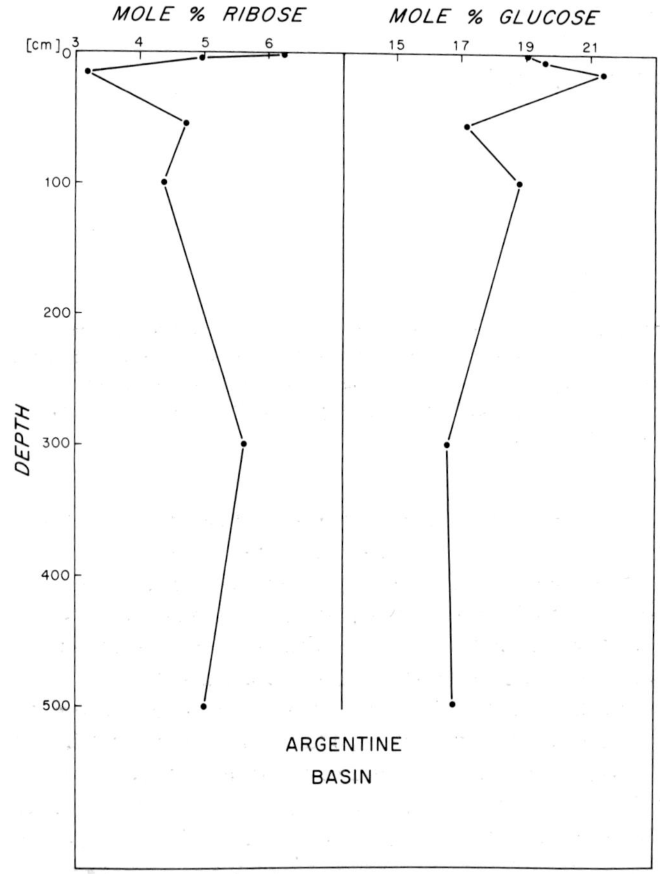

Fig. 4-4. Variation in ribose and glucose in Argentine Basin sediments with depth. The inverse relationship between contents of these sugars is noteworthy. The glucose to ribose ratio in sediments may reflect relative changes in the terrigenous and marine organic inputs.

metal-bound monosaccharides. Fucose, ribose, arabinose, and xylose are dominantly present as metal-bound monomers (Table 4-VI, line a). These similarities suggest that in deep- and shallow-sea oxygenated environments the end-products of biological degradation at the sediment—water interface are remarkably similar.

Fig. 4-5 indicates that the total sugar content relative to the total organic carbon content remains fairly uniform with increasing depth with the exception of the surface samples, which show an enrichment in sugar content. This enrichment may simply reflect time-dependent changes in benthic consump-

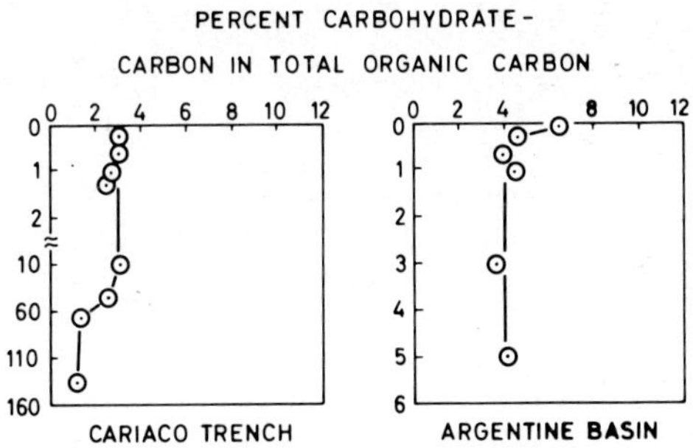

Fig. 4-5. Changes in carbohydrate carbon content relative to total organic carbon with depth. The change in the Cariaco Trench sediments at 40—60 cm is probably due to a change in the composition of the organic input. The reason for the change in the Argentine Basin surface sediments is probably related to biological degradation of carbohydrates in the surface sediments.

tion and sedimentation rates. With an increase in sedimentation rate carbohydrates may be buried at a faster rate than the biological consumption rate at the sediment—water interface. Alternatively, the surface enrichment in sugars may reflect the presence of breakdown products of readily consumable carbohydrates plus contributions from the organisms present alive in the top sediment; the residue, which is presumably inedible, is buried.

The results of sewage sludge analyses strongly suggest that metal-bound organic matter is resistant to the residue of biological degradation. Along with an increase in metal complexation with degradation, a decrease in sugar carbon content (relative to total organic carbon) is observed. Thus, if either of the explanations mentioned in the previous paragraph is true, then one would expect to observe in the Argentine Basin sediments an inverse relationship between the degree of carbohydrate complexation and the percent of sugar carbon in the total organic carbon content. That is, as the content of sugar carbon decreases with depth (Fig. 4-5), an increase in metal association is expected.

Fig. 4-6(a) shows the temporal change of the hydrolyzed EDTA extraction (as percent of the "total"); the average value is 64%. In contrast to what was expected, there is an actual decrease with depth of EDTA-extracted sugars. The possible explanations for this unexpected decrease include the following:

(1) The assumption that an inverse relationship exists between sugar carbon content and EDTA-extractable sugars only applies to recent degradation in surface sediments.

(2) The enrichment in sugar carbon and EDTA-extractable sugars at the surface may reflect a recent change in nature of the organic input.

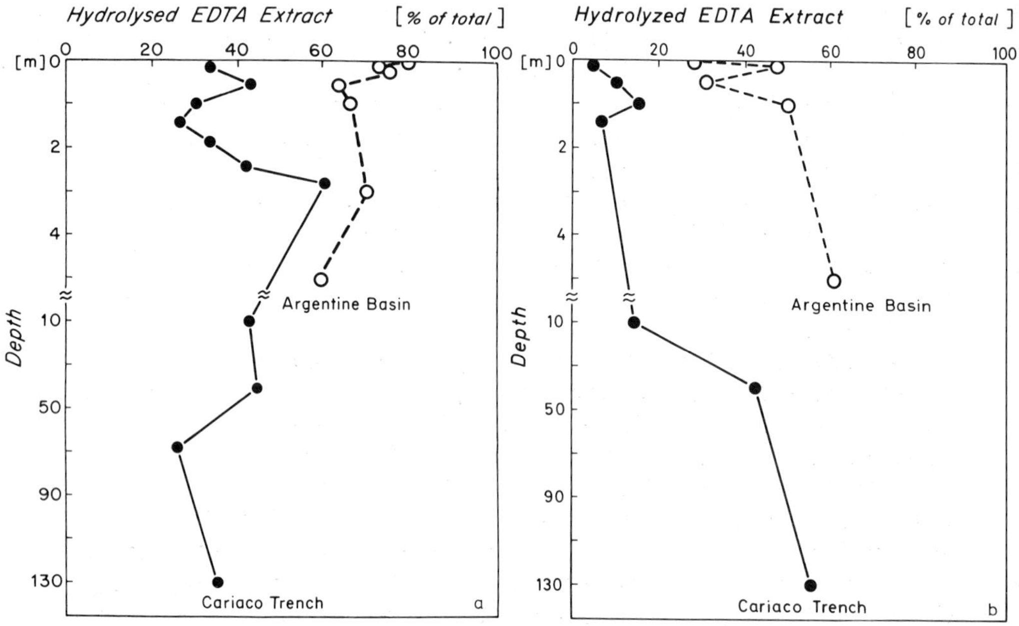

Fig. 4-6. Comparison of the temporal changes in the degree of metal association of carbohydrates (a) and amino acids (b) in Cariaco Trench (reducing) and Argentine Basin (oxidizing) sediments. The peak at a depth of 2.65 m in the Cariaco Trench carbohydrate data (a) represents a band of oxic sediment. The degree of complexation of carbohydrates in these sediments appears to be related to the Eh at the time of deposition. The degree of metal association of amino acids (b) in the reducing Cariaco Trench sediments temporally approaches that of recent oxic sediments.

(3) A diagenetic decrease in metal complexation with depth has occurred: metal ions may have been extracted from organic complexes by diagenetic incorporation into clays.

(4) The diagenetic formation of H_2O and EDTA-insoluble organic condensation products has occurred; thus, although the degree of association may not have decreased, the degree of intermolecular cross-linking among organic molecules may have increased. Inasmuch as EDTA is a weak acid, it is probably not capable of hydrolyzing these cross-linkages.

At present, the available data is insufficient to resolve this problem. Based on the experiments on sewage sludge (Mopper and Degens, 1972) where chelators stronger than EDTA (e.g., sulfur-containing amino acids) substantially increased the yield in metals bound to organic matter, the present writers tentatively suggest that enhancement of intermolecular cross-linking is the prime factor.

Amino acids in Argentine Basin sediments. The contents of amino acids and hexosamines in sediments of the Argentine Basin are shown in Tables 4-III and 4-VIII. In comparison to the sugar contents (Table 4-VI), the amino acid compositions of the various extracts are generally less uniform with depth. A few diagenetic relationships, however, may exist in the case of HCl extract: alanine and valine generally increase in relative abundance, whereas lysine generally decreases. Fig. 4-7 shows temporal changes in amino acid subgroups within the HCl (or "total") extract. Fluctuations in the alkyl amino acids correspond inversely to fluctuations in acidic and alcoholic amino acids; the contents of basic and cyclic amino acids also appear to show an inverse relationship.

The relative abundance of the more significant amino acids within the HCl extract (Table 4-VIII) is as follows: Gly > Ala ~ Asp ~ Val > Ser ~ Lys > Thr > Glu > Leu > Iso > Pro > Arg, i.e.: (alkyl > acidic > alcoholic > basic > cyclic > S-containing) [1]. In the hydrolyzed EDTA extract the order is: Gly > Lys > Asp > Ser > Thr > Ala > Glu > Val, i.e.: (alkyl > basic > acidic > alcoholic > S-containing > cyclic). In the unhydrolyzed EDTA extract the order is: Asp > Gly > Ser > Lys > Thr (acidic > alkyl > alcholic > basic).

As with the sugars, an average of 60% of the "total" amino acids are released with EDTA (Table 4-III) and, hence, in some fashion are metal-bound. About 20% of the "total" amino acids are present as metal-associated monomeric amino acids. Aspartic acid is dominantly present in the sediment in a metal-bound monomeric state (Table 4-IV, line a), whereas glycine and lysine are dominantly present in the metal-bound polymeric state (Table 4-VI, line b). In terms of relative abundance, aspartic acid, glycine, and lysine are also the dominant metal-bound amino acids (Table 4-VIII).

Similar to the result on the sugar contents, the total amino acid content relative to the total organic carbon content is higher in the surface sediments than in the older sediments. In contrast with the results on sugar content, however, the degree of metal-binding is lower in the surface sediments than in the older sediments (Fig. 4-6b). Thus, an inverse relationship exists between the percent amino acid carbon content in the total organic carbon and the percent of metal association in the Argentine Basin sediments. Inasmuch as the sugars do not show this inverse relation, they apparently follow a different diagenetic pathway.

Amino sugar concentrations are unusually high in all Argentine Basin samples and exceed amino acid concentrations by a factor of 2—5. The dom-

---

[1] *Alkyl* = glycine (Gly) + alanine (Ala) + valine (Val) + isoleucine (Iso) + leucine (Leu); *acidic* = aspartic acid (Asp) + glutamic acid (Glu); *alcoholic* = serine (Ser) + threonine (Thr); *basic* = lysine (Lys) + histidine (His) + arginine (Arg); *S-containing* = cystine (Cys) + methionine (Met); *cyclic* = proline (Pro) + tyrosine (Tyr) + phenylalanine (Phe).

TABLE 4-VIII

Distribution of amino acids and hexosamines in Argentine Basin sediments (residues/1000)

Amino acids depth (cm):	0			5		
	EDTA	EDTA + HCl	Total	EDTA	EDTA + HCl	Total
Aspartic acid	484	166	152	543	124	52
Threonine	23	32	54		8	71
Serine	147	57	55		10	146
Glutamic acid		48	57		15	43
Proline			29			16
Glycine	327	370	166	457	251	170
Alanine	18	38	151		7	147
Cystine			5			2
Valine		47	136		9	148
Methionine		8	21			3
Isoleucine		8	21			25
Leucine		16	31			37
Tyrosine			3			13
Phenylalanine			10			25
Lysine		211	84		576	70
Histidine			1			1
Arginine			26			32
Total amino acids (μg/g dry wt.)	43	129	451	32	205	440
Glucosamine (μg/g dry wt.)		} 52	693		} 43	46
Galactosamine (μg/g)			230			17
Ammonia (μg/g dry wt.)	39	16	36	29	68	35
Recovered N%			35			
C%			0.313			
N%			0.048			
H_2O%			44			
C/N			6.5			
AA/HA ratio		2.4	0.49		4.8	0.70

inance of hexosamine material is probably due to its resistant nature. Hexosamine material is dominantly unextracted by EDTA (Table 4-III) indicating that it is present as a high-molecular-weight water and EDTA-insoluble substance such as chitin.

Deep-sea sediments, such as in the Argentine Basin, contain more hexosamine material (relative to amino acids) than shallow-sea sediments (Degens, 1970), thus indicating that water depth plays a significant role. The low sedimentation rates and paucity of edible organic matter in deep oceanic areas apparently lead to more efficient utilization of the organic input which, in turn, leads to a preferential concentration of resistant hexosamines.

ABLE 4-VIII (continued)

			100			500		
DTA	EDTA + HCl	Total	EDTA	EDTA + HCl	Total	EDTA	EDTA + HCl	Total
9	186	143	511	260	152	279	107	123
3	32	60	11	76	80	37	25	66
72	68	67	44	107	67	56	28	83
	62	66	22	31	62		18	56
	tr.	20			58			14
7	539	183	252	437	103	520	576	89
	50	100	24	49	112		26	85
		1			5			3
	6	100	18	22	112		9	78
	9	22			12		10	33
	6	32	2	2	53		7	46
	9	48	3	7	61		10	72
		6			8			8
		17			23			23
	34	97	112	8	67	108	184	141
		1			6			1
		37			20			80
0	83	265	55	75	150	39	196	318
	}27	346		}26	}73		}91	600
		115						362
3	28	36	42	47	34	27	88	41
		14			7			35
		0.580			0.606			0.483
		0.075			0.073			0.044
		45			45			45
		7.7			8.3			11
	3.1	0.57		2.9	2.1		2.2	0.33

Cariaco Trench: anoxic oceanic trench

General description. The Cariaco Trench is a marine anoxic basin off the coast of Venezuela (Richards, 1965). Surface sediments (0—3 m) were sampled with a piston coring device and deeply buried sediment cores were obtained from JOIDES cores from drilling site 147.

Three sedimentary units within the piston cores were identified. The top unit (~120 cm long) is a dark green-grey, H_2S-rich, laminated sediment. Wood was encountered at a sediment depth of about 60—100 cm, indicating that a log buried in the sediment was penetrated. The age of the wood as determined by C-14 methods is 5400 years B.P., indicating an estimated

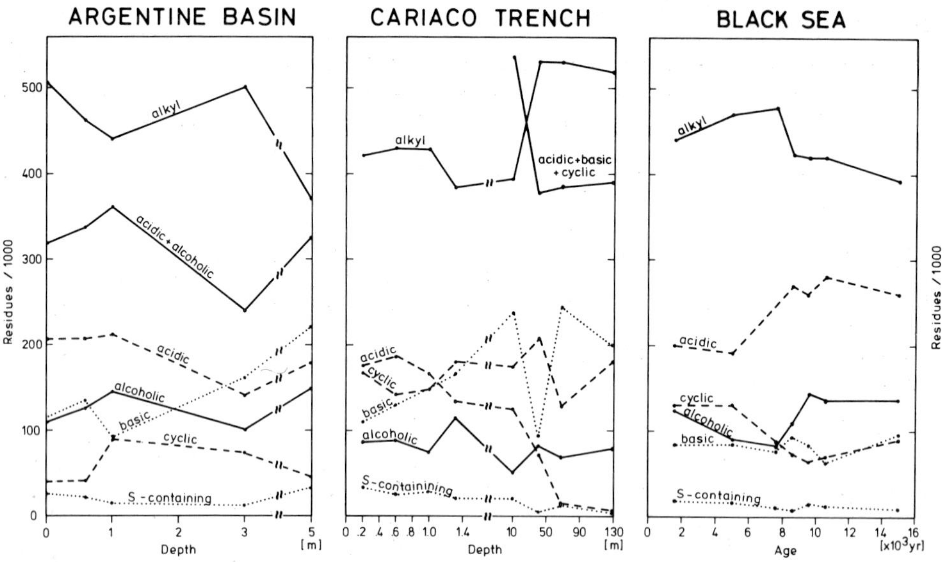

Fig. 4-7. Depth distribution patterns of amino acid subgroups in sediments of different marine environments. Alkyl and acidic groups are dominant in most sediments, whereas S-containing amino acids are always least abundant.

sedimentation rate of 10—20 cm/1000 years. Excluding the wood, the average organic carbon content of this unit is 5% and the $CaCO_3$ content is about 35%.

The second unit (~100 cm long) is also laminated and H_2S-rich; however, generally it has a lighter green-grey color. The average organic carbon content of this unit is 4% and the $CaCO_3$ content is about 15%.

A third unit consists of oxidized sediments as indicated by metal analysis (Price, personal communication). This unit has a uniformly light grey color with no apparent laminations. The organic carbon content is 1—2% and the $CaCO_3$ content is ~25%.

A detailed description of the JOIDES cores is given in the Initial Report No. 15 (1973), and the total depth of sediment penetration was 189 m. With the exception of a few narrow oxic bands, the sediments were almost entirely deposited under reducing conditions.

Sugars in Cariaco Trench sediments. The carbohydrate contents of twelve sediment samples from the reducing zone of the Cariaco Trench are listed in Tables 4-VI and 4-IX. Excluding the wood sample, which will be discussed below, the carbohydrate compositions for all three extracts (HCl, EDTA + HCl, and EDTA) are remarkably uniform down to 40 m even though the

TABLE 4-IX

Carbohydrate composition of Cariaco Trench sediments (mole%)

Depth, m	Sugar: Rh	Fu	Ri	A	X	M	Ga	Gl	Total (μmoles/g)
A. HCl									
0.15—0.20	11.0	8.0	4.2	5.9	10.1	19.4	27.0	13.9	23.7
0.60—0.64	10.9	12.4	3.1	4.9	9.8	18.1	27.9	12.4	19.3
0.64—0.68 (Wood-HCl)	6.5	1.8	tr.	15.1	23.1	4.4	13.7	35.4	258.9
0.64—0.68 (Wood-conc. H_2SO_4)	3.9	0.73	tr.	7.0	12.5	2.6	7.6	65.7	422.2
1.0—1.05	10.3	8.2	4.1	5.2	10.3	17.9	29.3	14.6	18.4
1.30—1.35	11.8	8.9	3.5	5.1	6.1	20.0	28.9	15.6	13.5
1.75—1.80	12.0	9.5	3.2	4.9	10.2	21.9	26.4	12.1	9.1
10	11.0	8.8	2.9	5.7	11.0	20.0	27.0	14.0	10.0
40	10.1	8.5	3.7	7.0	10.4	17.9	29.8	12.4	6.7
67	8.9	4.4	0.5	10.0	12.2	25.6	22.2	17.8	2.7
130	8.9	5.1	0.8	9.8	12.3	22.1	24.3	16.6	2.4
B. EDTA + HCl									
0.15—0.20	12.2	14.6	7.4	7.3	9.6	10.9	28.0	10.1	8.2
0.60—0.64	12.0	13.3	7.7	7.3	11.2	10.7	26.5	11.8	8.3
0.64—0.68 (Wood)	10.8	4.9	0.6	44.9	18.8	2.0	15.1	2.9	65.5
1.0—1.05	13.3	16.9	3.3	7.1	10.4	12.2	29.1	8.0	5.5
1.30—1.35	13.6	16.1	4.2	7.5	6.9	7.2	27.8	16.9	3.6
1.75—1.80	12.7	14.1	5.5	7.0	12.0	12.1	25.0	12.5	5.6
10	11.4	13.7	5.6	7.4	12.8	11.4	27.9	10.9	4.3
40	15.2	11.9	2.4	8.1	11.6	12.6	28.1	11.6	3.1
67	11.6	5.1	1.8	6.4	15.1	21.9	23.3	15.1	0.73
130	13.6	8.3	1.5	8.3	14.8	15.9	21.6	15.9	0.88
C. EDTA									
0.15—0.20	6.4	15.8	30.3	12.4	12.1	2.7	19.4	0.7	3.3
0.60—0.64	7.0	20.0	28.3	10.4	14.3	1.4	17.4	1.0	2.3
0.64—0.68 (Wood-EDTA ext.)	4.3	4.6	—	54.9	21.7	0.9	11.0	2.5	63.0
0.64—0.68 (Wood-H_2O ext.)	–	—	—	—	—	—	—	—	0
1.0—1.05	nd.	nd.	nd.	nd.	nd.	nd.	nd.	nd.	nd.
1.30—1.35	7.9	13.6	16.4	11.4	9.3	2.4	27.9	8.6	1.4
1.75—1.80	nd.	nd.	nd.	nd.	nd.	nd.	nd.	nd.	nd.
10	8.5	18.3	17.6	13.1	16.3	4.4	24.8	5.1	1.5
40	10.0	15.4	13.2	14.3	14.3	5.5	24.2	3.1	0.91
67	10.9	13.6	3.2	9.1	21.4	7.7	20.0	15.0	0.22
130	7.7	10.0	4.2	10.6	28.1	14.2	19.7	7.1	0.31

organic carbon content of the sediment drops from 6% to 1.9%. Figs. 4-5 and 4-8 indicate that the total sugar concentration correlates well with the drop in organic carbon content. These results suggest that the effects of diagenesis down to a depth of 40 m is either negligible (variations in the total carbon being due to variations of organic input rates) or remarkably nonselective.

The break in the sugar composition between 40 m and 67 m (Table 4-IX) along with the drop in sugar content relative to total organic carbon below 40 m (Figs. 4-5 and 4-8) may be explained by one of the following causes: (1) a sudden diagenetic change; (2) a change in the composition of the organic input; and (3) a change in the sediment-surface microbial population. The inverse correlation between the glucose and ribose as shown in Fig. 4-9 strongly suggests that a change in the composition of the organic input occurred below a depth of 40 m. These sediments appear to have a larger terrigenous input as evidenced by the increased glucose and decreased ribose contents.

The relative abundance of the sugars in the HCl extract at all depths down to 40 m is invariant: Ga > M > Gl > X ~ Rh > Fu > A > Ri. The order of abundance is similar to that of the Argentine Basin sediments from which it differs only in the inversion of mannose and glucose and the lower abundance of arabinose. The order of abundance is less similar to that of the plankton sample immediately above the trench (Table 4-I). The orders of abundance within the unhydrolyzed and hydrolyzed EDTA extracts also more closely resemble that of the Argentine Basin sediments than the Cariaco plankton. Fucose, ribose, arabinose, xylose, and galactose are the dominant metal-bound monomers (Table 4-XI). Ribose is dominantly in the metal-bound monomeric state (Table 4-VI, line a).

The relative abundance of sugars in the HCl extract below 40 m is somewhat different from that of the overlying sediment: M $\gtrsim$ Ga > Gl > X > A >

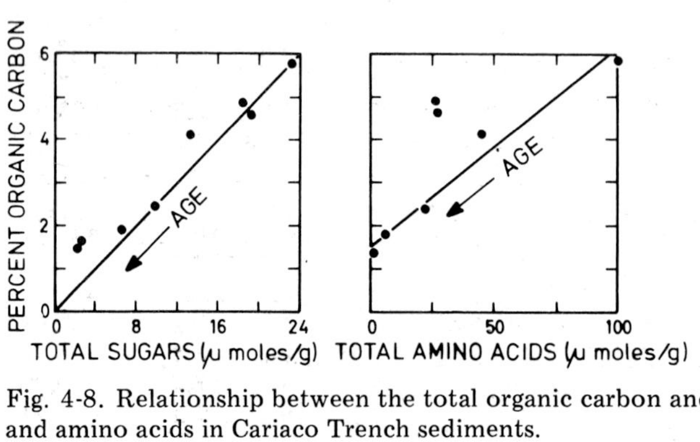

Fig. 4-8. Relationship between the total organic carbon and total extractable (HCl) sugars and amino acids in Cariaco Trench sediments.

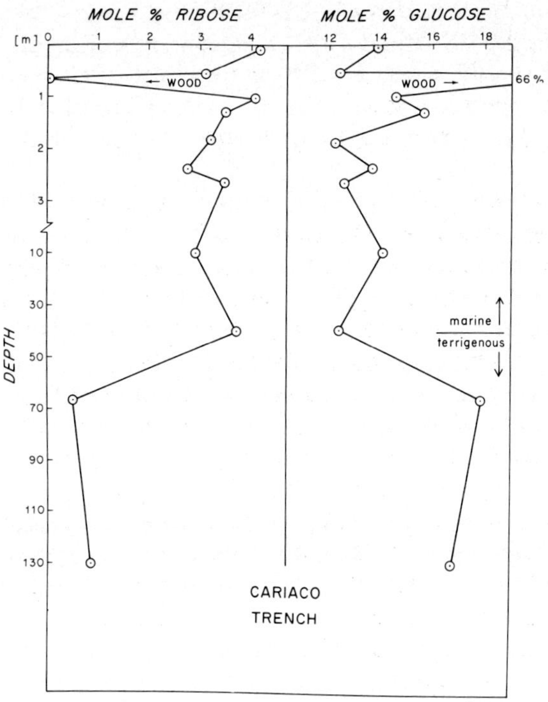

Fig. 4-9. Ribose and glucose variations with depth in Cariaco Trench sediments. The inverse relationship between contents of these sugars is striking. Contents of these sugars in the lower sediments (> 40 m) probably reflect an increased terrigenous organic input.

Rh > Fu > Ri, which again suggests a change in the source material and/or a change in the sediment-surface microbial population.

The 2.65-m sample represents sediment deposited under oxidizing conditions. Relative abundance of sugars within the HCl extract and the hydrolyzed EDTA extract (Table 4-IX) is nearly identical to those of the overlying and underlying reducing sediments. If it is assumed that the carbohydrate composition of the organic input during this interval (reducing—oxidizing—reducing) remained fairly constant, then these results imply that a change in the sedimentary environment from reducing to oxidizing, with the concomitant change in micro- and macrofauna, has little or no effect upon the composition of the sedimented carbohydrates.

The main difference in the carbohydrate extracts between the oxic and reducing zones lies in their degree of metal-complexation. Fig. 4-6 shows that for the reducing sediments the hydrolyzed EDTA extraction (metal-bound monomers and polymers) releases about 40% of the "total"; whereas for the oxidizing band at 2.65 m, the hydrolyzed EDTA extraction releases about 60%, a value which is similar to that of other oxidizing sediments, such as those of the Argentine Basin (Table 4-VI). These results strongly suggest that a correlation exists between the degree of metal-association of sedimentary carbohydrates and the Eh at the sediment—water interface at the time of deposition.

The wood sample is over 5000 years old; however, it has been remarkably well preserved and structural details such as intact cell walls still exist. This structure is actually very fragile and is easily destroyed in the dried specimen, however, indicating that some in situ hydrolysis did occur.

Approximately 14% of the "total" is released by the unhydrolyzed EDTA extraction; hydrolysis of this extract released an additional 2% of the "total" (Table 4-VI). Arabinose is the dominant metal-complexed monomer (Table 4-IX). Although 16% of the wood carbohydrates can be brought into solution by removal of the metals by EDTA, no equivalent process appears to have occurred in situ. The carbohydrate composition of the surrounding sediment [1] is completely unaffected by the presence of this carbohydrate-rich wood (Table 4-IX). These observations indicate that metal complexation of soluble carbohydrates in the wood fixes that fraction in situ, thereby inhibiting its diffusion. This conclusion undercuts the previous belief (Prashnowsky et al., 1961; Degens, 1965) that chromatographic separation of carbohydrates and other organic compounds within a sediment column is a significant diagenetic process.

Amino acids in Cariaco Trench sediments. The amino acid and hexosamine data for nine Cariaco Trench core samples are presented in Tables 4-III, X, XI. Fig. 4-8 shows that the amino acid yields roughly correlate with the decrease in organic carbon content with depth. Table 4-X shows that the C/N ratio stays within the narrow range of 9.7—11.4 with no obvious depth trends.

The sugar results indicated the presence of a major break between a depth of 40 m and 67 m. Examination of the amino acid data (Table 4-X) does not indicate this as clearly. For example, methionine, valine, serine, isoleucine, leucine, proline, ornithine and lysine show a depth break between a depth of 10 m and 40 m, whereas tyrosine, ornithine, and lysine show a break between 40 m and 67 m. Thus the amino acids either follow a different diagenetic pathway than the sugars or reflect source materials of a more variable nature.

[1] The wood actually extended from a depth of 0.64 to 1.0 m and presumably had a horizontal extension greater than the core diameter (~8 cm). The over- and underlying sediment samples approached the wood within a centimeter.

TABLE 4-X

Distribution of amino acids and hexosamines in Cariaco Trench sediments *

Amino acids	Depth (m) 0.15—0.20 (total)	0.60—0.64 (total)	0.64 (wood)	1.00—1.05 (total)	1.30—1.35 (total)	10 (total)	40 (total)	67 (total)	130 (total)
Aspartic acid	105	105	97	92	75	55	31	8	4
Threonine	38	38	55	32	18	27	33	24	25
Serine	47	50	79	44	96	26	50	45	53
Glutamic acid	72	81	67	74	58	65	43	5	2
Proline	71	73	90	69	61	63	91	41	21
Glycine	130	167	93	136	116	142	125	156	135
Alanine	132	111	86	94	94	95	123	140	111
Cystine	7	4	4	8	2	3	1	5	1
Valine	54	47	69	62	48	50	87	81	89
Methionine	28	21	6	20	19	18	5	6	3
Isoleucine	33	32	40	48	38	30	66	57	67
Leucine	74	76	87	92	88	77	135	101	117
Tyrosine	32	17	20	20	41	27	19	11	9
Phenylalanine	66	51	51	60	79	84	98	76	152
Ornithine	—	—	—	—	—	48	16	43	47
Lysine	80	75	98	87	102	146	41	174	140
Histidine	7	11	14	11	5	13	3	12	17
Arginine	24	43	44	51	60	32	34	16	6
Total amino acids (mg/g dry weight)	12.48	3.23	9.37	3.22	5.51	2.93	0.36	0.15	0.16
Glucosamine (mg/g)	7.83	1.75	4.08	1.93	0.88	4.02	0.022	0.058	0.127
Galactosamine (mg/g)	0.96		0.03				0.009	0.026	0.076
Ammonia (mg/g)	0.21	0.21	0.23	0.21	nd.	0.12	0.011	0.016	0.009
Total N (mg/g)	2.42	0.77	1.83	0.79	nd.	0.84	0.06	0.04	0.046
H_2O (%)	61.4	58.7	86.7	59.2	83.8	65.5	41.9	28.4	29.4
Organic C(%)	5.83	4.59	36.12	4.89	4.11	2.45	1.94	1.61	1.46
Organic N(%)	0.57	0.45	0.27	0.48	0.36	0.25	0.17	0.17	0.14
C/N ratio	10.2	10.2	134	10.2	11.4	9.9	11.2	9.7	10.2
$CaCO_3$ (%)	41.1	35.6	0.0	23.2	2.4	32.2	10.6	33.8	21.0

* Residues/1000.

TABLE 4-XI

Surface and deep sediment amino acid distribution in Cariaco Trench sediments (residues/1000)

	EDTA		EDTA + HCl		Total	
Amino acids	0—10 m	40—130 m	0—10 m	40—130 m	0—10 m	40—130 m
Aspartic acid	196	122	183	68	86	14
Threonine	—	18	36	8	31	17
Serine	5	44	60	38	53	58
Glutamic acid	—	—	50	4	70	17
Proline	—	—	24	15	67	51
Glycine	712	658	349	620	138	139
Alanine	—	—	49	30	105	125
Cystine	—	—	3	—	5	2
Valine	—	—	40	42	52	86
Methionine	—	—	4	—	21	5
Isoleucine	—	—	16	—	36	63
Leucine	—	—	15	—	81	118
Tyrosine	—	—	5	—	27	13
Phenylalanine	—	—	9	—	68	109
Ornithine	—	—	—	—	10	35
Lysine	88	159	168	176	98	118
Histidine	—	—	—	—	9	11
Arginine	—	—	—	—	42	19

To distinguish between early and advanced stages of diagenesis, the amino acid data are divided into two groups, i.e., 0—10 m and 40—130 m, as shown in Table 4-XI. In the HCl extract the order of abundance in the upper sediment (0—10 m) is: Gly > Ala > Lys > Asp $\gtrsim$ Leu > Glu $\gtrsim$ Phe ~ Pro > Ser ~ Val > Arg > Iso (alkyl > acidic ~ basic ~ cyclic > alcoholic > S-containing). It is of interest to note that glycine, alanine and aspartic acid are also dominant within the HCl extract of oxic Argentine Basin sediments. In the deeper Cariaco Trench sediments (40—130 m), the order of abundance is: Gly > Ala > Lys ~ Leu > Phe > Val > Iso > Ser > Pro > Thr > Arg > Glu > Asp, i.e.: (alkyl > basic $\gtrsim$ cyclic > alcoholic > acidic > S-containing). Thus, within the HCl extract (see Table 4-X and Fig. 4-7) the following trends can be discerned:

(1) the contents of acidic amino acids decrease from 15 to 20% of the "total" in modern sediments to about 1% in the oldest sediments;

(2) S-containing amino acids show a gradual decrease in time, whereas basic amino acids, in particular lysine, show substantial relative increases with time;

(3) within the upper sediments (<1.3 m) an inverse relationship appears to exist between acidic and cyclic amino acids and alkyl and alcoholic amino acids; and

(4) ammonia levels are reduced in the oldest samples by a factor of 10—20 relative to modern sediments.

The decrease in ammonia content with increasing geologic age may be due to (1) loss to interstitial solutions and eventual migration to the marine environment, and (2) tight fixation of ammonia to clay minerals. The inverse relationship (see item 3 above) in the younger sediments (Fig. 4-7) does not exist in the older sediments. The older sediments are dominated by a diagenetic decrease in acidic amino acids. This decrease is largely compensated for by a relative increase in alkyl amino acids. Fig. 4-7 shows that the summed fluctuations of the sum of acidic + basic + cyclic amino acids almost exactly balances the fluctuations of the alkyl amino acids in the older strata.

The order of abundance in the hydrolyzed EDTA extract (metal-associated polymers and monomers) within the upper sediment (0—10 m) is: Gly > Asp > Lys > Ser > Glu ~ Ala > Thr > Val > Pro > Iso ≳ Leu > Phe (alkyl > acidic > basic > alcoholic > cyclic > S-containing). In the lower sediment (40—130 m), the order is: Gly > Lys > Asp > Val ≳ Ser > Ala > Pro > Thr ≳ Glu, i.e.: (alkyl > basic > acidic > alcoholic > cyclic > S-containing). As in the case of the oxidizing Argentine Basin sediment samples, glycine, aspartic acid, and lysine are the dominant metal-associated species. In contrast to the modern Argentine Basin sediments, the amino acids in the upper Cariaco Trench sediments are dominantly unassociated with metals. For example, the hydrolyzed EDTA extraction (metal-bound monomers + polymers) releases less than 10% of the "total" from the Cariaco Trench sediments, whereas about 60% is released from the Argentine Basin sediments (Table 4-III).

In the deeper Cariaco sediments the degree of metal association of the amino acids greatly increases. This increase is illustrated in Fig. 4-6b, which shows that the oldest sediments are not distinguished in degree of metal association from modern oxidizing sediments. Three amino acids, notably aspartic acid, glycine and lysine, are mainly responsible for the trend towards progressive metal association (Table 4-III). In fact, metal binding becomes so strong that extraction with EDTA increases their absolute yields by about a factor of 10 relative to 6 *N* HCl hydrolysis, the "total" extract (Mopper and Degens, 1972).

In contrast, the degree of metal association of the carbohydrates as shown in Fig. 4-6a remains relatively constant with depth (with the exception of the oxic band) indicating again that sugars and amino acids follow different diagenetic pathways.

In the unhydrolyzed EDTA extract the order of abundance for the

0—10 m sediment is: Gly > Asp > Lys > Ser (alkyl > acidic > basic > alcoholic). For 40—130 m sediment, the order is: Gly > Lys > Asp > Ser > Thr (alkyl > basic > acidic > alcoholic). Interestingly, these are also the most abundant metal-bound monomeric amino acids present in the Argentine Basin sediments (Table 4-VIII).

Black Sea: restricted inland basin

General description. The environment at the sediment—water interface in the Black Sea has changed several times from oxic fresh-water to anoxic marine during the last 25,000 years (Degens and Ross, 1974). The frequency of marine spills and fresh-water dominance is recorded by the fossils assemblages, whereas the redox conditions at the sediment—water interface are reflected in the mineralogy of the sediments. The organic carbon content also mirrors these changes as shown in Fig. 4-3.

Degens (1971) proposed that between 17,000 and 9300 B.P. the Black Sea constituted an oxygenated fresh-water body for its entire depth. The sediments deposited during this interval consisted of alternating light and dark lutite bands with an organic carbon content of about 1%. Around 9300 years B.P., the first invasion of Mediterranean waters produced reducing conditions at the sediment—water interface for about 200 years. The frequency of saline spills, with concurrent reducing conditions at the sea bottom, over the next 200 years was reflected in the sedimentary organic carbon fluctuations (Fig. 4-3). Finally, at 7300 years B.P. the influx of Mediterranean water became so pronounced that saline (brackish) and reducing conditions were permanently established at the sediment—water interface. From 7300 to about 3000 years B.P., the organic content of the sediments was extremely high, as much as 40% on a dry-weight basis. This sediment has a dark brown, sapropelic appearance. From 3000 years B.P. to present the sediment is characterized by alternating thin sapropelic and coccolith layers.

Sugars in Black Sea sediments. The carbohydrate contents of three samples, representing the three major sedimentary units of the Black Sea, are presenred in Tables 4-VI and 4-XII. The deepest sample (120—130 cm, B.S. fresh) represents the fresh-water, oxic period of the Black Sea. The order of abundance in the HCl extract (Table 4-XII) is: Ga > Gl > M > Rh > X ~ A > Fu > Ri. This order is similar to that of the Argentine Basin and Cariaco Trench sediments. Table 4-VI indicates that about 65% of the "total" sugars are metal-complexed (EDTA + HCl), a value which compares well with sediments from other oxic environments (e.g., Argentine Basin, and Hudson Canyon, U.S.A.); 14% of the "total" sugars is extracted as metal-bound

TABLE 4-XII

Carbohydrate composition of Black Sea sediment (mole %)

Depth	Rh	Fu	Ri	A	X	M	Ga	Gl	Total (μ moles/g)
15 cm (HCl)	4.8	7.6	9.6	4.6	16.3	23.7	20.0	13.3	45.9
15 cm (EDTA + HCl)	11.9	12.4	10.3	7.0	10.3	10.8	26.5	10.8	18.5
15 cm (EDTA)	3.5	14.2	33.3	9.6	19.3	1.5	17.5	1.2	5.7
65—70 cm (HCl)	15.1	10.9	5.5	8.9	12.5	10.0	18.3	18.8	250.8
65—70 cm (EDTA + HCl)	19.3	13.0	5.5	6.8	14.0	8.7	15.6	17.0	137.5
65—70 cm (EDTA)	14.9	15.9	21.1	9.6	19.2	1.4	15.9	2.2	51.1
120—130 cm (HCl)	12.5	6.3	1.8	10.4	10.0	14.6	22.9	21.9	9.6
120—130 cm (EDTA + HCl)	15.5	11.8	4.3	11.2	11.2	12.6	18.5	15.4	6.5
120—130 cm (EDTA)	12.3	11.5	10.0	24.6	10.8	3.4	18.5	5.5	1.3
120—130 cm (H_2O)	—	—	—	—	—	—	—	—	—
120—130 cm (H_2O + HCl)	19.7	16.7	2.0	7.1	8.0	6.7	13.5	27.8	0.66
120—130 cm (H_2O + EDTA)	23.9	16.7	—	33.3	7.2	—	6.1	12.2	0.18

monosaccharides. Ribose, fucose, rhamnose, arabinose, and xylose are dominantly in metal-complexed forms (Table 4-VI).

The 65—70 cm interval sample (B.S. transition) is from the organic-rich unit and represents the transition period from oxic fresh water to anoxic marine (4000—7000 B.P.). The hydrolyzed EDTA extraction releases 54% of the "total" sugars, a value which lies between those for oxic and anoxic sediments. The high organic content of this sample (Fig. 4-3) suggests a terrigenous source. The $\delta^{13}C$ values of −24 to −26 for the organic matter, however, are between those of marine and terrigenous organic matter (Degens, 1969; Deuser, 1972). Furthermore, a "normal" order of sugar abundance in the HCl extract (Ga ~ Gl > Rh > X > M ~ Fu > A > Ri) and the low glucose concentration indicate an indigenous source for the organic matter.

Table 4-XIII compares the Black Sea organic-rich sediment sample with a submerged peat deposit from George's Bank dated at about 10,000 years B.P. The contrast in sugar compositions is striking. Glucose accounts for only about 20—25% of the total sugars of the Black Sea sample (even after concentrated sulfuric acid treatment), whereas in the peat deposit glucose represents 70% of the total sugars. Furthermore, the peat sample contains only traces of rhamnose, fucose and ribose, whereas in the Black Sea sample these sugars constitute 25—30% of the total. The relatively high abundance of these latter sugars suggests a planktonic source (see Table 4-I for typical plankton analyses) for the organic matter in the Black Sea, which, in turn, implies that the primary marine productivity must have been extremely high. This factor, combined with slow sedimentation rates (10 cm in 1000 years) due to a transgressive stage could account for a 30—40% organic matter content in the sediment.

Electron micrographs of the organic matter in the organic-rich zone (Fig. 4-10) reveal intact membranes and cell wall fragments (Degens et al., 1970). No terrigenous plant detritus was observed. It is interesting to note that the organic matter shown in Fig. 4-10 did not need to be stained with a heavy metal (e.g., Os or U-salts) prior to microscopy. In situ heavy-metal "staining" made visualization possible (Degens et al., 1970). Thus, electron microscopy provides direct evidence for existence of metal—organic-matter association in this sediment.

The third sample (15 cm, B.S. marine) is representative of the marine anoxic coccolith-rich zone. The hydrolyzed EDTA extraction releases 40% of the "total" sugars, a value which compares well with that for anoxic sediment from the Cariaco Trench and Walvis Bay (Table 4-V). Rhamnose, fucose, arabinose, and galactose are present dominantly in metal-bound states (Table 4-VI). The order of abundance within the HCl extract is: M > Ga > X > Gl > Ri > Fu > Rh ~ A, which is somewhat different from that of sediments from the lower zones. The particularly high contents of ribose and

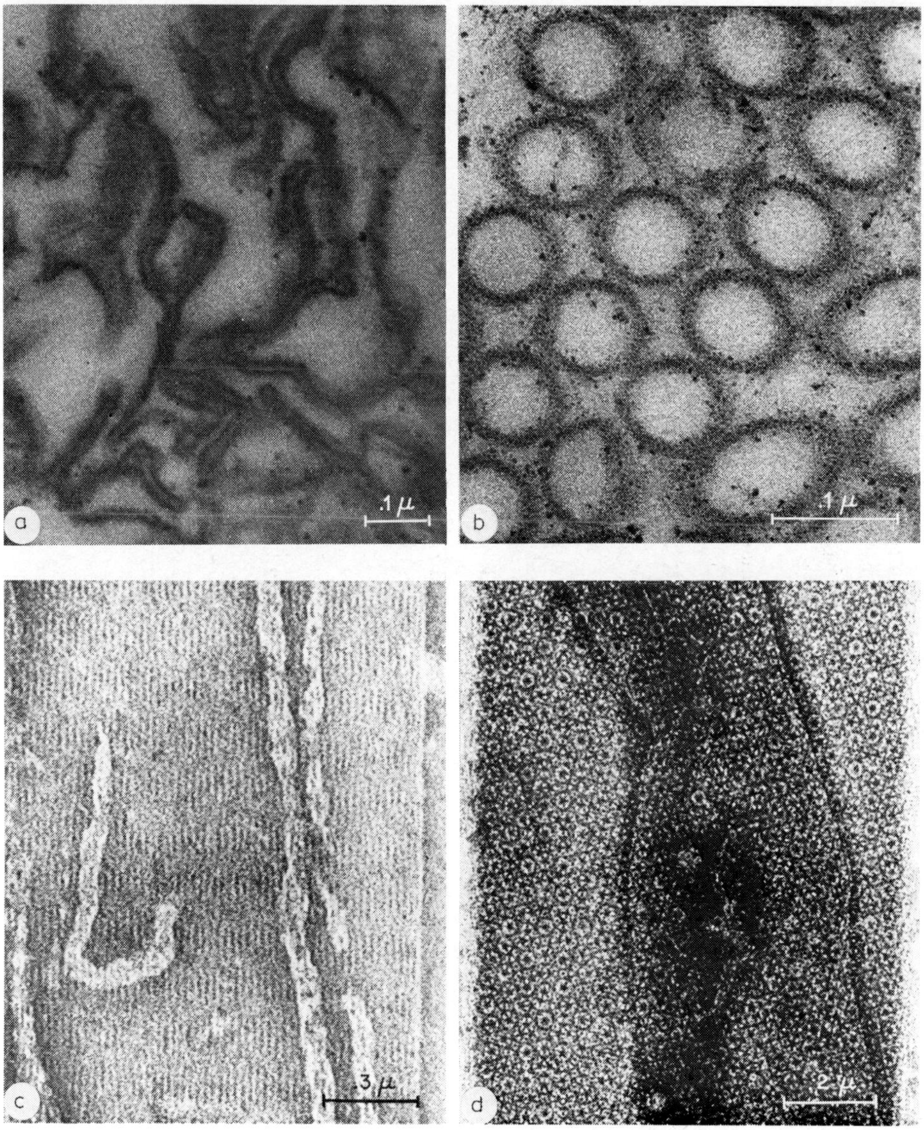

Fig. 4-10 (a) Branched tubular membranes; (b) large tubular membranes having a diameter of 700—800 Å and consisting of unit-membranes having a width of 80 Å; (c) organic substances resembling a bacterial cell wall and showing a unique pattern of subunits which have a 40 Å periodicity; and (d) portion of an organic substance resembling a bacterial cell wall and exhibiting a crystalline arrangement of subunits. All four electron micrographs were of Black Sea sediments (core 1474 K) 20—70 cm in the sapropel layer. The samples did not need to be stained prior to microscopy because in situ heavy-metal staining made visualization possible. (After Degens et al., 1970.)

TABLE 4-XIII

Comparison of the carbohydrate composition of true land-derived marine sediment and possible land-derived sediment (mole%)

	Rh	Fu	Ri	A	X	M	Ga	Gl	Total (μ moles/g)
Sample (conc. H_2SO_4 treatment):									
Peat, George's Bank, 10,000 years B.P.	3.1	1.0	0	2.3	7.3	8.4	8.9	68.9	354.8
Wood, Cariaco Trench, 5000 years B.P.	3.9	0.73	0	7.0	12.5	2.6	7.6	65.7	422.2
Black Sea, sapropel layer, 7000 years B.P.	14.9	8.9	4.5	6.1	12.4	9.2	19.5	24.5	246.1
Sample (1.8 *N* HCl treatment):									
Peat, George's Bank, 10,000 years B.P.	4.8	1.6	0	3.0	8.7	12.3	17.0	52.6	253.5
Wood, Cariaco Trench, 5000 years B.P.	6.5	1.8	0	15.1	23.1	4.4	13.7	35.4	258.9
Black Sea sapropel layer, 7000 years B.P.	15.1	10.9	5.5	8.9	12.6	10.0	18.3	18.8	250.8

xylose is noteworthy (Table 4-XII). This unusual distribution may be related to the dominant coccolith input.

Amino acids in Black Sea sediments. Inasmuch as the Black Sea sediments were deposited under varying environmental conditions, fluctuations in the amino acid composition attributable to these environmental changes may be clearly discerned. Table 4-XIV shows that within the marine sediments ($0–7 \cdot 10^3$ years) the order of amino acid abundance in the HCl extract is: Gly > Ala > Asp > Glu > Leu > Val > Lys ~ Pro ~ Thr > Ser ~ Iso > Phe (alkyl > acidic > cyclic > alcoholic > basic > S-containing). In the transitional zone ($7–9 \cdot 10^3$ years) the order is: Asp $\gtrsim$ Gly > Ala > Glu > Val >

TABLE 4-XIV

Total amino acids, hexosamines, and ammonia in Black Sea sediments (residues/1000)

Years $\cdot 10^3$: Number of samples:	0–3 (4)	3–7 (9)	7–8 (7)	8–9 (9)	9–10 (6)	10–11 (9)	13–17 (4)
Aspartic acid	109	105	145	177	159	166	151
Threonine	64	48	42	56	69	63	67
Serine	61	44	40	56	74	71	70
Glutamic acid	86	86	103	92	99	117	110
Proline	60	52	40	37	40	39	47
Glycine	138	127	160	155	151	157	150
Alanine	115	121	124	110	113	112	104
Cystine (half)	7	6	7	6	10	9	6
Valine	67	78	76	68	64	66	59
Methionine	12	11	5	3	5	4	3
Alloisoleucine	1.2	3.1	3.7	1.9	1.3	2.4	2.9
Isoleucine	45	57	46	36	35	31	28
Leucine	76	91	71	55	58	54	52
Tyrosine	26	25	12	8	6	9	9
Phenylalanine	42	54	36	30	20	23	36
Beta-alanine	9	4	13	13	14	13	16
Lysine	61	53	58	69	61	45	67
Histidine	1	1	1	1	1	1	1
Arginine	21	32	19	25	21	18	21
Total amino acids (mg/g)	16.7	36.8	2.99	1.64	1.73	0.66	0.40
Hexosamines (mg/g)	1.81	1.13	0.30	0.40	0.39	0.25	0.16
Ammonia (mg/g)	0.22	0.49	0.20	0.13	0.12	0.08	0.11
Organic carbon (%)	7.1	16.4	1.6	1.1	0.70	0.67	0.54
Organic nitrogen (%)	0.62	1.23	0.15	0.09	0.07	0.06	0.05
C/N ratio	11.5	13.3	11	12	11	11	11
AA/HA ratio	9	33	10	4	4	3	3
N recovered (%)	38	44	42	41	36	29	29

Leu ~ Lys > Thr $\gtrsim$ Ser > Iso > Pro > Phe, i.e.: (alkyl > acidic > alcoholic > basic $\gtrsim$ cyclic > S-containing). In the fresh-water sediments ($9–25 \cdot 10^3$ years) the order is: Asp $\gtrsim$ Gly > Ala > Glu > Ser > Thr > Val > Lys > Leu > Pro > Iso > Phe, i.e.: (alkyl > acidic > alcoholic > basic $\gtrsim$ cyclic > S-containing).

Thus, in comparison to the anoxic marine sediments (Table 4-XIV), the fresh-water oxic sediments are slightly enriched in aspartic acid, glutamic acid, and β-alanine, but are slightly depleted in isoleucine, leucine, and aromatic amino acids. Glycine, alanine, and aspartic acids are the dominant amino acids in all three sediment types. It is significant to note that these are also the principal amino acids in the Argentine Basin and modern Cariaco Trench sediments.

Fig. 4-7 shows the fluctuations in the relative abundance of amino acid subgroups with depth. An inverse relationship is suggested between (1) acidic and cyclic amino acids, and (2) alkyl and alcoholic amino acids. Sulfur-containing amino acids show a slight relative decrease with depth. Whether these fluctuations are due to changes in the nature of the organic input or to changes in relative preservation resulting from environmental shifts at the sediment—water interface can not be clearly established with the available data on the Black Sea sediments alone. Comparison with the Cariaco Trench data is somewhat more revealing. Similar relationships within amino acid subgroups were observed in these modern sediments (Fig. 4-7). Relative to the Black Sea, however, the environment at the sediment—water interface in the Cariaco Trench has remained constant during the time interval in question. Thus, this similarity in subgroup trends suggests that these fluctuations mainly reflect changes in the composition of the organic input, e.g., changes in the relative proportions of marine planktonic versus terrigenous sources. For example, cyclic (aromatic) amino acids may be dominantly land-derived, whereas acidic amino acids may be dominantly marine-derived.

The fluctuations within the sediment subgroups of the Black Sea, however, are minor when compared to the 100-fold reduction in the total amino acid concentration between the brackish-water anoxic sediments and fresh-water oxic sediments as shown in Fig. 4-11. Comparison of Figs. 4-11 and 4-3 shows that fluctuations in the total amino acid content closely follow fluctuation in total organic carbon.

Fig. 4-12 shows the distribution of total hexosamines and amino acid/hexosamine ratios in the Black Sea sediments. During the time interval between 3500 and 7500 years B.P., the hexosamine content was negligible. Preceding and following this period the hexosamine concentration represented a substantial fraction of the nitrogenous organic matter. It was suggested that this sudden drop in hexosamine concentration is related to the onset of extremely high productivity, perhaps eutrophication conditions in

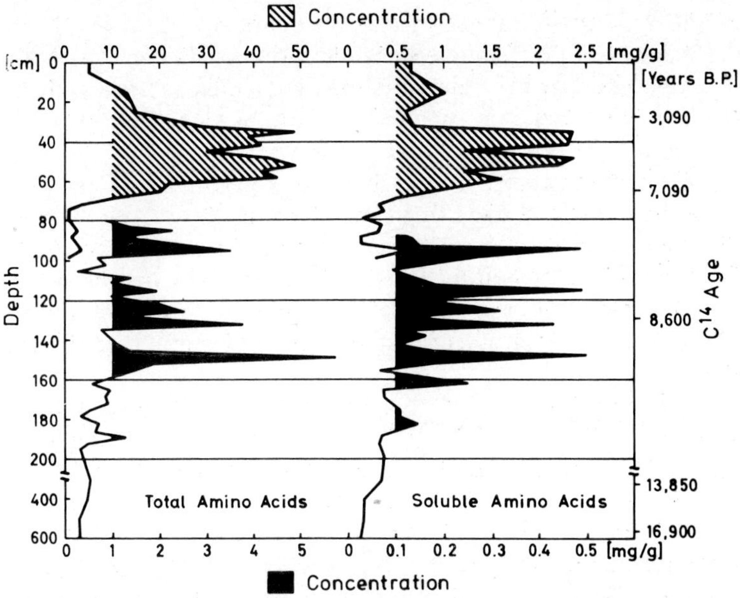

Fig. 4-11. Distribution of total and soluble amino acids in a Black Sea core (Degens and Ross, 1974). The spacing between the individual samples is about 3 cm. The lower values are plotted on an expanded scale. The observed fluctuations in the concentration levels of amino acids are principally determined by changes in deposition rates and the redox potential at the sediment—water interface at the time of deposition.

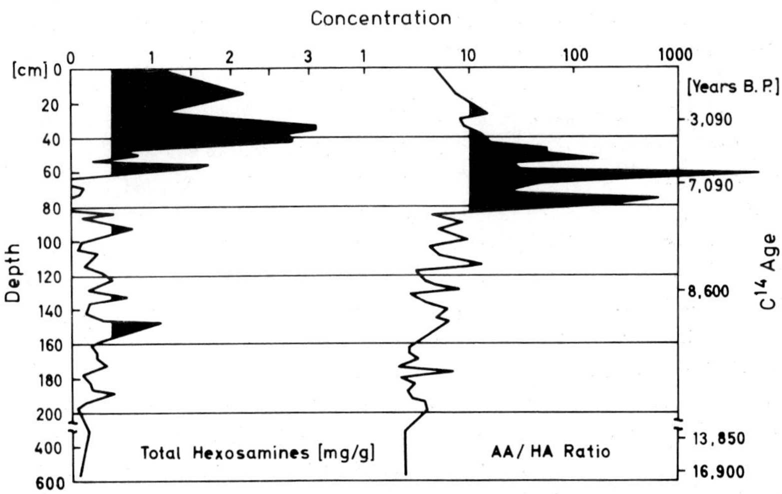

Fig. 4-12. Variation in the total hexosamines content and amino acid/hexosamine ratios (AA/HA) with depth in a Black Sea core. The spacing between individual samples is about 3 cm. The AA/HA ratios are plotted on a logarithmic scale.

the surface waters and resulting changes in the zooplankton/phytoplankton ratio (Degens and Hecky, 1974). During eutrophication conditions, the algal biomass, which contains little or no chitinous material (hexosamines), overwhelms the zooplankton biomass, that generally contains significant quantities of chitinous material. This gives rise to a high AA/HA ratio in the sediments. The data on sugar contents (preceding section) also indicate an extremely high algal productivity during this time interval.

Three samples, representing the fresh-water, transition, and present-day marine environments were treated with EDTA and the results are shown in Table 4-III. Fig. 4-13 shows that the percent of EDTA-extractable amino acids increases with depth, which implies that the degree of metal association

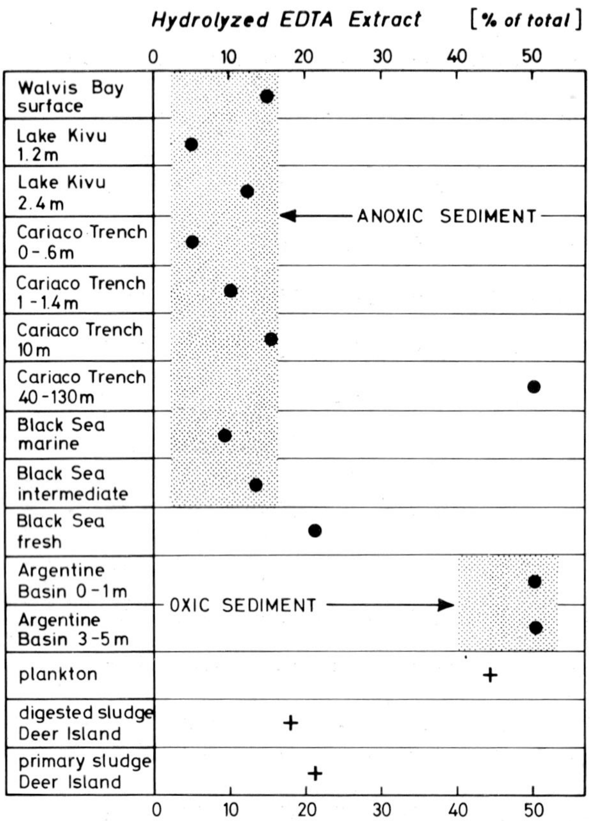

Fig. 4-13. Comparison of the degree of metal association of amino acids from various sedimentary environments. Reducing sediments are less metal bound than oxidizing sediments. Within a reducing sediment, however, the degree of metal binding increases rapidly with depth as shown for the Cariaco Trench sediments.

also increases. Although this increase is not as pronounced as in the case of sugars (Fig. 4-17), it probably also reflects a change in the depositional environment from anoxic to oxic. As in the case of Argentine Basin and Cariaco Trench sediments, aspartic acid, glycine and lysine are significantly metal-bound.

General trends

Table 4-XV presents a summary of carbohydrate composition (in mole%) of HCl ("total") extracts of sediments and plankton from different environments. Generally, there is no predominance of any sugar in any of these

TABLE XV

Summary of mole% composition of carbohydrates in sediment and plankton (HCl hydrolysis)

Area Sugar:	Rh	Fu	Ri	A	X	M	Ga	Gl
Argentine Basin <5 m (7) *	8.4	9.6	4.8	10.2	10.5	13.7	24.2	18.4
Bermuda, surface	10.7	9.8	4.1	8.9	12.6	16.8	23.8	13.1
N.Y. Bight, surface	9.1	8.1	1.6	8.6	12.9	15.1	21.5	23.1
Black Sea, marine, 15 cm	4.8	7.6	9.6	4.6	16.3	23.7	20.0	13.3
Black Sea, transition 65—70 cm	15.1	10.9	5.5	8.9	12.6	10.0	18.3	18.8
Black Sea, fresh 1.30 cm	12.5	6.3	1.8	10.4	10.0	14.6	22.9	21.9
Cariaco Trench <40 m (7)	11.0	9.2	3.5	5.5	9.7	19.3	28.0	13.6
Cariaco Trench 67—130 m (2)	8.9	4.8	0.7	9.9	12.3	23.9	23.2	17.2
Lake Kivu 2.0—9.3 m (3)	11.3	8.0	1.0	7.0	12.0	15.6	20.3	25.1
Oyster Pond, surface	11.1	6.8	0.9	10.2	11.3	13.3	20.2	26.1
Walvis Bay, surface	15.7	12.7	2.3	6.1	9.4	8.2	26.9	18.6
Santa Barbara Basin (2)	11.7	8.3	4.5	6.8	11.2	15.2	25.9	16.3
Average of above	10.4	8.1	3.1	7.7	11.1	15.0	24.0	20.6
Average of plankton (5)	5.4	6.7	10.0	2.7	8.2	14.5	29.5	22.9

* Number of samples represented.

extracts. The general order of abundance of various sugars in sediments is: Ga > Gl > M > X $\gtrsim$ Rh > Fu > A > Ri; the order in plankton is: Ga > Gl > M > Ri > X > Fu $\gtrsim$ Rh > A. These orders of abundance are almost identical with the exception of ribose, which shows a considerable depletion in sediments. As explained in the previous section, this relative depletion is attributable to the instability of ribose-containing organo-phosphate compounds. The loss of ribose probably occurred in the water column shortly after cell lysis.

Table 4-XVI gives a summary of the amino acid composition (residues/1000) of HCl extracts of sediment and plankton samples. Again, it can be seen that the average amino acid sediment composition is nearly identical to that of plankton. Glycine, alanine, aspartic acid, and glutamic acid are the dominant species in both cases.

TABLE 4-XVI

Summary of residues/1000 composition of amino acids in sediment and plankton (HCl hydrolysis)

Amino acids	Woods Hole Bermuda (14) *	N.Y. Bight, Hudson Canyon (3)	Cariaco Trench (3)	Walvis Bay (1)	Lake Kivu (3)
Aspartic acid	83	108	101	121	82
Threonine	75	81	36	55	67
Serine	87	96	47	97	61
Glutamic acid	67	95	76	73	79
Proline	39	55	71	50	65
Glycine	228	173	144	171	130
Alanine	111	90	112	95	131
Cystine (half)	7	7	6	2	2
Valine	62	53	54	45	78
Methionine	11	10	23	20	13
Isoleucine	27	29	38	23	56
Leucine	35	46	81	49	92
Tyrosine	9	29	23	24	35
Phenylalanine	15	43	59	31	53
Lysine	47	41	81	89	32
Histidine	14	3	10	3	6
Arginine	13	41	39	51	18
C/N ratio	5.2	11.1	10.2	7.3	11.9
AA/HA ratio	14.1	8.4	1.5	3.3	2.7

* Number of samples represented.

If it is assumed that the organic input of sediment is completely reworked by bottom fauna, then the above results indicate that this reworking is either remarkably nonselective or that the composition of the metabolic wastes (resynthesis) is remarkably similar to that of plankton. Data on the sugar and amino acid compositions of benthic organisms are lacking.

The similarities between sediment and plankton compositions (Tables 4-XV and 4-XVI) are, however, more consistent with the simple hypothesis that plankton is the source material for extracted sedimentary proteins and carbohydrates. This hypothesis is further supported by stable carbon isotope analyses. In temperate regions, marine plankton has a $\delta^{13}C$ value of about −20‰ (relative to PDB standard) with no apparent differences for zoo- and phytoplankton populations. Common land plants are 5—10‰ lighter (Craig, 1953; Degens, 1969). In a number of studies it has been shown that the

Argentine Basin (4)	Arabian Sea (5)	Black Sea, marine (4)	Black Sea, trans. (9)	Black Sea, fresh (6)	Average sediments (52)	Average plankton (22)
125	78	109	105	159	107	101
66	57	64	48	69	62	58
84	68	61	44	74	72	58
57	71	86	86	99	79	116
31	74	60	52	40	54	52
156	247	138	127	151	166	131
128	107	115	121	113	112	103
3	18	7	6	10	7	10
124	58	67	78	64	68	59
15	9	12	11	5	13	23
33	33	46	57	35	38	40
44	45	76	91	58	62	73
8	9	26	25	6	20	28
19	15	42	54	20	35	30
80	37	61	53	61	58	61
2	9	1	1	1	5	13
29	2	21	32	21	27	44
7.7	9.0	11.5	13.3	11	9.8	7.3
0.7	7.6	9	33	4	10.9	17

isotopic composition of organic matter in Recent marine sediments is identical to that of marine plankton (e.g., Hunt, 1962; Sackett and Thompson, 1963). Only sediments deposited in river estuaries and close to shorelines reveal a terrigenous isotopic influence.

The glucose and ribose concentrations in marine sediments also may be used to determine the relative terrigenous organic input of sediment. Table 4-XIII and Figs. 4-4, 4-9, and 4-14 reveal that terrigenous material contains a high percentage of glucose and a low percentage of ribose, whereas in marine plankton the reverse is true. Thus, in sediment where glucose >> galactose and where ribose is negligible, the presence of a major terrigenous fraction should be suspected. Fig. 4-14 shows that the relative abundances of glucose and ribose in most sediments are greatly reduced in comparison to terrigenous and marine sources, respectively.

From the viewpoint of characterizing sedimentary environments, the uniformity of the results in Tables 4-XV and 4-XVI is rather disappointing. To gain a better handle on environmental and diagenetic fluctuations, analyses of metal-associated (EDTA-extracted) amino acids and sugars were performed. Summaries of the sugar compositions (in mole%) of hydrolyzed and unhydrolyzed EDTA extracts of sediment are presented in Tables 4-XVII and 4-XVIII. Fig. 4-15 illustrates the composition changes in the averages of the different sugar extracts from Tables 4-XV, 4-XVII and 4-XVIII. With each extraction (HCl, HCl + EDTA, EDTA) the compositions generally become progressively different from that of plankton, which, presumably, is the source of sedimentary carbohydrates. The hexoses, glucose, mannose, and galactose, decrease in relative abundance, with galactose decreasing to a lesser extent. These decreases appear to be compensated for by increases in the pentoses and deoxyhexoses. The initially high content of ribose in plankton relative to sediments was explained previously. The reason for the drop in rhamnose between the EDTA + HCl and EDTA extracts is presently unclear. The trends mentioned above hold on an individual basis for nearly every sediment sample analyzed.

The trends in Fig. 4-15 might be explained in terms of differences in metal-ion affinities; thus pentoses and deoxyhexoses would have a greater affinity than hexoses for metals. In light of the fact that hexoses have more functional groups than the other sugars, however, this explanation is unlikely. Alternatively, it might be argued that hexoses are more strongly bound to mineral phases than are pentoses. Experiments with slurries of monosaccharide solutions and clays, however, revealed no detectable preferential absorption (Mopper, 1973).

The reasons for the compositional changes shown in Fig. 4-15 are most likely related to differences in the molecular-weight distribution. For example, it seems probable that mannose, glucose, and galactose are more associ-

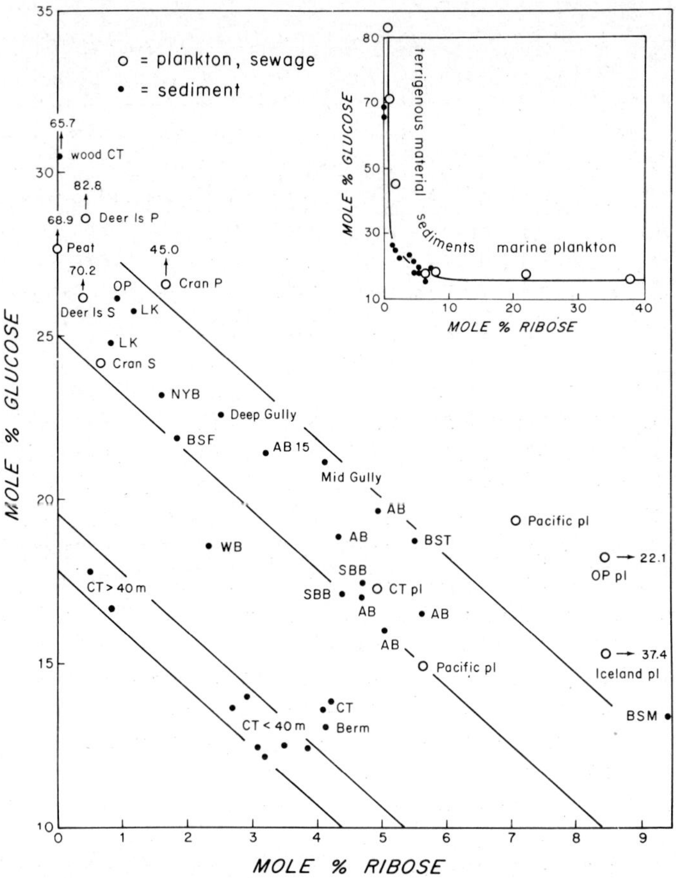

Fig. 4-14. Relationship (inverse) between glucose and ribose contents in marine sediments (Mopper, 1973). Terrigenous material is enriched in glucose and depleted in ribose, whereas the marine plankton shows the reverse trend. The following notations are used: AB = Argentine Basin; CT = Cariaco Trench; NYB = New York Bight; OP = Oyster Pond; BSF, T, M = Black Sea fresh water, transition, marine; Berm = Bermuda; WB = Walvis Bay; SBB = Santa Barbara Basin; LK = Lake Kivu; Cran = Cranston sewage sludge; Dear Is = Dear Island sewage sludge; P = primary sludge; S = secondary sludge; and pl = plankton.

ated with high-molecular-weight water-insoluble polysaccharides than other sugars. Thus, after EDTA treatment these polysaccharides are centrifuged down along with the sediment residue. Table 4-V shows that glucose, galactose, and mannose are indeed dominantly associated with the EDTA-insoluble residue.

Summaries of the amino acid compositions (in terms of residues per 1000)

TABLE 4-XVII

Summary of mole % composition of carbohydrates in sediments (hydrolyzed EDTA extract)

Area	Sugar: Rh	Fu	Ri	A	X	M	Ga	Gl
Argentine Basin, <5 m (7) *	10.2	12.0	5.7	8.7	11.6	12.3	25.1	14.6
N.Y. Bight Bouy St., surface	10.8	10.8	2.5	10.8	12.5	12.5	22.5	17.5
Hudson Canyon Deep Gully, surface	12.4	12.9	6.1	8.3	12.2	10.4	22.6	15.1
Black Sea, marine, 0.15 m	11.9	12.4	10.3	7.0	10.3	10.8	26.5	10.8
Black Sea, transition, 0.65—0.70 m	19.3	13.0	5.5	6.8	14.0	8.7	15.6	17.0
Black Sea, fresh, 1.3 m	15.5	11.8	4.3	11.2	11.2	12.6	18.5	15.4
Cariaco Trench, <40 m (9)	13.2	13.9	5.6	7.4	11.3	10.9	26.8	11.3
Cariaco Trench, 67—130 m (2)	12.6	6.7	1.7	7.4	14.9	18.9	22.5	15.5
Lake Kivu 2.0—9.3 m (3)	14.3	12.3	3.8	7.9	9.4	13.4	20.6	18.3
Oyster Pond, surface	13.6	10.3	2.0	15.0	12.5	9.5	21.4	15.6
Walvis Bay, surface	16.6	10.8	3.8	5.4	11.5	12.4	24.5	15.0
Santa Barbara Basin (50 years)	13.3	12.5	7.8	7.8	14.1	10.2	24.2	10.2
Average (mole %)	13.6	11.6	4.9	8.6	12.1	11.9	22.6	14.7

* Number of samples.

of hydrolyzed and unhydrolyzed EDTA extracts of sediments are given in Table 4-XIX. Fig. 4-16 illustrates the major compositional trends in the averages from Tables 4-XVI and 4-XIX. As with the sugars, with each extraction the compositions become progressively different from that of plankton.

TABLE 4-XVIII

Summary of mole % composition of carbohydrates in sediments (unhydrolyzed EDTA extract)

Area	Sugar: Rh	Fu	Ri	A	X	M	Ga	Gl
Argentine Basin <5 m (2) *	8.7	15.8	15.3	14.3	15.3	4.8	17.4	8.4
N.Y. Bight Bouy St., surface	6.7	14.4	9.2	21.5	21.5	3.5	17.8	5.4
Black Sea, marine, 1.5 m	3.5	14.2	33.3	9.6	19.3	1.5	17.5	1.2
Black Sea, transition, 0.65—0.70 m	14.9	15.9	21.1	9.6	19.2	1.4	15.9	2.2
Black Sea, fresh, 1.3 m	12.3	11.5	10.0	24.6	10.8	3.4	18.5	5.5
Cariaco Trench, <40 m (5)	8.0	16.6	21.2	12.3	13.3	3.3	22.7	3.7
Cariaco Trench, 67—130 m (2)	9.3	11.8	3.7	9.9	24.7	11.1	19.8	11.0
Lake Kivu, 2.0—9.3 m	13.8	17.3	5.8	14.0	15.2	5.8	20.4	7.2
Oyster Pond, surface	9.9	15.6	5.2	32.3	16.2	2.7	15.0	3.7
Walvis Bay, surface	11.3	18.1	15.9	12.5	17.0	3.6	19.3	2.4
Average (mole %)	9.8	15.1	14.1	16.1	17.3	4.1	18.4	5.1

* Number of samples.

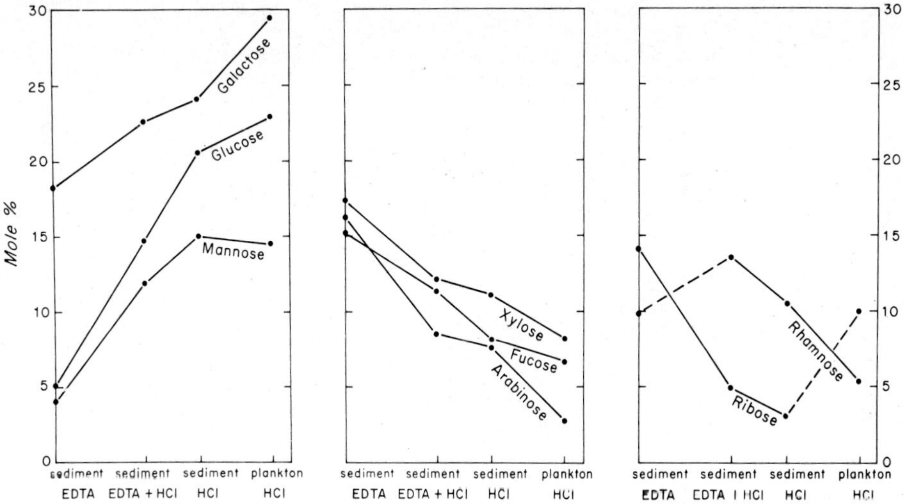

Fig. 4-15. Relationships among the relative sugar compositions in different sediment and plankton extracts; data was drawn from the averages in Tables 15, 17 and 18; these relations hold for individual samples as well as the averages; EDTA = unhydrolyzed EDTA extract, EDTA + HCl = hydrolyzed EDTA extract, and HCl = "total" hydrolyzable carbohydrates.

Glycine, aspartic acid and lysine increase in relative abundance, whereas concentrations of all the other amino acids decrease (Fig. 4-16a). In terms of amino acid subgroups (Fig. 4-16b), only the acidic group shows a marked progressive increase in abundance relative to the plankton extract. This increase is almost exactly balanced by the summed fluctuations of alkyl + cyclic + alcoholic groups as shown in Fig. 4-16c. It is significant to note that the trends observed in Figs. 4-16a, b, c also are followed by most of the samples on an individual basis.

As opposed to the sugars, the amino acids are more varied with respect to functionality and, hence, reactivity towards metal ions. For example, aspartic acid, because of its high degree of functionality and small molecular size, is a far better complexing agent than a bulky alkyl amino acid, such as valine or leucine. In fact, aspartic acid has been demonstrated to play a key role in calcification processes (as discussed in the plankton section). The high metal affinity of glycine (Fig. 4-16a) is also attributable to its small molecular size (it is the smallest amino acid). Consequently, glycine is easily incorporated into metal-ion—organic complexes.

Differences in the molecular-weight distribution of the amino acids probably also contribute to compositional variations between the different

TABLE 4-XIX

Summary of residues/1000 composition of amino acids in sediments

Amino acids	EDTA + HCl					EDTA				
	Cariaco Trench (3) *	Walvis Bay (1)	Lake Kivu (1)	Argentine Basin (4)	Average	Cariaco Trench (3)	Walvis Bay (1)	Lake Kivu (1)	Argentine Basin (4)	Average
Aspartic acid	206	241	210	184	210	250	495	761	502	502
Threonine	39	66	61	37	51	—	10	3	11	6
Serine	68	93	111	61	83	7	22	8	66	26
Glutamic acid	51	68	74	39	58	—	6	3	6	4
Proline	18	54	37	—	27	—	—	3	—	1
Glycine	337	194	136	399	267	596	79	181	371	307
Alanine	62	78	94	36	68	—	24	16	11	13
Cystine	7	—	2	—	2	—	—	—	—	—
Valine	40	47	73	21	45	—	4	5	5	4
Methionine	4	1	3	4	3	—	—	—	—	—
Isoleucine	12	11	33	4	15	—	6	—	—	1
Leucine	17	12	41	8	20	—	6	—	1	2
Tyrosine	6	3	38	—	12	—	—	—	—	—
Phenylalanine	15	5	52	—	18	—	—	—	—	—
Lysine	117	73	45	207	111	147	281	21	28	119
Histidine	—	13	15	—	7	—	63	—	—	16
Arginine	—	53	8	—	15	—	2	—	—	—

* Number of samples represented.

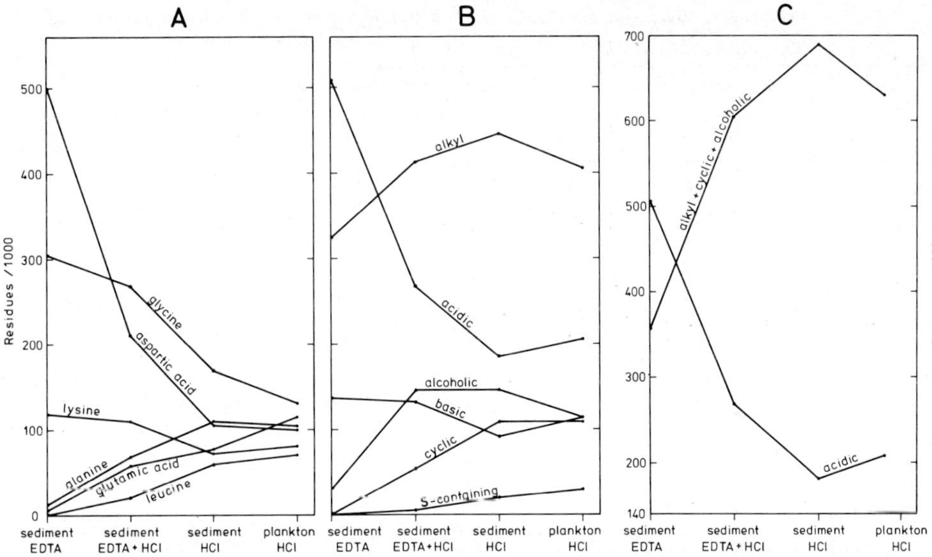

Fig. 4-16. Relationships among the relative amino acid compositions in different sediment and plankton extracts; data was drawn from the averages in Tables 4-XVI and 4-XIX.

extracts. This factor especially applies to the hexosamines which are approximately 95% unextractable by EDTA (Table 4-III). As discussed in the section on the Argentine Basin sediments, these compounds may accumulate preferentially as a biologically resistant material called chitin. Thus, in areas with high rates of benthic consumption, low AA/HA ratios can be expected. In areas with low rates of benthic consumption (anoxic sediments), however, changes in the sediment AA/HA ratio probably reflect changes in the algal/zooplankton ratio of the organic input as discussed in the section on Black Sea sediments.

Although the organic compositions of modern oxic and anoxic sediments are similar (Tables 4-XV and 4-XVI), differences between these depositional environments can be seen in terms of the degree of metal association. In Figs. 4-13 and 4-17, which are drawn from data presented in Tables 4-III and 4-VI, the percent of the "total" sugars released by the hydrolyzed EDTA extraction is plotted against the different sample types. For example, EDTA extracts 60—70% of the "total" sugars from oxidizing sediments and 35—40% of the "total" sugars from reducing sediments. The corresponding values for the amino acids are 40—60% and 5—15%. These trends indicate that carbohydrates and proteins are considerably more complexed in oxidizing sediments than in reducing sediments.

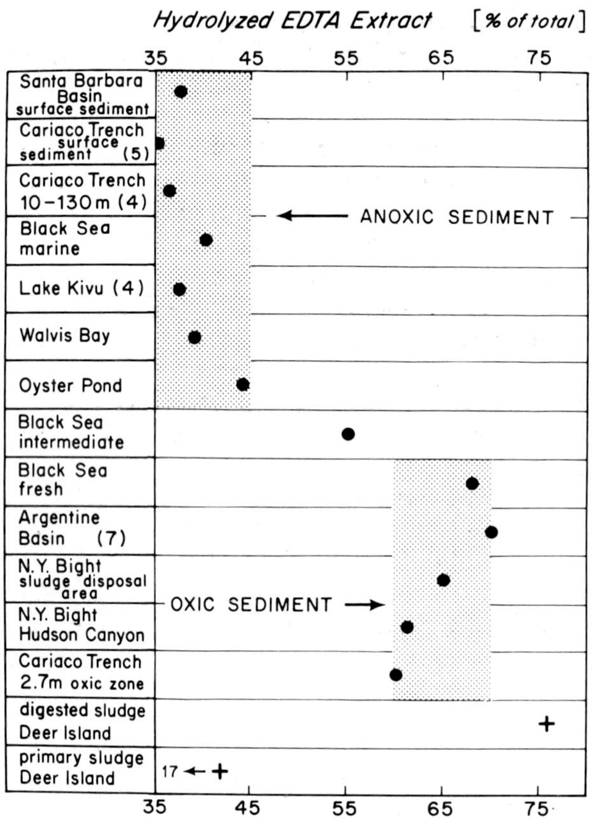

Fig. 4-17. Comparison of the degree of metal complexation of carbohydrates from various sedimentary environments. The % of HCl-extractable sugars which is released by EDTA is 25—45% for reducing sediments and 55—75% for oxidizing sediments. The "Black Sea intermediate" sample represents the transition period from oxidizing to reducing. Digestion of primary sewage sludge has the effect of drastically increasing the degree of metal complexation of the carbohydrate residue. Number in parenthesis represents number of samples averaged.

Differences in the degree of metal association are most likely related to differences in biological activity as discussed in the section entitled "General concepts". Analyses of primary and digested sewage sludges suggest that metal-bound proteins and carbohydrates are resistant to biological degradation, because they are the end-product of intensive biological degradation. For example, Fig. 4-17 shows that the percent of metal-bound (EDTA-extracted) sugars in the "total" sugars increases from 17% in Deer Island primary sludge to 77% in the microbially digested Deer Island secondary sludge. The degree of metal complexation by digestion of sewage sludge is similar to that of organic matter in oxic sediments. Thus, by analogy, the high degree of metal complexation of organic matter in oxic sediments is probably attributable to intensive biological consumption.

The low degree of complexation in reducing sediments implies that large fractions of carbohydrates and proteins in these sediments are potentially biodegradable and, hence, are consumed to a lesser degree than in oxidizing sediments. The low rate of consumption in reducing environments reflects the low biological activity associated with the presence of an anaerobic population and with the absence of a metazoan biomass. As discussed in the section "General concepts", the metazoan biomass appears to be mainly responsible for degradation in oxic sediments.

Fig. 4-18 shows that relationships exist among the degree of metal complexation, the sugar carbon content (expressed as % of the total organic carbon), and the degree of biological degradation. At the extremes of this plot are the Deer Island primary and secondary sewage sludges. The primary sludge has a high content of sugar carbon, which corresponds to the large cellulose content, and the degree of metal complexation is low. After intensive microbial degradation (secondary treatment), the sugar carbon greatly decreases and the degree of metal complexation correspondingly increases.

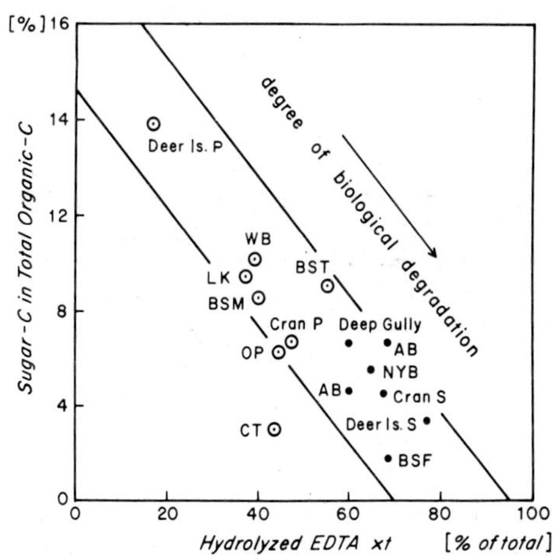

Fig. 4-18. Relationship (inverse) between sugar carbon (as % of total carbon) and hydrolyzed EDTA extracts of various sediment samples (Mopper, 1973). Reducing sediments have high sugar carbon values and low degree of metal complexation; oxic sediments show the reverse trends; the degree of metal complexation appears to be related to the degree of biological degradation. Notation as in Fig. 4-14; ⊙ = reducing sediments and primary sewage sludges; ● = oxic sediments and secondary sewage sludges.

Most sediments lie in-between these extremes; reducing sediments lie closer to the undigested sludge, whereas oxidizing sediments lie closer to the digested sludge.

PALEOENVIRONMENTAL CRITERIA

Sugars

Fig. 4-17 indicates that the degree of metal association of sedimentary carbohydrates is related to the Eh value at the sediment—water interface. Furthermore, Fig. 4-6 shows that the degree of metal binding appears to remain fairly unaltered after burial; therefore, changes in environment at the sediment—water interface from oxidizing to reducing may be discerned as shown for the Cariaco Trench.

Figs. 4-4, 4-9, and 4-14 indicate that the glucose and ribose contents of sediments may reflect relative changes in the terrigenous organic input as a result of changes in land runoff or sea level. First analyses show a good correlation between glucose/ribose ratio and the ratio of stable carbon isotopes.

Paleoeutrophication conditions also may be discerned from carbohydrate analyses. For example, organic-rich sediments (such as those of the Black Sea), which contain a "normal" algal sugar distribution, probably reflect extremely high productivity in the surface waters.

Amino acids

The degree of metal association of amino acids in modern sediments is also related to the Eh at the sediment—water interface. In contrast to the sugars, however, the amino acids undergo a temporal increase in metal association, probably as a result of their higher degree of reactivity. Thus, the amino acids are not as dependable as the sugars for determining paleochanges in the Eh at the sediment—water interface. The rate of increase in metal association may be time-dependent and, thus, may be used to estimate roughly the age of sediment deposited in a fairly uniform depositional environment, such as an anoxic basin. This hypothesis is presently under examination.

Changes in the amino acid/hexosamine (AA/HA) ratio reflect: (1) changes in the degree of biological degradation at the sediment—water interface, and (2) changes in the algae/zooplankton ratio of the organic input. Thus, a sediment with a very low AA/HA ratio (<1) probably reflects efficient benthic consumption, such as in the one in the Argentine Basin. A sediment with a very high AA/HA ratio (>20) probably corresponds to an unusually high algal input. Extremely high ratios (>100), such as those in the Black Sea sediments, may even reflect eutrophication in the surface waters.

CONCLUSIONS

Organic molecules in a living cell have a very complex structural order and are also far removed from thermodynamic equilibrium. Diagenesis of organic matter in sediments is expected to proceed in the direction of increasing thermodynamic stability. Initially, the organic input of sediment might be characterized as a loosely structured conglomeration of various biomolecules in different states of preservation. Diagenesis leads to complex interactions between organic molecules, metal ions, and minerals. The final product of late diagenesis, graphite, has achieved a very simple structural order. This order was obtained by the elimination of functional groups through condensation, deaminization, decarboxylation, etc.; CO_2, CH_4, NH_3, H_2O, and small organic molecules are released as byproducts.

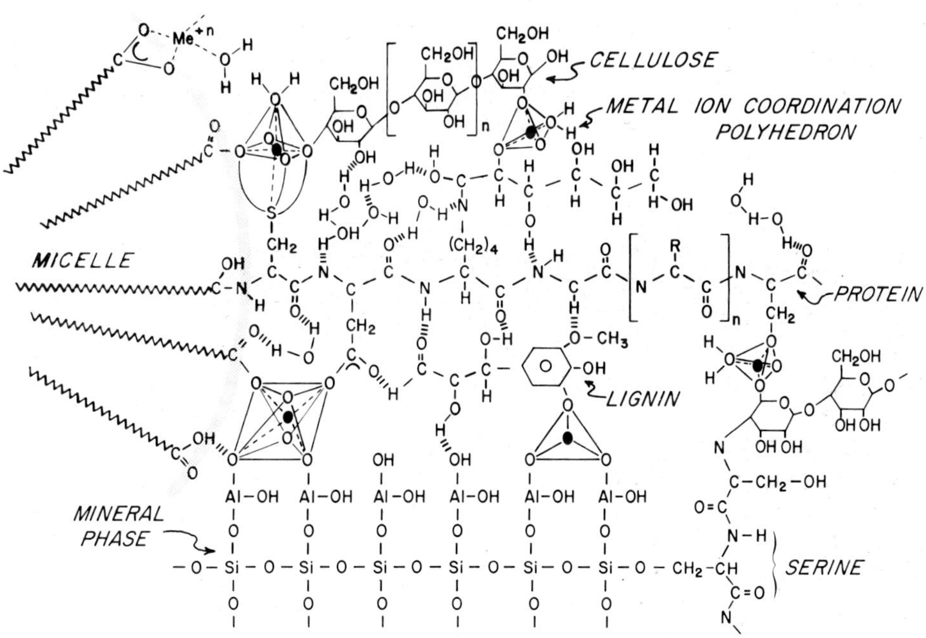

Fig. 4-19. A hypothetical schematic of the carbohydrate- and protein-containing residue of sediment. Various possible interactions among organic molecules, metal ions, and minerals are shown.

Although the structural nature of the carbohydrate- and protein-containing residue of Recent sediment has not been examined, a hypothetical schematic model of this residue is presented in Fig. 4-19. Although at the present stage of research this model is highly hypothetical, this type of model is useful in depicting possible interactions among organic molecules, metal ions, and minerals in sediments. Fig. 4-19 shows:

(1) Organic—organic condensations among carbohydrate, protein, lipid and lignin substances (the hashed lines represent hydrogen bonds).

(2) Organic—mineral interactions, such as condensation and adsorption of organic compounds onto surface of a kaolinite-type mineral.

(3) Organic—metal ion interactions; this type of interaction increases the structural order of the residue through the formation of metal ion coordination polyhedra. The functional groups of different organic molecules might participate in the coordination. In one of the polyhedra shown, a sulfide group (from cystine) coordinates with a metal ion, and the larger ionic radius of the sulfide ion distorts the polyhedron.

(4) Mineral—metal ion—organic interactions; oxygens on the mineral surface might also participate in metal ion coordination polyhedra.

(5) Micelle formation; molecules which contain both hydrophobic and hydrophilic portions (e.g., fatty acids) might tend to aggregate so that the hydrophilic portion interacts with the aqueous environment and the hydrophobic portion is protected from the aqueous environment.

REFERENCES

Baas Becking, L.G., Kaplan, I.R. and Moore, D., 1960. Limits of the natural environment in terms of pH and oxidation—reduction potentials. *J. Geol.*, 68: 243—284.

Bernal, J.D., 1965. Molecular matrices for living systems. In: S.W. Fox (Editor), *The Origins of Prebiological Systems.* Academic Press, New York and London, pp. 65—88.

Blumer, M., 1974. The geochemical significance of fossil porphyrins. *Acad. Cien. Brazil,* (in press).

Bordovskiy, O.K., 1965. Accumulation and transformation of organic substances in marine basins. *Mar. Geol.*, 3: 5—31.

Burton, F.G. and Neuman, W.F., 1971. On the possible role of crystals in the origins of life, 5. The polymerizations of glycine. *Curr. Mod. Biol.*, 4: 47—54.

Craig, H., 1953. The geochemistry of the stable carbon isotopes. *Geochim. Cosmochim. Acta,* 3: 53—92.

Dayhoff, M.O., 1971. Origin of organics from inorganics. In: S.J. Faust and J.V. Hunter (Editors), *Organic Compounds in Aquatic Environments.* Marcel Dekker, New York, N.Y., pp. 1—28

Deacon, G.E.R., 1963. The southern ocean. In: M.N. Hill (Editor), *The Sea.* Interscience, New York, N.Y., pp. 281—296.

Degens, E.T., 1965. *Geochemistry of Sediments.* Prentice Hall, Englewood Cliffs, N.J., 352 pp.

Degens, E.T., 1969. Biogeochemistry of stable carbon isotopes. In: G. Eglinton and M.T.J. Murphy (Editors), *Organic Geochemistry*. Springer, New York, N.Y., pp. 304—329.
Degens, E.T., 1970. Molecular nature of nitrogenous compounds in seawater and Recent sediments. In: D.W. Hood (Editor), *Organic Matter in Natural Waters—Inst. Mar. Sci., Univ. Alaska, Publ.*, 1: 77—106.
Degens, E.T., 1971. Sedimentological history of the Black Sea over the last 25,000 years. In: A.S. Campbell (Editor), *Geology and History of Turkey*. Petroleum Exploration Society of Libya.
Degens, E.T., 1974. Metal ion coordination in biogeochemical systems. In: *Actes du 6ème Congres International de Geochimie Organique*. Editions Technip, Paris, pp. 819—827.
Degens, E.T. and Bajor, M., 1960. Die Verteilung von Aminosäuren in bituminösen Sedimenten und ihre Bedeutung für die Kohlen- und Erdölgeologie. *Glückauf*, 96: 1525—1534.
Degens, E.T. and Matheja, J., 1968. Origin, development, and diagenesis of biogeochemical compounds. *J. Br. Interplanet. Soc.*, 21: 52—82.
Degens, E.T. and Mopper, K., 1975. Decay mechanisms of organic matter in marine soils. *Soil Sci.*, 119: 65—72.
Degens, E.T. and Mopper, K., 1976. Factors controlling the distribution and early diagenesis of organic material in marine sediments. In: J.P. Riley and R. Chester (Editors), *Chemical Oceanography*, 6. Academic Press, New York and London, pp. 59—113.
Degens, E.T. and Ross, D.A., 1974. The Black Sea—geology, chemistry, and biology. *Am. Assoc. Pet. Geol.*, Mem., 20: 630 pp.
Degens, E.T., Watson, S.W. and Remsen, C.C., 1970. Fossil membranes and cell wall fragments from a 7000-year old Black Sea sediment. *Science*, 168: 1207—1208.
Deuser, W.G., 1972. Late-Pleistocene and Holocene history of the Black Sea as indicated by stable-isotope studies. *J. Geophys. Res.*, 77: 1071—1077.
Emery, K.O. and Rittenberg, S.C., 1952. Early diagenesis of California Basin sediments in relation to origin of oil. *Bull. Am. Assoc. Pet. Geol.*, 36: 735—806.
Gehman, Jr., H.M., 1962. Organic matter in limestones. *Geochim. Cosmochim. Acta*, 26: 885—897.
Ghiselin, M.T., Degens, E.T., Spencer, D.W. and Parker, R.H., 1967. A phylogenetic survey of molluscan shell matrix proteins. *Breviora*, 262: 1—35.
Goldberg, E.D., 1961. Marine geochemistry. *Ann. Rev. Phys. Chem.*, 12: 29—48.
Handa, N. and Yanagi, K., 1969. Studies on the water-extractable carbohydrates of the particulate matter from the northwest Pacific Ocean. *Mar. Biol.*, 4: 197—207.
Hare, P.E., 1969. Geochemistry of proteins, peptides, and amino acids. In: G. Eglinton and M.T.J. Murphy (Editors), *Organic Geochemistry: Methods and Results*. Springer, New York, N.Y., pp. 438—463.
Harvey, G.R., Degens, E.T. and Mopper, K., 1971. Synthesis of nitrogen heterocycles on kaolinite. *Naturwissenschaften*, 58: 624—625.
Harvey, G.R., Mopper, K. and Degens, E.T., 1972. Synthesis of carbohydrates and lipids on kaolinite. *Chem. Geol.*, 9: 79—87.
Harvey, H.W., 1960. *The Chemistry and Fertility of Sea Waters*. University Press, Cambridge, pp. 1—240.
Hunt, J.M., 1962. *Some Observations on Organic Matter in Sediments*. (Paper presented at the Oil Scientific Session, 25 Years Hungarian Oil, 8—13 Oct. 1962, Budapest.)
Initial Reports of the Deep-Sea Drilling Project, 1973. U.S. Gov. Printing Office, Washington, D.C.
Jannasch, H.W. and Wirsen, C.O., 1973. Deep-sea microorganisms: In situ response to nutrient enrichment. *Science*, 180: 641—643.

Jannasch, H.W., Eimhjellen, K., Wirsen, C.O. and Farmanfarmaian, A., 1971. Microbial degradation of organic matter in the deep sea. *Science*, 171: 672—675.

King, K. and Hare, P.E., 1972. Amino acid composition of planktonic foraminifera: A paleobiochemical approach to evolution. *Science*, 175: 1461.

Lindblom, G. and Lupton, M.D., 1961. Microbial aspects of organic geochemistry. *Dev. Ind. Microbiol.*, 2: 9—22.

Lipman, C.B., 1928. The discovery of living microorganisms in ancient rocks. *Science*, 68: 272.

Matheja, J. and Degens, E.T., 1971. *Structural Molecular Biology of Phosphates.* Gustav Fischer, Stuttgart, pp. 1—180.

Mitterer, R.M., 1972. Calcified protein in the sedimentary environment. In: H.R. von Gaertner and H. Wehner (Editors), *Advances in Organic Chemistry, 1971.* Pergamon Press, Oxford, pp. 441—452.

Mopper, K., 1973. *Aspects of the Biogeochemistry of Carbohydrates in Aquatic Environments.* Ph. D. thesis, Joint-Program Mass. Inst. Technol. and Woods Hole Oceanogr. Inst.

Mopper, K. and Degens, E.T., 1972. Aspects of the biogeochemistry of carbohydrates and proteins in aquatic environments. *Tech. Rep. Woods Hole Oceanogr. Inst.*, WHOI, pp. 72—68.

Mulik, J.D. and Erdman, J. G., 1963. Genesis of hydrocarbons of low molecular weight in organic-rich aquatic systems. *Science*, 141.

Neher, J. and Rohrer, E., 1959. Dolomitbildung unter Mitwirkung von Bakterien. *Eclogae Geol. Helv.*, 51: 213—215.

Oppenheimer, C.H., 1961. Bacterial activity in sediments of shallow marine bays. *Geochim. Cosmochim. Acta*, 19: 244—260.

Porter, J.R., 1946. *Bacterial Chemistry and Physiology.* Wiley, New York, N.Y., 1073 pp.

Prashnowsky, A., Degens, E.T., Emery, K.O. and Pimenta. J., 1961. Organic materials in recent and ancient sediments, 1. Sugars in marine sediments of Santa Barbara Basin, California. *Neues Jahrb. Geol. Paläontol., Monatsh.*, 1961: 400—413.

Richards, F.A., 1965. Anoxic basins and fjords. In: J.P. Riley and G. Skirrow (Editors), *Chemical Oceanography.* Academic Press, London, pp. 611—645.

Rittenberg, S.C., 1940. Bacteriological analysis of some log cores of marine sediments. *J. Mar. Res.* (*Sears Found. Mar. Res.*), 3: 191—201.

Rittenberg, S.C., Emery, K.O., Hülsemann, J., Degens, E.T., Fay, R.C., Reuter, J.H., Grady, J.R., Richardson, S.H. and Bray, E.E., 1963. Biogeochemistry of sediments in Experimental Mohole. *J. Sediment. Petrol.*, 33: 140—172.

Rossini, F.D., 1960. Hydrocarbons in petroleum. *J. Chem. Educ.*, 37: 554—561.

Rowe, G.T., 1971. Benthic biomass and surface productivity. In: J.D. Costlow (Editor), *Fertility of the Sea.* Gordon and Breach, New York-London-Paris, pp. 441—454.

Ryther, J.H., 1963. Geographic variations in productivity. In: M.N. Hill (Editor), *The Sea.* Interscience, New York, N.Y., pp. 347—380.

Sackett, W.M. and Thompson, R.R., 1963. Isotopic organic carbon composition of Recent continental-derived clastic sediments of eastern Gulf Coast of Mexico. *Bull. Am. Assoc. Pet. Geol.*, 47: 525—528.

Sanders, H.L. and Hessler, R.R., 1969. Ecology of the deep-sea benthos. *Science*, 163: 1419—1424.

Sanders, H.L., Hessler, R.R. and Hampson G.R., 1965. An introduction to the study of deep-sea benthic faunal assemblages along the Gay Head-Bermuda transect. *Deep-Sea Res.*, 12: 845—867.

Sisler, F.D., 1961. Organic matter and life in meteorites. *Proc. Lunar Planet Explor. Coll.*, 2: 67—73.

Smith, K.L. and Teal, J.M., 1973. Deep-sea benthic community respiration: an in situ study at 1850 meters. *Science,* 179: 282—283.

Sørensen, L.H., 1972. Stabilization of newly formed amino acid metabolites in soil by clay minerals. *Soil Sci.,* 114: 15—11.

Stevenson, F.J. and Cheng, C.N., 1972. Organic geochemistry of the Argentine Basin sediments: carbon—nitrogen relationships and Quaternary correlations. *Geochim. Cosmochim.* Acta, 36: 653—671.

Swain, F.M., 1969. Fossil carbohydrates In: G. Eglinton and M.T.J. Murphy (Editors), *Organic Geochemistry: Methods and Results.* Springer, New York, N.Y., pp. 374—401.

Treibs, A., 1934. Chlorophyll und Häminderivate in bituminösen Gesteinen, Erdölen, Erdwachsen und Asphalten. Ein Beitrag zur Entstehung des Erdöls. *Ann. Chem. Liebigs,* 510: 42—62.

Treibs, A., 1936. Chlorophyll- und Häminderivate in organischen Mineralstoffen. *Angew. Chem.,* 49: 682—686.

Waksman, S.A., Carey, C.L. and Reuszer, H.W., 1933. Marine bacteria and their role in the cycles of life in the sea. *Biol. Bull.,* 65: 57—79.

Weeks, L.F., 1958. Habitat of oil and factors that control it. In: L.G. Weeks (Editor), *Habitat of Oil.* Am. Assoc. Pet. Geol., Tulsa, Okla, pp. 1—61.

Weiss, A., 1969. Organic derivatives of clay minerals, zeolites, and related minerals. In: G. Eglinton and M.T.J. Murphy (Editors), *Organic Geochemistry: Methods and Results,* Springer, New York, N.Y., pp. 737—781.

Welte, D.H., 1965. Relation between petroleum and source rock. *Bull. Am. Assoc. Pet. Geol.,* 49: 2246—2268.

White, A., Handler, P., Smith, E.L. and De Stetten, W., 1959. *Principles of Biochemistry.* McGraw-Hill, New York, N.Y., 1149 pp.

Whitehead, W.L. and Breger, I.A., 1950. The origin of petroleum: effects of low-temperature pyrolysis on the organic extract of a Recent marine sediment. *Science,* 111: 335—337.

ZoBell, E., 1946a. Studies on redox potential of marine sediments. *Bull. Am. Assoc. Pet. Geol.,* 30: 477—513.

ZoBell, E., 1946b. *Marine Microbiology.* Chronica Botanica, Waldham, Mass., 240 pp.

ZoBell, E., 1952. Bacterial life at the bottom of the Philippine Trench. *Science,* 115: 507.

ZoBell, E., 1963. Geochemical aspects of the microbial modification of carbon and its compounds. In: U. Colombo (Editor), *Advances in Organic Geochemistry.* Pergamon Press, London, pp. 330—356.

Reference added in proof

Degens, E.T. and Hecky, 1974. Paleoclimatic reconstruction of Late Pleistocene and Holocene based on biogenic sediments from the Black Sea and a tropical African lake. *Coll. Int. C.N.R.S.,* No. 219 — *Les Méthodes Quantitatives d'Etude des Variations du Climat au Cours du Pleistocène.*

Chapter 5

DIAGENESIS OF COAL (COALIFICATION)

MARLIES TEICHMÜLLER and ROLF TEICHMÜLLER

INTRODUCTION

In the stage of hard brown coal (subbituminous coal) there are already such severe physical and chemical changes that despite the prevailing low temperatures and pressures, many authors consider these changes as being metamorphic. They place the termination of coal diagenesis at the boundary between lignite and subbituminous coal (cf. Table 5-I). This diagenesis of coal is often referred to as "biochemical coalification". The later, more severe alterations commonly are called "geochemical coalification" or metamorphism of coal, although the temperature and pressure conditions do not lead yet to mineral transformations which are indicative of the metamorphism of rocks other than coal.

According to most mineralogists, the beginning of rock metamorphism (greenschist facies) coincides with the transformation of meta-anthracite into graphite (Schüller, 1961).

The coalification (or "carbonification") begins with peat and ends with meta-anthracite. Inasmuch as this development is a very complex process, one must use different properties to measure the degree of diagenesis and metamorphism, i.e., the "rank" of coal. The values of the different rank parameters change with the degree of coalification.

THE PARENT MATTER OF COAL

Most coals originate from peats of low moors with plant associations of forests or reeds. Plants usually decompose after their death, i.e., under the influence of oxygen they are converted into gaseous compounds and water. In swamps with a high water table and lack of aeration, however, the plant residues are protected from total decomposition by a reducing environment, which is conducive to the process of peat formation.

One can differentiate between humic coals and sapropelic coals. Humic coals originate from true peats, which develop through the accumulation of dead plant matter at the site of the peat-forming plants. Sapropelic coals, on the other hand, are formed from organic muds, which are deposited on the floor of poorly aerated quiet-water lakes and ponds. They contain many

TABLE 5-I

The main stages of coalification distinguished by chemical, physical and microscopical characteristics and the applicability of various measures of rank (after Patteisky and M. Teichmüller, 1960; *International Handbook of Coal Petrography*, 1963, 1971; and M. Teichmüller, 1974a)

Rank: German		Rank: USA	Refl. Rm_{oil}	Vol. M. d. a. f. %	Carbon d. a. f. Vitrite	Bed Moisture	Cal. Value Btu/lb (kcal/kg)	Microscopic Characteristics	Applicability of Different Rank Parameters
Torf		Peat	0,2	68; 64	ca. 60	ca. 75		free cellulose, details of initial plant material often recognizable, large pores	bed moisture (ash-free); calorific value (moist, ash-free)
Weich-	Braunkohle	Lignite	0,3	60; 56		ca. 35	7200 (4000)	no free cellulose, plant structures still recognisable, cell cavities frequently empty	bed moisture (ash-free); calorific value (moist, ash-free)
Matt-	Braunkohle	Sub-Bit. C, B	0,4	52; 48	ca. 71	ca. 25	9900 (5500)	geochemical gelification and compaction takes place, vitrinite is formed,	bed moisture (ash-free); calorific value (moist, ash-free)
Glanz-	Braunkohle	Sub-Bit. A; C	0,5; 0,6	44	ca. 77	ca. 8-10	12600 (7000)	formation of exudatinite; beginning of 1st coalification jump, formation of micrinite	carbon (dry, ash-free); reflectance of vitrinite; bed moisture (ash-free); calorific value (moist, ash-free)
Flamm-	Steinkohle	B (High Vol. Bituminous)	0,7	40					carbon (dry, ash-free); reflectance of vitrinite; calorific value (moist, ash-free)
Gasflamm-	Steinkohle	A (High Vol. Bituminous)	0,8	36				rapid rise of red/green quotient of sporinite fluorescence	carbon (dry, ash-free); reflectance of vitrinite; calorific value (moist, ash-free)
Gas-	Steinkohle		1,0	32					carbon (dry, ash-free); reflectance of vitrinite; calorific value (moist, ash-free)
Fett-	Steinkohle	Medium Volatile Bituminous	1,2; 1,4	28; 24	ca. 87		15500 (8650)	beginning of 2nd coalification jump, rapid rise of liptinite reflectance; Rm sporinite = Rm vitrinite	volatile matter (dry, ash-free); reflectance of vitrinite
Ess-	Steinkohle	Low Volatile Bituminous	1,6; 1,8	20; 16					volatile matter (dry, ash-free); reflectance of vitrinite
Mager-	Steinkohle	Semi-Anthracite	2,0	12				Rmax liptinite > Rmax vitrinite	volatile matter (dry, ash-free); reflectance of vitrinite
Anthrazit		Anthracite	3,0; 4,0	8; 4	ca. 91		15500 (8650)		hydrogen (d. a. f.); reflectance of vitrinite
Meta-Anthr.		Meta-A.							hydrogen (d. a. f.); moist.; X-ray; diffr.

The data refer to the German classification (1st column). They do not refer precisely to the North American (ASTM, USA) commercial classification (2nd column).

allochthonous elements. The sapropels do not undergo peatification, but pass through a "saprofication" stage, which is characterized by putrefaction processes under anaerobic conditions.

Most humic coals originate from forest peats and, therefore, mainly from wood, bark, leaves and roots of swamp vegetation. In swamps with herbaceous plant associations (e.g., reed marshes), the roots of the sedges and grasses play an important role in the formation of peat. During the process of peat formation, wood, bark, leaves and roots are usually almost completely transformed into humic substances, which eventually form the maceral-groups [1] of *huminites* (peat and lignite stages) and *vitrinites* (bituminous coal and anthracite stages). *Liptinites* or *exinites* are formed from chemically resistant and relatively hydrogen-rich plant components such as pollen grains and spores, leaf epiderms (cuticles), resins, and waxes. These substances commonly are of smaller importance in coals than is humic matter. Besides, relatively carbon-rich macerals occur in coals such as *fusinite*, *semifusinite* and *macrinite*, which are grouped under the *inertinite* maceral group. These components commonly originate from a relatively strong aerobic decomposition of plant residues at the peat surface or from swamp fires (fossil charcoal) (cf. Stach et al., 1975).

The parent material of coal consists mainly of celluloses, hemicelluloses, and lignins, with minor amounts of proteins, sugars, pentosanes, pectines, tannins, and bituminous substances. The latter comprise such substances as fats, oils, waxes, resins, sterins, sporopollenins, cutine, and suberine. The inorganic components of the coal originate partially from the plants. Most of them (i.e., clay, silt and sand), however, were transported by water or air into the swamps. Some were precipitated during syngenesis, diagenesis, and/or epigenesis from solutions in the peat or coal (pyrite, quartz, carbonates, etc.).

THE FORMATION OF PEAT

A prerequisite for the formation of peat is stagnant ground water, in which the plant residues are protected from total decomposition. The peat formation begins at the surface under oxidizing conditions. Fungi and aerobic bacteria play an important role in this process. Gradual sinking and covering by younger peat layers, however, produce reducing conditions, and fungi and aerobic bacteria are substituted increasingly by actinomycetae and anaerobic bacteria. The total microbial activity decreases with increasing

[1] Maceral = smallest petrographic unit of coal, comparable to the mineral of rocks (see Table 5-IV).

depth. Certain bacteria of the fluorescent group which are pressure-resistant, however, seem to survive up to the soft brown coal stage (Beck and Poschenrieder, 1957). This phase of coalification is, therefore, called biochemical coalification.

Typical products of peat formation are the humic substances which are responsible for the dark color of peat. One can divide peat formation into:

(1) a primary phase, which is effective in the "peatigenic layer" at and immediately below the surface, and which is characterized by relatively fast oxidation processes; and

(2) a secondary phase with much slower conversions in a reducing environment (Kurbatov, 1963).

The primary phase controls the degree of decomposition or of "humification" of a recent peat. A more or less great part of the original plant structures disappears, depending on the intensity of the microbial and chemical decomposition and on the type of parent material. The amount of colloidal humic matter increases. The chemically resistant bituminous substances are usually preserved and exhibit their original shapes and structures (pollen-exines, suberine, etc.); however, their optical properties (e.g., fluorescence) are changed. The lignin-poor parenchymous tissues, which are not protected by antiseptic agents, are more easily destroyed because of their high cellulose content than are lignin-rich wood tissues. A great part of the cell structure of the latter is usually conserved, even if the lignin fraction is subsequently completely converted into humic matter. One can, therefore, find well-preserved plant tissues even in coals.

The more easily hydrolyzable substances of the swamp vegetation, such as cellulose, hemicellulose, sugar, pentosanes, pectines, and proteins, are decomposed first. Lignin, tannins, and bituminous substances are thereby concentrated. But even lignin is attacked, mainly under the influence of wood-rotting fungi and aerobic bacteria, and is gradually converted into humic substances.

The most important chemical process during the formation of peat is undoubtedly the formation of humic acids, which is enhanced by oxygen supply (e.g., caused by a lowering of the water table and a temporary drying of the peat surface), by higher peat temperatures (e.g., in tropical climates), and by alkaline environments (e.g., as a result of the influx of calcium-rich water). The degree of humification thus depends to a large degree on facies conditions. Humic acids, however, are not formed only at the peat surface in the presence of air oxygen, but according to Welte (1952), the formation of humic acids, e.g., from lignin, can also be an auto-oxidative condensation process, which may take place in a weakly acid environment under microbiological influence. In neutral or alkaline environments, however, it can also proceed as a result of purely chemical reactions.

Lignin is not the only parent substance of humic matter, as has often been assumed. Tannins, decomposition products of cellulose, other carbohydrates, and proteins and metabolic products of microorganisms, are also involved in the formation of humic acids (Flaig, 1968). According to Rakowski et al. (1963), a large part of the humic acids seems to have been formed from carbohydrates. Thus, reed peats, for example, may contain a large amount of humic acids (>55%), although the original vegetable matter contains very little lignin. Recent peats generally contain, depending on their degree of humification, 15—50% of humic acids in their dry substance. The humic acid content increases or decreases with depth, depending on the primary degree of decomposition of the individual peat layers.

The average moisture content of recent peats is about 90%. Their water-retaining capacity may even reach 98%. Tropical and subtropical peats contain less water than do peats from moderate climates (Van der Molen and Smits, 1962). Only a relatively small part of the water fills the large pores and can be squeezed out. By far the greatest part is adsorbed and the capillary water can only be removed by drying. Air-dried peat still contains about 12% of water.

Relatively little is known about the diagenesis of peat with increasing depth. Undoubtedly the pore volume decreases and the density increases with depth of burial, whereas the moisture content decreases. In a 200 m thick, Macedonian peat section, there is a decrease in moisture content from 89% to 69% (i.e., 10% per 100 m versus 3—4%/100 m in the soft brown coal stage) (M. Teichmüller, 1968). In the same section the change of optical properties of peat macerals with depth are surprisingly small, whereas they are great in the peatigenic layer. According to Zailer and Wilk (1911), the water content of air-dried peats increases with increasing degree of decomposition. The chemical changes in the anaerobic zone are relatively insignificant. On the other hand, the carbon content (dry, ash-free) in the "peatigenic layer" near the surface may increase from 45 up to 60% as a result of the decomposition of cellulose and hemicellulose and the concentration of lignin and humic matter; however, there is only a small increase in carbon content at greater depths (Pigulevskaja and Rakowski, 1963). Rakowski et al. (1963) observed up to 40% undestroyed carbohydrates in primarily weakly decomposed interglacial peats.

BROWN COAL

The transition from peat to brown coal ("Braunkohle") is a gradual one. Kayser (1952) set the boundary at a moisture content of 75%. More important than the moisture content is the manner of moisture adsorption which

TABLE 5-II

Distinction between peat and brown coal

Distinguishing feature	Peat	Brown coal
Moisture (%)	>75	<75
Carbon (d.a.f.) (%)	mostly <60	mostly >60
Carbohydrates	present	not present
Free cellulose	present	not present
Ease of cutting	great	none

in brown coals is very much stronger than in peats (M. Techmüller, 1968). The limit for the carbon content is often given as 60%. Rakowski et al. (1963) set the boundary between peat and brown coal at the point where the carbohydrates, which can amount up to 70% of the original plant material of the peats, disappear. Others employ the presence of "free cellulose" (not intermixed with lignin) and the ease of cutting as parameters for the distinction between peat and brown coal (Table 5-II).

According to petrographic properties, in Middle Europe the brown coals are differentiated into soft brown coal ("Weichbraunkohle"), dull brown coal ("Mattbraunkohle"), and bright brown coal ("Glanzbraunkohle") (Table 5-I). Dull brown coal and bright brown coal are combined in the term hard brown coal ("Hartbraunkohle"). This classification corresponds to the ASTM classification of North America as presented in Table 5-I.

Soft brown coal

Soft brown coal ("Weichbraunkohle") is light brown to dark brown in color, dull, and earthy. Petrographically, soft brown coal still resembles peat, but is more solid and is denser. In many soft brown coals, individual plant remains, such as wood remnants, leaves and fruits, are macroscopically recognizable. The microscope reveals a relatively good preservation of many plant tissues. As a result of their thin sediment cover (overburden), these structures are only slightly deformed. The cell lumens are often empty or filled with water. The humic detritus surrounding the recognizable plant remains, is commonly loosely packed. Microbedding usually is not, or only obscurely, developed; subaquatic deposits such as gyttjae and other organic muds constitute an exception. The optical reflectance of the humic substance on polished surfaces is very weak: generally <0.35% R_m (oil) [1], and their absorption in thin sections is accordingly low.

[1] R_m (oil) = mean reflectance of huminite or vitrinite under immersion oil.

A pressure-induced anisotropy, which is oriented parallel to the bedding plane, does not yet exist. On the other hand, the residual cellulose can be identified in thin sections by its optical anisotropy. It has been preserved in such tissues where incrustations of the cell wall with lignin, resin or cutin protected it from microbial and chemical decomposition. The fluorescence properties of spores and pollen grains (see Fig. 5-9) resemble those which are characteristic of peats (Ottenjann et al., 1974).

Hard brown coal

Dull brown coal. Soft brown coal ("Weichbraunkohle") is similar to peat, whereas dull brown coal ("Mattbraunkohle") is more similar to bituminous coal ("Steinkohle"). Dull brown coal is more solid, darker in color and already well-bedded. Plant remains are only occasionally recognizable megascopically. Under the microscope one can observe a distinct compaction and homogenization of the humic substances. This phenomenon is especially striking in the case of xylites (stems): in soft brown coal they hardly differ from recent stems, whereas in hard brown coal they have become layers of vitrite. In dull brown coal the empty cell cavities disappear and the formerly rigid cell walls become homogenized, plasticly deformed, and compressed. One can simulate these processes to a great extent by compressing soft brown coals at room temperature and squeezing the water out (M. Teichmüller, 1962). The reflectivity of huminites in dull brown coals varies between ca. 0.45 and 0.55% R_m (oil).

The homogenization and compaction of the humic substance (petrographic term: *huminite*) is called *gelification* (Vergelung in German) (Teichmüller and Teichmüller, 1954). During diagenetic gelification, the huminites of peat and lignite are converted into the vitrinites of subbituminous and bituminous coal (*vitrinitization*). It is not yet understood whether or not peptization of the humic colloids occurs during this striking process, possibly induced by the neoformation of bituminous substances which act as the solvent (M. Teichmüller, 1974a, b) (cf. Fig. 5-1). The experiments of H. Stach (1948) indicate that peptization does occur, because he was able to convert soft brown coal, by peptization with alkali and subsequent drying of the coal gel, into bright brown coal ("Glanzbraunkohle"). These colloidal—physical processes in the dull brown coal stage lead to a significant compaction of the humic constituents and a consequent reduction of pore volume. The softened individual constituents are easily arranged into the bedding plane by the overburden pressure. This pressure also rearranges the smallest elements of the humus colloids, the micelles, and thus gives rise to birefringence of huminites—vitrinites. These petrographic changes in the stage of hard brown coal (subbituminous coal) are more striking than those between other coal ranks (Teichmüller and Teichmüller, 1954).

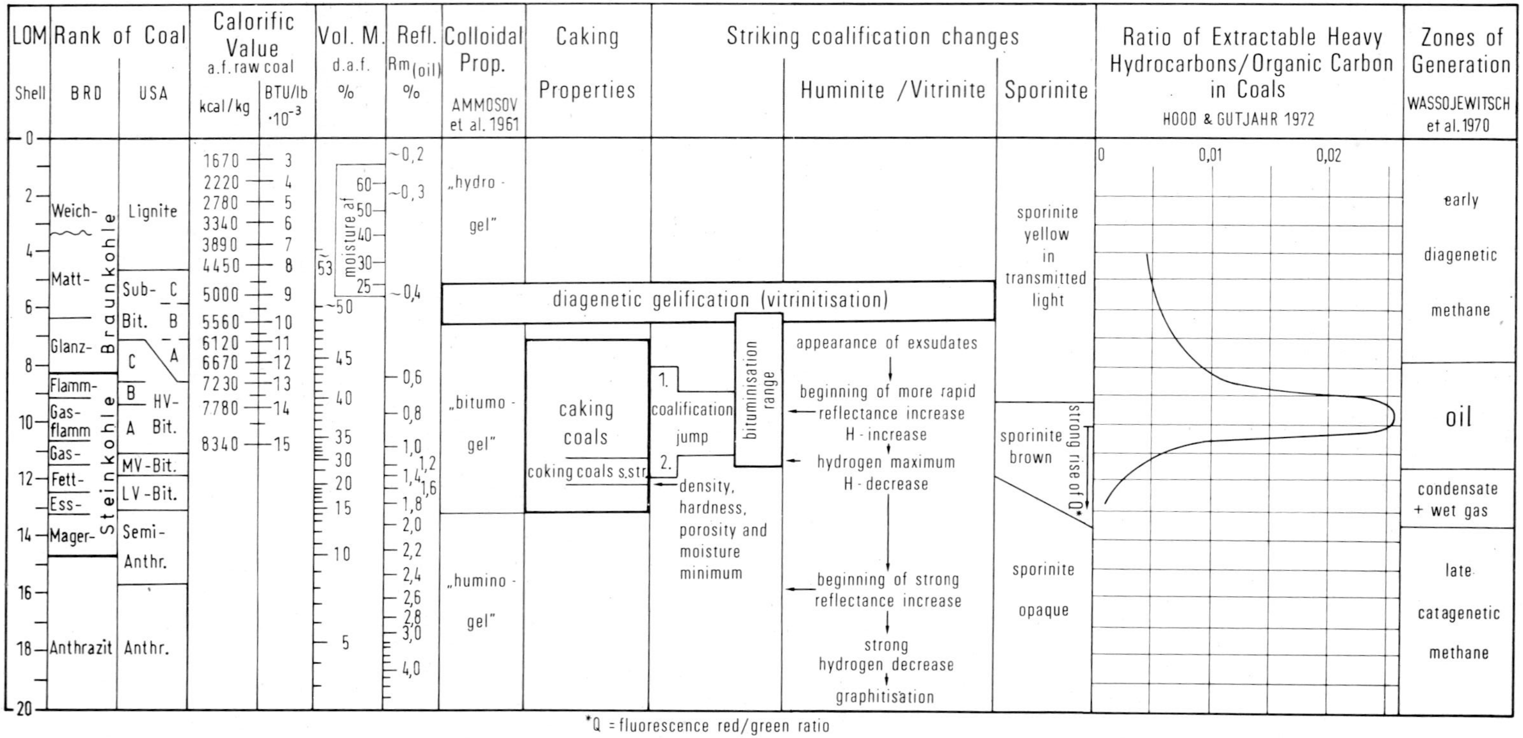

Fig. 5-1. Relationship between coalification and bituminization: changes of coal properties and generation of hydrocarbons from kerogen. (After Hood and Gutjahr, 1972, and M. Teichmüller, 1974a, b.)

Bright brown coal. The gelification is completed during the stage of bright brown coal ("Glanzbraunkohle"). Bright brown coals cannot be distinguished petrographically from bituminous coals and, therefore, in North America (ASTM classification) they are included among the subbituminous and high-volatile bituminous *C* coals (Table 5-I). Bright brown coal is solid, black, bright, and under the microscope shows the typical microlithotypes of the bituminous coals which are different in the Tertiary and the Carboniferous coals. Wood and bark have been converted into layers of vitrite and the originally loose detritus has been compressed into compact layers of clarite and durite. The reflectance of vitrinite varies from ca. 0.45 to 0.6% R_m (oil). Some liptinite macerals change distinctly, entering into a first coalification jump, which goes parallel with the beginning of oil formation in oil source rocks (cf. Fig. 5-1) (M. Teichmüller 1974a, b).

Chemical changes during the brown coal stage

On the basis of ultimate and proximate analyses, the chemical changes which accompany the significant petrographic changes at the boundary between dull and bright brown coal are relatively small. The average carbon content (d.a.f. = dry, ash-free) of 60—69% of soft brown coals increases to 71—77% in bright brown coals. The yield of volatile matter [1] varies considerably in brown coals (Fig. 5-2). It decreases roughly from ca. 65 to 56% (d.a.f.) in soft brown coals to ca. 42% at the boundary between bright brown coal and bituminous coal. The decrease in moisture and the corresponding increase in calorific value of the "as mined" coal are quite pronounced. When mined, a soft brown coal contains 75—35% water, a dull brown coal has 35—25% water, and a bright brown coal contains 25 to about 8% water (Table 5-I). This decrease in moisture content is partially caused by the reduction in porosity. In addition, there is a progressive splitting of hydrophilic groups from the humic molecules, which are very sensitive to temperature rises. The decrease in moisture content slows down with increasing coal rank: in soft brown coals, with about 60% water, the moisture of the coal as mined decreases by 1% per 25—30 m of increase in depth; in dull brown coals, by 1% per 100 m; and in bright brown coals, by 1% per 100—200 m.

At present little is known about the chemical changes which become explicit in the functional group analysis. According to Blom et al. (1957), the concentration of methoxyl ($-OCH_3$), carboxyl ($-COOH$), and carbonyl ($\geq C=O$) groups decrease rapidly with increasing carbon content during the different brown coal stages. The hydroxyl($-OH$) concentration, on the

[1] Volatile matter given off during crucible-coking of the proximate analysis.

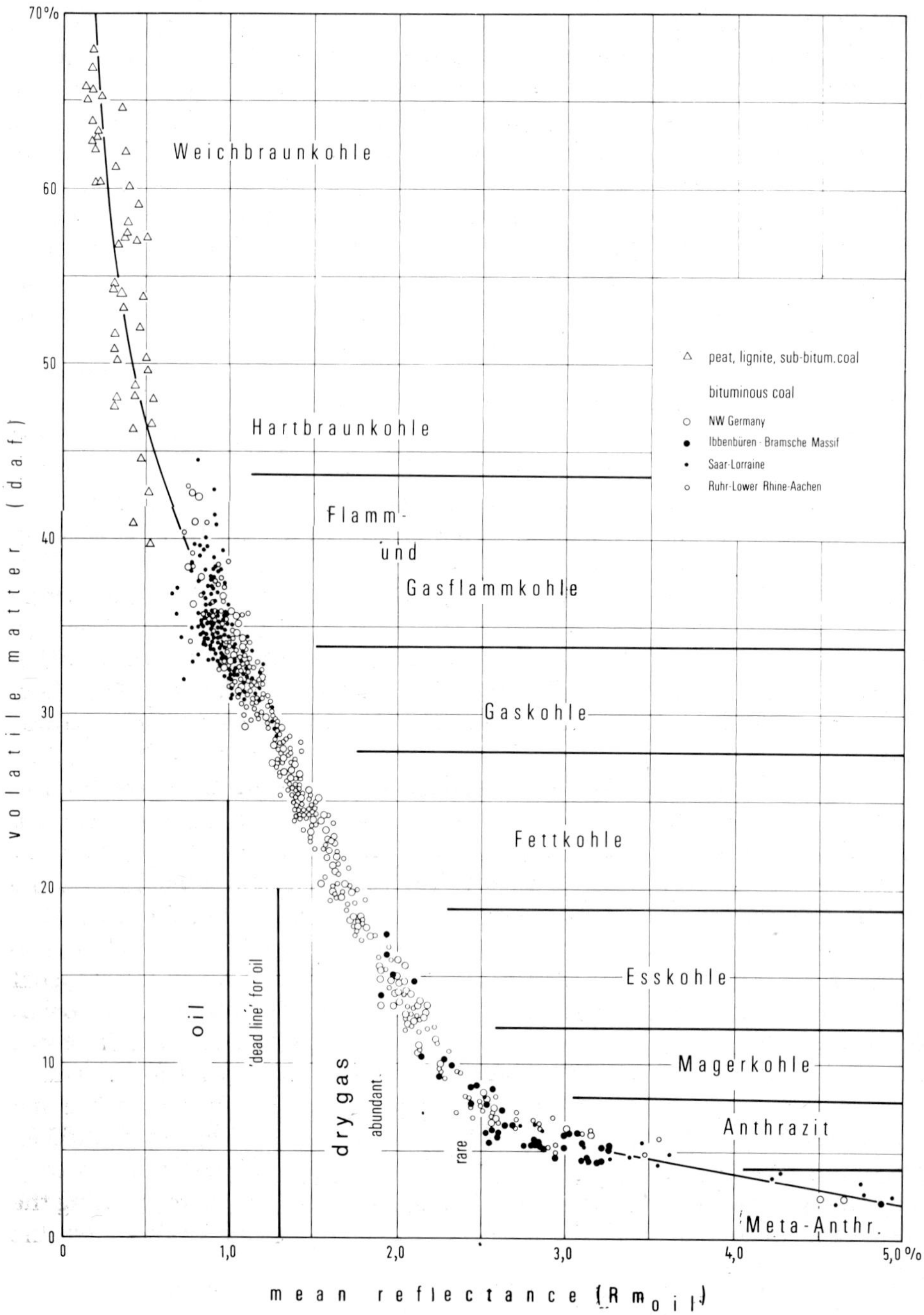

Fig. 5-2. Relationship between volatile matter of vitrite and mean reflectance of huminite—vitrinite for the different rank stages (soft brown coal to meta-anthracite) of German coal deposits and the occurrence of oil and natural gas.

other hand, varies and shows a significant drop only during the very first stages of brown coal formation and in the early stage of the bituminous coal formation. Lignin is gradually converted completely into humic substances. In addition, cellulose, originally present in small concentrations during the soft brown coal stage, eventually disappears. Starting from the hard brown coal stage, the humic acids gradually lose their acid character and, thus, become insoluble in alkali. The molecular complexes of the humic substance are enlarged and, consequently, color and streak of the coal become darker. The oxidizability by nitric acid decreases.

Boundary between brown coal and bituminous coal

The transition from brown coal to bituminous coal is, like the transition from peat to brown coal, a gradual one. The boundary between brown coal and bituminous coal is drawn differently. In central Europe three characteristics are used in distinguishing between brown coals and bituminous coals. Two of these distinguishing features must be present in order to name a coal "brown coal" or "bituminous coal" (Table 5-III).

On the basis of these characteristics, Patteisky and Teichmüller (1960) have drawn the boundary between brown coal and bituminous coal at concentrations of about 8—10% moisture, 77% carbon (d.a.f.), and 16% oxygen (d.a.f.), and a heating value of 7000 kcal/kg (ash-free), all values being based on vitrites. This classification has been accepted in the *International Handbook of Coal Petrography, 1963.* In the European commercial classification of coals (E.C.E. system), the boundary is drawn at a calorific value of 5700 kcal/kg (ash-free). Thus, using this classification, the bright brown coals ("Glanzbraunkohlen") would belong to the bituminous coals. According to the American ASTM classification, lignites and bituminous coals are separated by the transitional group of sub-bituminous coals (cf. Table 5-I).

TABLE 5-III

Distinction between brown coal and bituminous coal

Distinguishing characteristic	Brown coal ("Braunkohle")	Bituminous coal ("Steinkohle")
Streak (color)	brown, seldom black	black, seldom brown
Behavior on boiling with KOH (detection of humic acids)	brown solution	no color
Behavior with dilute HNO_3 (oxidizability)	red solution	no color

BITUMINOUS COAL AND ANTHRACITE

The stages of bituminous coal ("Steinkohle") and of anthracite are also defined differently. The German and the North American (ASTM) classifications are shown in Table 5-I. Depending on the range of rank, the commercial classifications for bituminous coals and anthracites as used in different countries are based on different rank parameters (calorific value and volatile matter content).

The microscopic picture of a low-rank bituminous coal resembles that of a bright brown coal ("Glanzbraunkohle"): the individual macerals show distinct differences in their reflectance, color, and relief on polished surfaces. The macerals of the liptinite (exinite) group reflect the incident light only weakly. The humic matter, which has been converted to vitrinite in bituminous coal, has a stronger reflectivity, appears light grey, and has a less pronounced relief. The macerals of the inertinite group (e.g., fusinite and macrinite) show the highest reflectance, appear white to yellowish, and have a strong relief. Some liptinites pass through a first coalification jump. New macerals are formed as a consequence of this jump, which can be correlated with the beginning of oil formation in oil source rocks (M. Teichmüller 1974a, b) (Fig. 5-1). Exudatinite and micrinite are such "secondary macerals".

The differences between the single macerals diminish with increasing rank of coal. The reflectance of vitrinite increases rather continuously (Table 5-I, Figs. 5-2, 5-6) and gradually, so that it is possible to determine the rank of coal from reflectance measurements. Inertinites, on the other hand, show little alteration. During the stage of coking coal, liptinites pass through a second coalification jump which corresponds to the disappearance of oil due to cracking (Fig. 5-1). This jump causes severe chemical and optical changes of liptinite macerals (Stach, 1953), which become visible under the microscope by a quick decrease and, finally, by the loss of transparency and of visible fluorescence under irradiation of ultraviolet light (cf. Fig. 5-9). In reflected light, liptinite becomes distinctly lighter, reaching the stage of high-rank coking coal, with about 22—20% volatile matter, the reflectivity of vitrinite (Table 5-I). Thus, liptinite (= exinite) becomes indistinguishable from vitrinite by optical and chemical means (cf. Fig. 5-4), and the microscopic picture of a high-rank bituminous coal shows much less variation than that of a low-rank bituminous coal.

With increasing rank of coal, especially in the anthracite stage, the reflectance of vitrinite greatly increases (cf. Figs. 5-2, 5-6 and 5-8) and approaches that of inertinite; the same holds true for the vitrinized exinite. In high-rank anthracites only the strongly reflecting fusinite is easily contrasted with vitrinite. In meta-anthracites, the maximum reflectivity of vitrinite even

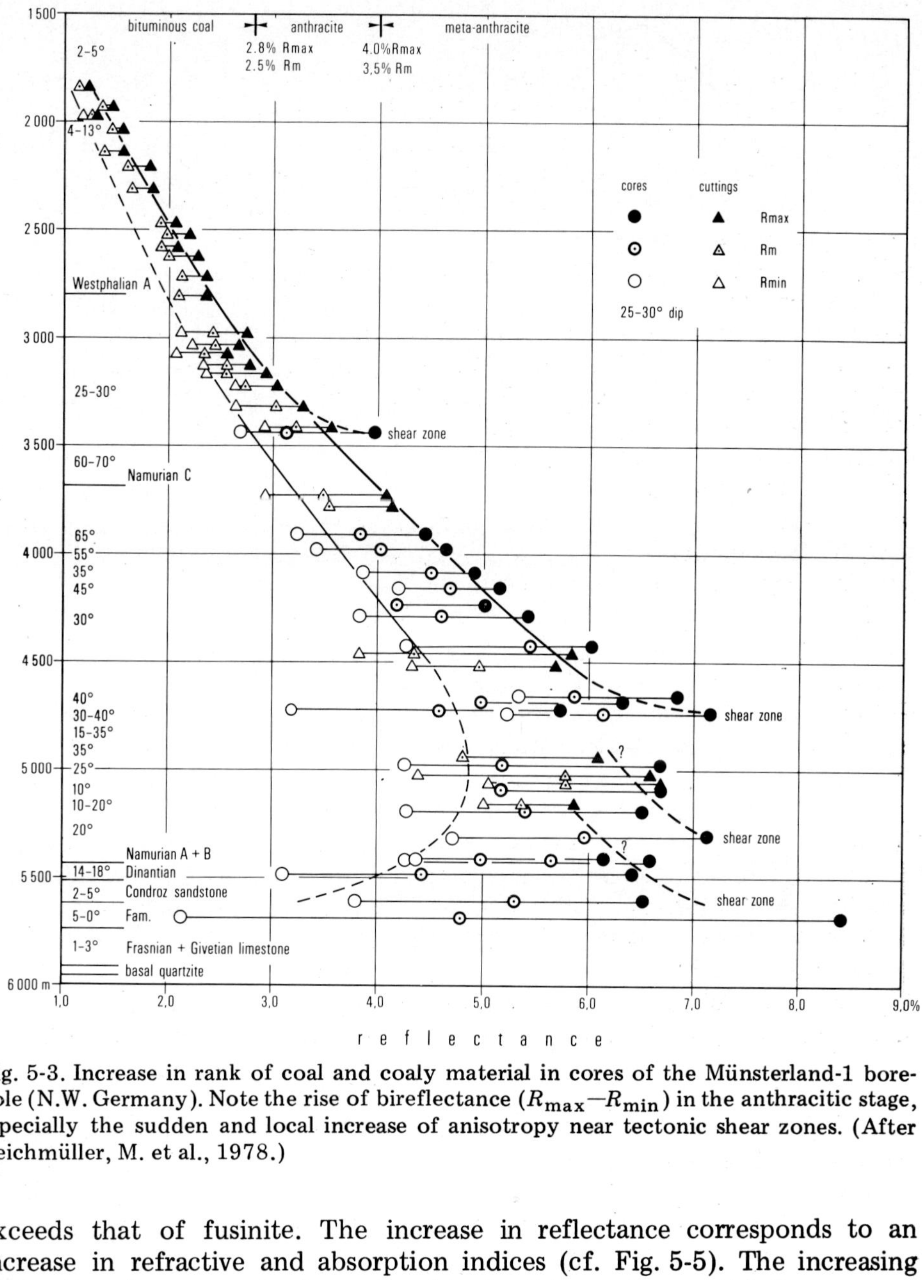

Fig. 5-3. Increase in rank of coal and coaly material in cores of the Münsterland-1 bore-hole (N.W. Germany). Note the rise of bireflectance ($R_{max}-R_{min}$) in the anthracitic stage, especially the sudden and local increase of anisotropy near tectonic shear zones. (After Teichmüller, M. et al., 1978.)

exceeds that of fusinite. The increase in reflectance corresponds to an increase in refractive and absorption indices (cf. Fig. 5-5). The increasing absorption becomes clearly noticeable in thin sections: vitrinite becomes increasingly darker, so that with increasing rank of the coal the sections must

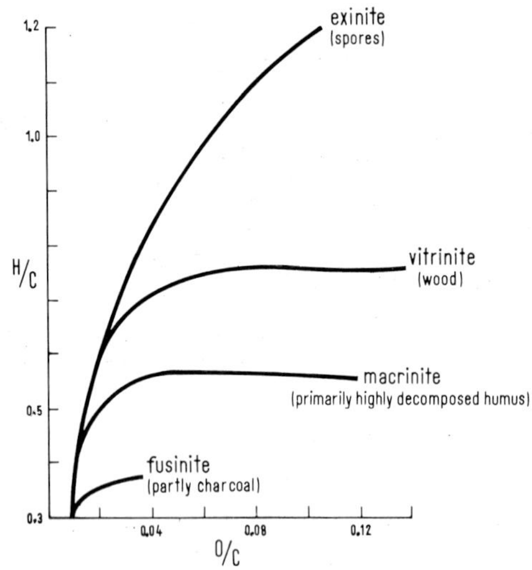

Fig. 5-4. Coalification tracks of different macerals based on H/C and O/C atomic ratios. (After Van Krevelen, 1961.)

be made continuously thinner, in order to obtain transparent thin sections.

These optical changes are correlated mainly with chemical changes, i.e., increasing aromatization and condensation of the molecular groups. Inasmuch as the volatile matter yield represents roughly the non-aromatic fraction of the humin molecules and because there exists close relationship between the degree of aromaticity and the reflectivity of vitrinite, the increase of reflectance in the bituminous coal range is more or less parallel with the decrease of volatile matter (Fig. 5-2). In the anthracite stage, the reflectivity rises especially quickly due to condensation processes and to a pronounced orientation of the large aromatic lamellae (cf. Fig. 5-5) into a graphite-like order.

In binary diagrams for different rank parameters, such as Figs. 5-3 and 5-8, and in graphs where rank is plotted versus depth (e.g., Figs. 5-6 and 5-7), four coalification "jumps" of vitrinite become explicit by breaks in the coalification curves. The first one occurs during the high-volatile bituminous coal stage [ca. 80% C, 43% volatile matter, and 0.6% R_m (oil)]; the second, in the medium-volatile bituminous coal stage [ca. 87% C, 29% v.m., and 1.3% R_m (oil)]; the third, at the boundary between semi-anthracite and anthracite [91% C, 8% v.m., and 2.5% R_m (oil)]; and the fourth, at the anthracite—meta-anthracite boundary [93.5% C, 4% v.m., and 3.5% R_m (oil)].

The rise of optical anisotropy with increasing degree of coalification must be attributed partly to physical changes: with increasing overburden pressure

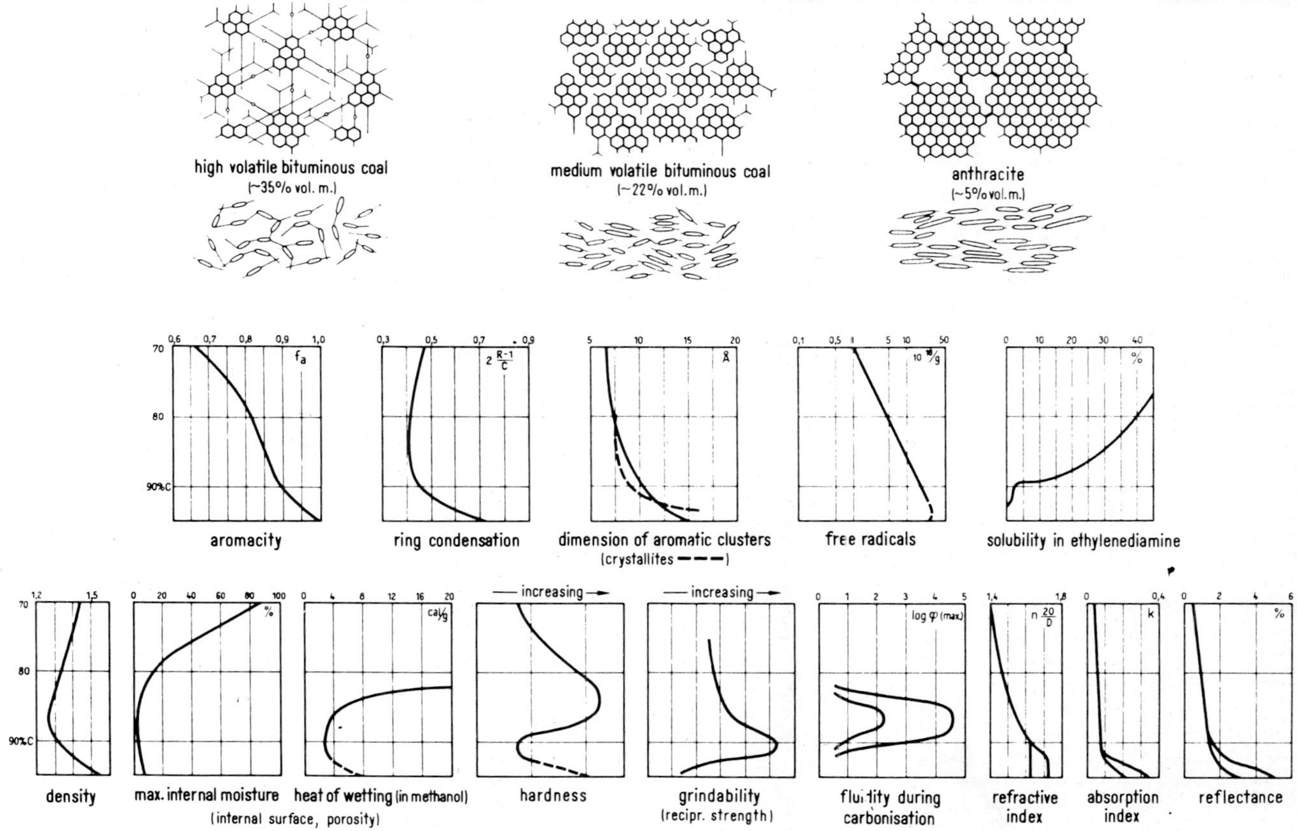

Fig. 5-5. Molecular, chemical and physical alterations of vitrite during coalification. (vol. m. = volatile matter).
First row: molecular structure; —○— = hydrogen bonding; — = molecular bonding.
Second row: orientation of molecules perpendicular to the bedding plane.
Third row: development of chemical properties in relation to carbon content (d.a.f.).
Fourth row: development of physical properties in relation to carbon content (d.a.f.).
(After Bangham and Maggs, 1944; Blayden et al., 1944; King and Wilkins, 1944; Dulhunty, 1950; Dryden, 1951; Dryden and Griffith, 1953; Huck and Karweil, 1953; Van Krevelen, 1953; Brown and Hirsch, 1955; Austen and Ingram, 1958; and others; from M. Teichmüller, 1962.)

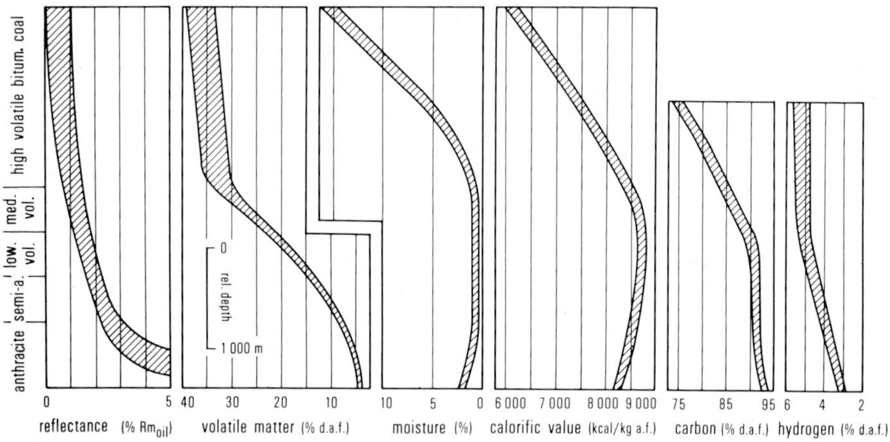

Fig. 5-6. The increase in degree of coalification with depth, based on vitrite analyses from deep boreholes and shafts, mainly in the Ruhr Basin.

the more or less laminated molecular complexes are arranged into the bedding planes. This pressure-induced anisotropy, which is at first only visible in transmitted light (in the stage of dull brown coal), starts to become noticeable in the medium-volatile bituminous coal stage also in reflected light. This becomes more and more pronounced with increasing coal rank, especially in anthracites (cf. Fig. 5-3): in a meta-anthracite with 3% volatile matter, the reflectance of vitrinite (in air) in polarized light may fluctuate between 18.1% (R_{max}) and 10.5% (R_{min}) on rotating the microscope stage (Dahme and Mackowsky, 1951). Figure 5-3 shows the sudden local increase of reflectance anisotropy ($R_{max(oil)} - R_{min(oil)}$) of anthracites in the vicinity of shearing zones in the borehole Münsterland 1. Shear movements obviously facilitate pre-graphitization, with accompanying strong anisotropy of anthracites.

Although meta-anthracite resembles graphite, as shown not only by reflectance anisotropy but also by X-ray and electron diffractions and by electric conductivity, it is still colloidal in nature. Macromolecules of meta-anthracite, consisting of aromatic lamellae, are merely well-oriented into the bedding plane and thus yield diffraction patterns similar to those of graphite. The formation of a true graphite lattice, however, requires pressures and temperatures which exist only during the metamorphic stage of phyllitization (Schüller, 1961; Landis, 1971).

The physical and chemical changes of coal, during the stages of bituminous coals and anthracites, are better known than the alterations in the different brown coal stages. New knowledge has been gained on using colloid-chemistry techniques, the structural analysis of petroleum, spectrometry

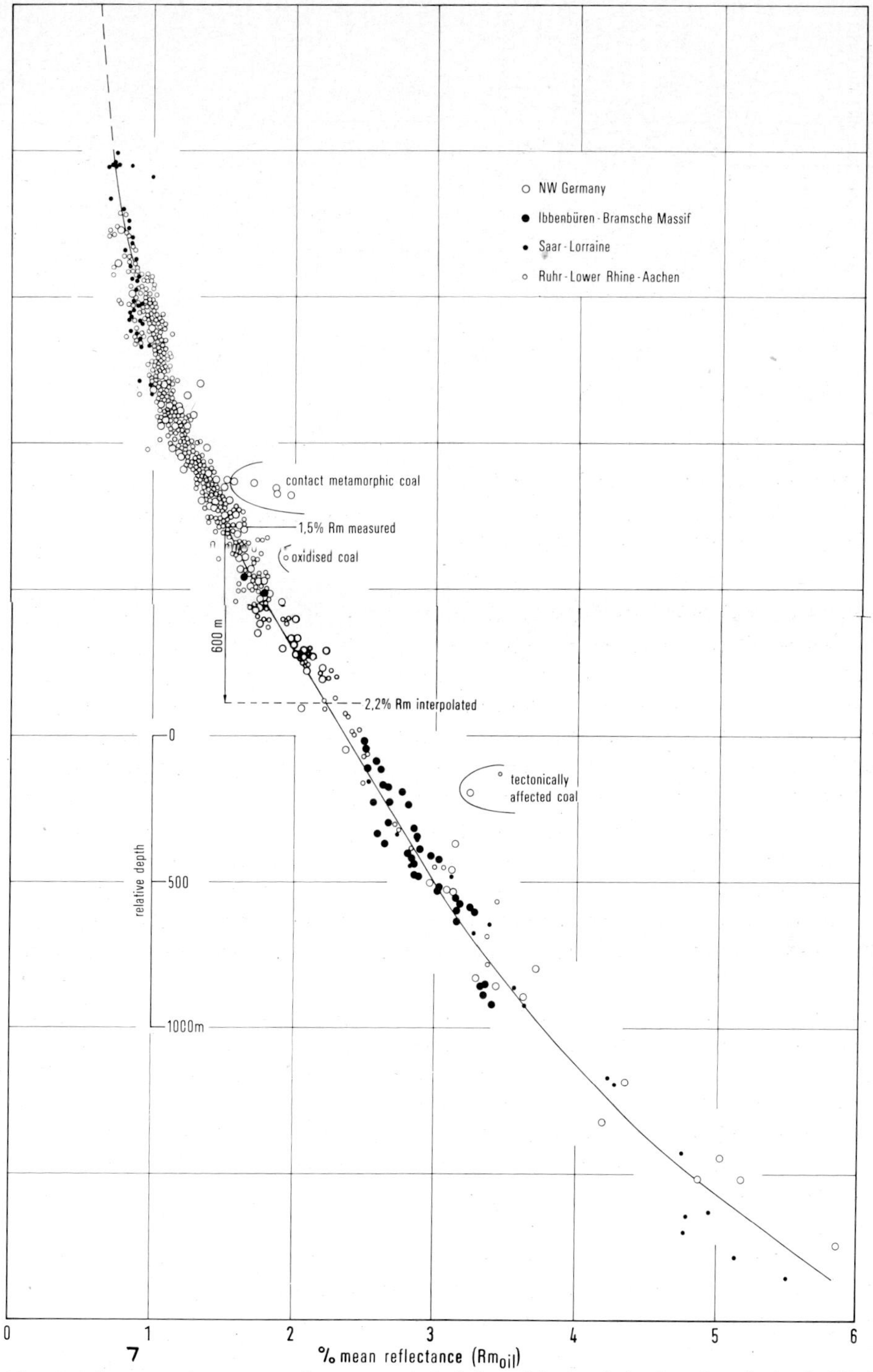

Fig. 5-7. Increase in mean reflectance of vitrinite with depth in deep boreholes in W. Germany. The reflectance values obtained in different boreholes are correlated on the basis of reflectance range. The overall mean curve allows predictions of coal rank at greater depths than those reached by a single borehole.

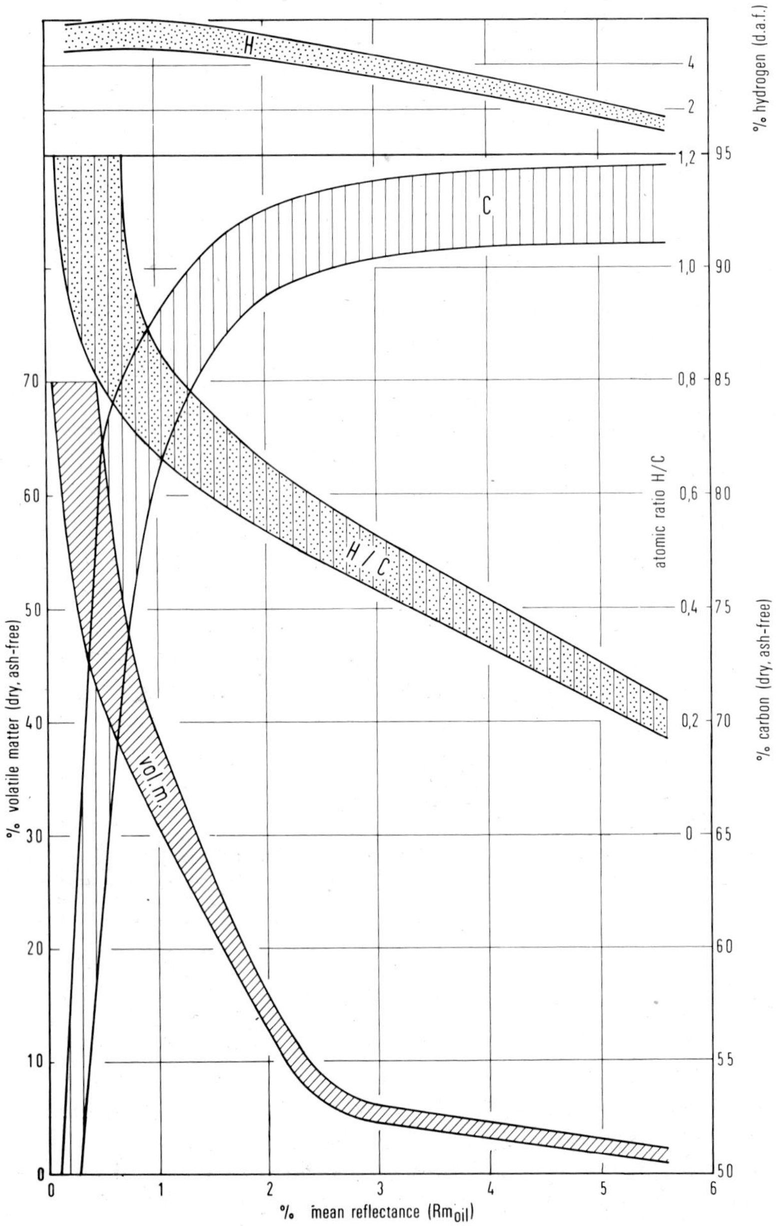

Fig. 5-8. Relationships between vitrinite reflectance and different chemical rank parameters. (After M. Teichmüller, 1971.)

(e.g., infrared, nuclear magnetic resonance, electron magnetic resonance), X-ray and electron diffraction, electron spin resonance, differential thermal analysis, electron microscopy, and solvent extraction (cf. Fig. 5-5).

The chemical alterations during the stage of bituminous coals and anthracites consist mainly of an increase in carbon content and a decrease in oxygen and, subsequently, also in hydrogen content (cf. Figs. 5-6 and 5-8). Fig. 5-4 shows these changes in terms of the H/C and O/C atomic ratios for different macerals. Moreover, the yield of volatile matter (which is liberated during the crucible-coking of the proximate analysis) decreases with increasing coal rank (cf. Figs. 5-2 and 5-6). As far as the chemical structure is concerned, the humic complexes become more and more aromatized and condensed (Fig. 5-5); the aromatic clusters of low-rank bituminous coals are still relatively small. Long chains and bridges of oxygen-rich and hydrogen-rich groups (—OH, —COOH, $-CH_3$) are attached to these aromatic clusters. With increasing rank, these groups are gradually split off, which at first yields mainly CO_2 and H_2O and then increasing quantities of CH_4. The concentration of aromatic carbon increases at the same time, and the aromatic clusters grow. At concentrations of 29% volatile matter and 87% carbon (d.a.f.), the rate of oxygen loss suddenly decreases. Instead, hydrogen is given off in increasing amounts as CH_4. At concentrations of about 20% volatile matter and 89% carbon (d.a.f.), many physical properties of the coal are changing rather rapidly. At this stage the lowest microporosity and internal surface area (as determined by the heat of wetting), the lowest density, and the minimum strength are attained, whereas plasticity and coking qualities become maximum (Figs. 5-1, 5-5). Also, the moisture content of the coal reaches a minimum of about 0.5% H_2O. The intermolecular spaces at this stage are particularly small, because bulky bridges and chains have split off to a great extent. The relatively heavy oxygen has been given off for the most part, whereas the light hydrogen has been concentrated. The low strength results from the lack of the original web of chains and bridges.

PHYSICAL AND CHEMICAL INDICATORS OF COAL RANK

One must use different physical and chemical parameters of coalification depending on the rank range which is to be investigated. It is always necessary to perform comparative rank investigations on the same constituent of coal, because the chemical changes proceed differently in the different macerals (Fig. 5-4). Vitrinite is best suited for comparative rank investigations, because it is the most abundant component of the coal and can be isolated relatively easily. In addition, it exhibits more continuous changes than the other macerals.

Table 5-I shows the applicability of different indicators of rank to the various rank ranges, as has been shown by binary diagrams of the individual parameters (Patteisky and Teichmüller, 1960; Teichmüller, 1971) and by rank gradients studied in deep boreholes (cf. Figs. 5-2, 5-5, 5-6 and 5-8).

For brown coal and low-rank bituminous coal (up to 30% volatile matter, d.a.f.), the moisture content or the calorific value of humite—vitrite, as mined, is the best measure of rank. Inasmuch as the calorific value of humite and low-rank vitrite is dependent on the moisture content, the determination of the moisture content is usually sufficient. The analysis should be performed under certain constant conditions (temperature, air humidity, etc.). Instead of the moisture content of the coal as mined, it is also possible to choose as a measure the maximum internal moisture content (moisture holding capacity). In the range from sub-bituminous to medium-volatile bituminous coal (with 30% volatile matter, d.a.f.), the carbon and oxygen [1] contents can also be used as rank parameters. The concentrations of volatile matter and hydrogen, on the other hand, are unsuitable as rank parameters in this range because their values vary considerably and show little correlation with depth (Figs. 5-2, 5-6, 5-8).

The rank of medium- and low-volatile bituminous coals (<30% volatile matter) is best determined by using the volatile matter content. As shown in Fig. 5-6, the volatile matter content of these coals decreases rather rapidly with depth down to the anthracite boundary (8% volatile matter). In the Carboniferous coals of the Ruhr Basin, this rate of decrease is equal to 1.9% volatile matter per 100 m stratigraphic depth. Volatile matter content is, therefore, used as a parameter in the commercial classifications of bituminous coals and anthracites (cf. Table 5-I and Fig. 5-2). The contents of carbon, hydrogen, and moisture show only small changes in coals having 30—8% volatile matter (cf. Fig. 5-8). Starting with the anthracite stage, the hydrogen content decreases considerably, whereas the carbon content increases slowly. The volatile matter yield decreases also relatively slowly over this range. Because of this, the hydrogen content is the best-suited chemical indicator of rank for anthracites. Other parameters for the degree of anthracitization are X-ray and electron diffraction diagrams which show an increasing pre-graphitization.

The reflectance of vitrinite on polished surfaces is a highly applicable rank measure for bituminous coals and anthracites, especially because it may be determined quickly and requires small samples, which would be insufficient for chemical analyses. The rank of even microscopically small coaly inclusions in various sedimentary rocks can be determined in this manner. Figure

[1] Inasmuch as oxygen usually is not determined directly in the ultimate analysis of coals and in this rank range the oxygen content is inversely proportional to the carbon content, the carbon content should be used preferentially.

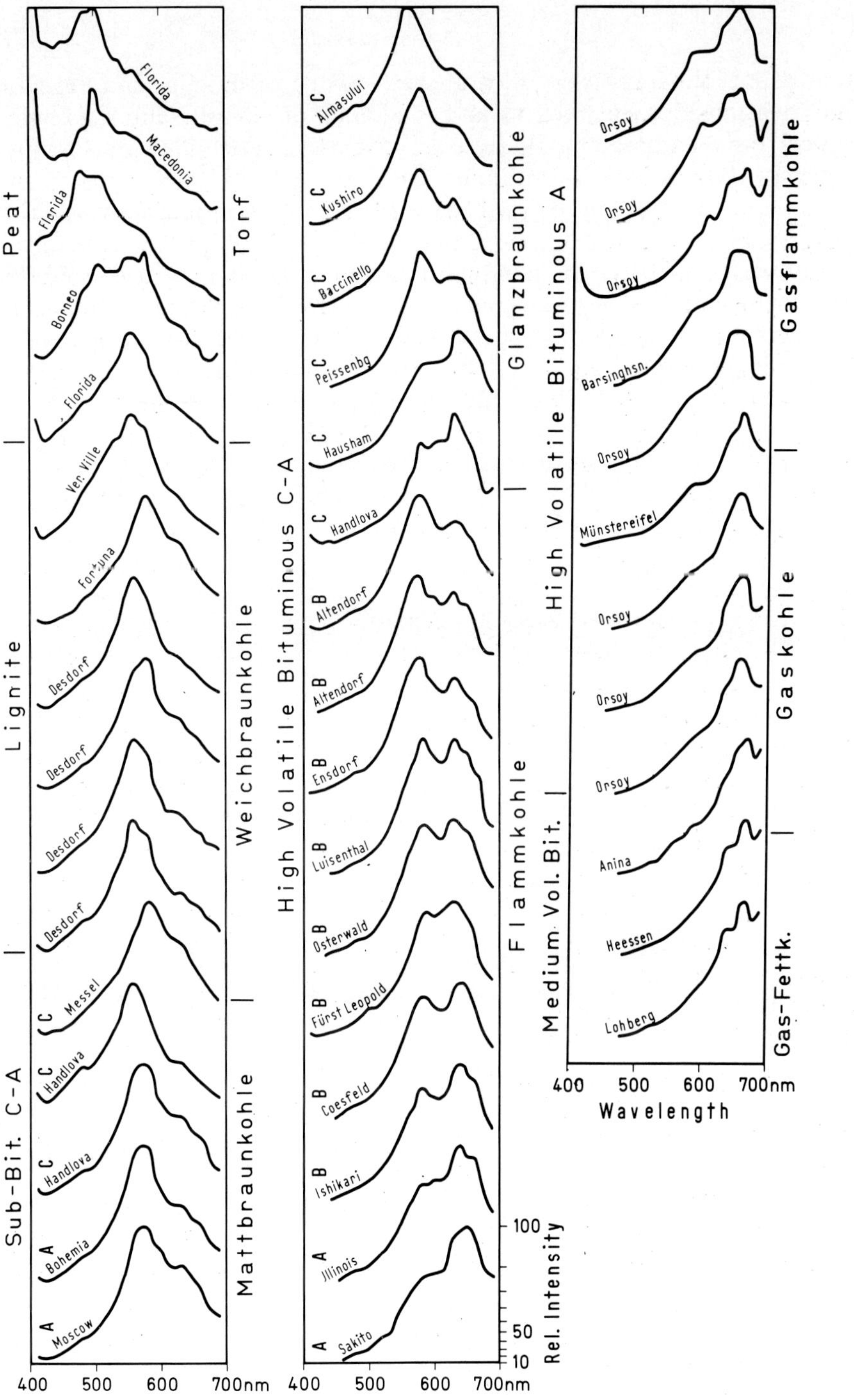

Fig. 5-9. Variation of the fluorescence spectra of sporinites with increasing degree of coalification (U.S.A. and German classifications). (After Ottenjann et al., 1974.)

5-7 demonstrates the increase of vitrinite reflectivity with depth and Fig. 5-8 shows the relationships between mean reflectance of vitrinite and the classical chemical rank parameters. Because of increasing reflectance anisotropy in the stages of high-rank bituminous coal and anthracite (cf. Fig. 5-3), it is best to use the maximum reflectivity as an optical rank parameter for such coals. [1]

The method of reflectance measurements on vitrinites in sedimentary rocks, other than coal, attained a great importance not only in the solution of paleogeographic, tectonic and geothermic problems but also in prospecting for ore, oil and natural gas (Teichmüller and Teichmüller, 1966b; Taylor, 1971; M. Teichmüller, 1971; Hood and Castaño, 1974).

Another optical rank parameter for the low-rank range (peat to medium-volatile bituminous coal) is the spectral fluorescence of sporinites of peats and coals (Ottenjann et al., 1974). Figure 5-9 shows the characteristic changes of fluorescence maxima with increasing rank. The increase in higher wavelength fluorescence with increasing rank of coal can be expressed numerically as follows:

$$\text{red/green ratio } Q = \frac{\text{relative intensity at 650 nm}}{\text{relative intensity at 500 nm}}$$

As in the case of vitrinite reflectance, the red/green ratio of sporinite fluorescence can be used to determine the rank of finely dispersed organic matter in sedimentary rocks and, thus, determine the degree of diagenesis of these rocks. Palynologists use the color, brightness, and structure of separated pollen grains in normal transmitted light as rough parameters for low-rank diagenesis (Gutjahr, 1966; Correia, 1971). Electron-spin-resonance (ESR) of finely divided organic matter is also applied to determine the rank of kerogenous and coaly material (Pusey, 1973).

SIGNIFICANCE OF PRESSURE, TEMPERATURE AND TIME IN COALIFICATION

Pressure

In the past it was assumed that pressure promotes the process of coalification. The following observations appeared to support this assumption:

(1) The coal rank increases with increasing depth of burial and, thus, with increasing overburden pressure (Hilt's rule).

(2) The rank of coal in the folded part of a basin is frequently higher than in the non-folded region. It was concluded, therefore, that the coalification

---

[1] In the stage of meta-anthracite (ASTM classification: >6% R_{max}), however, the maximum reflectance varies greatly, whereas the minimum reflectance decreases rapidly (cf. Fig. 5-3), reaching values of less than 0.5% R_{min} (oil) when graphite is formed (Ragot, 1977; Teichmüller, M. et al., 1978).

process was accelerated by the tangential (folding) pressure.

The experiments of Huck and Patteisky (1964; cf. Huck and Karweil, 1962), Davis and Spackman (1964), and Bostick (1973) have shown, however, that the static pressure not only does not enhance but, instead, retards chemical processes during coalification, because removal of gas is made more difficult.

The increase of coal rank with depth of burial is also readily explained by the increase of temperature with depth. Jüntgen and Karweil (1962) succeeded in obtaining a distinct increase in rank on applying dynamic pressure by grinding coal in a rocking ball mill. The authors themselves, however, emphasized that this experiment does not reproduce natural conditions, for, as a rule, the tectonic movements proceed so slowly that the term "dynamic pressure" (i.e., pressure which changes continuously in magnitude, direction, or even acceleration) is not applicable. The small increase in rank, which occasionally and locally occurs in the immediate vicinity of tectonic thrust planes, could have been caused by concentrated frictional heat generated during rapid tectonic movements.

It can be shown that in the Carboniferous of the Subvariscan Foredeep (Ruhr Basin, Belgium) and in other coal-seam-bearing foredeeps, the coalification was largely completed before initiation of folding (R. Teichmüller, 1952; Damberger, 1974), because the isorank lines run parallel to the folded bedding planes (Fig. 5-10). Consequently, the increase in degree of coalification cannot be due to the folding. Inasmuch as the layers in the folded part of the foredeeps generally were buried to a greater depth, more intense coalification of the folded coal seams can be easily explained by the higher rock temperature. The coals of the most intensely folded southern margin of the Ruhr Carboniferous basin exhibit a remarkably low degree of coalification. The coal seams here, i.e., at the basin margin, sank to shallower depths before folding than did the same seams further north. Likewise, a

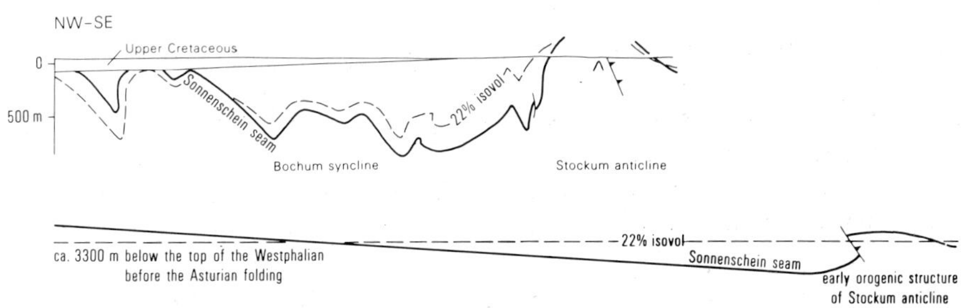

Fig. 5-10. Cross-section through the Bochum megasyncline (Ruhr Basin) showing the 22% isovol running approximately parallel with folds in the Sonnenschein seam (above) and the stage at the beginning of orogenic folding (below).

particular seam will be less coalified in the large epeirogenic swells than in the adjacent large depressions, where the seams sank more deeply before folding (Patteisky and Teichmüller, 1962).

Whereas pressure obviously does not have a significant effect on the chemical process of coalification, it can undoubtedly change the *physical* structure of coal. Pressure can close the pores and rearrange the micelles. Intensely folded coals are sometimes anomalously poor in moisture content because of their relatively low porosity (Dunningham, 1944; Edwards, 1948; Berkowitz and Schein, 1952). In addition, their optical anisotropy may have been affected by tangential pressure (Williams, 1953; Petrascheck, 1954; Monomakhoff, 1961).

The influence of pressure is most pronounced during compaction and, thus, is most obvious during the peat (up to the early bituminous coal) stage, as exhibited by a decrase in porosity and consequent reduction of the moisture content with depth (Fig. 5-6). The direct influence of tangential pressure becomes obvious during the metamorphic transition of meta-anthracite to graphite. This process certainly also requires high temperatures; however, inasmuch as it is accelerated by strong tectonic shearing stresses, it can also take place at lower temperatures. Taylor (1971) even quoted temperatures of the order of 300°C necessary for the formation of graphite in rocks under shear.

Temperature

The effect of temperature rise on the chemical reactions during coalification is undeniable (see, e.g., Dulhunty, 1954; Huck and Karweil, 1955; Dryden, 1956; Van Krevelen, 1961). Examination of coal at magmatic contacts demonstrates this as well as the experimental work. The effect is more pronounced the larger the intrusive body and the higher its temperature. Narrow vertical magmatic dikes dissipate relatively little heat and, consequently, their influence extends only from a few inches to a few feet (M. Teichmüller, 1973). The heating of coals above large intrusive bodies (batholiths), on the other hand, is still distinctly noticeable over distances of several thousand meters, even where the host rock is not any longer affected (Teichmüller and Teichmüller, 1950; Stadler and R. Teichmüller, 1971; Stadler and M. Teichmüller, 1971; Damberger, 1974). The Bramsche Massif in N.W. Germany is a classical example of an intrusive body that was detected by geophysical methods and which is overlain by meta-anthracites and anthracites of Carboniferous snd Lower Cretaceous ages (Fig. 5-11) (Teichmüller and Teichmüller, 1968b). Probably many other anthracite occurrences are caused by additional heating of this kind, for the high temperatures which lead to anthracitization are seldom reached through subsidence in regions of "normal" geothermal gradients (3—4°C/100 m).

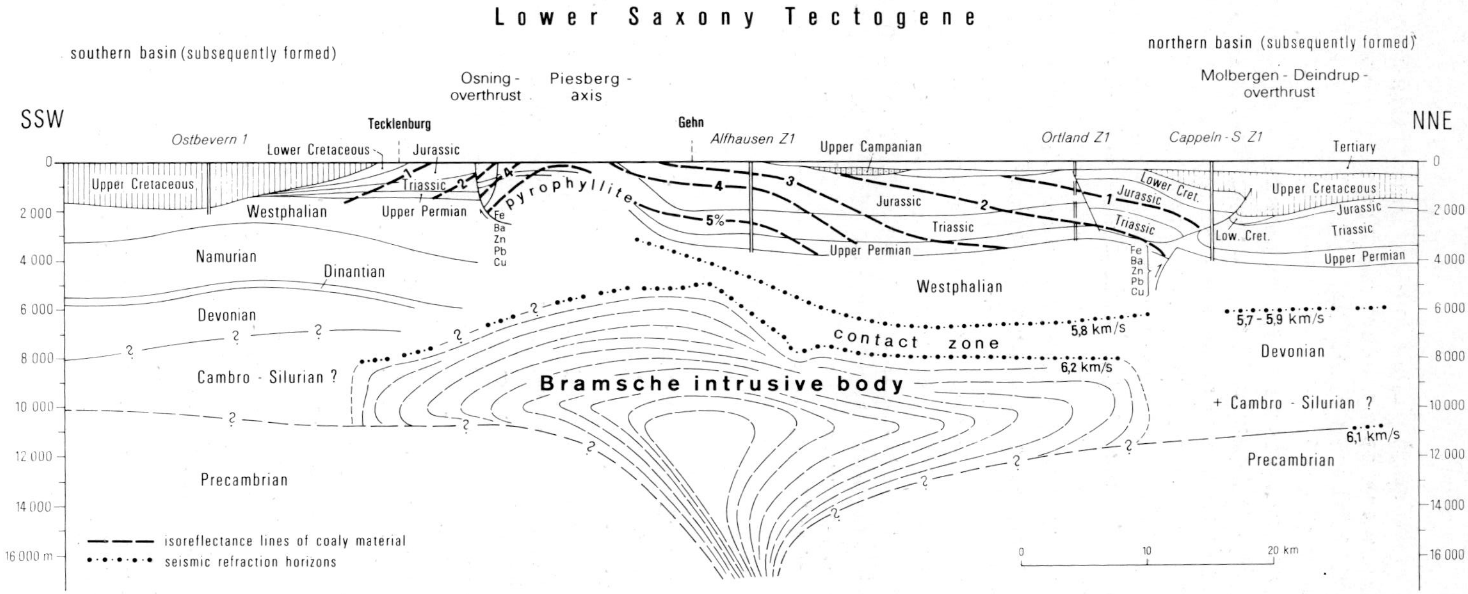

Fig. 5-11. Influence of the coalification pattern above the Bramsche Massif in N.W. Germany. Note the doming of isoreflectance lines and the occurrence of pyrophyllite above the intrusive body. (After Stadler and R. Teichmüller, 1971.)

The effect of temperature plainly follows "Hilt's rule", i.e., the increase of coal rank with depth. The maximum depth of burial and the corresponding maximum temperature, to which the coal was exposed for a sufficient length of time, determines its rank. Based on experience, commonly this maximum temperature in the large foredeeps is attained prior to the initiation of folding ("pre-orogenic"). This, for example, is the case in: (1) the Ruhr Basin (Teichmüller and Teichmüller, 1949, 1968a); (2) most coalfields of the eastern United States (Damberger, 1974), and (3) most coalfields of the Canadian Rocky Mountains and the inner foothill belt (Hacquebard and Donaldson, 1974). The rate of rank increase with depth, i.e., the "rank gradient", is dependent on the geothermal gradient and on the heat conductivity of the rocks. In the Tertiary deposits of the Upper Rhine Graben, for instance, the rank gradient is high in the "warm" borehole Landau 2 (present-day gradient: 7.7°C/100 m), whereas it is relatively low in the "cold" borehole Sandhausen 1 (present-day gradient: 4.2°C/100 m). As shown in Fig. 5-12, bituminous rank is already attained at a depth of 1500 m in Landau, whereas in Sandhausen it is only reached at 2600 m depth. Especially "cool" areas, such as the subduction zones of the Great Valley in California and the northern foreland of the Alps, are characterized by a very low rank gradient (Bostick, 1974; Castaño and Sparks, 1974; Teichmüller and Teichmüller, 1975). These observations allow inferences on geothermal regimes and tectonics of a particular region from rank gradients.

The influence of heat conductivity on the increase of coalification with depth (smaller in sandstones than in claystones) was observed in boreholes of the Saar and the Lower Rhine district (Damberger, 1968; Teichmüller and Teichmüller, 1968a).

Coalification temperatures are much lower than has been earlier inferred from experiments. Bituminous coal formation takes place at temperatures of the order of 100—150°C. Based on experimental data and other evidence, to attain a certain rank, higher temperatures are necessary with more rapid heating rates (contact metamorphism) than with slower heating rates (gradual subsidence).

The knowledge of the rank gradient and of the subsidence history allows evaluation of the paleogeothermal regime of a certain area (Ammosov, 1968; Robert, 1971; Doebl et al., 1974; Bostick, 1974; Castaño and Sparks, 1974). For example, for the Variscan foredeep a much higher geothermal gradient must be assumed (5°C/100 m) (Teichmüller, M. et al., 1978) than for the Alpine foredeep (ca. 2.5°C/100 m) (Teichmüller and Teichmüller, 1975).

Time

Inasmuch as the diagenesis of the Lower Carboniferous coals in the Moscow Basin, USSR, is still in the lignite stage, the conclusion has been

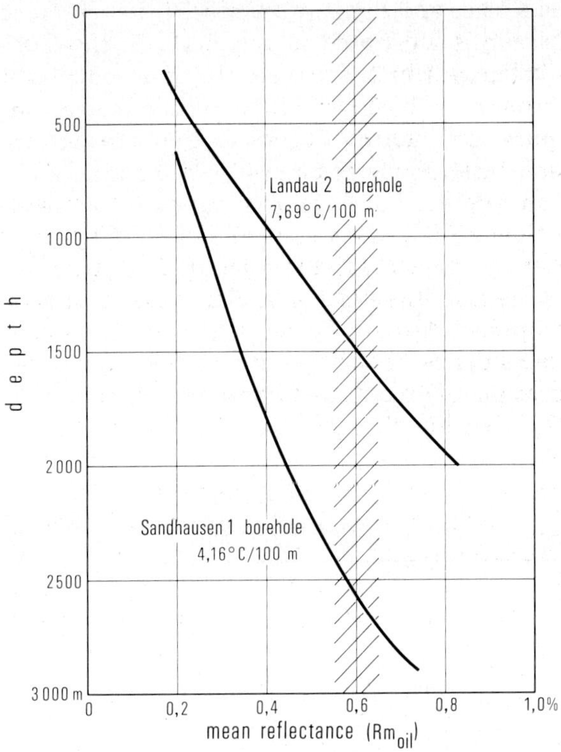

Fig. 5-12. Increase of vitrinite reflectance (measured on finely dispersed particles in Tertiary sediments) with depth in the "warm" borehole Landau 2 and the "cool" borehole Sandhausen 1 in the Upper Rhine Graben. The hatched band indicates the boundary range between subbituminous and bituminous coal.

drawn that time does not play an important role in the coalification process. This undoubtedly applies to those coals which have never been buried deep enough during their geological history and, therefore, have never been subjected to considerable heating. The Lower Carboniferous lignites of the Moscow Basin never reached temperatures of more than 20—25° C (Lopatin, 1971) and, thus, the minimum temperature required for the formation of bituminous coals has not been attained. Some originally low-rank Carboniferous coals, however, were later buried in Mesozoic and Cainozoic times at greater depths (2000—4000 m), as in the Lower Saxony Basin of northwestern Germany. In this case the coals have been subjected to temperatures of 80—140° C for periods in excess of 100 million years. This resulted in an intense subsequent "post-orogenic" coalification; the high-volatile bituminous coals, which formed during Westphalian C time, were subsequently converted into low-volatile bituminous coals and semi-anthracites.

The effect of time on coalification is illustrated in Fig. 5-13. In the Gulf Coast area of Louisiana, coaly inclusions in the Upper Miocene sediments reached the stage of low-rank bituminous coal (35—40% volatile matter) at a depth of 5440 m and a maximum temperature of 140°C. In northwestern Germany, Carboniferous high-rank bituminous coals (14—16% volatile matter) sank to approximately the same depth, but have been exposed to a similar temperature over a far longer time ($270 \cdot 10^6$ yr instead of $17 \cdot 10^6$ yr).

The influence of time is greater the higher is the temperature. Using reaction kinetics, Karweil (1956) was the first investigator to clarify this relationship. His basic work has been applied and partly revised by many other authors in order to clarify the dependence of coalification on paleotemperature and burial history, i.e., duration of exposure to various temperatures (e.g. Lopatin, 1971; Bostick, 1973; Hood et al., 1975).

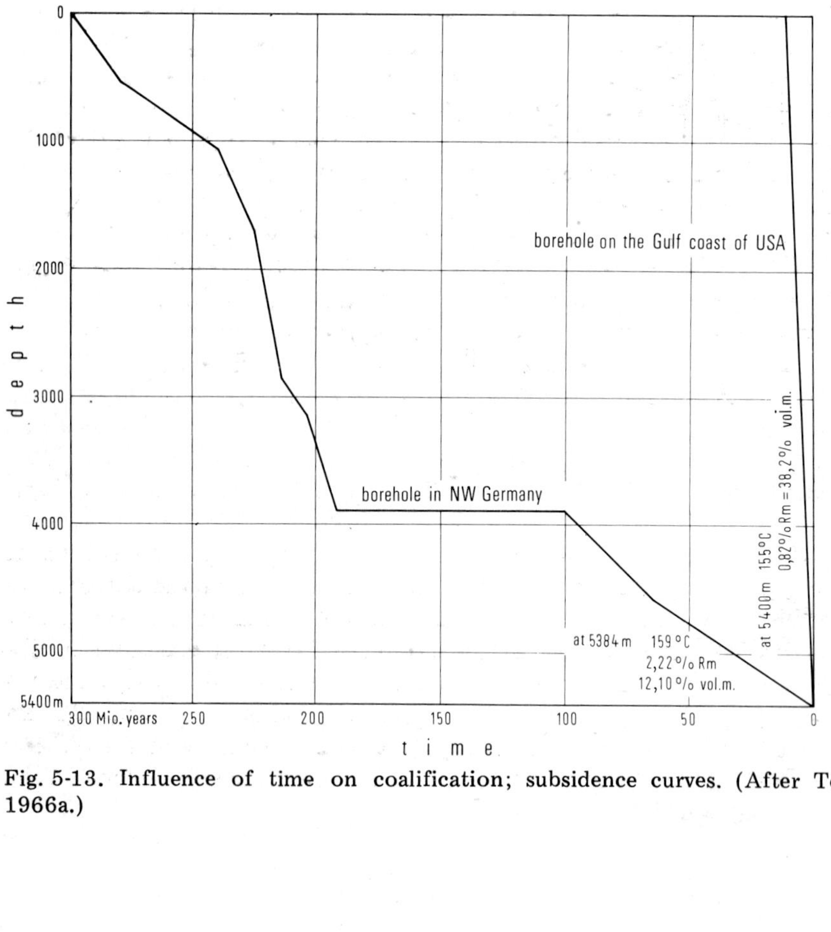

Fig. 5-13. Influence of time on coalification; subsidence curves. (After Techmüller, 1966a.)

GASEOUS PRODUCTS OF COALIFICATION

The coalification of lignite results mainly in release of water and carbon dioxide. In the stage of low-rank bituminous coal (up to more than 29% volatile matter) mainly carbon dioxide is produced, whereas subsequently methane is the predominant gas [1] produced as a result of coalification. Furthermore, small amounts of heavier hydrocarbons, carbon monoxide, and nitrogen are split off. These findings are partly based on experiments (see Jüntgen and Karweil, 1962; Van Heek et al., 1971; Klein, 1971). On grinding coal samples, Gedenk et al. (1964) found that heavier hydrocarbons are present as adsorbed gases. Furthermore, they found that with increasing rank, hexane, pentane, butane, propane, and ethane disappear successively from the coal. Anthracite contains only methane as an adsorbed hydrocarbon gas.

The quantities of gas which are released during the coalification process are considerable. According to Jüntgen and Karweil (1962, 1966), starting from a high-volatile bituminous coal (yielding 40% volatile matter) up to an anthracite with 5% volatile matter, about 100 m^3 of carbon dioxide and about twice as much methane are produced per ton (1000 kg) of final coal. Patteisky (1950) obtained even higher values in his calculations. The carbon dioxide is dissolved in water and rises with it upward along joints and cracks. The greater part of the methane likewise migrates away; it accumulates in pores and joints of the coal host rocks or in porous sandstones of the overlying sediments and can form natural gas reservoirs of commercial significance. Part of the methane, however, remains adsorbed on the coal and is released only on decompression, which occurs in boreholes or during the mining of coal seams. In the mining district of Karwin, Czechoslovakia (coal basin of Upper Silesia), more than 100 m^3 of methane were measured per ton (1000 kg) of mined coal (Patteisky, 1950). Such large quantities of gas, however, are rarely found in other places.

COALIFICATION AND BITUMINIZATION

Both coalification and bituminization are diagenetic processes during which mobile products (gas or crude oil) are given off and aromatization and condensation of the residual products take place. Bituminization which leads to the formation of crude oil (from the kerogen of source rocks) runs paral-

[1] Carbon-isotope ($^{12}C/^{13}C$) studies of coals and adsorbed methane lead to the conclusion that during coalification the light carbon isotope (^{12}C) is preferentially split off (Teichmüller et al., 1970).

lel with coalification. As to which one of these two processes will take place is dependent on the parent substances and their depositional milieu, as well as on the later diagenetic environment. Inasmuch as there are transitions between coaly and bituminous materials and because coal seams usually contain some lipid constituents, bituminization also takes place in coals. Extractable, petroleum-like hydrocarbons are formed during the subbituminous to high-volatile bituminous coal stage (Brooks, 1970; Hood and Gutjahr, 1972) (cf. Fig. 5-1). This "bituminization range" is part of the coalification process. The newly formed bitumen (expelled from liptinites and hydrogen-rich molecular groups of vitrinites) cannot migrate because the ultrafine pore system of coal functions as a molecular sieve. The bituminous matter is adsorbed on and chemically incorporated into the humic substance, with only a small part being deposited in empty cavities as "exsudatinite". On further coalification the newly formed bituminous matter is cracked into hydrocarbons of smaller molecular size. During this process residual solid products attain a strong reflectance. This decomposition begins in the medium-volatile bituminous coal stage and is known as the second coalification jump of liptinites (cf. Fig. 5-1).

Petroleum formation in source rocks begins when the coals pass through the subbituminous A-stage and ends when the coals have attained medium-volatile bituminous rank (26% volatile matter). This has led to extensive studies on coalification for improving oil and gas prospecting methods, mainly on the basis of reflectance measurements of finely divided vitrinite particles in borehole rock samples (Teichmüller and Teichmüller, 1966b; Ammosov, 1968; M. Teichmüller, 1971; Shibaoka et al., 1973; Castaño and Sparks, 1974; Hood et al., 1975).

COMPARISON OF DIAGENESIS OF COALS WITH DIAGENESIS OF THE ASSOCIATED ROCKS

Soft brown coals are associated with weakly consolidated clays and sands, whereas hard brown coals occur with more indurated sediments. Bituminous coals are associated with consolidated host rocks; therefore, the diagenesis of coal commonly runs parallel to the diagenesis of the associated sediments. The porosity of sand and clay deposits decreases as the rank of coal increases (Doebl et al., 1974). Apart from the transformation of montmorillonite into mixed-layered minerals during the subbituminous coal stage (R_m(oil) = 0.4—0.5%) (Heling and M. Teichmüller, 1974), more pronounced transformations in shales, claystones and sandstones occur only when the coal attains anthracite rank. An example of this is presented in Fig. 5-14 for the range of bituminous coals and anthracites found in a deep borehole in N.W. Germany, in

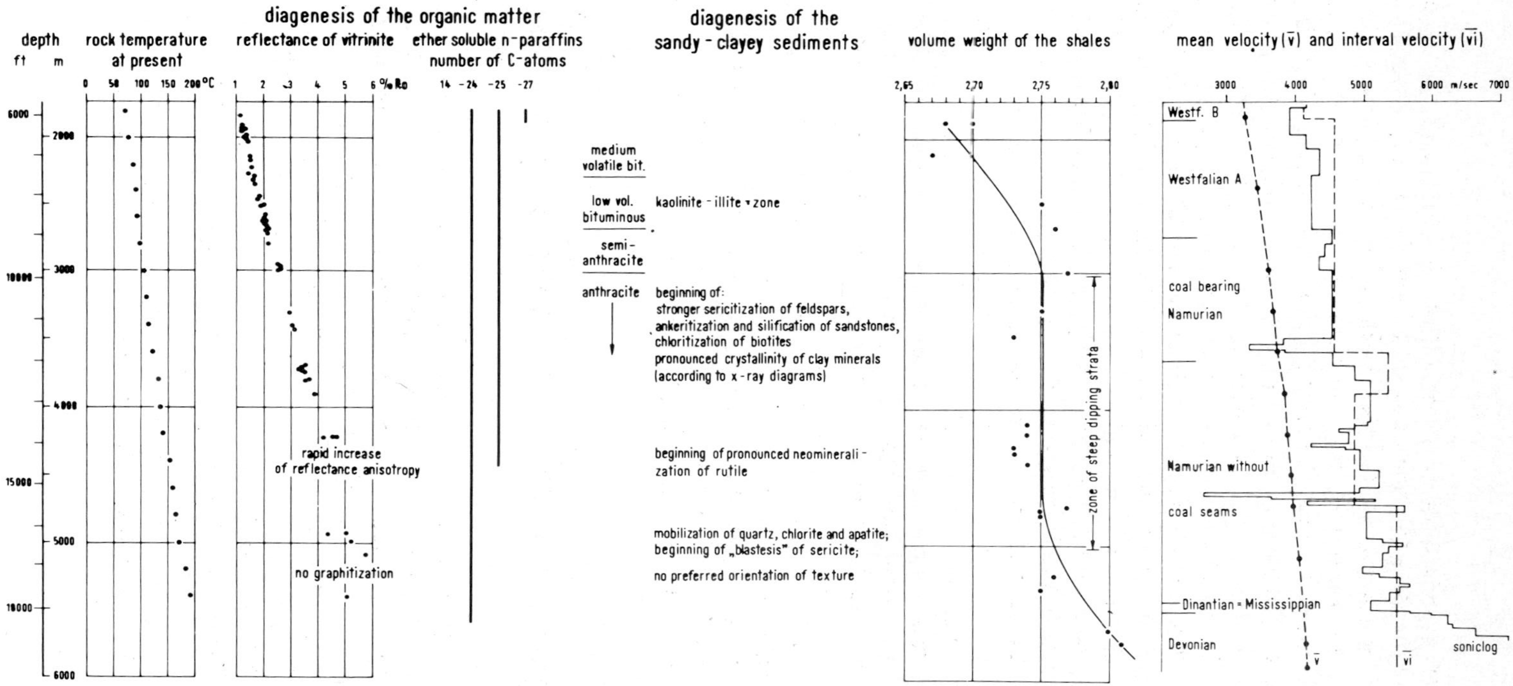

Fig. 5-14. Increase of present subsurface temperature, degree of coalification (vitrinite reflectance), and diagenesis of sand and clay sediments with depth in the Münsterland-1 borehole, N.W. Germany. (After Lensch, 1963; Scherp, 1963; Stadler, 1963; and Hedemann and R. Teichmüller, 1966).

which the present depth of the Carboniferous coal seams approximately corresponds to the original one. As shown in Fig. 5-14, a stronger sericitization of feldspars and an increasing ankeritization and silicification of sandstones begin during the anthracite stage.

The reaction of coals to temperature rise is much more sensitive and rapid than that of minerals. This is the reason why over relatively short periods of time coalification often proceeds in advance of mineral transformation in the contact zones of intrusive igneous bodies (Stadler and M. Teichmüller, 1971; Stadler and R. Teichmüller, 1971). The same is valid for the transformation of montmorillonite into mixed-layered minerals in young sediments of rapidly subsiding areas (Heling and M. Teichmüller, 1974). For this and other reasons (see below), *precise* correlations between degrees of coalification and mineral transformation are not possible.

In contrast to coalification, mineral transformation depends not only on temperature, time, and pressure, but also on a complex interaction of other physical and of chemical factors, e.g., on supply of cations. The replacement of kaolinite or mixed-layered minerals by illite—muscovite or chlorite, for instance, requires K, Mg, and Fe cations. If cations (from unstable detritus or pore solutions) are not available, kaolinite will be replaced by pyrophyllite, but not unless temperatures of 200—300° C are attained. This large temperature range corresponds to the entire coalification stage of anthracites and meta-anthracites.

Moreover, in the transitional stage between diagenesis and metamorphism, the increasing crystallinity of illite, as well as the neoformation of chlorite, cannot be precisely correlated with individual stages of anthracitization. Generally, low-grade metamorphism is first reached when the relative half-band width of the (001) X-ray reflection from the illite lattice is 120—100 (Weber, 1972). This stage corresponds to a vitrinite reflectivity of 5—10% R_{max}(oil), i.e., to highly graphitized meta-anthracite or semi-graphite (Fig. 5-15). Furthermore, it must be considered that pre-graphitization of meta-anthracites, as well as the formation of true graphite, is strongly enhanced by shear movements (Fig. 5-3).

In the case of magmatites, it has been attempted to correlate the formation of different zeolites with rank of coal (Kisch, 1974; Castaño and Sparks, 1974). Laumontite may already appear with coaly material of low rank, depending on the time of heating, with vitrinite reflectivities of 0.45—0.63% R_{max}(oil). The pumpellyite-prehnite-quartz facies, i.e., the range of anchimetamorphism, corresponds to the meta-anthracite stage (>4% R_{max}(oil)) (Kisch, 1974; Teichmüller et al., 1978) (Fig. 5-15).

The transition to epimetamorphism is characterized by a stepwise graphitization, depending on the parent matter of meta-anthracite. According to Diessel and Offler (1975), graphite inclusions show an increasing lattice

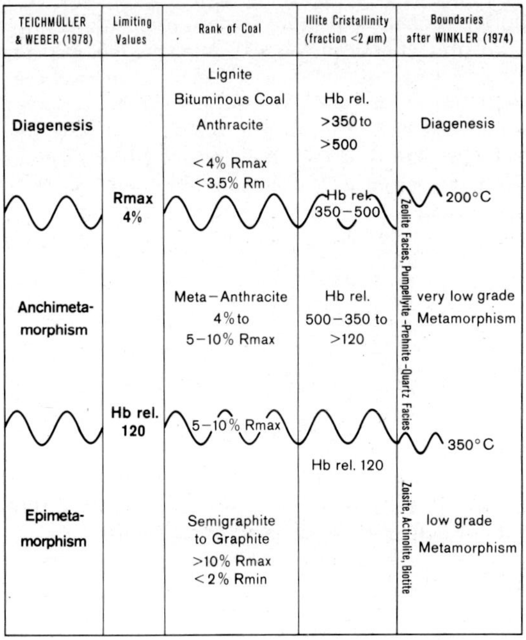

Fig. 5-15. Distinction between diagenesis and metamorphism on the basis of vitrinite reflectivity and illite cristallinity. (Proposed by Teichmüller, M. et al., 1978).

perfection with increasing metamorphism of the host rock from the greenschist to the granulite facies. Graphite, representing the final stage of coalification, is used as a marker for rock metamorphism (Schüller, 1961). It commonly forms at temperatures above 400° C.

Finally, it should be stressed that coal is a much better geological thermometer than minerals and that diagenetic stages can be best studied by means of coal petrological methods, i.e., reflectance measurements of coaly inclusions, which are present in most sedimentary rocks.

ACKNOWLEDGEMENT

The authors are greatly indebted to Dr. H. Reuter for the translation of the German text of the 1st edition and to Dr. D.G. Murchison for the translation of additional texts of the 2nd edition.

APPENDIX

Petrographic Coal Classification

Stopes (1919) classified coal into four *lithotypes*, which may be distinguished megascopically: vitrain, clarain, durain, and fusain. Each is composed of microscopic entities termed *macerals*, such as telinite, collinite, sporinite, resinite, macrinite, and fusinite. Macerals which are similar in their optical and chemical behavior are divided into the three *maceral groups:* vitrinite, liptinite (= exinite), and inertinite. Under the microscope, it is possible to distinguish layers of characteristic maceral associations which are called *microlithotypes* and *microlithotype-groups* (vitrite, clarite, durite, fusite, etc.). The precursors of the vitrinite group in the peat and lignite stages are the huminites with the macerals textinite, ulminite, attrinite, densinite, gelinite, and corpohuminite. (See, *International Handbook of Coal Petrography*, 1971, 1975.)

The petrographic classification of the Stopes-Heerlen nomenclature for bituminous coals is presented in Table 5-IV.

TABLE 5-IV

Classification of macerals, microlithotypes, and lithotypes of bituminous coals (after *International Handbook of Coal Petrography*, 1963, 1971)

Maceral	Maceral group	Microlithotype-group (and its principal maceral group constituents)	Lithotype
Telinite		vitrite (vitrinite)	vitrain
Collinite	vitrinite	vitrinertite (vitrinite + inertinite)	
Vitrodetrinite		clarite (vitrinite + liptinite)	
Sporinite			clarain
Cutinite	liptinite	duroclarite (vitrinite + liptinite + inertinite)	
Resinite	(= exinite)	clarodurite (inertinite + liptinite + vitrinite)	
Alginite		liptite (liptinite)	
Liptodetrinite		durite (inertinite + liptinite)	durain
Micrinite			
Macrinite			
Sclerotinite	inertinite	inertite (inertinite)	
Inertodetrinite			
Semifusinite		fusite (fusinite + semifusinite)	fusain
Fusinite			

GLOSSARY

Anthraxylon: humic matter of coal, largely derived from leaves, roots, stems and bark (lignin and cellulose). Because of gelification (transformation to colloidal humic matter), the cell structures are more or less flattened as a result of the overburden pressure. Megascopically bright, glossy, or jet-like bands or lenses (vitrains).

Attritus: component of coal consisting of finely divided organic matter. According to the classification of Thiessen-Bureau of Mines divided into translucent attritus (humic matter, protobitumen) and opaque attritus (fusinite, macrinite), megascopically dull.

Bone coal: argillaceous coal or carbonaceous shale.

Cannel coal: coalification product of sapropelic organic muds, deposited in quiet-water lakes and ponds. Coals composed of a very fine attritus (with anthraxylon largely absent) having a pronounced microbedding, may be differentiated into (1) *cannel coals* s.str. (composed of spores and similar organic materials, vitrinite and inertinite); (2) *humic* or *pseudocannels* (largely composed of vitrinite); and (3) *boghead cannels* (fatty algae are the characteristic ingredients).

Clarite: microlithotype (maceral association) of banded bituminous coals, consisting of vitrinite and exinite.

Durite: microlithotype (maceral association) of banded bituminous coals, consisting of inertinite and exinite.

Fusinite: maceral of coal, relatively high in carbon content and having high reflectance. It has no coking power and has lower contents of volatile matter and hydrogen than all other components of coal. Derived from wood or other plant tissues which are highly charred or underwent condensation processes by dehydration, etc. (*Fusain* = fossil charcoal, mineral charcoal or mother-of-coal = lithotype of coal, which is megascopically visible).

Gyttja: black organic mud deposited in poorly aerated lakes and ponds (Recent).

Huminite: maceral group of peat and lignite; precursor of vitrinite.

Inertinite: group of coal macerals, including fusinite, semifusinite, macrinite and sclerotinite (fungal remains). Characterized by a relatively high carbon content and a high reflectance in polished sections. More or less inert during coking processes.

Liptinite (= *exinite*): group of coal macerals, comprising mainly sporinite (spore exines), cutinite (surface layer of leaves) resinite (resins and waxes), and alginite. Characterized by a relatively high hydrogen content and a low reflectance on polished surfaces.

Maceral: smallest micropetrographic unit of coal. Recognized by virtue of physical, mainly optical, similarity.

Macrinite: maceral of coal, gel-like, without structure, high in carbon content and in reflectance. Derived from highly decomposed plant matter.

Micrinite: maceral of coal, consisting of very fine grains of strong reflectance. In part coalification product of liptinite.

Phyteral: coal particles of microscopic size based on recognition of plant fossil parts and pieces (e.g., spore, cuticle, bark, etc.).

Vitrinite: major maceral of most bituminous coals, relatively high in oxygen. Coalified humic matter, largely derived from lignin and cellulose of plants.

REFERENCES

Ammosov, I.I., 1968. Coal organic matter as a parameter of the degree of sedimentary rock lithification. *Int. Geol. Congr., 23rd, Prague, 1968, Proc.*, Sect. 11: 23—30.

Ammosov, I.I. and Tan Sju, I., 1961. *The Coalification Stages and the Paragenetic Behaviour of Fossil Fuels.* Akad. Nauk. U.S.S.R., Moscow, 117 pp. (in Russian).

Austen, D.E.G. and Ingram, O.J.E., 1958. Electron resonance in coals. *Brennstoff-Chemie*, 39: 25—30.

Bangham, D.H. and Maggs, F.A.P., 1944. The strength and elastic constants of coals in relation to their ultra-fine structure. *Proc. Conf. Ultrafine Struct. Coals Cokes, London, 1944*, pp. 118—130.

Beck, Th. and Poschenrieder, H., 1957. Drucktoleranz, ein Kriterium für den autochthonen Charakter der Braunkohlenmikroflora. *Zentralbl. Bakteriol. Parasitenkd. Infektionskr. Hyg., Abt. 2*, 110: 534—539.

Berkowitz, N. and Schein, G., 1952. Some aspects of the ultrafine structure of lignites. *Fuel*, 31: 19—32.

Blayden, H.E., Gibson, J. and Riley, H.L., 1944. An X-ray study of the structure of coals, cokes and chars. *Proc. Conf. Ultrafine Struct. Coals, Cokes, London, 1944*, pp. 176—231.

Blom, L., Edelhausen, L. and Van Krevelen, D.W., 1957. Chemical structure and properties of coal, 18. Oxygen groups in coal and related products. *Fuel*, 36: 135—153.

Bostick, N.H., 1973. Time as a factor in thermal metamorphism of phytoclasts (coal particles). *C.R. 7me Congr. Int. Stratigr. Géol. Carbonifère*, 2: 183—193.

Bostick, N.H., 1974. Phytoclasts as indicators of thermal metamorphism, Franciscan assemblage and Great Valley sequence (Upper Mesozoic), California. *Geol. Soc. Am., Spec. Pap.*, 153: 1—17.

Brooks, J.D., 1970. The use of coals as indicators of the occurrence of oil and gas. *Aust. Pet. Explor. Assoc. J.*, 10: 35—50.

Brown, J.K. and Hirsch, P.B., 1955. Recent infra-red and X-ray studies of coal. *Nature*, 175: 229—242.

Castaño, J.R. and Sparks, D.M., 1974. Interpretation of vitrinite reflectance measurements in sedimentary rocks and determination of burial history using vitrinite reflectance and authigenic minerals. *Geol. Soc. Am., Spec. Pap.*, 153: 31—52.

Correia, M., 1971. Diagenesis of sporopollenin. In: J. Brooks, P.R. Grant, M. Muir, P. van Gijzel and G. Shaw (Editors), *Sporopollenin.* Academic Press, London, pp. 569—620.

Dahme, A. and Mackowsky, M.Th., 1951. Mikroskopische, chemische und röntgenographische Untersuchungen an Anthraziten. *Brennstoff-Chemie*, 32: 175—186.

Damberger, H., 1968. Ein Nachweis der Abhängigkeit der Inkohlung von der Temperatur. *Brennstoff-Chemie*, 49: 73—77.

Damberger, H., 1974. Coalification patterns of Pennsylvania coal basins of the eastern United States. *Geol. Soc. Am., Spec. Pap.*, 153: 53—74.

Davis, A. and Spackman, W., 1964. The role of the cellulosic and lignitic components of wood in artifical coalification. *Fuel*, 43: 215—224.

Diessel, C.F.K. and Offler, R., 1975. Change in physical properties of coalified and graphitised phytoclasts with grade of metamorphism. *Neues Jahrb. Mineral. Monatsh.*, 1: 11—26.

Doebl, F., Heling, D., Homann, W., Karweil, J., Teichmüller, M. and Welte, D., 1974. Diagenesis of Tertiary clayey sediments and included dispersed organic matter in relationship to geothermics in the Upper Rhine Graben. In: J.H. Illies and K. Fuchs (Editors), *Approaches to Taphrogenesis.* Schweizerbart, Stuttgart, pp. 192—207.

Dryden, I.G.C., 1951. Einige neue Fortschritte zum Verständnis der physikalischen Struktur und chemischen Natur von Glanzkohlen. *Brennstoff-Chemie,* 32: 321—324.

Dryden, I.G.C., 1956. How was coal formed? *Coke Gas,* 18: 1—11.

Dryden, I.G.C. and Griffith, M., 1953. Quantitative estimation of the changes in chemical structure of coals during metamorphism. *Fuel,* 32: 199—210.

Dulhunty, J.A., 1950. Relations of rank to inherent moisture of vitrain and permanent moisture reduction on drying. *Proc. R. Soc. N.S. Wales,* 82: 286—293.

Dulhunty, J.A., 1954. Geological factors in the metamorphic development of coal. *Fuel,* 33: 145—152.

Dunningham, A.G., 1944. The inherent moisture in coal. *Proc. Conf. Ultrafine Struct. Coals Cokes, London, 1944,* pp. 57—70.

Edwards, A.B., 1948. Some effect of folding on the moisture content of brown coal. *Australas. Inst. Mining Metall., Proc.,* 150/151: 101—112.

Flaig, W., 1968. Biochemical factors in coal formation. In: D.G. Murchison and T.S. Westoll (Editors), *Coal and Coal-bearing Strata.* Oliver and Boyd, Edinburgh, pp. 197—232.

Gedenk, R., Hedemann, H.A. and Rühl, W., 1964. Oberkarbongase, ihr Chemismus und ihre Beziehungen zur Steinkohle (Untersuchungsergebnisse aus Nord- und Westdeutschland). *Congr. Av. Étud. Stratigr. Géol. Carbonifère, C.R., Paris, 1963,* 2: 431—450.

Gutjahr, C.C.M., 1966. Carbonization measurements of pollen grains and spores and their application. *Leidse Geol. Meded.,* 38: 29 pp.

Hacquebard, P.A. and Donaldson, J.R., 1974. Rank studies of coals in the Rocky Mountains and inner foothill belt, Canada. *Geol. Soc. Am., Spec. Pap.,* 153: 75—94.

Hedemann, H.A., 1963. Die Gebirgstemperaturen in der Bohrung Münsterland 1 und die geothermische Tiefenstufe. In: *Aufschlussbohrung Münsterland 1—Fortschr. Geol. Rheinl. Westfalen,* 11: 403—418.

Hedemann, H.A. and Teichmüller, R., 1966. Stratigraphie und Diagenese des Oberkarbons in der Bohrung Münsterland 1. *Z. Dtsch. Geol. Ges.,* 115: 787—825.

Heling, D. and Teichmüller, M., 1974. Die Grenze Montmorillonit/Mixed-layer Minerale und ihre Beziehung zur Inkohlung in der Grauen Schichtenfolge des Oligozäns im Oberrheingraben. *Fortschr. Geol. Rheinl. Westfalen,* 24: 113—128.

Hood, A. and Castaño, J.R., 1974. Organic metamorphism: its relationship to petroleum generation and application to studies of authigenic minerals. U.N. ESCAP, *C.C.O.P. Tech. Bull.,* 8: 85—118.

Hood, A. and Gutjahr, C.C.M., 1972. Organic metamorphism and the generation of petroleum. (Paper presented Annual Meeting Geol. Soc. Am., Nov. 1972, Minneapolis.)

Hood, A., Gutjahr, C.C.M. and Heacock, R.L., 1975. Organic metamorphism and the generation of petroleum. *Bull. Am. Assoc. Pet. Geol.,* 59: 986—996.

Huck, G. and Karweil, J., 1953. Versuch einer Modellvorstellung vom Feinbau der Kohle. *Brennstoff-Chemie,* 34: 97—102, 129—135.

Huck, G. and Karweil, J., 1955. Physikalisch-chemische Probleme der Inkohlung. *Brennstoff-Chemie,* 36: 1—11.

Huck, G. and Karweil, J., 1962. Probleme und Ergebnisse der künstlichen Inkohlung im Bereich der Steinkohlen. *Fortschr. Geol. Rheinl. Westfalen,* 3 (2): 717—724.

Huck, G. and Patteisky, K., 1964. Inkohlungsreaktionen unter Druck. *Fortschr. Geol. Rheinl. Westfalen,* 12: 551—558.

International Handbook of Coal Petrography, 1963, 2nd ed., edited by Int. Comm. Coal Petrol. Centre National Recherche Scientifique, Paris.

International Handbook of Coal Petrography, 1971, 1975, suppl. to 2nd ed., edited by Int. Comm. Coal. Petrol. Centre National Recherche Scientifique, Paris.

Jüntgen, H. and Karweil, J., 1962. Künstliche Inkohlung von Steinkohlen. *Freiberg. Forschungsh., A.,* 229: 27—36.

Jüntgen, H. and Karweil, J., 1966. Gasbildung und Gasspeicherung in Steinkohlenflözen. *Erdöl Kohle,* 19: 251—258; 339—344.

Karweil, J., 1956. Die Metamorphose der Kohlen vom Standpunkt der physikalischen Chemie. *Z. Dtsch. Geol. Ges.,* 107: 132—139.

Kayser, H., 1952. Die Veredlung der Braunkohle und der geologisch jüngeren Brennstoffe. In: K. Winnacker and E. Weingaertner, *Chemische Technologie, 3. Organische Technologie.* Hanser, München, pp. 123—124.

King, J.G. and Wilkins, B.T., 1944. The internal structure of coals. *Proc. Conf. Ultrafine Struct. Coals, Cokes, London, 1944,* pp. 46—57.

Kisch, H.J., 1974. Anthracite and meta-anthracite coal ranks associated with "anchimetamorphism" and "very-low-stage" metamorphism. *K. Ned. Akad. Wet. Proc., Ser. B,* 77 (2): 81—118.

Klein, J., 1971. *Untersuchungen zur Abspaltung von Wasserdampf, CO, CO_2, N_2 und H_2 bei der nicht-isothermen Steinkohlenpyrolyse unter inerter und oxidierender Atmosphäre.* Dissertion Technische Hochschule Aachen, 117 pp.

Kurbatov, J.M., 1963. Zur Frage über die Genesis des Torfes und der Torfhuminsäuren. *Int. Peat Congr., Leningrad, 1963,* pp. 1—8.

Landis, C.A., 1971. Graphitization of dispersed carbonaceous material in metamorphic rocks. *Contrib. Mineral. Petrol.,* 30: 34—45.

Lensch, G., 1963. Die Metamorphose der Kohle in der Bohrung Münsterland 1 auf Grund des optischen Reflexionsvermögens der Vitrinite. In: *Die Aufschlussbohrung Münsterland 1—Fortschr. Geol. Rheinl. Westfalen,* 11: 197—203.

Lopatin, N.V., 1971. Temperature and geologic time as factors in coalification. *Izv. Akad. Nauk S.S.S.R., Ser. Geol.* (3): 95—106 (translated by N.W. Bostick).

Monomakhoff, C., 1961. La tectonique tangentielle dans les bassins houillers de la France et sa répercussion sur la continuité et le comportement de ces gîsements. *C.R. Congr. Av. Étud. Stratigr. Géol. Carbonifère, 4, Heerlen, 1958,* 2: 423—435.

Ottenjann, K., Teichmüller, M. and Wolf, M., 1974. Spektrale Fluoreszens-Messungen an Sporiniten mit Auflicht-Anregung, eine mikroskopische Methode zur Bestimmung des Inkohlungsgrades gering inkohlter Kohlen. *Fortschr. Geol. Rheinl. Westfalen,* 24: 1—36.

Patteisky, K., 1950. Die Entstehung des Grubengases. *Bergbau Arch.,* 11/12: 5—24.

Patteisky, K. and Teichmüller, M., 1960. Inkohlungs-Verlauf, Inkohlungs-Masstäbe und Klassifikation der Kohlen auf Grund von Vitrit-Analysen. *Brennstoff-Chemie,* 41: 79—84; 97—104; 133—137.

Patteisky, K., Teichmüller, M. and Teichmüller, R., 1962. Das Inkohlungsbild des Steinkohlengebirges an Rhein und Ruhr, dargestellt im Niveau von Flöz Sonnenschein. *Fortschr. Geol. Rheinl. Westfalen,* 3 (2): 687—700.

Petrascheck, W.E., 1954. Zur optischen Regelung tektonisch beanspruchter Kohlen. *Mineral. Petrogr. Mitt.,* 4: 232—239.

Pigulevskaja, L.V. and Rakowski, V.E., 1963. Die Änderung der chemischen Zusammensetzung einiger Torfarten in Abhängigkeit von ihrem Alter, 1. Alter und Änderung der Torfkomponenten. *Tr. Inst. Torf. Akad. Nauk, Beloruss. S.S.R.,* 6: 12—31.

Pusey III, W.C., 1973. The ESR-kerogen method—a new technique of estimating the organic maturity of sedimentary rocks. *Petrol. Times,* 1973: 21—26.

Ragot, J.P., 1977. *Contribution à l'Etude de l'Evolution des Substances carbonées dans les Formations géologiques.* Thésis Univ. P. Sabatier, Toulouse, 150 pp.

Rakowski, W., Baturo, W. and Pigulewskaja, L., 1963. Die Humusbrennstoffe und ihre Bildung. *Int. Peat Congr., Leningrad, 1963:* 35 pp.

Robert, P., 1971. Étude pétrographique des matières organiques insolubles par la mesure de leur pouvoir réflecteur. Contribution à l'exploration pétrolière et à la connaissance des bassins sédimentaires. *Rev. Inst. Fr. Pét. Ann. Combust. Liq.*, 26: 105—135.

Scherp, A., 1963. Die Petrographie der paläozoischen Sandsteine in der Bohrung Münsterland 1 und ihre Diagenese in Abhängigkeit von der Teufe. In: *Die Aufschlussbohrung Münsterland 1—Fortschr. Geol. Rheinl. Westfalen,* 11: 251—282.

Schüller, A., 1961. Die Druck-, Temperatur- und Energiefelder der Metamorphose. *Neues Jahrb. Mineral., Abh.*, 96: 250—290.

Shibaoka, M., Bennett, A.J.R. and Gould, K.W., 1973. Diagenesis of organic matter and occurrence of hydrocarbons in some Australian sedimentary basins. *APEA* J., 13: 73—80.

Stach, E., 1953. Der Inkohlungssprung im Ruhrkarbon. *Brennstoff-Chemie,* 34: 353—355.

Stach, E., Chandra, D., Mackowsky, M.Th., Taylor, G.H., Teichmüller, M. and Teichmüller, R., 1975. *Stach's Textbook of Coal Petrology.* Borntraeger, Stuttgart, 428 pp.

Stach, H., 1948. Experimentele Beiträge zur Frage der Brikettierbarkeit von Weich- und Hartbraunkohlen und der Quellung und des Zerfalls von Braunkohlenbriketts. *Braunkohle,* 1: 35—44.

Stadler, G., 1963. Die Petrographie und Diagenese der oberkarbonischen Tonsteine in der Bohrung Münsterland 1. In: *Die Aufschlussbohrung Münsterland* 1. *Fortschr. Geol. Rhein. Westfalen,* 11: 283—292.

Stadler, G. and Teichmüller, M., 1971. Die Umwandlung der Kohlen und die Diagenese der Ton- und Sandsteine in der Untertagebohrung 150 der Steinkohlenbergwerke Ibbenbüren. *Fortschr. Geol. Rheinl. Westfalen,* 18: 125—146.

Stadler, G. and Teichmüller, R., 1971. Zusammenfassender Überblick über die Entwicklung des Bramscher Massivs und des Niedersächsischen Tektogens. *Fortschr. Geol. Rheinl. Westfalen,* 18: 547—564.

Stopes, M.C., 1919. On the four visible ingredients in banded bituminous coal. Studies on the compositions of coal. *Proc. R. Soc., London, Ser. B,* 190: 470—487.

Taylor, G.H., 1971. Carbonaceous matter: a guide to the genesis and history of ores. *Soc. Min. Geol. Jpn.*, Spec. Issue, 3: 283—288.

Teichmüller, M., 1962. Die Genese der Kohle. *C.R. Congr. Av. Étud. Stratigr. Géol. Carbon., 4, Heerlen, 1958,* 3: 699—722.

Teichmüller, M., 1968. Zur Petrographie und Diagenese eines fast 200 m mächtigen Torfprofils (mit Übergängen zur Weichbraunkohle?) im Quartär von Philippi (Mazedonien). *Geol. Mitt.*, 8: 65—110.

Teichmüller, M., 1971. Anwendung kohlenpetrographischer Methoden bei der Erdöl- und Erdgasprospektion. *Erdöl Kohle,* 24: 69—76.

Teichmüller, M., 1973. Zur Petrographie und Genese von Naturkoksen im Flöz Präsident/ Helene der Zeche Friedrich Heinrich bei Kamp-Lintfort (Linker Niederrhein). *Geol. Mitt.*, 12: 219—254.

Teichmüller, M., 1974a. Entstehung und Veränderung bituminöser Substanzen in Kohlen in Beziehung zur Entstehung und Umwandlung des Erdöls. *Fortschr. Geol. Rheinl. Westfalen,* 24: 65—112.

Teichmüller, M., 1974b. Generation of petroleum-like substances in coal seams as seen under the microscope. In: *Advances Organic Geochemistry 1973.* Technip., Paris, pp. 321—348.

Teichmüller, M. and Teichmüller, R., 1949. Inkohlungsfragen im Ruhrkarbon. *Z. Dtsch. Geol. Ges.*, 99: 40—77.

Teichmüller, M. and Teichmüller, R., 1950. Das Inkohlungsbild des niedersächsischen Wealdenbeckens. *Z. Dtsch. Geol. Ges.*, 100: 498—517.

Teichmüller, M. and Teichmüller, R., 1954. Die stoffliche und strukturelle Metamorphose der Kohle. *Geol. Rundsch.*, 42: 265—296.

Teichmüller, M. and Teichmüller, R., 1966a. Geological causes of coalification. *Coal Sci., Adv. Chem. Ser.*, 55: 133—155.

Teichmüller, M. and Teichmüller, R., 1966b. Inkohlungsuntersuchungen im Dienst der angewandten Geologie. *Freiberg. Forschungsh.*, C 210: 155—195.

Teichmüller, M. and Teichmüller, R., 1968a. Geological aspects of coal metamorphism. In: D.G. Murchison and T.S. Westoll (Editors), *Coal and Coal-bearing Strata.* Oliver and Boyd, Edinburgh, pp. 233—267.

Teichmüller, M. and Teichmüller, R., 1968b. Cainozoic and Mesozoic coal deposits of Germany. In: D.G. Murchison and T.S. Westoll (Editors), *Coal and Coal-bearing Strata.* Oliver and Boyd, Edinburgh, pp. 347—379.

Teichmüller, M. and Teichmüller, R., 1975. Untersuchungen in der Molasse des Alpenvorlandes. *Geol. Bavarica,* 73: 123—142.

Teichmüller, M., Teichmüller, R. and Weber, K., 1978. Inkohlung und Illit-Kristallinität. *Fortschr. Geol. Rheinl. Westfalen,* 27 (in press).

Teichmüller, R., 1952. Zur Metamorphose der Kohle. *C.R. Congr. Av. Étud. Stratigr. Geol. Carbon. 3, Heerlen, 1951,* 2: 615—623.

Teichmüller, R., 1962. Zusammenfassende Bemerkungen über die Diagenese des Ruhrkarbons und ihre Ursachen. *Fortschr. Geol. Rheinl. Westfalen,* 3 (2): 725—734.

Teichmüller, R., Teichmüller, M., Colombo, U., Gazzarrini, F., Gonfiantini, R. and Kneuper, G., 1970. Das Kohlenstoff-Isotopen-Verhältnis im Methan von Grubengas und Flözgas und seine Abhängigkeit von den geologischen Verhältnissen. *Geol. Mitt.*, 9: 181—206.

Van der Molen, W.H. and Smits, H., 1962. Die Sackung in einem Moorgebiet in Nord-Griechenland. *Int. Torf Kongr., Bremen, 1962, Ber.*, 11 pp.

Van Heek. K.H., Jüntgen, H., Luft, K.F. and Teichmüller, M., 1971. Aussagen zur Gasbildung in frühen Inkohlungsstadien auf Grund von Pyrolyseversuchen. *Erdöl Kohle,* 24: 566—572.

Van Krevelen, D.W., 1953. Physikalische Eigenschaften und chemische Struktur der Steinkohle. *Brennstoff-Chemie,* 34: 167—182.

Van Krevelen, D.W., 1961. *Coal.* Elsevier, Amsterdam, 514 pp.

Wassojewitsch, N.B., Korchagina, Ju.I., Lopatin, N.V. and Chernyshev, V.V., 1970. Principal phase of oil formation. *Int. Geol. Rev.*, 12: 1276—1296.

Weber, K., 1972. Notes on determination of illite crystallinity. *Neues Jahrb. Mineral., Monatsh.*, 6: 267—276.

Welte, E., 1952. Über die Entstehung von Huminsäuren und Wege ihrer Reindarstellung. *Z. Pflanzenernähr. Düng. Bodenkd.*, 56 (101): 105—139.

Williams, E., 1953. Anisotropy of the vitrain of South Wales coals. *Fuel,* 32: 89—99.

Zailer, V. and Wilk, L., 1911. Der Einfluss des Vertorfungsprozesses auf die Zusammensetzung von Carextorf. *Z. Moorkult. Torfverwert.*, 9: 153—168.

Chapter 6

DIAGENESIS OF CARBONATE SEDIMENTS AND EPIGENESIS (OR CATAGENESIS) OF LIMESTONES

GEORGE V. CHILINGAR, HAROLD J. BISSELL and KARL H. WOLF

INTRODUCTION

Certain factors will initiate diagenesis, whereas the same or other factors will perpetuate the old and/or cause commencement of new diagenetic processes. The sediments have a tendency to adjust to new physical and chemical conditions and would, theoretically, reach equilibrium. The micro- and macro-environmental conditions above and within the sediments, however, change continuously. Sometimes, equilibrium may be established, as, for example, in cases where limestones are completely replaced by iron oxide, silica or dolomite. In many cases, however, the physical and chemical conditions shift so rapidly that only a small fraction of the reactions involving the limestone framework reach equilibrium. In particular during the early diagenetic stages numerous successive and overlapping processes will be acting at a relatively fast rate on both micro- and macro-scales, when movements of interstitial fluids are at a maximum, biological activity is producing chemically-active substances, maximum pore space is available, temperature change is more or less sudden due to diurnal exposure, and so forth.

The following list of factors influence diagenesis of carbonate sediments: (1) geographic factors (e.g., climate, humidity, rainfall, type of terrestrial weathering, surface water chemistry); (2) geotectonism (e.g., rate of erosion and accumulation, coastal morphology, emergence and subsidence, whether eugeosynclinal or miogeosynclinal); (3) geomorphologic position (e.g., basinal versus lagoonal sediments, current velocity, particle size, sorting, flushing of sediments); (4) geochemical factors in a regional sense (e.g., supersaline versus marine water, volcanic fluids and gases); (5) rate of sediment accumulation (e.g., halmyrolysis, ion transfer, preservation of organic matter, biochemical zonation); (6) initial composition of the sediments (e.g., aragonite versus high-Mg and low-Mg calcite, isotope and trace element content); (7) grain size (e.g., content of organic matter, number of bacteria, rates of diffusion); (8) purity of the sediments (e.g., percentage of clay and organic matter, base exchange of clays altering interstitial fluids); (9) accessibility of limestone framework to surface (e.g., cavity systems permit replacements); (10) interstitial fluids and gases (e.g., composition, rate of flow, exchange of ions); (11) physicochemical conditions (e.g., pH, Eh, partial pressures of gases, CO_2 content); (12) previous diagenetic history of the sediment (e.g., previous expulsion of trace elements will determine subsequent diagenesis).

The numerous large-scale environmental parameters listed above influence in one way or another the more local environments and these in turn influence the micro-environments. There is a complete gradation and overlap of these macro- and micro-factors as one example below illustrates (Wolf, 1963b):

Climate ↓ Geomorphology	{ particle size	→	amount and type of bacteria	→	rate of diagenesis	→	pH and Eh	→	type of replacement

The actual processes that lead to diagenetic alterations and modifications of limestones are divisible as follows (among other):

(*1*) *Physicochemical processes:* solution, corrosion, leaching, bleaching, oxidation, reduction, reprecipitation, inversion, recrystallization, cementation, decementation, authigenic mineral genesis, overgrowth, crystal enlargement, replacements, chemical internal sedimentation, aggregation, and accretion.

(*2*) *Biochemical and organic processes:* accretion and aggregation, particle-size reduction, corrosion, corrasion, mixing of sediments, boring, burrowing, gas-bubbling, breaking down and synthesizing of organic and inorganic compounds.

(*3*) *Physical processes:* compaction, desiccation, shrinkage, penecontemporaneous internal deformation and corrasion, and mechanical internal sedimentation.

Many of the above processes are commonly considered syngenetic. As they can occur within the sediments and directly alter and influence diagenesis, however, they must be considered as part of diagenesis. It is the total or collective influence of all factors that must be examined in a final analysis. As Krumbein (1942) pointed out, variations in the diagenetic end-products may occur either with different sediments in the same environment, or with the same kind of sediment in different environments.

COMPACTION

Compaction of sediments is the process of volume reduction expressed as a percentage of the original voids present. The process affects mainly loose, unlithified limestones and, of course, other sediments not considered here. Autochthonous limestones such as reefs do not undergo much compaction. The intergranular spaces of allochthonous deposits are eliminated by closer packing, crushing, deformation, expulsion of interstitial fluids, and possibly corrosion of the grains. Krumbein (1942) gave the following values of porosities or amounts of fluid content of freshly deposited material: sand = 45%, silt = 50—65%, mud = 80—90%, and colloids (less than 1 μ) = approximately 98% water. Lime-mud apparently behaves similarly to clay minerals. The degree of compaction, in general, depends largely on the ratio of fine to coarse material and on the character of the sediment framework.

Fine-grained sediments undergo the highest degree of compaction in the first foot (Ginsburg, 1957). As he suggested, the negligible weight of overlying sediments cannot cause compaction during this early stage. Ginsburg believed that mixing by organisms, the gel-like character of the sediment and the escape of bacterial gases contribute to rapid packing. The burial pressure of sediment accumulation becomes effective somewhat later to produce some sort of physical cohesion between the particles.

The expulsion of interstitial fluids and gases during compaction may be predominantly vertical or horizontal. Even freshly deposited sediments have different degrees of permeability and, although unlithified, some may act as "cap" rocks to cause very early horizontal fluid movements. Differential compaction may determine the direction and rate of fluid and gas migration. Newell et al. (1953) proposed, for example, compaction-induced lateral movement of $CaCO_3$-rich basinal fluids toward the reef-talus and reef, with a subsequent precipitation of carbonate cement, to explain the well-lithified state of these sediments.

Interbedded carbonate and clay-mineral accumulations may both have at the beginning interstitial water that is identical in composition. Differential adjustment of the fluids to their new micro-environments will soon take place. In particular the clay minerals will motivate chemical changes. For example, the Ca—Na relationships seem to depend partly on the base exchange properties of clay minerals (Von Engelhardt, 1961). During compaction, the interstitial fluids of a number of different physicochemical horizons intermingle and cause reactions that may have significant results. Clay-mineral layers between limestone beds may cause filtration of certain cations when fluids pass through them. Werner (1961) believed that during compaction the fluid-movements from a clay into iron—oolite deposits caused filtration along the clay—oolite boundary. Certain cations were held back and remained below the boundary, whereas others passed freely until the concentration was sufficiently high within the oolite sediment to result in precipitation. The poorly lithified oolite beds exhibit compression features in contrast to the well-cemented oolites which lack any signs of deformation. This seems to indicate to Werner that the precipitation of the $CaCO_3$ cement must have been early diagenetic and occurred during compaction when the interstitial fluids assured sufficient quantities of chemicals. Werner's explanation may well apply to limestone deposits that contain clay-rich beds. Shelf-to-basin facies of Pennsylvanian and Permian carbonates in the Cordilleran miogeosyncline are examples.

Compaction processes can alter textures and structures of carbonate rocks. Poorly cemented faecal pellets, for example, have been reported to form a textureless lime-mud a few inches or feet below the surface due to merging of the individual grains. Movements of connate waters during compaction may form tubes, channels and bubbles (Cloud et al., 1962) which, when filled by carbonate cement, resemble the so-called birdseyes (dismicrite of Folk, 1959; dispellet of Wolf, 1960). An excellent example in the geologic record is found in lower limestones of the Ely Group in the Hamilton district of White Pine County, Nevada. On the other hand, lime-mud, more so than its clayey or muddy terrigenous counterpart, may lack compaction especially if early cementation took place. The result may be a loose, sponge-like dismicrite with minute sparite-filled voids. However, relatively large open-space structures in micrite and pellet limestones, for instance, cannot be explained by lack of compaction. It has been suggested that soft-bodied organisms became buried and upon decomposition left voids. Although this is possible in some cases, most cavities in micrite limestones are probably of inorganic origin and Algae were responsible only in an indirect way (Wolf, 1965a).

Rate of cementation, degree of compaction and pressure-solution are closely related. If the former varies on a regional scale, the latter may follow the same pattern. For example, loosely packed, birdseye-rich shallow-water Nubrigyn algal calcarenites of New South Wales must have undergone early cementation in contrast to the basinal algal, graded-bedded deposits that exhibit tight packing and extensive pressure-solution.

Structures believed to have been the product of compaction were described by Terzaghi (1940); and early diagenetic formation of cone-in-cone structures may be related to compaction of clayey micrite limestones (Usdowski, 1963; see also the section on recrystallization). For a detailed treatment on compaction of fine- and coarse-grained sediments, the reader is referred to the recent books by Rieke and Chilingarian (1974) and Chilingarian and Wolf (1975, 1976).

LITHIFICATION

Lithification is the process that changes unconsolidated sediments into weakly to strongly consolidated rocks. Lithification may occur through cementation, recrystallization, replacements (i.e., dolomitization), crystallographic welding of lime-mud, and by other processes such as desiccation.

Only cementation is considered in this section. Cementation, as understood here, is the process of open-space filling by physicochemical and biochemical authigenic precipitates, and excludes allogenic internal sediments. In cases of allochthonous limestones, cementation causes lithification; but autochthonous carbonate rocks may merely undergo a decrease in porosity and permeability without marked consolidation. [1] Cementation of limestones is very often associated with solution, corrosion, leaching and replacement phenomena, and can form a number of generations until the available open space is completely eliminated. Some limestone bodies are cemented stratum by stratum, whereas others are cemented "en bloc" (Kaye, 1959). Precipitation of carbonate cement can take place: (1) in littoral environments and subaerially; (2) within sediments but above the water-table; (3) at or near the water-table; (4) below the water-table; (5) in zones where fresh water mixes with marine water, or normal marine with supersaturated waters; and (6) under a thick overburden. According to Jaanusson (1961), however, no undisputable evidence of recent submarine calcium carbonate cementation at or close to the sediment—water interface has been reported in contrast to the widely occurring cementation processes above low-tide level.

Calcium carbonate is the principal cement in limestones and occurs as aragonite and various types of calcite in Recent and near-Recent sediments, and as calcite in older rocks (see inversion). The $CaCO_3$ source for cement may be either endogenic or exogenic, i.e., the carbonate may be derived from within the formation or brought to the site of precipitation from an outside source. Calcite is also one of the most significant cement types in terrigenous sediments. It is interesting to note, therefore, that whatever the original conditions of the depositional environment may have been, subsequent convergence of the physicochemical conditions appears to permit the formation of $CaCO_3$ cement. The explanation of calcite cement in a wide variety of rocks may lie in the independence of both Ca^{2+} and CO_3^{2-} of the Eh parameter (Krumbein and Garrels, 1942). Precipitation of calcium carbonate can be brought about by numerous factors discussed below (Niggli and Niggli, 1952). When considering the problems of $CaCO_3$ solution and precipitation one has to distinguish between: (1) changes in solubility when the CO_2 content remains constant or CO_2 is absent, and (2) changes in solubility when there is a possibility of CO_2 decrease or increase.

(1) The solubility decreases if the free CO_2 content remains constant or if free CO_2 is absent (and when the pressure of carbonic acid remains constant), i.e., $CaCO_3$ is precipitated from a saturated solution:

(a) With increasing temperature, if CO_2 is not present. (For example, in

[1] See discussion by Crickmay (1945).

CO_2-free sea water the solubility product constant of $CaCO_3$ at 0°C is $8.3 \cdot 10^{-7}$ and at 30°C it is $4.4 \cdot 10^{-7}$.)

(b) With decrease in the hydrostatic pressure associated with a decrease in dissociation of carbonic acid at constant CO_2 content.

(c) With a decrease in soluble NaCl or Na_2SO_4, etc. (so-called salt content) at constant gas pressure of the carbonic acid in the gaseous phase. This is due to the changes of the dissociation constant of the carbonic acid and the constant K (solubility product), where $(Ca^{2+})(CO_3^{2-}) = K$. (At 20°C and a salinity of 35‰, $K = 6.2 \cdot 10^{-7}$; at 20°C and no salt content, $K = 0.5 \cdot 10^{-8}$.)

(d) With addition of Ca ions (for example, bonded to SO_4, Cl), or if these are present and not bonded to carbonate and will form new $(NH_4)_2CO_3$ as a result of organic decomposition.

(e) When only calcite can exist and not the unstable aragonite or vaterite.

(f) When the water evaporates.

(2) Water, to which CO_2 has been added, increases the solubility of $CaCO_3$ because of formation and dissociation of H_2CO_3 into H^+ and HCO_3^-. The added H^+ will combine with the CO_3^{2-} ions already present to form the more stable HCO_3^-. In order to reach equilibrium at a constant solubility product, therefore, more Ca has to go into solution. On the other hand, because of this phenomenon a decrease in $CaCO_3$ solubility occurs, i.e., $CaCO_3$ is deposited, when the carbonic acid content decreases. That is the case:

(a) When the partial pressure of CO_2 in sea water (in equilibrium with the CO_2 of the atmosphere) decreases. The CO_2 content in the atmosphere is increased by volcanic activity, respiration of animals, decomposition of organic substances, etc. On the other hand, CO_2 is removed from the atmosphere by photosynthesis.

(b) When the pressure decreases (at constant temperature, CO_2 escapes into the atmosphere).

(c) When the temperature increases (at constant pressure and constant partial pressure of CO_2, with increasing temperature CO_2 is freed into the gaseous phase).

(d) When organic substances are formed by plants in the marine waters (in contrast to the animals which exhale CO_2).

(e) When the formation of CO_2 through organic decomposition is reduced or made impossible.

(f) When the salinity increases, because less CO_2 can be dissolved in marine than in fresh water.

In general, the above conditions conducive to $CaCO_3$ cement precipitation are controlled by three main processes discussed further below, namely, physicochemical, bacterial and decompositional, and algal processes. Little work has been done on rock cementation and much of the following presentation is confined to theories that attempt to explain the formation of

Recent and Pleistocene beach-rocks. Least of all is known about the factors that control genesis of fibrous, drusy and granular carbonate cement (see carbonate types). Very little precise information is available on the parameters that control precipitation of aragonite in preference to high-Mg and low-Mg calcite. Many of the experimental results appear to be contradictory. Goto's (1961) experiments suggest that slow reaction, higher pH value of solution, diminished solvation effect of water, and balanced proportions of Ca^{2+} in relation to CO_3^{2-} are responsible to some extent for the formation of aragonite and are less favorable to calcite genesis. The presence of Mg^{2+}, Sr^{2+}, and Ba^{2+} seems to be unfavorable for aragonite formation. On the other hand, experiments by Zeller and Wray (1956) suggest that aragonite formation is favored by low Mn^{2+} but also by high Sr^{2+}, Ba^{2+} and Pb^{2+} contents, which differs from Goto's conclusions. Other factors evidently are responsible to cause such seemingly contradictory results. In the present-day calcareous sediments aragonite is formed especially in shallow-water environments in tropical and semitropical regions suggesting that temperature is significantly influential in the genesis of aragonite.

Physicochemical precipitation

Ginsburg (1957) mentioned that extensive cementation of beach-rocks occurs in those young carbonates that are subaerially exposed or located in zones of meteoric waters. Those sediments still in a marine environment or above the groundwater table are very friable. This agrees in general with the observation made by Kaye (1959) who stated that cementation of beach-rocks of Puerto Rico is not coincidental with high tide but extends up to 3 ft above it. The upper limit is possibly controlled by capillary action or by splash. The lower limit lies slightly below low spring tide and is probably controlled by the lowest level of wave trough at low spring tide. These beach-rocks are mainly cemented by calcite rather than aragonite. As Kaye examined very recently formed sediments, it seems that inversion or recrystallization from aragonite to calcite is unlikely. In many other localities, however, beach-rock is cemented by aragonite. Illing (1954) suggested that calcite is precipitated from fresh water and aragonite from salt water. Some of the Puerto Rican beach-rocks were locally cemented by iron oxide which was derived from a rubbish dump containing iron objects. Kaye (1959) pointed out that this dump is not more than 80--100 years old, which is indicative of relatively rapid alterations.

Emery et al. (1954) reported that lithification of beach-rocks may be due to the consolidation of interstitial pasty lime-mud matrix. They believed that consolidation results from precipitation of microcrystalline carbonate within the paste, or from crystallization of the paste probably during daytime periods at low tide, when the tidal waters are warm and algal processes utilize much CO_2. Inasmuch as most beach-rocks appear to lack a matrix, however, this explanation is of restricted application.

Cementation of littoral, beach, and dune limestones has also been explained by the action of fresh water that dissolves part of the carbonate framework and later precipitates it upon evaporation, aeration and possibly with the help of organisms such as bacteria. Russell (1962) supported this theory by showing different degrees of corrosion of carbonate grains in relation to the fresh ground-water table (see corrosion). Cementation of these Puerto Rican beach-rocks investigated by Russell seems to take place in the vicinity of the fresh-water table. The beach-rock is thickest where seasonal contrasts in sea level are most pronounced. It seems that with the changes of sea water level, the fresh—salt water table is displaced correspondingly, with the lighter fresh (or brackish) water floating above more saline water. Thus, the zone of cementation is shifted causing thickening of the beach-rock. The cement is almost wholly calcite with subordinate amounts of aragonite. Although an endogenic origin of cement is likely where there are signs of internal corrosion, this theory is not applicable in cases where no solution of the carbonate sediments has taken place. In some instances, therefore, the calcium carbonate must have come from an exogenic source. This is supported by the carbonate cementation of beach-rocks composed wholly of terrigenous material, i.e., quartz, which could not have supplied any endogenic $CaCO_3$. Crickmay (1945) discussed some interesting examples in limestones of Lau, Fiji.

Related to the above is the theory that weak acids, in particular humic acids, percolating down into the fresh ground-water lower its pH. Thus, the fresh water dissolves carbonate from the surrounding medium and, at its contact with the underlying salt water, precipitation of the calcium carbonate cement may occur. Seasonal and yearly fluctuations of the fresh—salt water interface may cause a thick zone of cementation. However, Maxwell (1962), for example, found this theory inapplicable to the beach-rocks of Heron Island, Great Barrier Reef.

Capillary action is a third likely process that supplies $CaCO_3$ in some cases. In particular, limestones subaerially exposed may be heated to the extent that fluids are brought up from a lower level by capillary forces and deposit carbonate cement on evaporation and aeration. Certain tufa, travertine and caliche deposits are the result of this process: for example, the areally-extensive tufa and travertine deposits of Pleistocene and Recent age within alluvial fans, slope-wash, and even lacustrine sediments of parts of the Great Basin area of western United States (in particular southern Nevada).

Another hypothesis is based on the premise that fresh ground-water brings dissolved calcium carbonate from the hinterland, which is followed by precipitation as this water seeps out through the coastal sediments. Various objections have been raised against this process because it is thought that it does not explain the sporadic occurrence of beach-rock, for example. Also, cementation of sediments takes place where the hinterland is devoid of carbonate source material.

It has been pointed out that present-day accumulations of tufa along the shores of Utah Lake west of Provo, Utah, consist of beach-rock, shell heaps and other materials more or less tightly cemented by calcium carbonate (Bissell, 1963). It was also noted that beach-rock and massive tufa deposits formed along the shores of Pleistocene Lake Bonneville through combined action of wave splash and Algae in releasing CO_2 and thus precipitating $CaCO_3$ (Bissell, 1963).

Ginsburg (1953a) believed that because the beach-rocks he investigated are exposed to saturated marine waters and provide abundant nuclei for $CaCO_3$ precipitation, cementation occurs due to heating and evaporation of interstitial fluids. At low tide, water remains as intergranular films and permits a more complete exchange of CO_2 between atmosphere and solution, inducing a more rapid equilibrium and precipitation from supersaturated solution. According to Revelle and Fairbridge (1957), the water temperature of splash pools just above tide limit in Western Australia varies from 13°C at night to 24°C in daytime, with a change of pH from 8.2 to 9.4. During the night the pH remains equal to 9 and is ample to account for large precipitation of carbonates. The above authors also mentioned the formation of superficial pelagosite crusts and pore-space fillings formed by the action of spray and evaporation. Wolf (1963a) has similarly explained the numerous open-space calcite patches in internal channels and cavities of Devonian littoral algal bioherms. The internal voids must have undergone a sharp temperature increase during low tide when the reef-structures were directly exposed to sunlight. The films and small patches of intrastratal fluids that remained behind at low tide then reached supersaturation and precipitated $CaCO_3$ on the cavity walls.

The restriction of penecontemporaneous carbonate cementation to particular localities, e.g., intertidal zones (it is absent below low tide), is explained by Ginsburg (1953a) by the sluggishness of the equilibrium between solid and dissolved calcium carbonate, and the inhibition of this equilibrium by organic matter. According to experiments, it may take 6—8 h under laboratory conditions for a system to reach equilibrium (Hindman, 1943; Miller, 1952; both in Ginsburg, 1953a).

Daly (1924, in Ginsburg, 1953a; and Kaye, 1959) thought that beach-rock cementation occurs in two stages. An initial precipitation of $CaCO_3$ from sea water was a result of ammonifying action of decaying organic matter originally incorporated in the sediments as detritus. The chemical reaction consists of ammonia combining with CO_2 to form ammonium carbonate, which then reacts with calcium salts in solution to form $CaCO_3$. Daly envisioned a second stage of cementation during which the precipitation of $CaCO_3$ from marine water is caused by aeration and surf agitation, and their effect on the CO_2 partial pressure in the water. The first stage provides nuclei essential to the deposition of $CaCO_3$ in the second stage. The varying distribution of detrital matter can, therefore, explain the localization of cementation and formation of beach-rock. As Kaye pointed out, however, numerous localities rich in organic matter lack beach-rock genesis. Kaye discussed at length the physicochemical factors and had to reject them as an explanation for beach-rock cementation. He believed that the Puerto Rican sediments were formed most likely by microbiological processes.

One has to conclude that the problems of physicochemical cementation have not been solved. Whatever the factors are, they cannot be of equal importance in all environments. It seems that temperature is one of the most important parameters and restricts beach-rock genesis to tropical and subtropical localities. Other factors, however, must be of equal significance as beach-rock formation does not take place at many localities where temperature, evaporation and other conditions seem to be favorable for cementation.

Bacterial processes and decomposition

Bacterial processes and decomposition of organic matter are closely linked and are inseparable in the study of organic influences on $CaCO_3$ precipitation, and other diagenetic processes. Recent calcareous sediments may contain bacteria in concentration from about 10 to 10,000 billion organisms/g in contrast to the water above the sediments that contains only 10 to 1,000 organisms/mm^3. ZoBell (1942) has reported even higher concentrations of bacteria in sediments. The quantity of bacteria is a function of sediment grain size and presence of organic matter. The heterotrophic bacteria depend on organic matter as a source of carbon or energy. Not much organic matter, however, is required to assure the presence of some bacteria. Small quantities of organic particles adsorbed on sand grains and cavity walls suffice.

The type of bacteria and availability of oxygen control the depth to which oxidation of organic matter to CO_2 can take place. For example, Ginsburg (1957) mentioned that in the reef and back-reef deposits investigated by him most of the organic matter is relatively rapidly removed and the sediments are, therefore, light colored and have only a slight odor of H_2S at depth. On the contrary, the shallow-water calcareous muds may contain three to six times as much organic compounds. Here, only the upper layer, one inch in thickness, is light colored and the sediments as far down as 8 ft smell strongly of H_2S. The limy muds of the Red Sea reef complex, although light colored, give off a slight odor of H_2S and, when placed in a closed bottle, become black and form minute pyrite crystals 1 mm long. Lagoonal white carbonate muds of Pacific atoll reefs become black and saturated with H_2S only to a depth of about one inch (Termier and Termier, 1963). All these occurrences are attributable to bacterial processes. Based on these observations, Ginsburg (1957), among others, concluded that shallow bays can form a type of barred or stagnant basin in which the original organic matter need not be diagenetically removed.

Bacterial decomposition of proteins leads to the formation of carbon dioxide, ammonia, hydrogen sulfide and a variety of intermediate products, and many carbohydrates are converted into carbon dioxide, carbon monoxide, methane and organic acids (ZoBell, 1942). Some species oxidize, for example, organic calcium salts, thereby increasing the Ca^{2+} concentration. On the other hand, the autotrophs obtain energy from oxidation of inorganic substances such as ferrous iron, manganous manganese, hydrogen sulfide, hydrogen, carbon monoxide, methane or ammonia. These latter bacteria are aerobes, i.e., require free oxygen, and thus predominate mainly in the surface sediments. The bacteria may also produce significant amounts of biocatalysts or enzymes which can activate numerous chemical reactions. Some of these catalysts continue to be functional after death of the bacteria (ZoBell, 1942). The biologically mobilized material in turn stimulates and controls diagenesis by changing the pH, Eh, partial pressure, composition of interstitial fluids, temperature, and so on. For example, when normal sea water (chlorinity 19‰ at 25°C) is in equilibrium with solid $CaCO_3$, the pH remains stable at approximately 7.5 due to bicarbonate—carbon dioxide reactions. This carbonate buffer system can be changed by bacterial decomposition of organic matter and by addition of acidic or basic components. Addition of CO_2 acidifies water, whereas sulfate reduction may produce either an acid or an alkaline effect depending on the organic matter and products of decomposition. Addition of ammonia increases the pH to alkaline conditions, whereas oxidation of ammonia to nitrate and nitrite causes a decrease in pH.

Cloud et al. (1959) reported that an increase in hydrogen ions, with an accompanying decrease in pH, is caused by the bacteria that produce CO_2. Together with HCO_3^- addition from the reactions: $CO_2 + H_2O \rightarrow H_2CO_3 \rightarrow H^+ + HCO_3^-$, the combination of new H^+ with CO_3^{2-} to produce still more HCO_3^- causes a reduction in the CO_3^{2-} component of alkalinity, and allows both Ca^{2+} and alkalinity to reach high values without precipitation. The interstitial fluids are, therefore, in a condition favoring precipitation whenever they may be exposed to an environment of higher pH and higher CO_3^{2-} content. This condition may be fulfilled when, for example, bacterial production of ammonia causes an increase in pH, waves or organisms stir up bottom sediments, burrowers transfer sediments, or when lateral movements of interstitial fluids bring them to a higher pH environment; $CaCO_3$ precipitation and cementation may then result. In the latter case of lateral movements of fluids, the transfer to a higher pH locality, results in loss of H^+ from HCO_3^- and in increase in CO_3^{2-} content which causes precipitation of $CaCO_3$. This may explain, as Cloud et al. (1959) pointed out, (1) the seaward increase (toward the bank margin) of aragonite lithification of pellets and algal grains, and (2) the cementation of pellets to form lumps (grapestones). Measurements by these authors indicate that this lithologic change from lime-muds to pellets and lumps is accompanied by rising pH and Eh as pore space and circulation of oxygen-bearing water increase. It seems, therefore, that lateral movements of interstitial waters, in addition to other factors, may cause regional variations in limestone lithology, textures and structures.

Purdy (1963) mentioned the possibility of cementation of bahamite sediments by decomposition of organic detritus by ammonifying and nitrate-reducing bacteria which produce amnonia. The latter reacts with the calcium bicarbonate in the immediate surrounding water and causes $CaCO_3$ precipitation. Such a process can occur on a very small scale within pellets, for example, and cause cementation and preservation of very friable material.

Revelle and Fairbridge (1957) concluded from the available published evidence that bacterial precipitation of calcium carbonate in Recent marine environments seems to be strictly limited in scope. This may be correct if one speaks of the total volume of carbonate sediments but, as the considerations above suggest, it does not exclude the possibility that bacterial activities lead to cementation and stimulate other diagenetic processes.

Algal cementation

Algal cementation is one of the most important lithification processes in shallow-water limestone genesis. The lime precipitates of blue-green, green and red Algae can occur as crusts and are then considered more or less of syngenetic origin. As it seems possible, however, that microscopic cells and filaments can exist for short periods of time to some depth within sediments and, as Algae in general cause chemical modifications in surface and interstitial waters, they must find a place in a discussion on cementation and diagenesis in general.

Calcareous Algae play a dual role: on one hand they dissolve lime possibly by use of oxalic acid or by indirectly acidifying the water and, on the other, the same organisms cause $CaCO_3$ precipitation. Ginsburg (1957) listed the corrosive action of Algae as one of the major diagenetic processes changing extensively Recent calcareous sediments on the Bahama Bank and in Florida.

Precipitation of $CaCO_3$ by Algae occurs through respiration processes. At night the Algae give off CO_2 into the microenvironment and this may result in corrosion and solution, whereas in daylight CO_2 is utilized during photosynthesis leading to alkaline conditions sufficiently strong to cause precipitation of $CaCO_3$ and pelagosite. It seems then that corrosion and precipitation may alternate, and the predominance of one over the other depends on local circumstances. Emery et al. (1954) mentioned the possibility that boring Algae which dissolve chemically surface layers might cause precipitation of calcium carbonate a distance down within the sediments.

Another indirect process of algal cementation may be due to the detritus-binding Algae. *Schizophyta* filament and cell colonies, for example, when covered by a sudden influx of detrital sediment are not necessarily killed but move upward through the layer and re-establish themselves above the sediment. Recent stromatolites are formed by similar processes. As Algae give off oxygen and use CO_2 during their life processes, it may be possible that they stimulate diagenetic changes within the uppermost sediment accumulations while they move upward. The microenvironment beneath algal mats may also be conducive to corrosion and precipitation. The waters beneath the mats are relatively isolated from the overlying sea water as indicated by emission of H_2S odor from the mats upon disturbance (Revelle and Fairbridge, 1957). The metabolic activity of Algae and the decay of organic matter cause marked changes in CO_2 content and other properties of the water. The pH variations from 6.5 to 8.7 in fluids collected from beneath algal mats in Tahiti are sufficient for solution and precipitation of $CaCO_3$.

In addition to the indirect algal influence, some genera are capable of precipitating carbonate on their surfaces and/or internally, which is partly due to absorption of CO_2 from the water medium. This biogenic carbonate commonly appears as dense cryptocrystalline material in thin-section (Wolf, 1965a).

It is quite conceivable that the so-called umbrophile, i.e., shade-adapted, Algae can exist within the upper layers of sedimentary frameworks and cause precipitation of thin crusts around detrital grains and on walls of voids. For example, Wolf (1963a) described beach-rocks of Portuguese Timor composed of skeletal and algal debris and volcanic rock fragments which are circumcrusted by layers of dark brown, nearly opaque, cryptocrystalline to very dense calcium carbonate (Plate 6-I). This carbonate is identical to the algal debris forming part of the detrital framework, and where this debris is encrusted by the brown layers, the two merge completely and can be distinguished only with difficulty in thin-section. Wolf has also shown that algal calcareous deposits of algal pisolites (Plate 6-II) and algal crusts (Plates 6-III, 6-IV) of very recent origin can change relatively quickly from a thin upper layer containing clear algal cellular structure to a dense textureless crypto-

PLATE 6-I

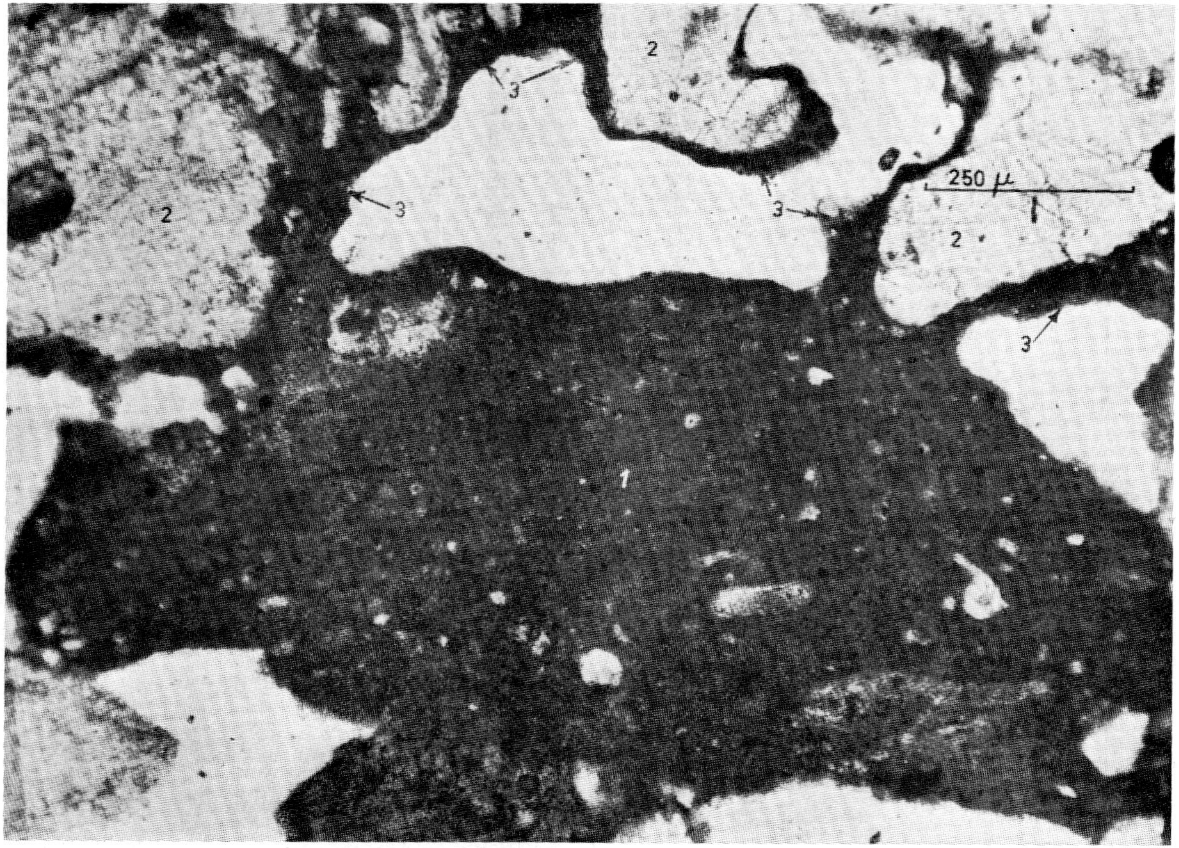

A section of a beach-rock of Portuguese Timor composed of a dense algal fragment (*1*) and skeletal fragments (*2*). All have been surrounded by a crust of micrite (*3*). The remaining white spaces are pores. Petrographic studies indicate that the dense algal grain has most likely formed by "degrading recrystallization" (see text). It is, therefore, a fragment of pseudomicrite. The origin of the micrite crust or "cement" is not certain—it may have been formed by direct biochemical precipitation or by degrading recrystallization of either physicochemical or algal calcite (see text). Note the merging of the dense fragment and crystalline calcium carbonate a few millimeters below. Intermediate cases with faintly preserved cellular features are also present. Remarkably similar changes have been reported in a Devonian algal reef complex (Plate 6-V). The exact mechanism of this alteration is not clear, but it seems, that the mere absence of algal features does not preclude a floral origin for the cryptocrystalline circumcrusts such as those that lead to the cementation of Portuguese Timor beach-rocks. Further research is required to confirm these

PLATE 6-II

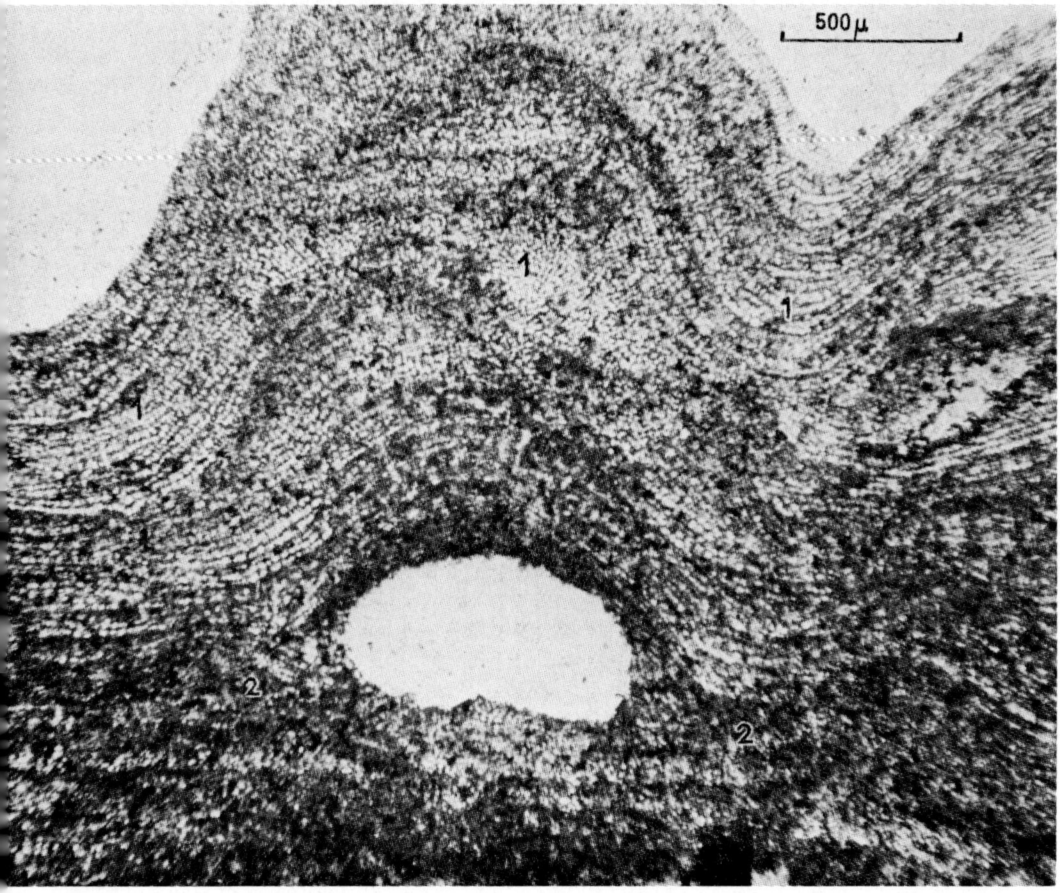

Section of a Recent algal pisolite exhibiting distinct cellular features (*1*) grading into dense cryptocrystalline calcium carbonate (*2*) toward the nucleus. The latter may be a product of "degrading recrystallization" or disintegration (see text). (Thin-section supplied by Mr. J. Standard.)

PLATE 6-III

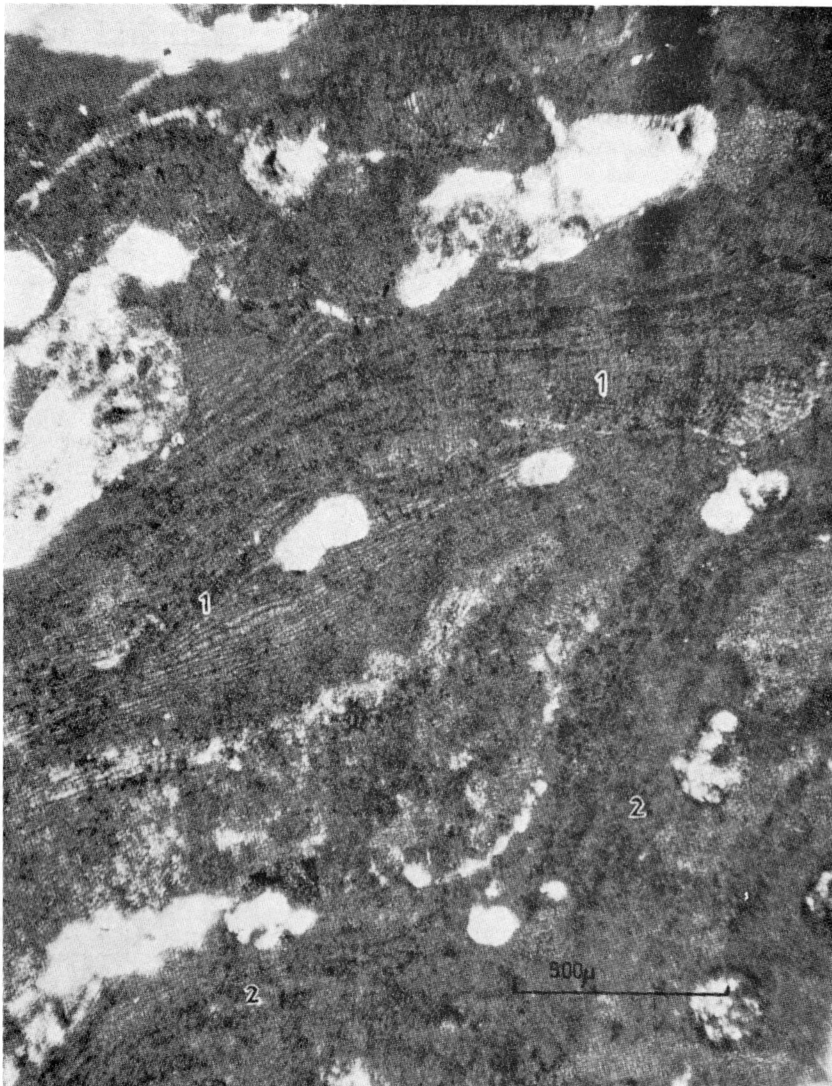

Recent algal biolithite composed of cells (*1*) and dense cryptocrystalline carbonate (*2*), which may be micrite formed directly by algal precipitation, by "degrading recrystallization", or disintegration of algal cellular material (see text). Note the resemblance between the material in Plates II and III of recent origin and that of the Devonian in Plate V.

PLATE 6-IV

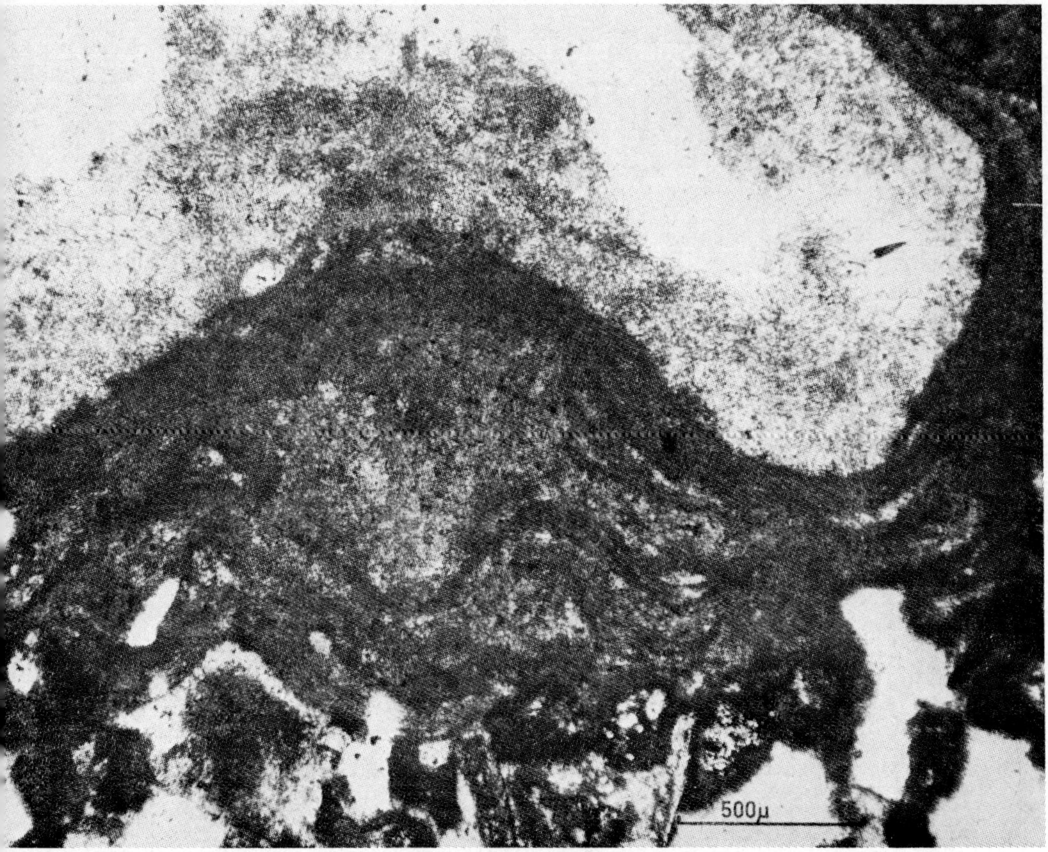

Recent algal micrite crust. It is a product of direct algal precipitation (= orthomicrite), or formed by "degrading recrystallization" or disintegration of an algal cellular colony (= pseudomicrite). Important to note is that whatever its origin, it is automicrite, i.e., formed in situ. Therefore, it is either an ortho-automicrite or a pseudo-automicrite (see text and Table 6-III). Many of the Devonian algal bioherms, Nubrigyn Formation, N.S.W., consist of identical dense algal automicrite-biolithites.

observations, in particular because it is quite likely that other processes can give rise to cryptocrystalline carbonate cement.

CORRASION, CORROSION, SOLUTION, DECEMENTATION, DISINTEGRATION

Several processes can alter and obliterate limestones during syngenetic, diagenetic and epigenetic stages. Some of the products may have (resemble) present-day, near-surface weathering features.

PLATE 6-V

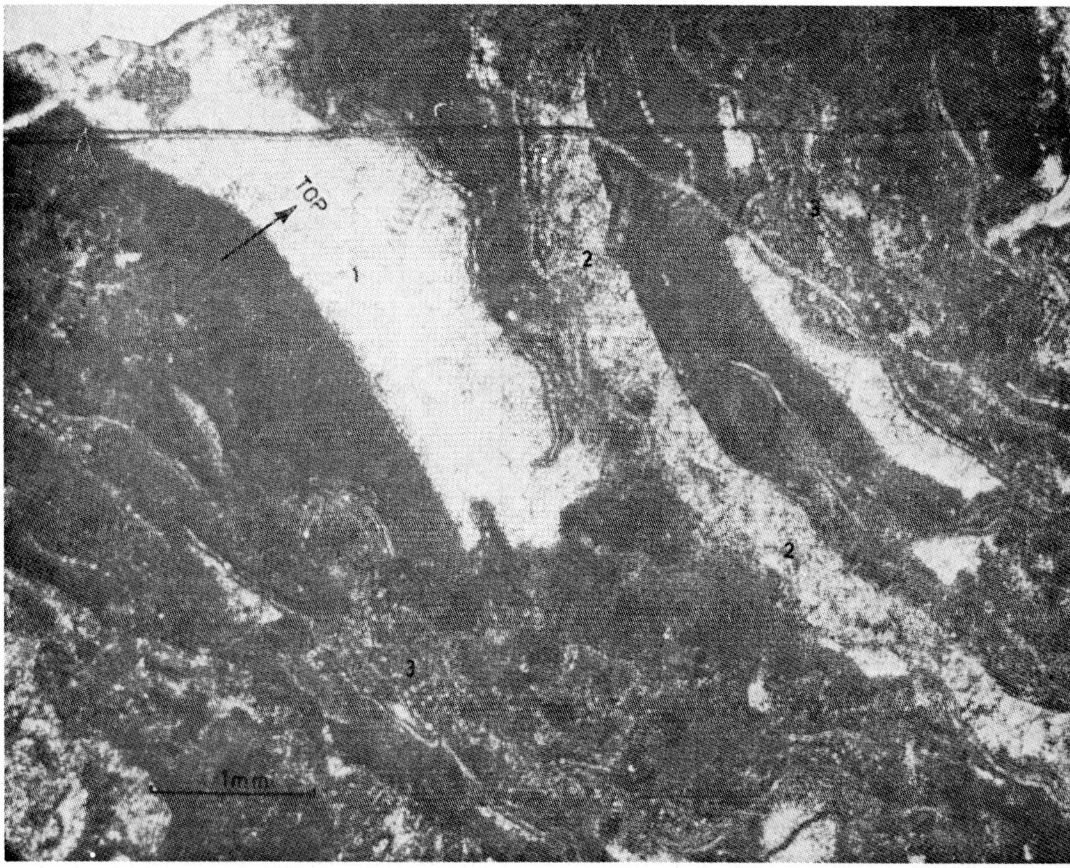

Micrite and algal filament-biolithite of a Devonian algal bioherm, Nubrigyn Formation, N.S.W. One stromatactis (*1*) and a recrystallized stromatoporoid crust (*2*) are present. The filaments are *Rothpletzella* (*3*). Most of the dense micrite is automicrite of which most of the 300 knoll-shaped bioherms are composed. Note the resemblance with the material in Plates III and IV. Compare it with the micrite circumcrusts in Plates I, XV, XVI and XIX; and with algal pellets and grains in Plates, I, X, XIII, XV, XX and XXIV. (The filaments have been identified by Professor J.H. Johnson.)

Corrasion and corrosion may occur very extensively in littoral and subaerially-exposed limestones and can form diagenetic "micro-karst" structures. Little is known as to how significant these processes are in sublittoral carbonate rock environments. They deserve considerably more attention because they may be excellent paleogeographic indicators, may illustrate formation and destruction of porosity and permeability, and may influence diagenesis to a large degree. Internal cavity systems originate in a number of ways. Many of them were originally surface depressions of various kinds which became part of the limestone framework. For example, the pitting of algal-encrusted limestone surfaces reported by Kaye (1959) resulted in depressions and irregularities from a few millimeters up to a few feet in size with shapes changing systematically corresponding to the environment. A relative rise of sea level causing spreading, or transgression, of the encrusting Algae over the smaller pits would result in an internal cavity system. A similar process is mentioned by J.W. Wells (1957) who described recent surge channels with upper "eaves" of calcareous algal deposits. These spread until the channels are completely roofed over and constitute part of a tubular labyrinth within the limestone body. From this stage onward, diurnally surging waters and solutions can penetrate the limestone to flow internally below the surface and cause internal corrosion and abrasion, internal sedimentation, replacements, and chemical precipitation, all described later. As many of the voids have been inherited from the surface, they are mostly horizontally oriented and are often concentrated along specific beds. Hence, the internal fluids will continue to flow predominantly horizontally and any further corrosion occurs mainly sideways and to a lesser degree downward, commonly resulting in flat-bottomed voids.

Internal open-space structures in Devonian algal bioherms (Plates 6-VI—XIV) have been reported by Wolf (1963a, 1965a, c), among others. Analogous to the recent occurrences, many have flat bottoms with irregular upper parts and are predominantly horizontally oriented. In serial thin-section and polished section studies it can be demonstrated that two or three horizons of cavities meet laterally to form part of one system (Plate 6-VII). Many of the cavities, all of which have been completely filled by secondarily introduced material, exhibit solution and/or abrasion features (Plates 6-VI, -VIII, -IX, -XII). Where extensive alterations occurred, it is impossible to determine the original factors that localized the fluids, for evidence of the last process only is present. In less altered void structures, however, it can be shown that many were originally inter-biolithite spaces (i.e., spaces between colonies) and channels with algal overgrowths. Some of the algal filaments and cells clearly line and follow more or less concentrically the outlines of the channels. The flat-bottomed character of these cavities is identical to the so-called stromatactis structures described, for example, by Bathurst (1959a) and Otte and Parks (1963). The above discussion suggests that no "soft-body burial" hypothesis is required to explain the genesis of stromatactis, and that these controversial structures are more likely caused by a combination of both syngenetic and diagenetic inorganic processes, and that Algae play only an indirect role. (A "soft-body burial" origin is not completely impossible in

PLATE 6-VI

1

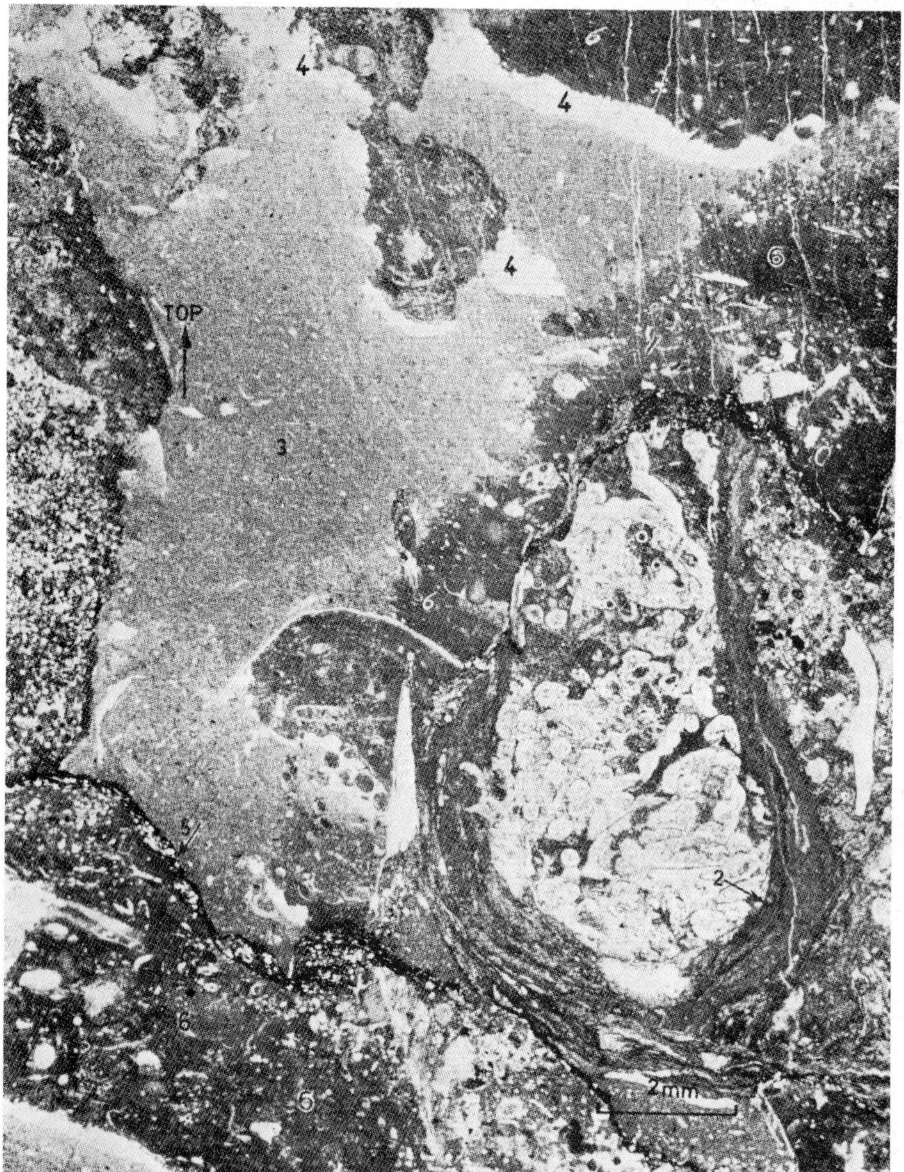

Complex section of a Nubrigyn algal bioherm showing algal colony (*1*) and detrital skeletal fragment enveloped by algal micrite layer (*2*). A large cavity, with some features of differential solution has been filled by allomicrite internal sediment (*3*) and some clear granular orthosparite (*4*). A stylolite cuts across the slide (*5*). Several patches of dense algal automicrite (*6*) are recognizable.

PLATE 6-VII

Part of a Devonian Nubrigyn algal bioherm consisting of lower filament colony (*1*) overlain by spongy algal growth and one coral fragment. The intra-biolithite cavity is filled by dense automicrite bottom sediment (*2*). Two horizons of open-space structures merge to form part of one cavity (*3—4*). Detrital internal sediment composed of algal pellets and micrite is distinctly recognizable (*3*). Most of the calcite cement is of the clear granular sparite type (*4*).

PLATE 6-VIII

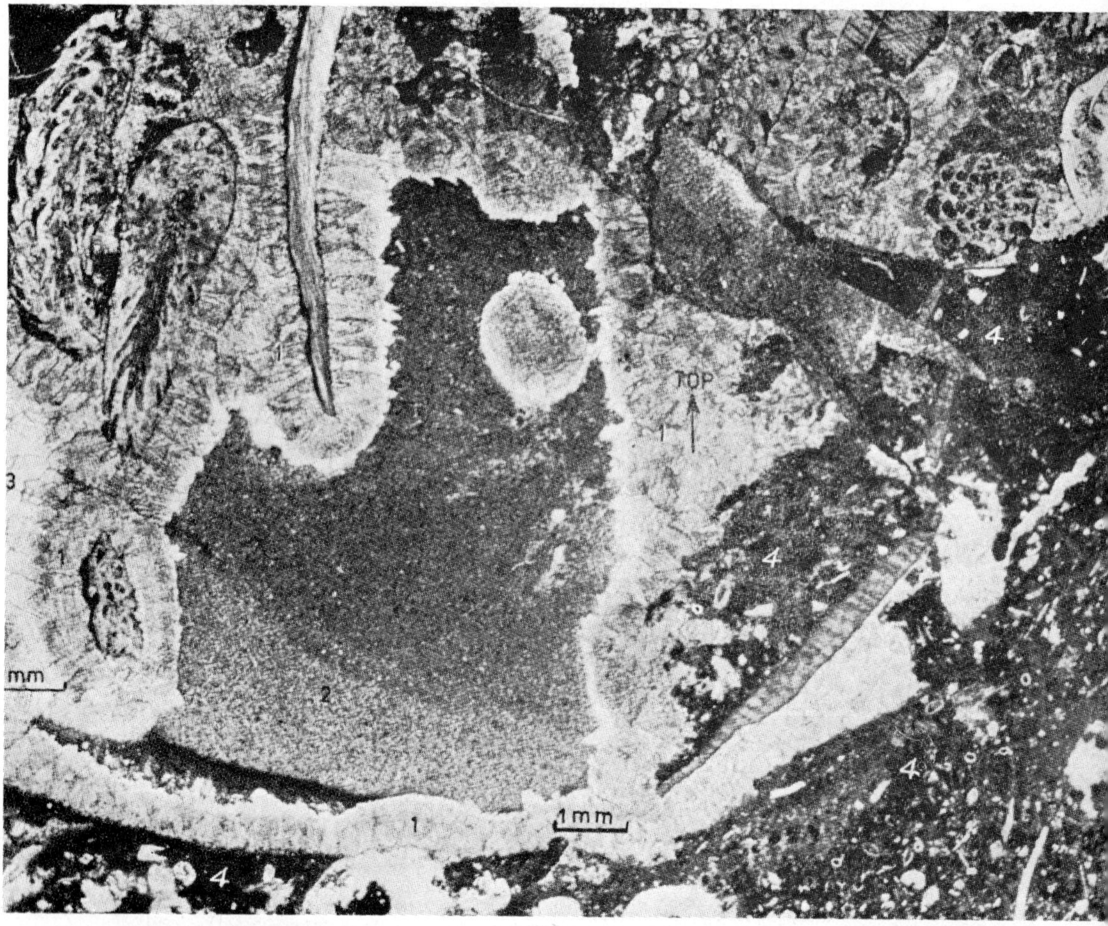

Open-space structure of a Nubrigyn bioherm composed of a rim of brown fibrous orthosparite (*1*) lining both the micrite framework and the skeletons extending into the cavity. The remainder of the space is occupied mainly by internal allomicrite (*2*) and clear granular orthosparite (*3*). The host rock is composed of algal automicrite (*4*) rich in detrital skeletal fragments, bound by filaments and cells, and some encrusting Foraminifera of the genus *Wetheredella* (identified by Professor J.H. Johnson).

the genesis of some other open-space structures, however.) It has been reported that the characteristics of the stromatactis change with location in a reef complex. This is in agreement with Kaye's (1959) observations of a systematic environmental change of the shape of surface pits before they become part of an internal cavity system.

PLATE 6-IX

Micrite limestone of an algal bioherm with detrital fragments. The stromatactis open spaces were clearly formed by differential solution of the micrite. The crinoid ossicles (*1*) and the shell (*2*) remained unaffected. Note the hematite impregnation of the bottom and the thin film of iron oxide on the exterior of one valve extending into the cavity.

Not all surface pits are incorporated into the sedimentary rocks as open spaces. Under conditions other than those described above, the depressions may be completely filled by detritus, especially if the "eaves" of encrusting Algae are not present. Jaanusson (1961) mentioned pits in limestone, some of which are flat-bottomed. All are completely occupied by fine-grained detritus forming part of the overlying bed. Interesting to note is the

PLATE 6-X

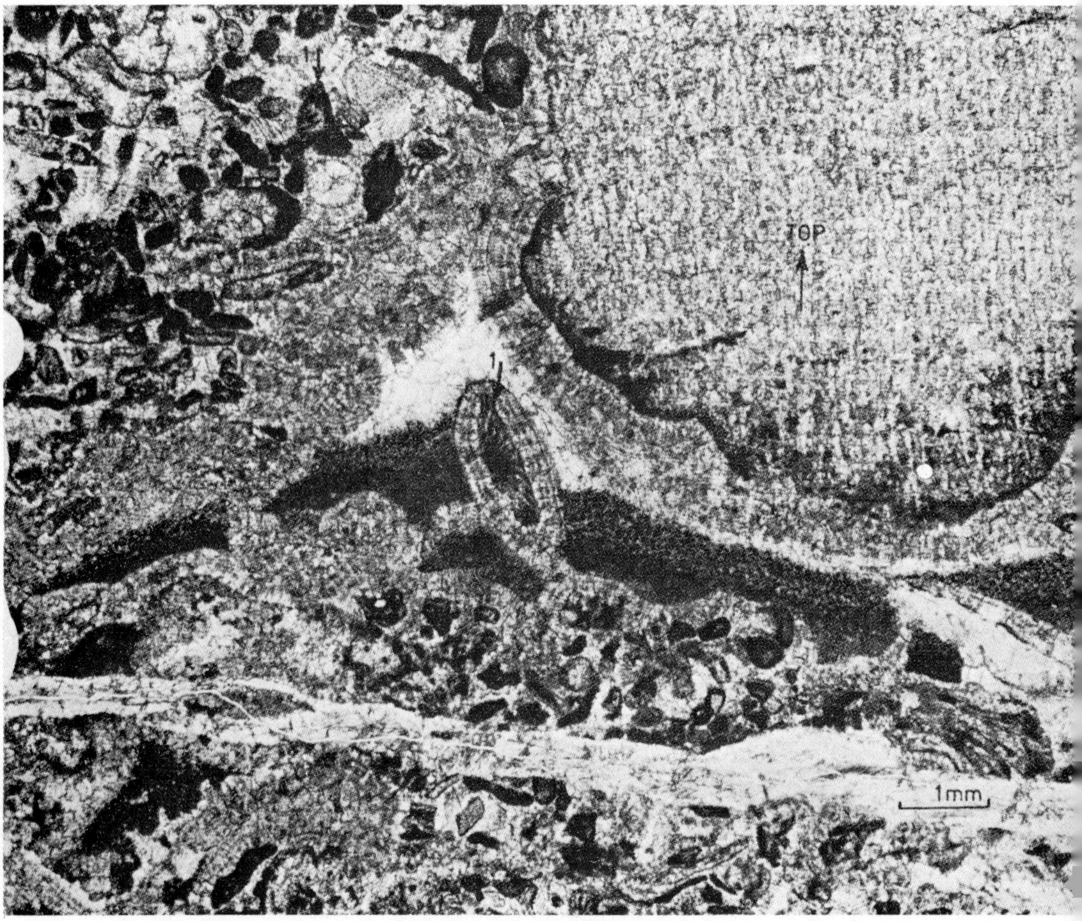

Detrital skeleton and algal grain limestone of the Nubrigyn Formation, N.S.W., cemented by brown fibrous orthosparite. A cavity left after fibrous sparite precipitation was partly filled by red iron oxide internal sediment and clear granular orthosparite. Note two well-preserved algal stem segments (*1*). The large skeletal fragment is thinly encrusted by dark algal micrite.

The paragenetic sequence is: detritus accumulation → solution (?) cavity → fibrous sparite → hematite internal sediment → clear granular sparite.

In this case it is not quite clear how the cavity has been formed. If it is of primary origin, it is difficult to see how the grains could have supported the fragments. It may be possible that a slightly cemented calcarenite underwent solution resulting in cavities similar to those shown in Plate XVIII.

PLATE 6-XI

Nubrigyn algal micrite limestone with some skeleton and algal fragments and flat-bottomed stromatactis. Note that the bottoms of two large stromatactis have been impregnated by red hematite. Some others have thin films of iron oxide. The cavities have been filled by orthosparite. The vertical fracture is filled by calcite, which is synchronous with that of the stromatactis cavity into which the fracture passes.

bleaching of one bituminous limestone parallel to the pitted surface. Probably diagenetic oxidation of the bituminous substance resulted in bleaching of the upper part of the sediment.

Corrosion, solution and leaching of argillaceous limestone subsequent to its accumulation can form patches, lenses, laminae and beds of marls or clay (Lindström, 1963). Removal of calcareous skeletons by solution may leave

PLATE 6-XII

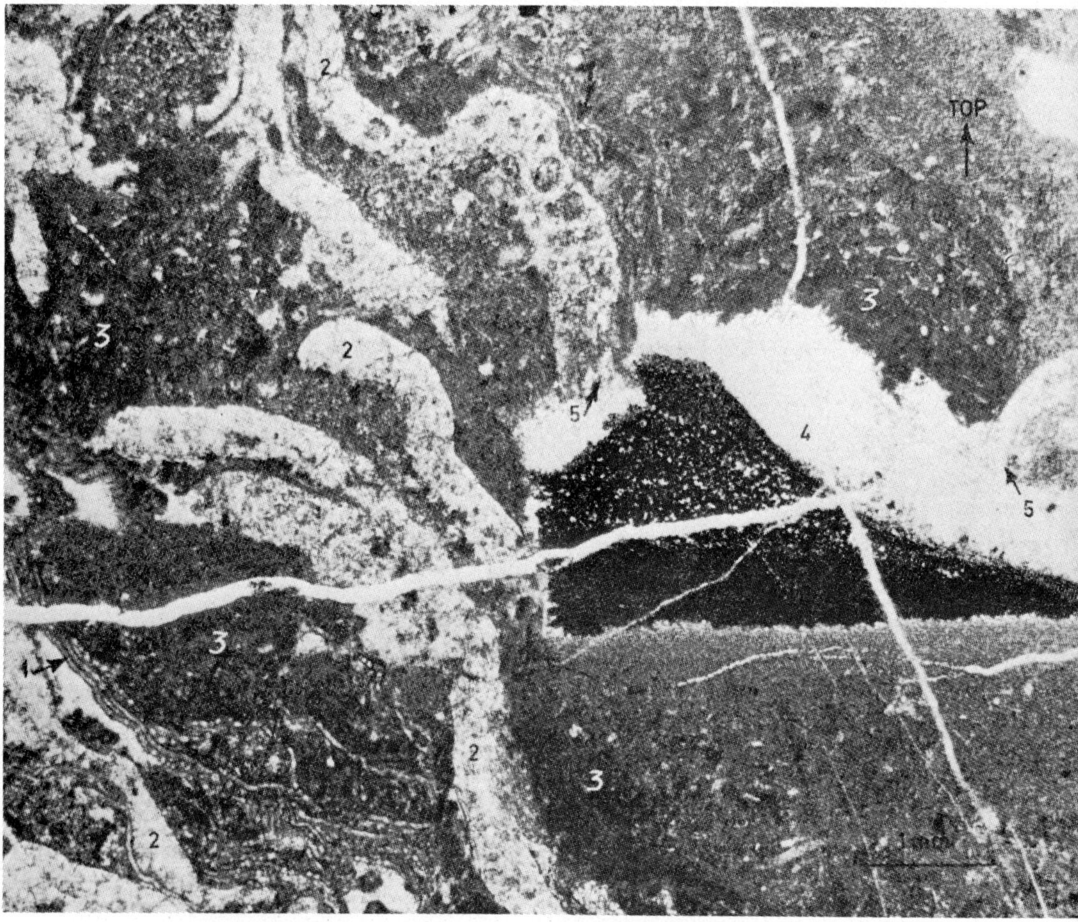

Section of a Nubrigyn bioherm with algal filaments (*1*), recrystallized stromatoporoid crusts (*2*), and automicrite (*3*). A flat-bottomed cavity, i.e., a stromatactis (*4*), is filled by detrital red-brown hematite and granular orthosparite. The cavity shows corrosion features (*5*). Note that the sparite of the fractures is contemporaneous with that of the cavity above the hematite. The longest dimension of that portion of the sample represented by the thin-section is equal to 8.25 mm.

internal casts if the internal parts of organisms were filled by less soluble or insoluble material.

Both direct and indirect organic processes may lead to significant corrosion, solution and disintegration of calcareous sediments. The bacterial processes leading to corrosion and solution of $CaCO_3$ have been reviewed by Re-

PLATE 6-XIII

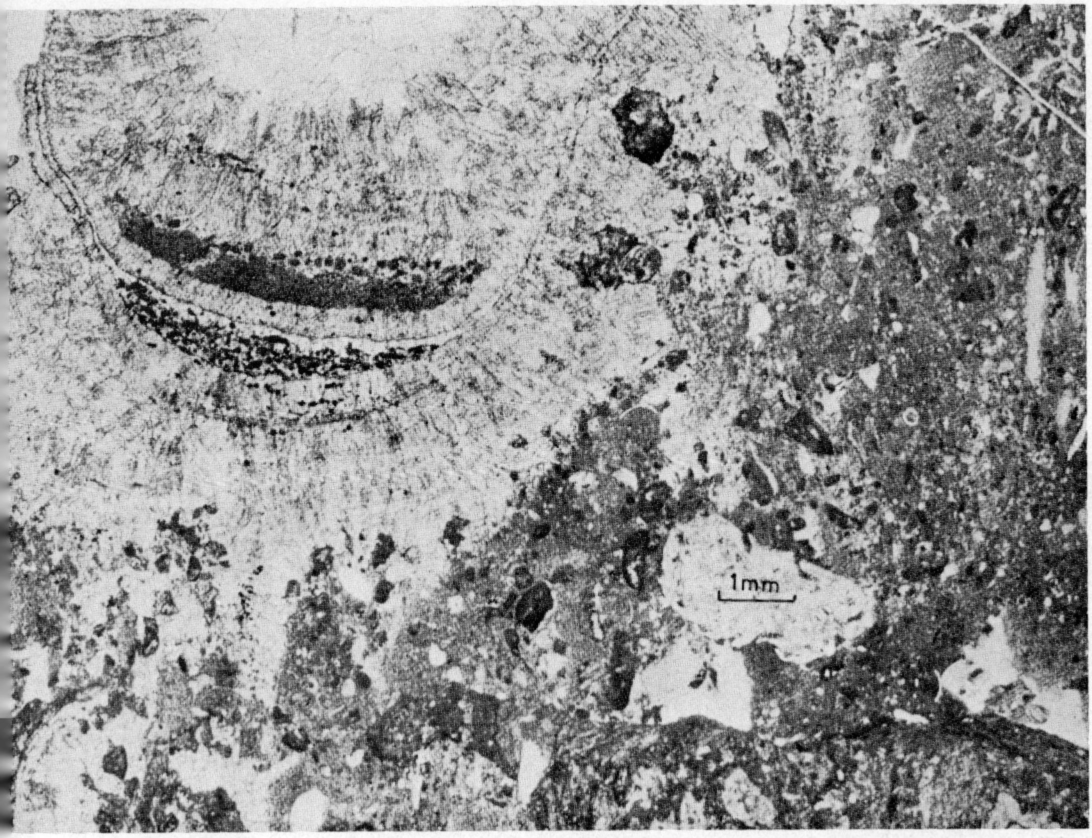

Open-space structure in an allomicrite limestone of the Nubrigyn Formation. Note the numerous brown fibrous orthosparite generations each separated by a layer of hematitic pellets, micrite and calcareous pellets, or films of hematite. The remaining space is filled by clear granular orthosparite. The allomicrite host rock contains numerous algal grains and pellets and skeletal fragments. It seems that the cavity was formed by solution. The paragenesis is as follows:
Syngenetic: micrite accumulation.
Precementation—diagenetic: solution cavity.
Syncementation—diagenetic: fibrous calcite and internal sediments.
Postcementation—diagenetic: granular sparite.

velle and Fairbridge (1957). As soon as organisms die and become buried, bacteria concentrate to decompose the soft parts. It has been illustrated, for example, that during the decompositional process shells lost 10—24% of their $CaCO_3$ in 1—2 months (in one case, 25% in 2 weeks); and merely traces

PLATE 6-XIV

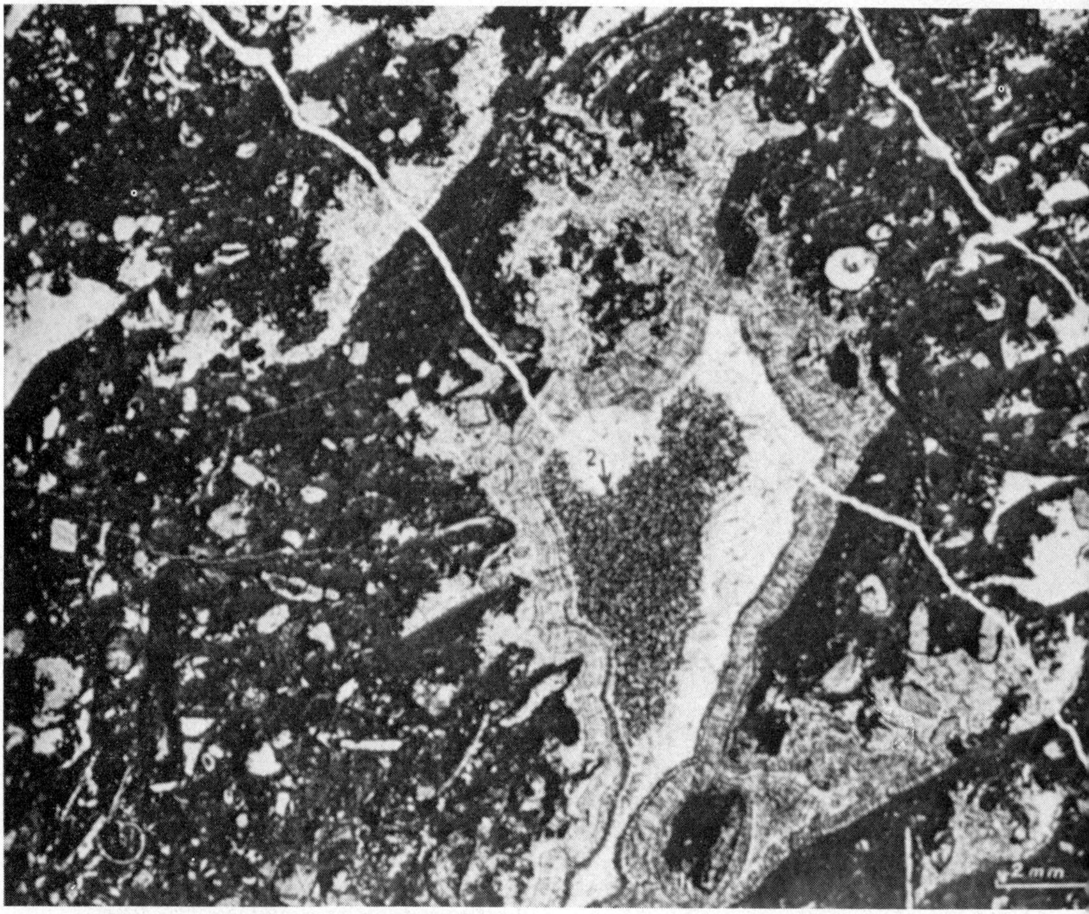

Algal micrite limestone of the Nubrigyn Formation, N.S.W., with skeletal and algal fragments and stromatactis. The large open-space structure is lined with numerous generations of brown fibrous orthosparite (*1*). The dark brown crystalline patch is dolomite (*2*) formed by internal chemical precipitation. Subsequently, fracturing of the host rock permitted solutions to precipitate clear granular orthosparite. The flat-bottomed stromatactis are filled with fibrous, granular, or both types of orthosparite. The paragenesis is as follows:
Syngenetic: framework.
Precementation—diagenetic: solution cavity.
Syncementation—diagenetic: fibrous calcite.
Postcementation—diagenetic: dolomite, fracture and granular calcite. There are numerous dark micrite circumcrusts around many crinoid ossicles and other skeletal debris. The layers are most likely formed by Algae.

of insoluble chitinous material remained after complete removal of the carbonate.

Etching, corrsosion and solution of calcareous material occur in mangrove environments with high organic content and rapid decay processes. The processes are not well known but it has been suggested that carbonic acid produced by decomposition of organic matter and other acids are the principal agents. Similar changes of calcareous components may take place in freshly accumulated sediments where bacterial oxidation of organic matter produces CO_2 and lowers the carbonate concentration and pH. Revelle and Fairbridge mentioned estuarine and lagoonal muds in France and Africa with a pH as low as 6.5 and 5, respectively. Particularly in the latter case, carbonate shells were found to dissolve with great rapidity. Both concentration of organic matter and porosity—permeability of the sediment influence greatly the amount and rate of $CaCO_3$ solution. The smaller the organic matter content and permeability of sediments, the greater is the possibility that shells remain unaffected.

Sulfate-reducing bacteria may cause an increase of pH up to 8.5 due to replacement of the strongly acid sulfate radical by sulfide, a weak Brönsted base. The $CaCO_3$ solution will not take place then, and the skeletons may be preserved. If free iron is not available to react with the sulfide, however, dissolved H_2S will diffuse upwards to the surface where it becomes reoxidized to the sulfuric acid and thus reduces the pH causing solution of $CaCO_3$ (Revelle and Fairbridge, 1957). Pyrite associated with the corrosion horizons may indicate that H_2SO_4 produced by the oxidation of H_2S may have been responsible for solution.

Corrosion and solution of calcareous particles can occur while sediments pass through the digestive system of organisms. Dapples (1942) mentioned that this is indicated by the pH of the fluids of the alimentary tract which may range from 4.75 to 7 before feeding and increases to 7 when the gut is filled with calcareous material. Limestones may be reduced up to 1 inch and more in thickness annually. Dapples gave examples where holothurians dissolved up to 414 g/year. On the Aua reef flat at Samoa, 290,000 individual holothurians destroy 104 tons of sand and lower the entire reef flat 0.2 mm/year. Ginsburg (1957) reported that material larger than sand in size is broken down by boring and burrowing organisms. Worms, mollusks, sponges, and Algae chemically bore into limestones from hightide level to a depth of a few hundred feet below sea level. Differential attack by Algae also has been reported from ancient sediments. In Devonian algal reefs of the Nubrigyn Formation, New South Wales, in particular crinoid ossicles exhibit algal corrosive surfaces (Plates 6-XV, 6-XVI) beneath thin cryptocrystalline calcite circumcrusts, and occasional shells are riddled with boreholes formed most likely by Algae (Plate 6-XVII) (Wolf, 1963a, 1965a).

PLATE 6-XV

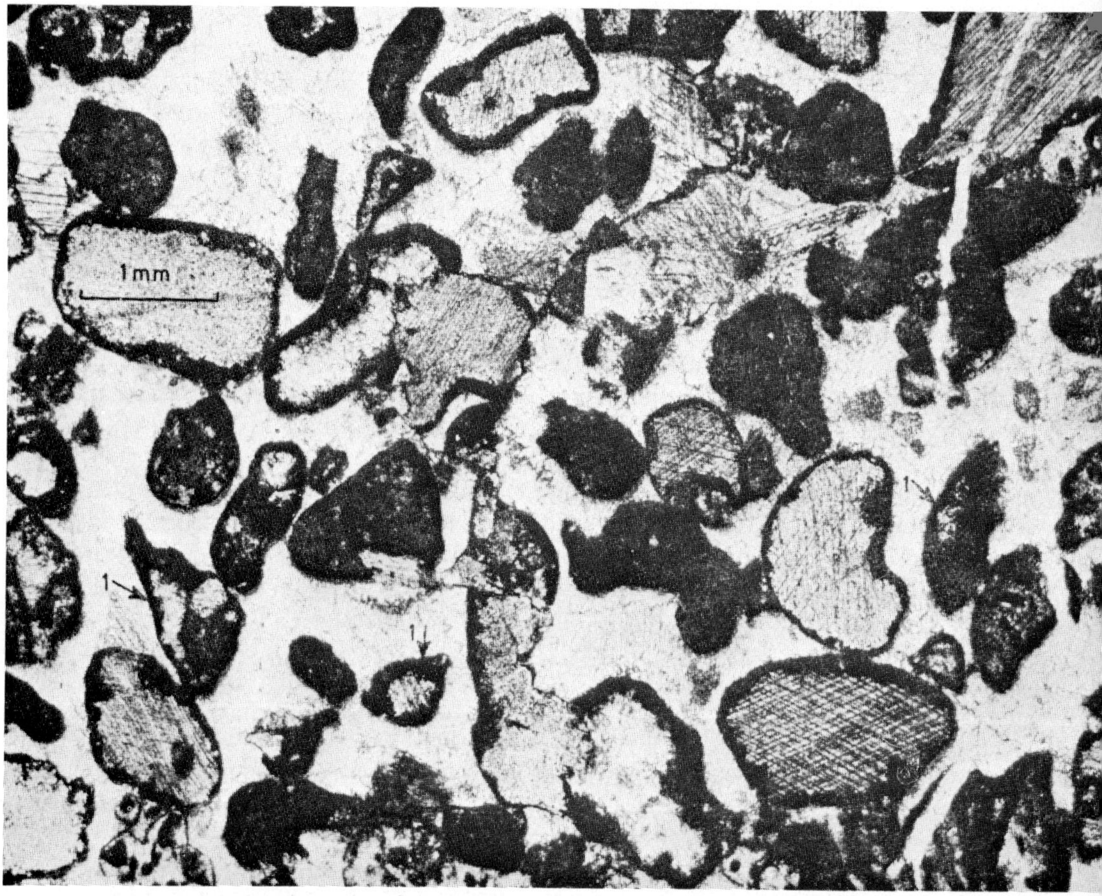

Algal circumcrusted crinoid—sparite—calcarenite of a Devonian reef complex. A number of the crinoid ossicles have been nearly completely destroyed by algal corrosion (see Plate XVI) as shown by minute specks of crinoid fragments (*1*) left in some of the dense micrite grains. Once a nucleus has been completely destroyed, the result of corrosion with simultaneous formation of a micrite crust is an algal pellet or grain composed of dense micrite, which resembles algal debris directly derived from abrasion of algal micrite bioherms.

The selected examples of organic corrasion and corrosion indicate that the cumulative diagenetic effects of organisms control to a considerable degree the growth, porosity and permeability of reefs. The constant organic corrosion and abrasion weaken the reef limestones and make them more suscep-

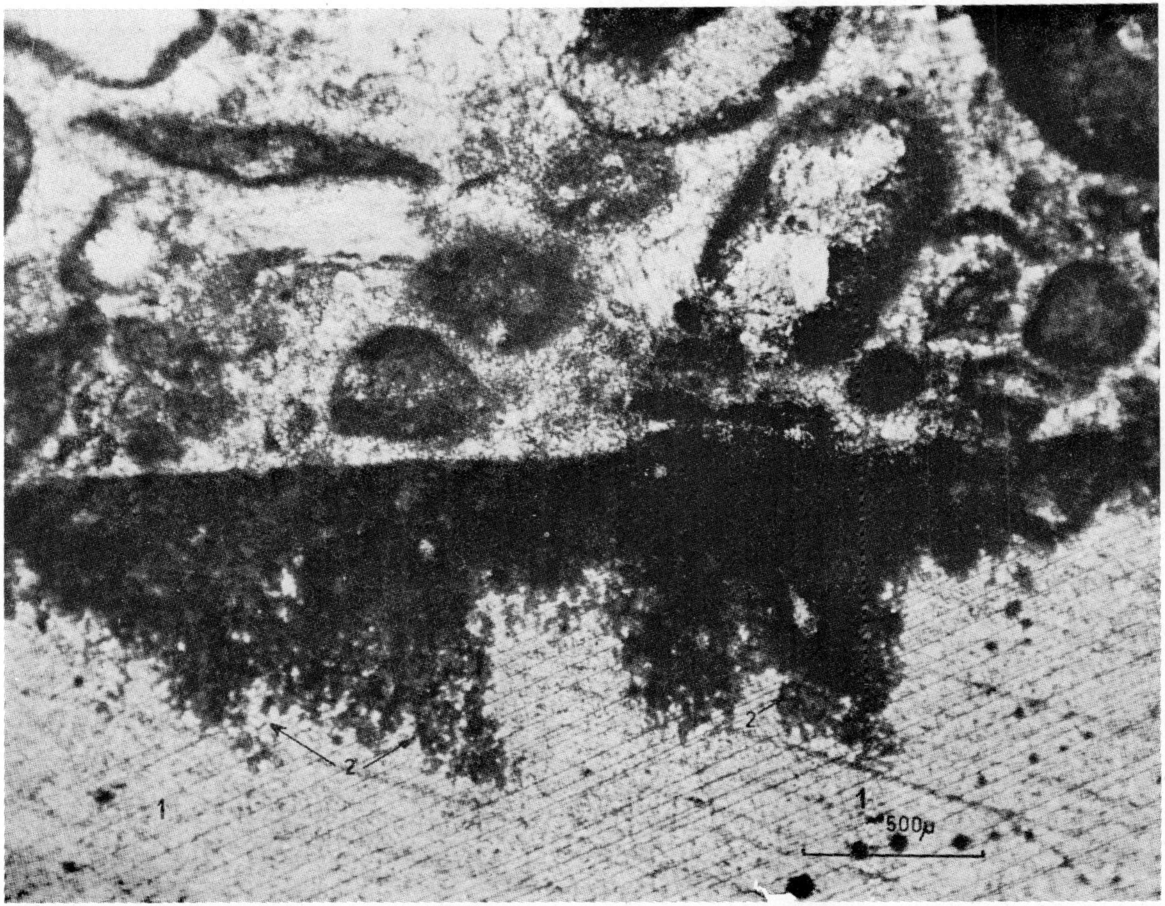

Section of a Nubrigyn calcarenite composed of a greatly enlarged crinoid ossicle (*1*) with a thick algal micrite circumcrust exhibiting irregular algal corrosion features (*2*).

PLATE 6-XVII

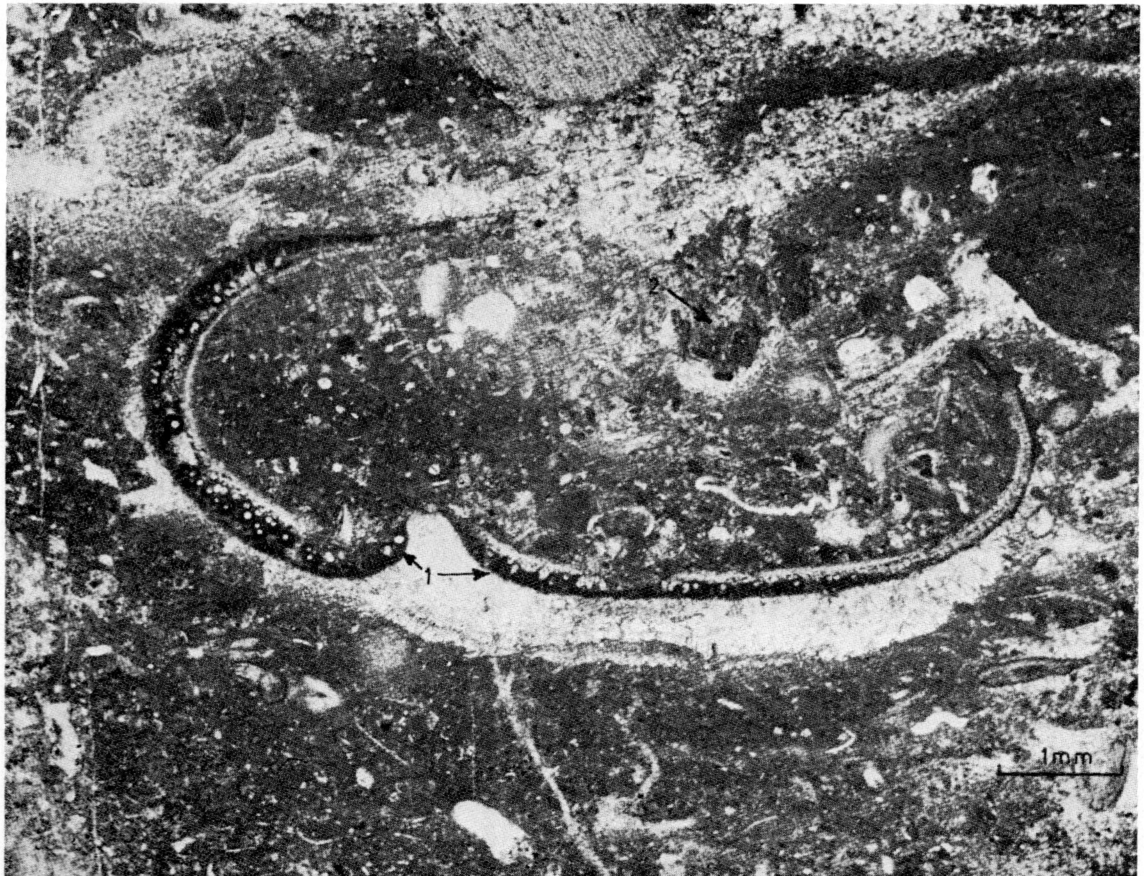

Boreholes (*1*), most likely of algal origin, in a shell surrounded by allomicrite. The latter contains a small patch of algal filaments (*2*).

tible to mechanical erosion by waves and currents. Further research on the textures and structures caused by diagenetic organic processes may reveal some useful environmental criteria.

Subsurface physicochemical solution and corrosion may be closely related to the water table as indicated by carbonate grain morphology. Russell (1962) has shown one example where the grains of beach sand above the ground-water table are polished and devoid of corrosion features, whereas the grains within the ground-water zone show pitting and various degrees of dissolution caused by the undersaturated fresh water. Precipitation of $CaCO_3$ apparently occurs in the vicinity of the water table resulting in coating of the detrital fragments.

Selective solution and leaching is widespread in some carbonate rocks. Lucia (1962) suggested the possibility that the presence of lime-mud in the detrital rocks studied by him inhibited the genesis of initial calcite cement. The lime-mud was later selectively leached to form pores. A similar selective matrix removal process has been postulated by Graf and Lamar (1950). Lucia (1962) presented evidence that the best porosity is found in those crinoidal limestones that originally contained 5—10% of micrite matrix. These rocks all had a supporting framework and the lime-mud merely occupied the available intergranular spaces; at these conditions solutions moved easily. As the lime-mud increases above 20%, the porosity decreases. Lucia concluded that once the matrix exceeded this value it supported more and more of the overburden and this resulted in compaction of the micrite. The sediment became less permeable to the fluids and prevented the removal of the matrix.

Selective removal by internal corrosion has also been shown to be widespread in the algal bioherms of the Nubrigyn reef complex (Wolf, 1963a). Relatively large proportions of the micrite have been dissolved to leave cavities and tubes. In many cases, Bryozoa and brachiopod fragments extend deeply into the solution voids (Plates 6-VIII, 6-IX). On the other hand, in some rare occurrences the micrite matrix and Bryozoa remained unaffected, whereas gastropod shells and crinoid ossicles were selectively removed leaving external molds.

The published material on decementation of carbonate rocks, a process suggested for terrigenous rocks by Pettijohn (1957), is not extensive. Murray (1960) described anhydrite cement which partially replaces both fossils and matrix of limestones. Leaching of the anhydrite left characteristically shaped vugs and increased the porosity of the sediments. If under similar conditions intergranular anhydrite cement is leached, it seems conceivable that a large section of a limestone may undergo decementation. Later, possibly during epigenesis, recementation by calcite would form a limestone without evidence of its previous cementation and decementation history. A case in point might be the reefal limestones of the Devonian Guilmette Formation in parts of western Utah and eastern Nevada of the Great Basin region, which illustrate this process in a superb fashion.

Pressure-solution of allochthonous carbonate grain accumulations is possible, especially in cases where the deposit is not exposed to warm, saturated water and, consequently, remains uncemented for a relatively long period. It can be demonstrated, for example, that in the Nubrigyn-Tolga reef complex, New South Wales, the shallow-water calcarenites exhibit pressure-solution features in contrast to the graded-bedded basinal deposits (Wolf, 1963a, 1965a). This paleoregional change is most likely a function of different degree of saturation and pH at the time of sedimentation: the near-shore waters were more saturated and aerated, caused early cementation and prevented pressure-solution; whereas the basinal waters were undersaturated and reducing, delayed cementation and permitted pressure-solution. The basinal fluids may have migrated upwards and reefwards during compaction and pressure-solution, thus moving the dissolved $CaCO_3$ and allowing a continuation of the pressure-solution process.

Emery et al. (1954) observed in the subsurface limestones at Bikini that coral and mollusk fragments have disintegrated to a chalky powder. They noticed a general disintegration of crystalline calcareous faunal and floral skeletons into microcrystalline and cryptocrystalline material without obvious change in the mineralogy and chemical composition. The original fibrous aragonite crystals of organisms can be seen in thin-sections to have altered to microgranular material. Especially *Halimeda* segments underwent alterations from a microcrystalline to a denser cryptocrystalline calcium carbonate. Unaffected *Halimeda* in thin-section exhibit a mat of minute aragonite needles approximately 1 μ in size, which changes into a brown isotropic matrix with scattered needles of high birefringence. A similar alteration occurs in the fibrous calcite tests of small Foraminifera, for example, which change to brown isotropic material. Interstitial microgranular lime-mud shows a similar change to brown, marly, isotropic micrite. These alterations increase with depth in an irregular to regular fashion. This change of organic carbonate structures to cryptocrystalline micritic material is identical to those observed in Australian algal pisolites and crusts mentioned earlier (Plates 6-I, 6-IV; Wolf, 1963a). Hadding (1958) suggested that bacteria may obliterate or destroy algal components. It is not known, however, to what extent bacteria are responsible for the disintegration mentioned above.

It appears, therefore, that crystalline substances can change relatively rapidly into dense calcium carbonate by inorganic and/or organic processes as yet poorly understood. An interesting question arises as to whether inorganically formed aragonite needles can also convert into a brown cryptocrystalline mass or not. If this is so, it may explain the controversy between those who have observed cryptocrystalline crusts surrounding carbonate fragments in beach-rocks and believe them to be of algal origin, versus those who recorded acicular or fibrous aragonite cement and insist that it is of physicochemical origin. If both algal and physicochemical aragonite needles are identical, as shown by Lowenstam (1955), and the former can change into a cryptocrystalline mass, then there seems to be a distinct possibility that the latter does the same as suggested previously (Wolf, 1963b). Hence, a modification in our concepts of diagenesis, and more refined techniques than thin-section studies only, will be necessary to solve these problems. Our present information, however, suggests that both algal and physicochemical, and possibly other, processes can cement carbonate sediments.

INVERSION, RECRYSTALLIZATION AND GRAIN GROWTH

Inversion is the process by which unstable minerals change to a more stable form of the same chemical composition (except for a possible change in content of trace elements and isotopes) but with a different lattice structure. Mayer (1932) mentioned that some organisms form a gel-like $CaCO_3$ which quickly changes into vaterite. The latter is very unstable (Termier and Termier, 1963), but may remain unaltered up to almost 1 year before inverting into aragonite. In the sequence gel—vaterite—aragonite—calcite the latter two are the most stable. Stehli and Hower (1961) reported that of recent high-Mg calcite, aragonite and low-Mg calcite, the first is very unstable, whereas the other two may persist for a long time under natural conditions. They suggested that the increase of volume during change from aragonite to low-Mg calcite and from high-Mg calcite to low-Mg calcite (the Mg content in the latter two is approximately 8% and 1%, respectively) may affect the porosity, cementation and dolomitization of the sediments. Inversion of aragonite may be relatively rapid or slow, i.e., it may occur within 12 months or require tens of thousands of years. Lowenstam (1954), for example, mentioned that more than 50% of the aragonite laid down by some of the marine invertebrates inverted to calcite within 1 year. On the other hand, aragonite has been identified from rocks as old as Late Paleozoic. Degens (1959) believed that aragonite may be found in traces in rocks as old as the Cambrian. Taft (1963) reported that aragonite and high-Mg calcite of Florida Bay sediments, determined by ^{14}C dating to be 3600 years old, exhibit no evidence of recrystallization. Taft stated that recrystallization rates appear to be controlled by concentration of a particular cation in the surrounding liquid. As experiments suggest, magnesium chloride solution and Mg in sea water seem to prevent recrystallization; whereas solutions of calcium and strontium chloride (and distilled water) cause recrystallization of aragonite and high-Mg calcite at different rates. Taft suggested, therefore, that marine carbonates tend to remain unstable for long periods until they are exposed to Mg-deficient water. It is quite obvious, then, that the inversion process is both early and late diagenetic as well as epigenetic, and may even be due to burial metamorphism.

In general, the exact causes that initiate and perpetuate inversion, recrystallization, and grain growth of limestones are not well known. In particular, calcite-to-calcite conversion is largely an unsolved enigma. Numerous parameters have been thought to be conducive to secondary alterations: trace elements, associated organic and inorganic impurities, unstable mineralogic composition, physical and chemical conditions of interstitial fluids, degree of compaction, degree of solubility, permeability, differential pressure and distortion, availability of nuclei or seeds, temperature variations, and others.

Near-recent fossils have been shown to be susceptible to recrystallization in a certain order (Crickmay, 1945; Emery et al., 1954), namely, corals, mollusks, *Halimeda*, thin-walled pelagic Foraminifera, thick-walled Foraminifera, larger Foraminifera, echinoids, and *Lithothamnion*. Some evidence suggests that corals may break down into microgranular aragonite before changing into a mosaic of calcite. The segments of *Halimeda* are rarely completely recrystallized and are altered first in the central areas (pores) and boundaries. The fibrous calcium carbonate of some organisms changes into coarsely crystalline calcite and may be more or less radially oriented. This indicates that fibrous aragonite may invert to granular calcite and need not form pseudomorphs as suggested by Usdowski's (1962) experiments. It is interesting to note here that the recrystallization or inversion to coarser calcite is contrary to the "disintegration" into cryptocrystalline material mentioned earlier.

The term recrystallization has been used loosely for a number of processes that commonly cause a change in crystal or grain size, predominantly an enlargement and occasionally a reduction in size, without causing a chemical alteration except for changes in isotope and trace element concentrations. Some prefer to include inversion and grain growth, whereas others prefer a restricted use of the term recrystallization. Bathurst (1958) stated that "grain growth s. str. is nowadays distinguished from primary recrystallization which may precede it and from secondary recrystallization which may follow it. Grain growth acts in monomineralic fabrics of low porosity. The intergranular boundaries migrate causing some grains to grow at the expense of their neighbors. The reaction takes place in the solid state, ions being transferred from one lattice to another without solution. Larger grains tend to replace smaller and a fine mosaic is gradually replaced by a coarser. As grain growth proceeds, many of the enlarged grains are themselves replaced by their more successful neighbors." Recrystallization, as defined by Bathurst occurs when "nuclei of new unstrained grains appear in or near the boundaries of the old, strained grains. These nuclei grow until the old mosaic has been wholly replaced by a new, relatively strain-free mosaic with a nearly uniform grain size. Its coarseness depends on the density of the initial nucleation. Where the nuclei are widely spaced there is an intermediate porphyroblastic stage." Grain growth and recrystallization should be accepted as two distinct processes wherever possible.

Bathurst (1958) considered the following processes that may cause grain enlargement: solution of supersoluble small grains with redeposition on larger grains (aragonite is 3—9% more soluble than calcite, Chilingar, 1956c), solution transfer, primary recrystallization, inversion of aragonite to calcite, and grain growth s. str. It is possible, however, that at least one of these can cause a relative decrease in crystal size. If recrystallization of a coarse crinoid ossicles accumulation occurs, starting with nucleation in the interior of the ossicles, minute calcite crystals will replace each larger crinoid crystal and may spread to consume the whole limestone. Wardlaw (1962) suggested, therefore, that under favorable conditions a calcarenite may be converted into a limestone composed of silt-sized calcite crystals, i.e., microsparite.

Bathurst (1958) mentioned syntaxial replacement rims which are similar in appearance to those formed by what he termed syntaxial rim cementation. For example, crinoid ossicles in contact with lime-mud may undergo grain growth at the expense of the fine matrix. The result is a calcite rim in optical continuity with the ossicles. Calcite deposited from interstitial solutions onto free surfaces of crinoids may give the same result. The two processes, however, are very different. A similar syntaxial replacement phenomenon has been described by Folk (1962a). He mentioned an oolite with an echinoderm fragment as nucleus, and rays of sparry calcite in optical continuity with the echinoderm.

To make a clear distinction between the so-called open-space calcium carbonate precipitated in voids on one hand and that formed by inversion, recrystallization, and grain growth on the other, the latter have been named pseudosparite by Folk (1959) and the former orthosparite by Wolf (1963b).

The term "sparite" is purely descriptive, therefore. The processes that cause crystal enlargement have been collectively called "aggrading recrystallization" by Folk (1956; 1959). Those that cause a decrease in crystal size were named "degrading recrystallization" by Folk (1956), "degenerative recrystallization" by Dunham (in Folk, 1956), and "grain diminution" by Orme and Brown (1963). The disintegration change of crystalline coral and algal material to brown cryptocrystalline calcium carbonate described in the foregoing sections may be considered as "grain diminution", although the actual causal factors are not known.

Inversion, recrystallization and grain growth vary not only in sign and extent but also in position and resultant grain or crystal morphology, as indicated by Folk (1956). The synthesis below is based on Folk's work and is presented here with slight alterations, with his permission:

Sign: (1) marked increase in crystal size, (2) marked decrease in crystal size, and (3) no or very little change in crystal size (e.g., formation of pseudomorphs). The extent and position of the processes discussed here can be divided into phases α, β, and so forth. This is useful as it may save time and eliminate repetitious descriptions in preparing logs and reports, for example. It is understood, of course, that all phases are completely gradational.

α phase. Limestone is unaffected by inversion, recrystallization, or grain growth.

β phase. Limestone is slightly affected. A few of the allochemical grains and possibly small portions of the micrite matrix or sparite cement are "recrystallized". Inversions of originally aragonite fossils, e.g., most pelecypods, many gastropods, and some Algae, to calcite have occurred.

γ phase. The limestone has undergone major alteration, but the original nature of the matrix is still discernible; the rock can still be described and classified according to Folk's scheme, the modified version of which is given in this chapter. The allochems are still recognizable and may range from unaltered to completely recrystallized. This phase passes into the next phase when the original matrix is completely recrystallized.

δ phase. Limestone is extensively altered. Inversion, recrystallization and/or grain growth of an original cryptocrystalline to microcrystalline calcite or argonite matrix and cement resulted in microsparite. This is probably the most common process according to Folk. It agrees with Bathurst's (1958, 1959b) idea of the existence of a "universal threshold state at which fabric evolution stops and beyond which it can, but need not, continue". The microspar consists of calcite crystals 5—20 μ in diameter in contrast to the original micrite size of 1—5 μ. The fragments, i.e., allochemical grains, or colonial growths in autochthonous limestones, remain largely unaffected in this phase. In handspecimens it is impossible to distinguish micrite from microsparite, but in thin-section they are easily discriminated. If recrystallization of the matrix is incomplete, patches of the matrix may "float" in the microsparite. The contact between micrite and microsparite may be very gradual or sharp. Sometimes the contact is oblique or vertical to bedding planes. Hence, the microspar is not merely a coarser crystalline detrital material as pointed out by Folk. If the alteration of the matrix is complete, the detrital grains may "float" in the microsparite. Open-space sparite may be more resistant to changes in contrast to micrite matrix and clear calcite cement may be found in a microsparite due to preferential alterations (Wolf, 1963a).

ϵ phase. Alterations affected matrix, cement and grains or framework of the limestone. The patches formed by recrystallization and/or grain growth may be very irregularly shaped, may occur as "fronts", veins, or transgress the whole rock. Faint relics ("ghosts") of grains are still recognizable. The criteria for partial recrystallization are the same as those employed to determine other replacement phenomena.

ζ phase. Limestone alteration is complete. None of the original textures and structures of allochemical grains or colonial growths, matrix and cement are recognizable. The rock is composed of microsparite or sparite only. The genetic nomenclature of different types of micrites and sparites is presented below (Wolf, 1963b).

Folk (1956) defended the viewpoint that the products listed in phases ϵ and ζ may be common locally, but their overall volumetric significance is small. Phases α—γ are the more common ones.

It must be noted that the above phases are based on the assumption that there is an increase in crystal size during alteration. Although this seems to be the case in the majority of recrystallization and grain growth occurrences, one should not lose sight of the "degradation recrystallization" and disintegration possibilities mentioned earlier. It is not known how significant they have been in the geologic past, but it seems possible that many of the problematic micrite knoll-reefs may have been formed by grain diminution of algal and faunal colonies.

The morphology of the crystals or grains after inversion, recrystallization and/or grain growth may be granular, drusy, fibrous and/or bladed. Usdowski's (1962) experiments indicate that inversion of aragonite may result in the formation of pseudomorphs and does not necessarily destroy the original fibrous nature. The small volume changes are apparently insignificant in obliterating the original texture. Hence, it seems reasonable to conclude that any primary textures and structures of aragonite, whether granular, drusy, fibrous, oolitic or spherulitic, may be preserved as suggested in Table 6-I. On the other hand, it has been observed that during inversion a change of crystal morphology can take place, and the likely possibilities have been given in Table 6-I. Changes in textures attributed to volume increase during inversion have been reported by Bathurst (1959b). More research is required on these aspects.

Recrystallization and grain growth have been shown to form granular, fibrous (Folk, 1962a; Orme and Brown, 1963) and blady (Harbaugh, 1961) calcite crystals or grains. Hence, the latter two forms are not always indicative of open-space formation.

The powerful displacing ability and forces of both open-space and replacement fibrous calcite have been demonstrated by Folk (1962a). He described fibrous calcite overgrowths in optical continuity with an ostracod shell. The spar began as open-space calcite growth which completely filled the cavity,

TABLE 6-I

Possible grain or crystal morphology changes during inversion, recrystallization and grain growth (After Wolf, 1963b)

Inversion		
originally		finally
Granular aragonite	→	granular calcite pseudomorphs [1]
Drusy aragonite	→	drusy calcite pseudomorphs [1]
Fibrous aragonite	→	fibrous calcite pseudomorphs [1]
Granular aragonite Drusy aragonite Fibrous aragonite	—[2]→	granular calcite fibrous calcite (?) bladed calcite (?)

Recrystallization and grain growth		
originally		finally
Granular aragonite or calcite Drusy aragonite or calcite Fibrous aragonite or calcite	—[2]→	granular calcite fibrous calcite bladed calcite

[1] These types of pseudomorphs are usually referred to as paramorphs because no change in composition occurs during inversion (except for possible changes in trace element and isotope contents).

[2] Referring to possible processes that cause change of drusy aragonite to either granular or fibrous calcite; fibrous to either granular or blady calcite; and so on.

but then appears to have continued to enlarge and force the shell apart. In other cases, articulate ostracods were originally filled by clay. Fibrous calcite overgrowths then grew inward from the upper and lower valves into the former body cavity forcing the clay to the center by the pressure of crystallization. The fan-like overgrowth of fibrous sparite caused in some cases an expansion of the sediment to two to three times its original volume. The spar must have crystallized when the sediment was at sea-floor level or upon very slight burial to permit such expansion. That some of the fibrous calcite is definitely of the replacement rather than the open-space type is further indicated by cases where fibers grew preferentially downward from both the upper and lower valves of fossils. The lower ones definitely replaced host-rock matrix material.

Folk (1962a) mentioned unaltered micrite intraclasts within a micro-sparite "matrix". He suggested that the differential recrystallization was due to differences in compaction and permeability. The micrite limestone fragments, which were derived from some locality within the depositional environment, were less permeable as they had undergone some compaction before dislocation and transportation. In the new area of deposition these clasts were mixed with a less compacted micrite matrix. The latter was more permeable to solutions and, possibly, had other features conducive to recrystallization. On the other hand, faecal pellets, possibly because of presence of organic matter, have recrystallized at the same rate and/or extent as the matrix and the end-product is a fairly uniform microsparite with "ghosts" of pellets outlined by organic specks and other impurities.

The degree of inversion and recrystallization may change on a regional scale. Newell et al. (1953) mentioned, for example, a regional trend in degree of recrystallization: the basinal sediments are least affected, whereas the reef and lagoonal limestones are most extensively recrystallized. The fore-reef talus deposits take an intermediate position and exhibit slight recrystallization near the basin and grade into more altered limestones close to the reef. Hence, the fossils are generally better preserved near the basin. Somewhat similar conditions typify some of the Pennsylvanian and Permian limestones (Callville and Pakoon) of the shelf facies near the shelf-to-basin transition along the Las Vegas Line of southern Nevada (Bissell, 1959). These micrites and algal to pelletal limestones that have a fine-textured matrix have been recrystallized (and dolomitized) on a much greater regional scale than have their basinal equivalents (Bird Spring Group, Nevada, U.S.A.). Furthermore, the fossils (and in particular fusulinids) are better preserved in the basinal sediments.

Some recent sediments show evidence of extensive recrystallization, whereas others lack it. The causes are poorly known. Inversion seems to be more rapid in subaerial environments and in zones of meteoric waters. According to Folk's (1962a) experience, it seems that brackish-water micrites recrystallize more readily than either normal marine or lacustrine fresh-water micrites. Stehli and Hower (1961) indicated that minerals of shallow- and deep-water environments are different and their diagenesis susceptibility must differ accordingly. In some recent shallow-water areas at least 70% of the carbonates consist of metastable $CaCO_3$: aragonite and high-Mg calcite. The deep-water sediments appear to be composed of low-Mg calcite ad are, therefore, more stable. Thus, diagenesis should in general be initiated earlier and be more pronounced in shallow-water type carbonates.

Changes in contents of trace elements and isotopes are commonly associated with inversion, recrystallization and, possibly, grain growth. These processes are accompanied by expulsion of trace elements (e.g., Mg^{2+}, Sr^{2+}, Mn^{2+}, Ba^{2+}) to the interstitial fluids, matrix or cement, where they are available for other diagenetic processes either in the immediate vicinity or at remote localities.

As high-Mg calcite is the least stable among the aragonite and calcite sediments, it seems that Mg is one of the earliest elements available. Quantitatively, it is also more significant than the other elements and this Mg may form, therefore, the raw material for early diagenetic dolomitization. In the sediments examined by Stehli and Hower (1961) the elements present in the calcium carbonate lattice are in the following order: $Mg^{2+} > Sr^{2+} > Mn^{2+} > Ba^{2+}$. In every case diagenetic alterations appear to have resulted in a marked decrease in trace element concentration.

Siegel (1960) stated that whenever the Sr/Ca ratios of Recent corals are compared, it becomes clear that when Sr is present in amounts that indicate that none or only very little of it has been lost from the original aragonite, the inversion to calcite has hardly begun. On the contrary, however, where the amount of Sr has been appreciably reduced, the alteration from aragonite to calcite has reached a point that appears to be directly related to the degree of Sr removal. Siegel suggested, therefore, that the presence of Sr, not as $SrCO_3$ but rather in substitution for Ca in the aragonite lattice, inhibits and, therefore, prevents or slows the rate of inversion under natural conditions. Siegel further proposed that the inversion occurs only when much of the Sr has been removed. Many scientists, however, maintain that inversion merely expels Sr, i.e., loss of Sr is an effect and not a cause of inversion. Lowenstam (1954) reported that a distinct decrease of the Sr/Ca ratio occurs during recrystallization. For example, unaltered corals with about 1.4% strontium carbonate content recrystallize to microgranular calcite having about 0.7% $SrCO_3$ and to more coarsely crystalline calcite containing 0.2% strontium carbonate.

Usdowski (1962) advanced an interesting theory to support the idea of inversion of aragonite oolites to calcite. This author assumed that the composition of the water medium remained the same from the time of oolite formation until cement precipitation. Therefore, both must have had the same trace element composition. Analyses indicate, however, that the cement has a much higher content of Mg, Fe, Mn, and Sr; Usdowski suggested that during inversion the oolites expelled the trace elements, which were either incorporated into the calcite cement or were removed by interstitial solutions. In a subsequent publication on early diagenetic cone-in-cone structures, Usdowski (1963) supported his theory of trace element expulsion during recrystallization. The limestone beds which underwent recrystallization, resulting in cone-in-cone features, have an Sr content of 247 p.p.m. The unrecrystallized beds are richer by a factor of 0.4. A similar relationship exists for the Mg content which is 0.7% and 4.7% for recrystallized and unaltered limestones, respectively.

Degens' (1959) studies show that recent fresh-water limestones have a lower content of strontium than marine carbonate sediments, caused presumably by the lower amounts of Sr in fresh water. With an increase in age of limestones, however, the difference in Ca/Sr ratio between the fresh- and marine-water sediments appears to diminish; and Paleozoic carbonates, independent of facies, do not deviate much from the average value of 500 p.p.m. of Sr. Hence, fresh-water limestones must have gained and marine limestones lost Sr during the diagenetic—epigenetic geologic history.

Ross and Oana (1961) concluded that both the environment of deposition and the diagenetic history of limestones determine the carbon-isotope distribution in carbonates. Their work indicates that biosparites (terminology of Folk, 1959) with a large amount of sparry calcite cement have $\delta^{13}C$ values between +1.0 and −1.0. On the other hand, biomicrites or biomicrosparites have either distinctly positive or negative $\delta^{13}C$ values. As the amount of sparry calcite decreases, the $\delta^{13}C$ value becomes either more positive or negative. This suggested to the two authors that limestones which underwent little recrystallization have a wider spread of $\delta^{13}C$ values than do those which exhibit evidence of considerable recrystallization or introduction of calcite cement. The effect of recrystallization is to shift the $\delta^{13}C$ values toward the range of +1 to −1. More research on different types of grains, micrites of diverse origins, and sparry calcites formed by different processes is necessary, however, in order to check the validity of these interpretations.

INTERNAL FILLING AND INTERNAL SEDIMENTATION

Internal filling and sedimentation processes of physicochemical, biochemical, and physical nature cause partial to complete filling of voids within sedimentary frameworks and form the so-called open-space structures (Plates 6-VI–XIV; Fig. 6-1, Wolf, 1965c). Many of the cavities that are formed diagenetically are interconnected and give the sediments a very high degree of primary permeability. Most of the larger systems are open at some points to the upper surface and tidal waters can penetrate the system to deposit detritus (Fig. 6-1, top). The same fluids may also chemically precipitate a number of substances or cause wall-rock alterations as, for example, in the Nubrigyn algal reef complex (Table 6-II; Wolf, 1963a). These components form a complex paragenesis due to cyclic deposition. The detrital internal sediments form minute lenses, patches and thin layers in the voids, and smooth out irregularities of the floor of the cavity (Plate 6-VII). In thin-section, the internal sediments are either dense and structureless, or are laminated and graded on a microscopic scale. In most cases, the internal sediments are different both in texture and/or composition from the host rock. In some occurrences, however, they blend. Occasionally, cavities are completely filled by internal sediments; but in the majority of internal sediments they are con-

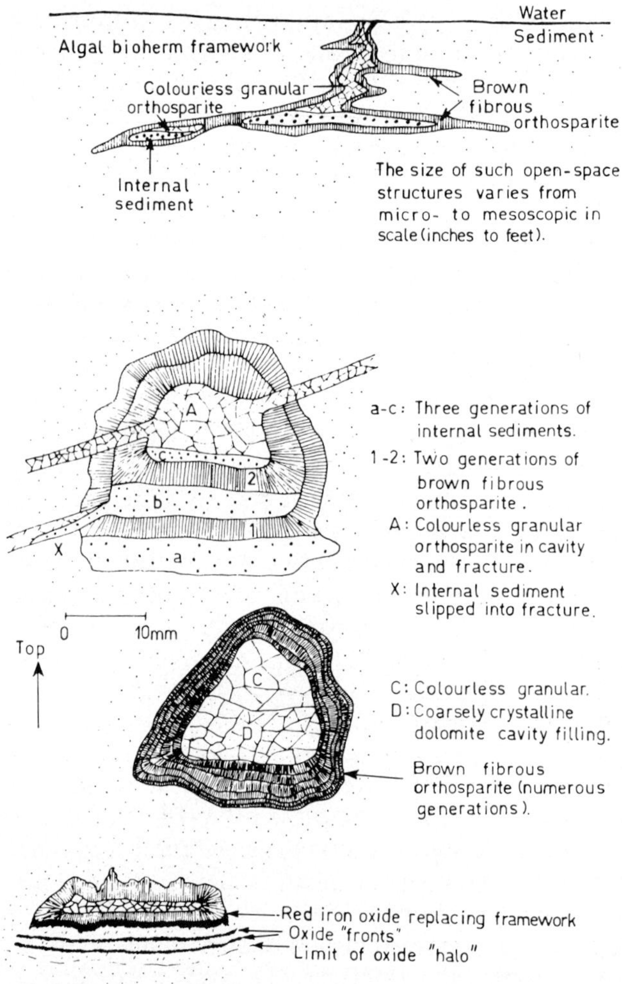

Fig. 6-1. Open-space structures in Devonian algal bioherms, Nubrigyn Formation, N.S.W. (After Wolf, 1963a.)

fined to the lower part of the voids and it is the subsequent deposition of clear colorless sparite that filled the upper spaces.

Numerous cavities show wall-rock alterations prior to internal sedimentation and calcite cement precipitation (Plates 6-IX, 6-XI). Either iron oxide replacement or leaching and bleaching occurred in a semiconcentric fashion around the voids or was limited to the area near the floor of the openings. Some of the lower parts of the cavities were differentially leached, corroded

TABLE 6-II

Diagenetic modifications in Devonian algal bioherms, New South Wales

Detrital internal sediments	Chemical internal fillings	Wall-rock alterations
Lime-mud	fibrous calcite	leaching
Pellets	drusy calcite	bleaching
Fine algal and skeletal debris	granular calcite	solution
Iron oxide	iron oxide	iron oxide replacement (irregular and as "fronts")
Clay	dolomite	

and oxidized, resulting in red iron oxide rich pellets that are easily mistaken for detrital internal sediments. They were formed in situ, however, and constitute a residual product on a microscopic scale. On the other hand, in most cases the internal open-space iron oxide was directly precipitated from solution and/or mechanically deposited. Although it may have originated at the same time as the iron oxide replacing the wall-rock, it is of a different origin. In numerous occurrences, fibrous calcite precipitation encrusted the walls of the cavities before the internal sediments, iron oxide, dolomite, and/or granular sparite were deposited (Plates 6-VIII, XIII, XIV; Fig. 6-1).

In addition to the above-mentioned cavities, minor open-space structures beneath large faunal fragments are common. They usually lack a complex paragenetic history, however, and are only filled by internal sediments, fibrous and/or granular sparite. It seems that they were isolated and out of reach of oxide- and dolomite-precipitating solutions.

MORPHOLOGIC AND GENETIC CALCIUM CARBONATE TYPES

As illustrated in Table 6-I, the three basic morphologic types of calcium carbonate can be formed by a number of primary and/or secondary processes. Recent research in carbonate petrology has resulted in valuable information that permits the discrimination of the numerous aragonite and calcite types formed by open-space precipitation, recrystallization and grain growth. Hence, it is possible to present in Table 6-III a scheme that attempts to cover all likely occurrences ranging from a simple descriptive to a more complex genetic nomenclature of cryptocrystalline to coarsely crystalline carbonate. Figure 6-2 gives a diagrammatic illustration of the numerous possible fabrics or textures. All have been listed also in Table 6-III except for the two types of syntaxial rims. They are either of open-space or grain

TABLE 6-III

Descriptive and genetic nomenclature for micrite-sparite range of aragonite and calcite [1] (after Wolf, 1963b)

Descriptive				Genetic descriptive		Origin
indicating crystal size	approx. size	indicating crystal morphology and size	approximate proportions			
Sparite	>0.02 mm	granular sparite	equidimensional	orthosparite	open-space versus recrystallization and grain growth	open-space precipitation, i.e., void fillings
		drusy sparite (size and morphology change distally)	elongate	pseudosparite [1]		recrystallization or grain growth (Bathurst, 1958)
		fibrous sparite [6]	elongate			
Microsparite	0.005—0.02 mm	granular microsparite	equidimensional	orthomicrosparite		open-space precipitation
		drusy microsparite [4]	elongate	pseudomicrosparite [1]		recrystallization or grain growth
		fibrous microsparite [6]	elongate			
Micrite [7] (often called calcilutite, ooze, lime-mud) cryptocrystalline [7]	<0.005 mm	too small to observe visually morphologic differences except by use of electron microscope		pseudomicrite [2,3]		degradation recrystallization (= grain or crystal diminution)
				orthomicrite [3,5]	detrital versus in situ	"genuine primary" micrite [5]
				allomicrite [3]		allochthonous micrite
				automicrite [3]		autochthonous micrite

[1] It is uncertain at the present time whether recrystallization or grain growth can form a drusy fabric.
[2] Recent algal colonies appear to change to cryptocrystalline material by an unknown diagenetic process (Wolf, 1965a, b).
[3] Commonly called "matrix" in contrast to sparite cement.
[4] Grades rapidly into sparite size grade.
[5] Orthomicrite is a collective term for unaltered micrite and includes both allo- and automicrite.
[6] Fibrous sparite has been called "drusy" by mistake. Fibrous carbonate consists of acicular needles of roughly uniform length. In contrast, drusy carbonate changes from small blady or acicular grains or crystals to larger ones toward the center of the cavity. This change is size is accompanied by a gradual change of morphology to equidimensional (granular) sparite (Bathurst, 1958).
[7] Micrite when not resolvable by a petrographic microscope is cryptocrystalline in appearance.

growth origin and need no special pigeonhole. It should be emphasized that the writers feel as strongly as others who oppose the "game" of semantics. Analogous to nuclear physics, however, the more we learn about the minute, often subtle, and yet important differences, the more terms will be required for precise, unambiguous communications, and in order to eliminate long repetitive descriptions. Also, both descriptive and genetic types of terms are necessary if confusion is to be avoided. The former can be easily changed to the latter by merely adding prefixes.

The following criteria and discussions represent a modified summary of the works of Bathurst (1958, 1959b), Harbaugh (1961), Folk (1962a), Orme and Brown (1963), and Wolf (1963a).

Granular, drusy, and fibrous open-space calcium carbonate

The following features may be characteristic (4 and 5 may also occur in grain-growth products):

(1) The crystals or grains are in contact with surfaces that were once free, i.e., surfaces of voids. The contact may be horizontal, vertical or oblique, and crystal growth may occur preferentially upward, downward, or in any other direction (Plates 6-VIII, X, XIII, XIV, XVIII, XIX; Fig. 6-1).

(2) If the cavity is not completely filled, the remaining space may be filled by succeeding generations of the same calcite type, any of the other two types, by detrital or chemical internal sediment, or any combination of the above. Euhedral terminations extending into the voids are frequent.

(3) Some of the cavities underwent pre-cement modifications: the host rock may have been slightly replaced or was leached, bleached, corroded; or internal detrital bottom sediments smoothed out irregularities before precipitation of cement (Plates 6-VI, VII, IX, XI; Fig. 6-1).

(4) There is usually an abrupt contact between calcite mosaic and host rock.

(5) The mosaic-filled region has the obvious form of a cavity, but may be very complex in shape, or too large to be recognized in one thin-section.

(6) The intergranular boundaries of the mosaic are usually planar (Plates 6-XVIII, XIX; Fig. 6-1).

(7) In many cases there is an increase in grain or crystal size of the mosaic away from the wall: this is the so-called drusy carbonate (Plate 6-XVIII). In other cases, fibrous calcite of uniform length forms a relatively wide crust on the surfaces of open spaces, and no changes of crystal size need occur (Plates 6-XIV, XIX). Similarly, granular sparite may fill open spaces without systematic grain size change.

(8) Drusy and fibrous calcite show a preferred orientation of the longest grain axis normal to the surface of the host-particle (Plates 6-XVIII, XIX).

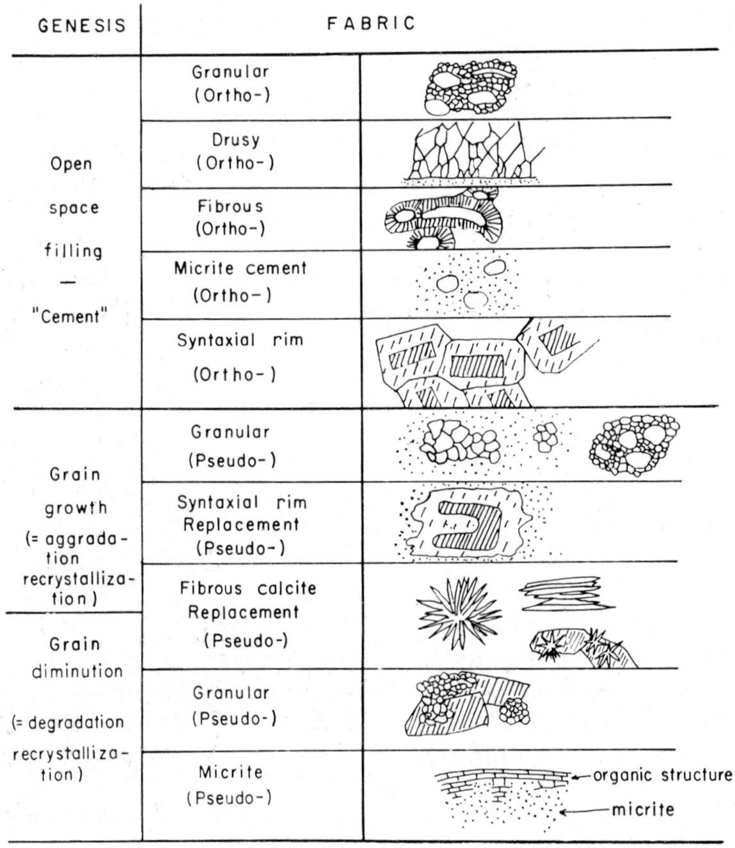

Fig. 6-2. Diagenetic fabrics (Wolf, 1963b). (Modified after Orme and Brown, 1963.)

(9) Drusy and fibrous calcite grains and crystals are preferentially oriented with the optical axes normal to the surface. Occasionally, it may be a type of overgrowth, e.g., fibrous calcite on a shell in optical continuity with the shell's surface.

(10) Most commonly the drusy and granular calcite is clear and colorless. The fibrous calcite, however, has been frequently reported as light brown in color (Plate 6-XVIII, XIX).

(11) In many cases, early diagenetically cemented limestones are reworked by intraformational processes. In the Nubrigyn Formation, N.S.W., fibrous sparite is present as intraclasts, indicating that it is of very early diagenetic origin.

PLATE 6-XVIII

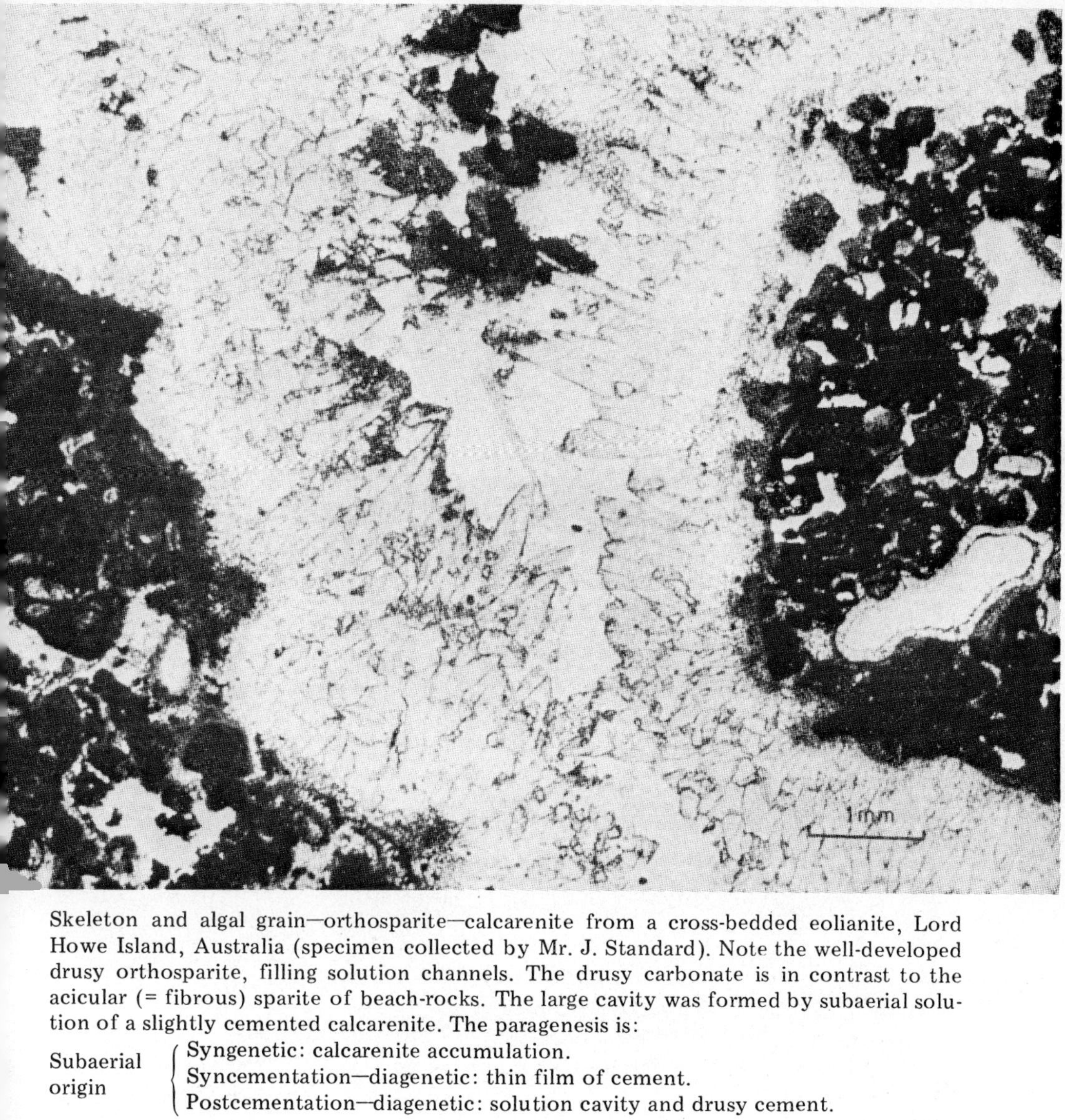

Skeleton and algal grain—orthosparite—calcarenite from a cross-bedded eolianite, Lord Howe Island, Australia (specimen collected by Mr. J. Standard). Note the well-developed drusy orthosparite, filling solution channels. The drusy carbonate is in contrast to the acicular (= fibrous) sparite of beach-rocks. The large cavity was formed by subaerial solution of a slightly cemented calcarenite. The paragenesis is:

Subaerial origin {
- Syngenetic: calcarenite accumulation.
- Syncementation—diagenetic: thin film of cement.
- Postcementation—diagenetic: solution cavity and drusy cement.

Syntaxial rim cementation

Syntaxial rim cementation is characterized by some or all of the following: (1) A detrital core is present. It is usually a single crystal, most commonly

PLATE 6-XIX

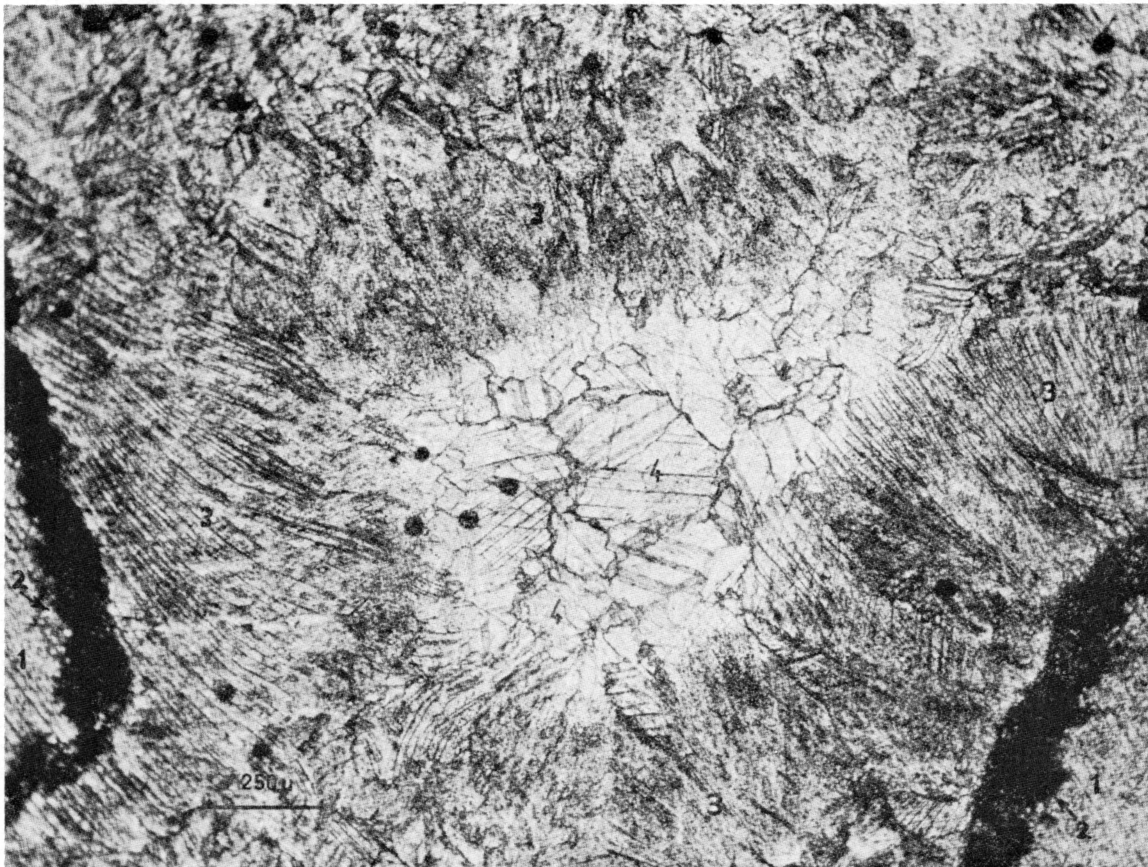

Two crinoid ossicles (*1*), circumcrusted by dense algal micrite (*2*), are surrounded by brown fibrous orthosparite (*3*). The central void is occupied by clear granular sparite (*4*). The fibrous morphology has been slightly obliterated, most likely as a result of diagenesis.

PLATE 6-XX

Nubrigyn calcarenite, N.S.W., composed of skeletal fragments and algal grains (*1*) surrounded by pseudosparite. A crinoid ossicle circumcrusted with micrite exhibits syntaxial rim cement (*2*).

a crinoid ossicle, but other fossils have served as nuclei.

(2) The core may be recognized by its inclusions or outer rim of impurities, in contrast to the clear outer overgrowth. In some cases, relatively thick algal circumcrusts around the ossicles did not prevent syntaxial growth (Plate 6-XX; Wolf, 1963a).

(3) Host and rim are syntaxial, i.e., in optical continuity. Lucia (1962) stated that rim cement grows on single-crystal fragments, such as crinoid ossicles, and multicrystalline (dog-tooth) cement grows on multicrystalline hosts. There is little doubt that this is possible, but it must be remembered that not all calcite cement deposited on uni- or multicrystalline components forms overgrowths or is in optical continuity. In the Nubrigyn Formation, N.S.W., much of the crinoid debris is cemented by open-space fibrous calcite that is not in optical continuity with it. It seems then, that more than the mere presence of suitable nuclei controls the genesis of syntaxial rims.

(4) The outer boundaries of the rims are mostly in contact either with other rims or granular cement or with detrital particles, but seldom with a micrite matrix. According to Bathurst (1958), contacts with a micrite matrix are typical of syntaxial rims formed by grain growth or replacement. The presence of lime-mud, however, need not exclude the possibility of open-space syntaxial deposition. For example, a crinoid ossicle calcarenite overlain by lime-mud has been shown to develop syntaxial rim cement preferentially downward, because the lime-mud prevented overgrowth along the upper ossicle—lime-mud contact. Also, as Folk (1962b) has shown, a matrix saturated with fluids may cause formation of fibrous calcite which, due to the crystallization force, may merely force the matrix aside without actually replacing it.

(5) Boundaries between the rim and adjacent cement are planar.

(6) Host grains are in contact with each other in three dimensions.

(7) The mosaic resulting from overgrowth has plane intergranular boundaries.

(8) The mosaic may be rather equidimensional resulting from the syntaxial growth on well-sorted detrital particles, which is in contrast to the rather more heterogeneous grain-size pattern of granular cement and grain growth on recrystallization mosaics.

(9) The mosaic may be arranged in layers which differ in composition and coarseness.

(10) The mosaic may contain patches of skeletal fragments, pellets, oolites, and so on, with components similar to those of the mosaic grains.

(11) The longer axes of the detrital particles plus their syntaxial rim may be arranged sub-parallel to the original substratum.

Grain growth

The calcium carbonate formed by grain growth has the following features:

(1) The grain or crystal diameter ranges upwards from about 5 μ. Diameters between 50 and 100 μ are common and larger grains occur. The coarse mosaic has often been confused with granular open-space cement (Plate 6-XXI).

(2) The contact between fine and coarse mosaic may be abrupt; it can also be very gradual and the intermingling of fine and coarse grains makes it difficult to draw a definite boundary (Plate 6-XXI).

(3) Grain growth may be very selective or preferential (Plate 6-XXI).

(4) The grain size in the coarse mosaic varies irregularly and may change from place to place even over distances of 0.5 mm. This irregular pattern of grain size is distinct from the vectorial variation in drusy and fibrous mosaics and from the well-sorted mosaics of rim-cemented detritus. Presence of porphyroblasts is possible. Grain growth can also form rather equidimensional grain mosaics.

(5) Boundaries between grain growth and unaffected material may cut depositional features, e.g., laminations.

(6) Grain boundaries in the coarse mosaic vary generally from curved to consertal. Implicate boundaries appear among the larger grains; the plane boundaries so typical of open-space sparite occur less frequently.

(7) Some large marginal grains in the coarse mosaic embay the adjacent fine mosaic causing it to have a "nibbed" appearance. Many embayments are plane sided. Once convex curved boundaries, e.g., of pellets, are now locally concave. Fine-grained mosaic may occur as wisps or threads in the coarse mosaic. Fossils may be extensively interrupted to leave only disconnected relics of the original skeletons. In addition, pseudobreccias may form (Bathurst, 1959b).

(8) Some detrital components, e.g., patches of lime-mud, oolites, pellets, sparite cement, are entirely surrounded in three dimensions by the grain growth mosaic (Plate 6-XXI).

(9) Although well and extensively developed drusy and fibrous carbonate is usually indicative of open-space precipitation, both fibrous and blady calcite may form by grain growth (Harbaugh, 1961; Orme and Brown, 1963).

(10) Unidirectional grain growth may occur under favorable conditions.

(11) Grain growth may cut textures formed during preceding generations of diagenesis.

(12) Presence of impurities between crystals is common.

(13) Spherulites may be formed by grain growth (Plate 6-XXII).

(14) Patches of pre-grain growth open-space sparite cement remain often unaffected as they are more stable (Plate 6-XXI).

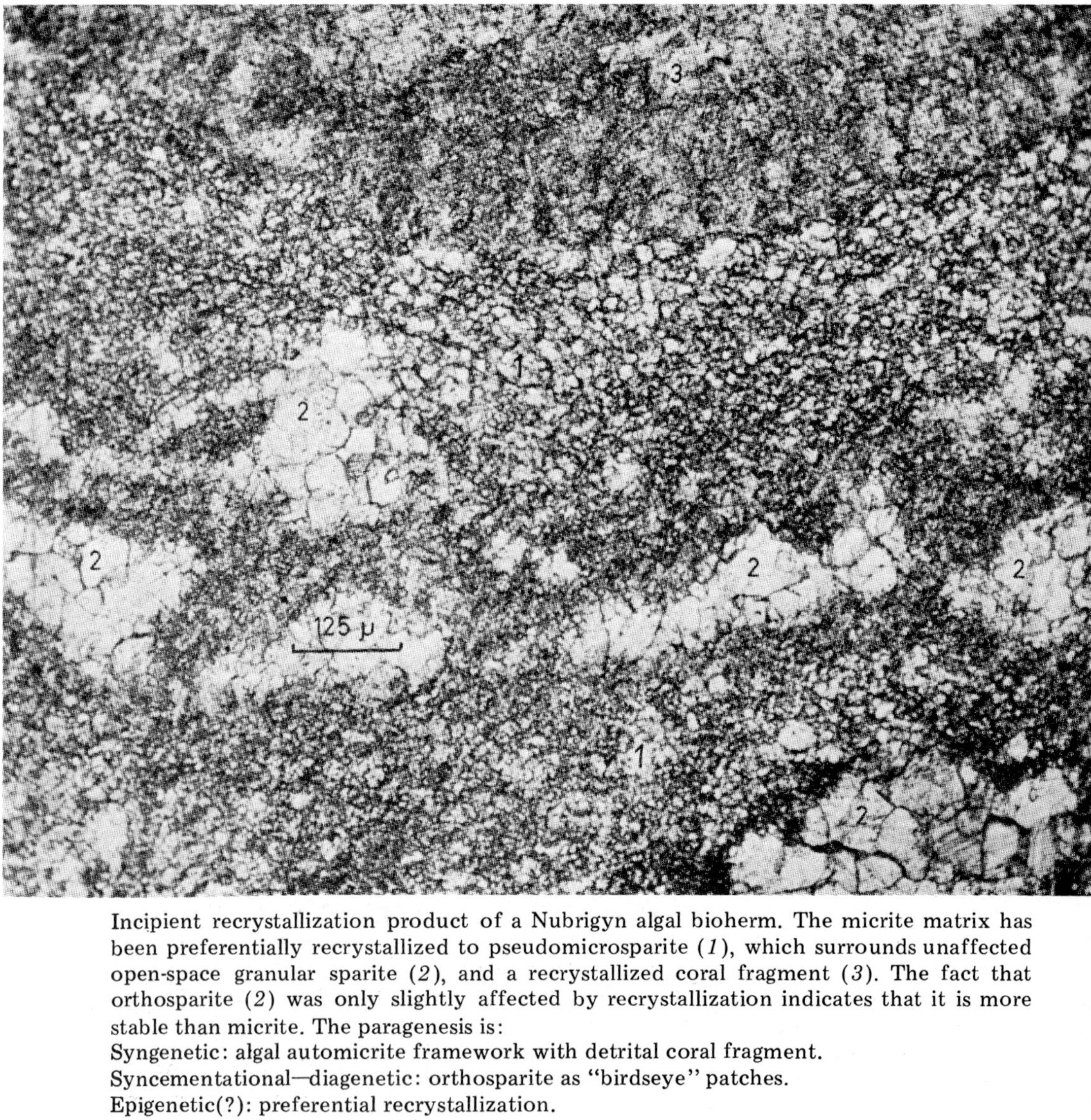

Incipient recrystallization product of a Nubrigyn algal bioherm. The micrite matrix has been preferentially recrystallized to pseudomicrosparite (*1*), which surrounds unaffected open-space granular sparite (*2*), and a recrystallized coral fragment (*3*). The fact that orthosparite (*2*) was only slightly affected by recrystallization indicates that it is more stable than micrite. The paragenesis is:
Syngenetic: algal automicrite framework with detrital coral fragment.
Syncementational—diagenetic: orthosparite as "birdseye" patches.
Epigenetic(?): preferential recrystallization.

Syntaxial grain-growth rims

Syntaxial grain-growth rims usually have some or all of the following distinct features:

PLATE 6-XXII

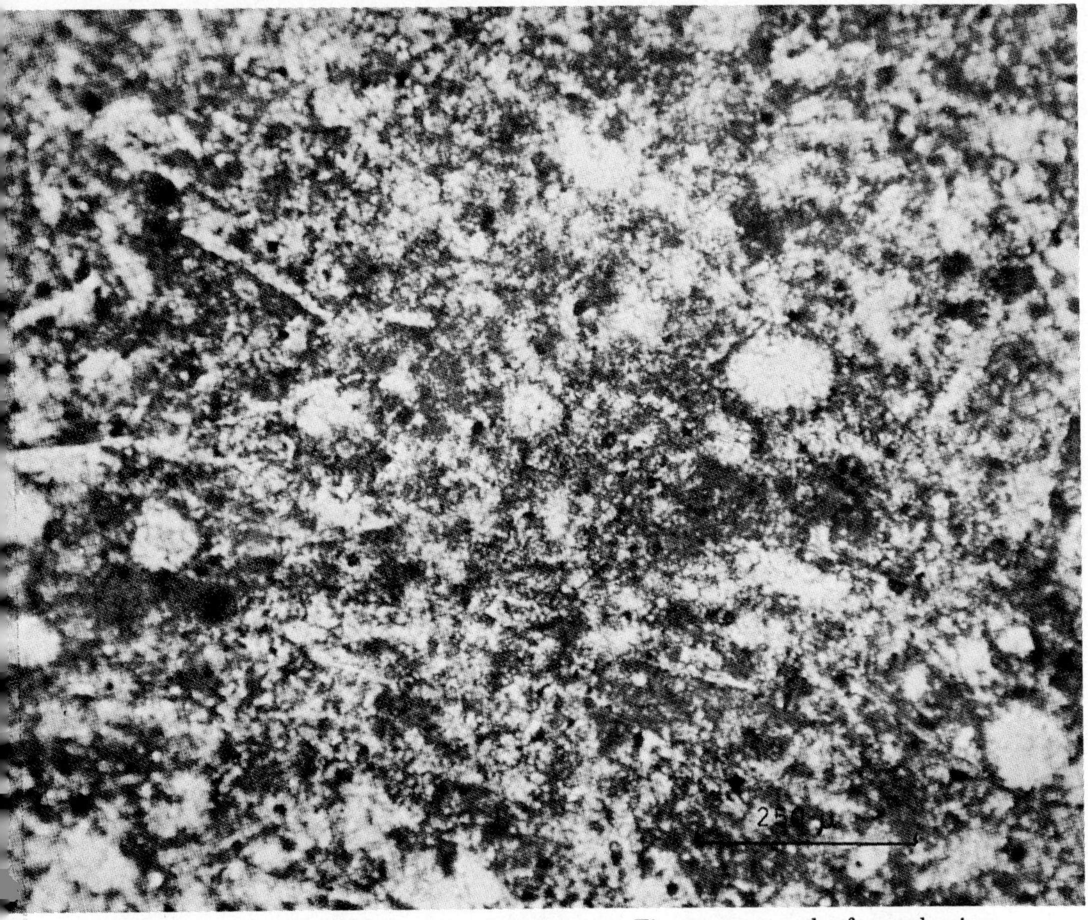

Recrystallized Tertiary(?) limestone of Portuguese Timor composed of pseudomicrosparite and pseudospherulites formed by grain growth. Note that the latter are quite distinct from the algal genus *Calcisphaera*. The pseudospherulites are composed of sparite formed by grain growth or recrystallization, whereas the latter are open-space structures filled by micro-drusy calcite in the Nubrigyn Formation. The longest dimension of that portion of the sample represented by the thin-section is equal to 1.3 mm.

(1) The syntaxial grain-growth rim resembles superficially a cement rim, but otherwise has quite different fabric relations. The host is most commonly a crinoid ossicle.

(2) The rim, or the hosts where no rimming occurred, is in contact with a

matrix of lime-mud (unlike a cement rim), with other rims, or other detrital particles.

(3) The adjacent lime-mud matrix may include detrital particles. It is this micrite which is interrupted by the rims.

(4) A rim may interrupt the fabric of a skeleton or embay the surface of pellets.

(5) Unlike the open-space syntaxial rim, the grain-growth rim has a highly irregular outer boundary, part of which may be produced into small spires often plane sided. These may be relatively wide, e.g., 10—30 μ in examples reported by Bathurst (1958), and taper generally distally to a point. Other extensions of the rim taper proximally, having swollen distal ends.

(6) This kind of syntaxial rim is commonly associated with the coarse grain-growth mosaic, and genuine open-space sparite cement may be absent from the rock.

(7) The nuclei that underwent syntaxial enlargement may "float" in the spar.

Grain diminution

Calcium carbonate can "recrystallize" at low temperatures and low pressures resulting in a relative decrease in crystal or grain size to form features such as the ones listed below:

(1) Small grains or crystals replacing coarse material of crinoids, Bryozoa, and algal colonies (Plates 6-II—IV). This process may explain the controversial knoll-reefs in various parts of the world, which are composed of micrite and fine-grained particles. If preferential grain diminution of an algal bioherm framework occurs and internal sediments and calcite cement remain unaffected, the resultant limestone is composed of dense material (lacking any evidence of previous organisms) and some patches of open-space fillings (Wolf, 1963a, b).

(2) Both granular and fibrous grains or crystals can form.

(3) Patches of fine grains may be irregularly distributed.

(4) Selective replacement may be common.

(5) The mosaic formed by grain diminution may increase by grain growth to form a coarser mosaic. If this occurs, many of the features associated with grain growth are applicable here also.

NON-CALCAREOUS REPLACEMENT

Non-calcareous replacement or substitution of one mineral by another of different composition in limestones may be: (1) absent, (2) very local, (3)

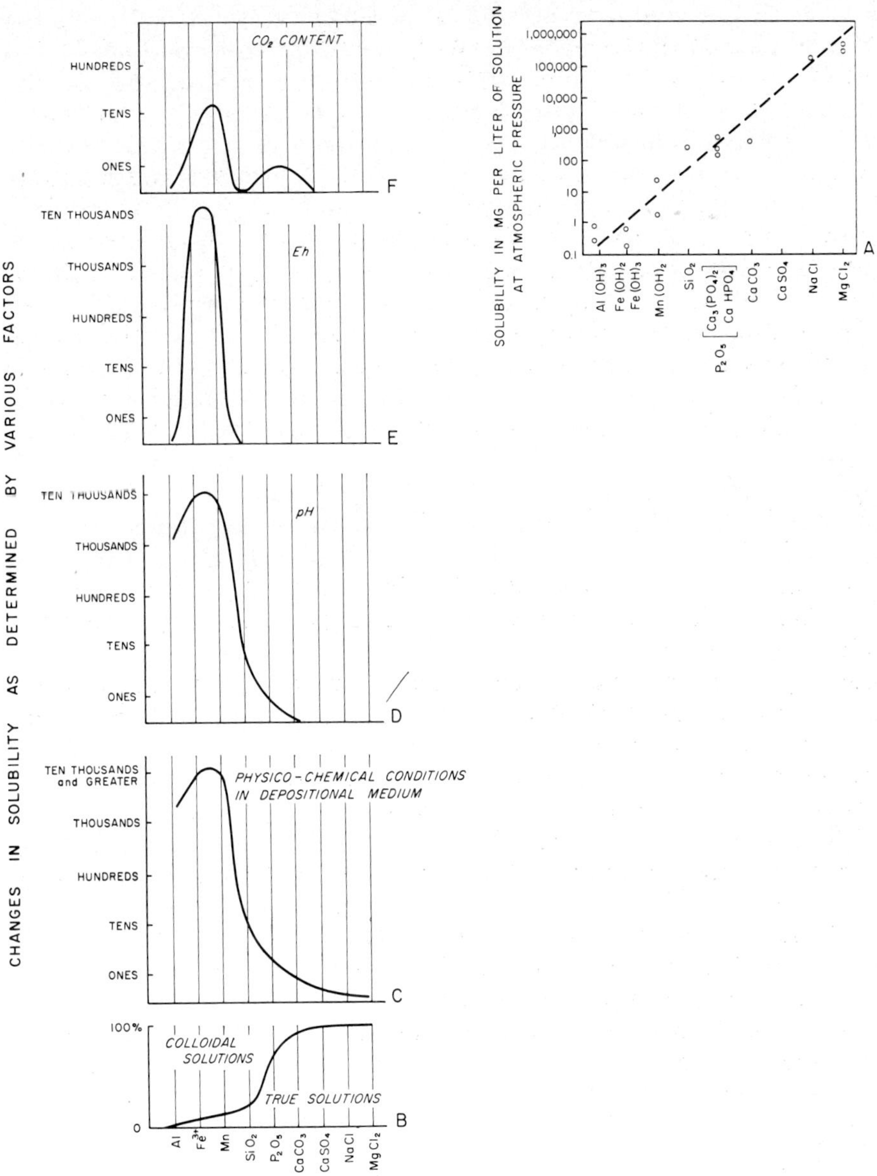

Fig. 6-3. Solubility of the most important chemical components of sediments (A); their type of solution (B); their general variability depending on physicochemical conditions of the depositional medium (C); and the specific factors pH (D), Eh (E), and CO_2 content (F). (After Rukhin, 1961, p. 275, 277.)

regional, (4) partial, (5) preferential, and (6) complete. Dolomite replacement, which is the most common of all, is treated in Chapter 7. As other replacement phenomena are considered in detail by other authors in this book, a few remarks on regional differentiation and local conditions that lead to silica, iron oxide and pyrite formation in limestones will suffice here.

The parameters responsible for non-carbonate replacements are similar to those listed for other diagenetic processes, but some have a more intense significance. For example, in the monomineralic limestone all or most of the raw material for diagenesis may be of endogenic origin. The components necessary for non-carbonate replacement, however, must have had an exogenic source in many cases. The iron and silica for extensive limestone replacements must have come from an outside source; this could have been from the continent or from intrabasin highs such as volcanic archipelagos (Bissell, 1959). Suitable climatic and geomorphic conditions are a prerequisite to assure a supply from an outside source. As indicated in Fig. 6-3, the components in solution react differently to physicochemical conditions, and inasmuch as these solutions pass through various natural environments (for example, from the continent through near-shore to deep-water environments), chemical differentiation is possible (Figs. 6-4—6). The precipitation of components from solution depends on their solubilities (Fig. 6-3A). In the sequence Al—Fe—Mn—SiO_2—P_2O_5—$CaCO_3$—$CaSO_4$—NaCl—$MgCl_2$, the solubility gradually increases from Al to $MgCl_2$: from a fraction of a milligram to 10^6 mg/l at atmospheric pressure. In normal, near-shore, marine environments concentrations may be high enough for dolomitization. Special conditions, however, are required for a high degree of supersaturation and deposition of sulfates and halides to permit diagenetic reaction between them and limestones. Under coastal conditions such as those depicted in Figs. 6-5 and 6-6, concentrations are usually not high enough and closed basinal environ-

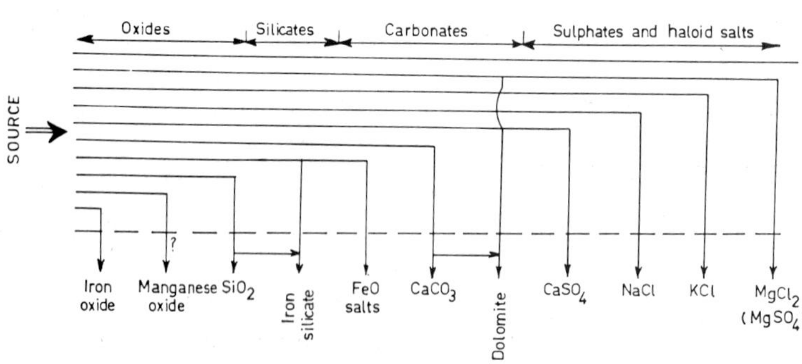

Fig. 6-4. Generalized deposition sequence during transportation leading to chemical differentiation. (After L.V. Pustovalov, in: Rukhin, 1958, p. 323.)

ments such as those in the Red Sea are necessary.

The differences in solubility of various components are related to a change from a colloidal state to a true solution (Fig. 6-3B), i.e., components which are only slightly soluble have a tendency to form colloidal solutions and coagulate readily, whereas highly soluble ones form true solutions. Sea water has a coagulating effect on certain constituents in solution, which explains the near-shore concentration of a number of minerals (Figs. 6-5 and 6-6). The CO_2 content greatly affects the solubility of colloids, but it causes less drastic relative effects on components in true solution (Fig. 6-3F). In addition, the pH, Eh and chemical composition of the solutions and environments are the most important factors influencing precipitation and, thus, replacement diagenesis.

Iron under certain conditions is predominantly transported as a bicarbonate and is primarily deposited in shallow water due to oxidation and early coagulation. Only small quantities may reach the deeper parts of the basin for pyrite genesis. After its precipitation in shallow water it will be subject to hydraulic factors, similar to clay; in other words, iron oxide may be just another detrital component undergoing mechanical reworking from the time of precipitation. This may explain the occurrence of red-brown hematite, predominantly as clay-sized detritus, in cavities within the Nubrigyn algal bioherms (Plates 6-X, XII; see also Wolf, 1965c). On the other hand, the small proportion of iron in true solution may have been responsible for the occasional wall-rock impregnations and replacement in the vicinity of the cavities (Plates 6-IX, XI) and of the matrix in detrital limestones (Plate 6-XXIII).

Silica is transported in true solution and as a colloid in natural waters

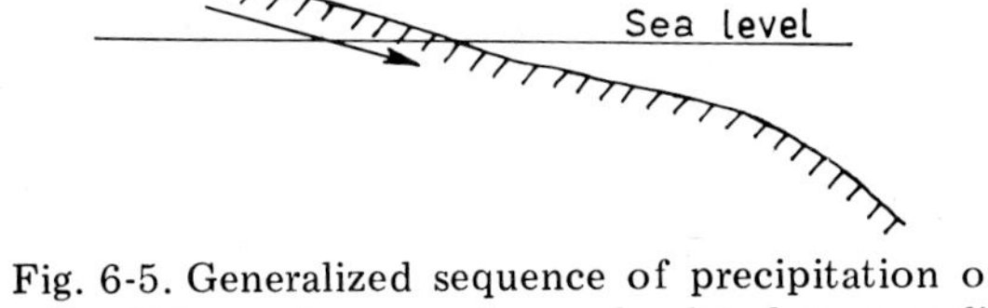

Fig. 6-5. Generalized sequence of precipitation of oxides of iron, manganese and silica with distance from source and related to coastline. (After N.M. Strakhov, in: Rukhin, 1961, p. 382.)

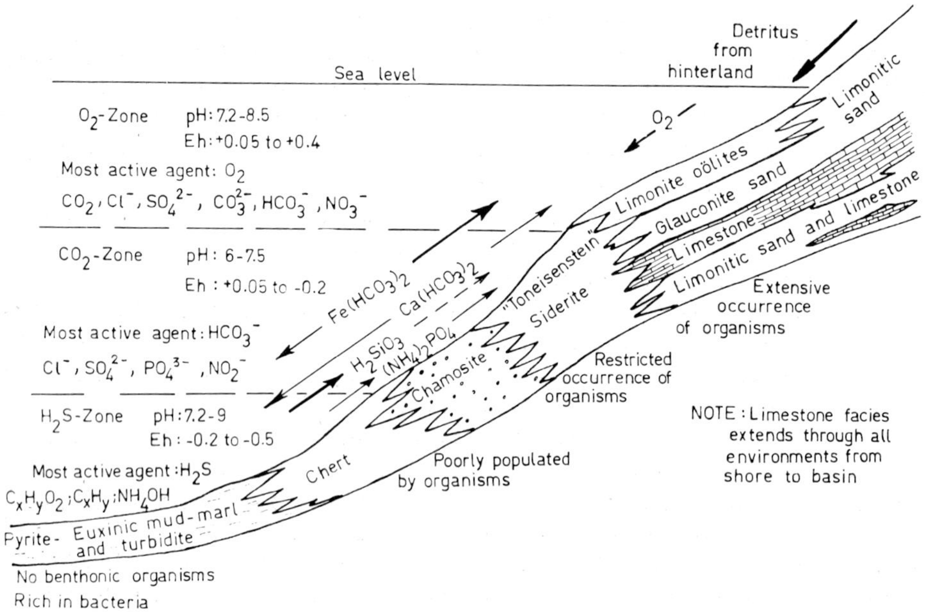

Fig. 6-6. Diagrammatic presentation of physico chemical zones related to diagenesis of carbonate sediments. (Diagram by Borchert, in: Braun, 1961, p. 472. Reprinted with permission of Drs. H. Braun and H. Borchert and *Z. Erzbergbau Metallhuettenw.*)

(Fig. 6-3B) and has, therefore, a better chance to reach deeper waters in contrast to iron oxide (Figs. 6-5 and 6-6). On the other hand, silica may have an endogenic origin, e.g., siliceous skeletal debris such as spicules, that upon solution may be precipitated as chert or chalcedony. Newell et al. (1953) suggested that horizons having different physicochemical attributes, in particular pH (some changes are possibly caused by bacteria), are conducive to the migration of SiO_2 and $CaCO_3$. The former moves to layers of lower pH and the latter to higher pH environments during diagenesis. In agreement with Figs. 6-4—6-6, Newell et al. (1953) and Wolf (1963a) report silicification in deeper water limestones. The former described fossils preferentially replaced by silica, and silica that occurs (1) as geode-like cavity fillings, postdating the sparry calcite cement, and (2) as nodules and crusts. These silica formations are typically absent from the reef and lagoonal limestones. The selective silicification of fossils is in particular obvious in the lower slopes of the reef talus, but may extend into small reefs buried within the basin sediments of the Capitan reef complex. Studies by one of the writers of this chapter (H.J. Bissell) on the Permian Kaibab Limestone in southern Nevada in the shelf-to-basin transition zone indicate that greatest silicification of limestones occurred near the hinge-line, particularly slightly basinward.

Selective silica replacements of calcareous fossils in preference to the limy matrix of the rock is not unusual. The sequence, from most readily to least readily silicified groups of fossils listed by Newell et al. (1953) seems to agree in general with observations made elsewhere and is as follows: (1) bryozoans, tetracorals, tabulate corals, punctate brachiopods; (2) impunctate brachiopods; (3) mollusks (replacement is usually spongy and imperfect), (4) echinoderms (replacement is usually limited to the surface); (5) Foraminifera; and (6) calcareous sponges and dasycladacean Algae.

Silicification beyond material (6) is not selective and will affect the matrix also. Selective replacement of fauna and flora is most probably analogous to other replacement processes. The relative solubility or rate of solution of the particular skeletons concerned may control the differential replacement. Space vacated by the dissolved carbonate is immediately, or sometimes later, occupied by silica (Newell et al., 1953). Although aragonite is more soluble than calcite under ordinary conditions, aragonite skeletons are not necessarily silicified prior to calcitic ones. Newell et al. suggested that the protective nitrogenous conchiolin which pervades the mollusk shells, for example, releases sufficient ammonia to raise the pH within the immediate microenvironment to prevent solution. Under certain conditions, therefore, aragonite may be more stable than calcite and can resist replacement.

The basinal Tolga calcarenite of New South Wales (Wolf, 1965a) has two very distinct chalcedony types: one detrital and derived from an older limestone source, and a second type formed authigenically by replacement. Only the latter is of interest here. Silicification is quite often selective to the extent that only a very small central portion of brachiopods is replaced by silica, whereas the outer portions are encrusted by "sooty" pyrite. The silica and pyrite are separated by recrystallized shell calcite (Plate 6-XXIV). On the other hand, along many laminae of the Tolga calcarenite beds there are irregular small patches of chalcedony up to 2—4 mm in thickness which replace grains and matrix. These chalcedony-rich layers are parallelled by upper and lower portions that have a distinct leached and recrystallized appearance in thin-section. The silicifying fluids apparently have also affected the host rock to some extent to form a "halo" parallel to the siliceous laminae.

The inverse physicochemical relationship between the silica and carbonate precipitation, i.e., the former precipitates whereas the latter dissolves, may explain the widespread pressure-solution in the basinal Tolga calcarenite in contrast to its absence in the shallow-water Nubrigyn detrital accumulations. The chemical conditions of the basinal interstitial fluids favored solution of $CaCO_3$ and precipitation of SiO_2, whereas possibly no such reactions could occur in the shelf environment.

PLATE 6-XXIII

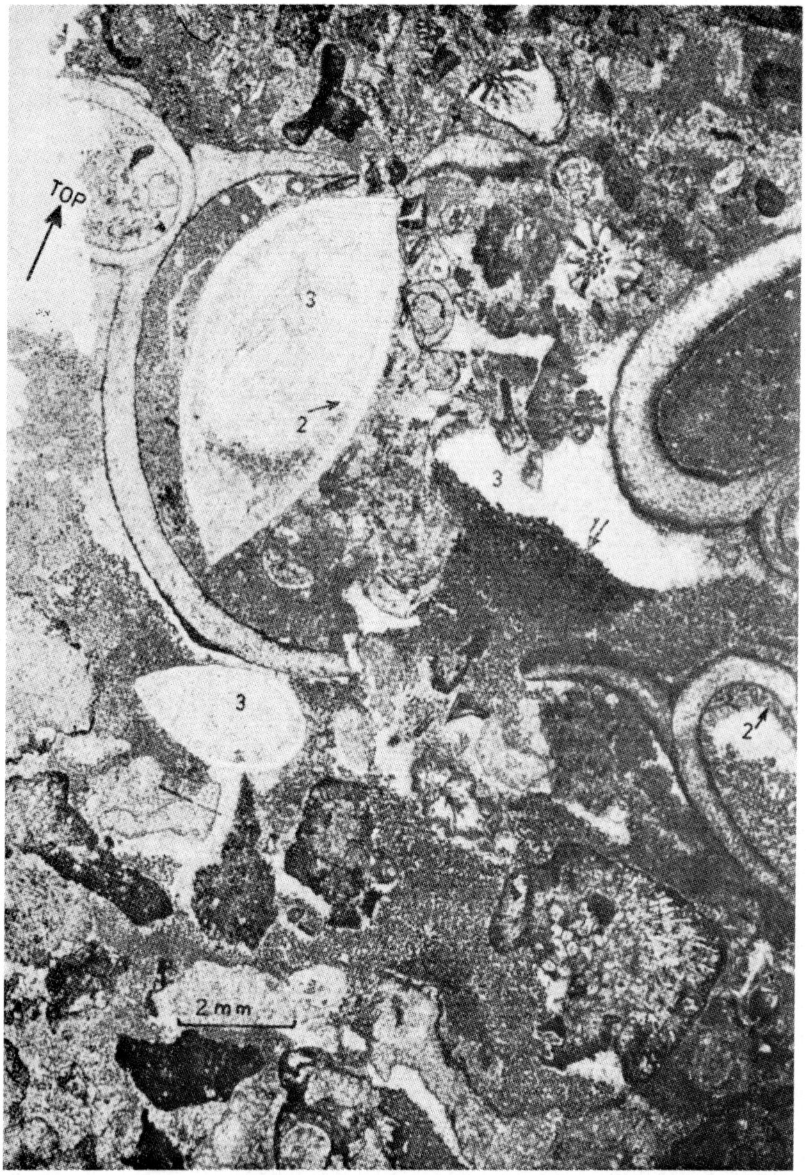

Nubrigyn skeletal-pellet limestone. The different color shades of the pellets are due to differential hematite impregnation. The pellet patch (*1*) shows a gradual downward decrease in degree of oxidation. The pellets in direct contact with the oxidizing fluids passing through the upper void became more intensely affected before cementation. The limestone consists of gastropod, coral, bryozoan and algal fragments and some brachiopods. Under high magnification, the pellets are composed of loosely compacted light-brown micrite unless impregnated with red iron oxide.

The cement is composed of thin layers of light-brown fibrous orthosparite (*2*) and clear granular orthosparite (*3*). The paragenesis is:

Syngenetic—syndepositional: skeletal and algal debris.

Syngenetic—prediagenetic: secondarily introduced pellets.

Precementation—diagenetic: hematite impregnation of pellets.

Syncementation—diagenetic: sparite cement.

PLATE 6-XXIV

A brachiopod shell fragment within the basinal Tolga calcarenite, which is the deep-water facies of the Nubrigyn-Tolga algal reef complex, N.S.W. The fragment is outlined by "sooty" pyrite (*1*), and is partly replaced by minute patches of chalcedony (*2*). One algal pellet (*3*) is present.

Minute (up to 8 μ) authigenic quartz crystals with well-developed hexagonal cross-sections are present in both shelf and basin limestones of the Nubrigyn-Tolga reef complex, N.S.W. Significant to note is their restriction to algal cryptocrystalline calcite and their absence in faunal products and detrital matrix. It is not possible to determine the cause and stage of formation of this authigenic quartz. Most likely it is of late diagenetic—epigenetic origin. On the other hand, Termier and Termier (1963) reported fairly early-formed euhedral quartz in recently emerged reefs buried in mud.

Similar to the SiO_2, pyrite is confined to the deep-water limestone facies (Fig. 6-6) of the Nubrigyn-Tolga complex. Three morphologic pyrite types have been noticed here: (1) the most widespread are the fine films of "sooty" pyrite on faunal skeletons penetrating even the most minute surface pores on brachiopods, for example; (2) minute pyrite spheres occurring in clusters; and (3) cubes. The formation of most of the pyrite must have been precementation—diagenetic, because it occurred distinctly before $CaCO_3$ cement precipitation in most cases. Only the cube-shaped pyrite is of postcementation origin as it replaces both detrital grains and calcite cement. The occurrence of silica with pyrite suggests that both formed in reducing, low-pH conditions characteristic of the basinal euxinic black limestone and marl facies. Although it is impossible to state with absolute certainty the origin of the pyrite, a bacterial origin is possible under euxinic conditions. Anaerobic bacteria attack organic matter, extract oxygen and release hydrogen. The latter combines with sulfur derived from sulfates to form H_2S, which is a toxic gas readily soluble in sea water. H_2S attacks soluble iron compounds to form FeS_2, which is highly insoluble and is precipitated in the form of pyrite or marcasite. Other anaerobic bacteria attack sulfates to obtain oxygen needed in metabolism and free sulfur for the genesis of H_2S required. The precipitation of iron sulfides by anaerobic bacteria may take place as finely divided dark pigment; sulfides may also replace shells or form nodules. Although pyrite genesis may be restricted in some localities to the deep-water facies, it is important to remember that euxinic shallow-water environments may also lead to pyrite and marcasite formation.

TEXTURES, STRUCTURES AND DIAGENESIS

Diagenesis may lead to formation or destruction of textures and structures. Some of these have been mentioned already in this chapter and others are so well known that a few remarks will suffice. Laminations of certain types, discontinuity-surfaces, stromatactis, birdseyes, club-shaped stromatolites, cone-in-cone, certain spherulites and oolites, faecal pellets, pseudobreccias, and early fractures may all be of syngenetic—diagenetic or purely diagenetic origin.

Botvinkina (1960) pointed out that lamination and stratification can be the result of iron transfer and differential precipitation of iron oxide, silica, carbonates, and others, at horizons with corresponding favorable pH and Eh values. Such laminations may be very similar to those formed by ordinary detrital accumulations. Other diagenetic products are concretions, lenses and beds. Stratification or bedding may also be the result of solution and corrosion, sometimes forming discontinuity surfaces (Jaanusson, 1961). Residual clay, iron oxide, phosphate, glauconite, corrosion and bore pits, burrows, and bleaching of the underlying sediments characterize hiatuses formed by near-surface diagenetic alterations.

Early cementation may control the shape and preservation of stromatolites. Logan (1961) suggested that the domed and club-shaped stromatolites are a function of acicular aragonite cement precipitation, which has to occur very early to prevent collapse of these high relief structures in the turbulent littoral environment.

Usdowski (1963) presented evidence that cone-in-cone structures, composed of fibrous calcite, are the result of early diagenetic recrystallization of lime-mud beds shortly after the sediment accumulated and was still in an unconsolidated state and saturated with interstitial fluids. If this interpretation is factual, further research may indicate that cone-in-cone structures are valuable paleoenvironmental criteria.

Spherulites, other than those shown in Plate 6-XXII and formed by recrystallization, have been reported as products of diagenetic bacterial processes (Monaghan and Lytle, 1956; Lalou, 1957). Kaye (1959) mentioned that precipitation of calcium carbonate may occur as a gel, and coagulation may mechanically entrap particles of non-colloidal size and form alternate bands of colloidal and entrapped material. Laboratory experiments indicate that spherulites composed of vaterite or one of the several hydrates of calcium carbonate form during initial crystallization. These have not been found in nature, however, and it is most likely that their unstable nature caused a change to aragonite and calcite spherulites soon after formation. Cloud et al. (1962) stated that the tendency of bacteria to adhere to surfaces may be conducive to the genesis of some types of oolites. Accretion of successive layers by aggregation of sedimentary particles around successive slimy or gelatinous bacterial sheaths surrounding the initial nucleus may be a likely process. It may conceivably occur up to a few feet within the sediments. The process, however, requires further study. Eardley (1938) believed that radial, in contrast to the concentrically laminated, structures of oolites in Great Salt Lake of Utah are a diagenetic feature formed during inversion of aragonite to calcite.

Early diagenetic fracturing of the algal micrite bioherms of N.S.W. resulted in calcite veins that post-date internal sedimentation and fibrous calcite, but predate or are contemporaneous with granular sparite. From the structural relations described elsewhere (Wolf, 1963a, 1965c), it appears that the solutions that precipitated the granular sparite reached the voids only after fracturing took place.

PARAGENESIS

In a general way, though not always, the sequence of diagenesis takes place in the following order: (1) biological and biochemical, (2) physico-chemical, and, (3) physical (Termier and Termier, 1963). These processes, of course, overlap to a large extent. With time, there is a decrease in rate of these processes. Little information is available that would permit a paragenetic reconstruction of diagenesis on a regional scale, although it may be a valuable tool for paleogeographic reconstructions. Most of the data are of local value, of meso- and microscopic dimensions.

The previously mentioned difficulty of making clear distinctions between syngenetic, diagenetic, and epigenetic processes and products even on a micro-scale is illustrated in the following paragenetic example. Diagenetically formed cavities first have been lined by diagenetically precipitated fibrous calcite. The central cavity was then filled by dolomite which was precipitated from saturated surface waters penetrating the limestone framework (Fig. 6-1; Plate 6-XIV; Wolf, 1963a). Both Schwarzacher (1961) and Folk (1962a) reported similar dolomite infillings. The former calls it syngenetic or primary dolomite. This, although correct, is confusing as it suggests that syngenetic products may be preceded by diagenetic cement, for example. An identical situation occurs when cavities are diagenetically encrusted by fibrous calcite and the central cavity is filled by detrital internal sediments, brought into the system from the surface that is exposed to syngenetic processes (Plates 6-VI—VIII). From these two examples it seems clear that under certain circumstances one has to expect cyclic formation of syngenetic and diagenetic products. Wolf (1963a) has distinguished, therefore, between syngenetic, diagenetic and epigenetic processes and products on one hand and stages on the other.

The same complexities occur on a regional scale. Within a limestone formation or a reef complex, one geomorphologic environment may be still in the syngenetic stage, whereas others are undergoing rigorous diagenetic alterations. Or, if all sections of a formation are exposed to similar diagenetic processes, then various parts may be characterized by distinct paragenetic histories as a synthesis in Table 6-IV indicates. If both are present, i.e., characteristic diagenetic products as well as complex paragenetic histories, then combination of the two will be a valuable tool for environmental studies.

Significant paragenetic relationships may exist between different types of sparry calcite cement. It has been noticed, for example, that light brown fibrous sparite always precedes colorless granular calcite, where both are present in the Nubrigyn-Tolga reef complex of New South Wales (Wolf, 1963a, 1965a, c). The fibrous calcite is restricted to the shallow-water shelf

TABLE 6-IV

Paragenesis of diagenetic features (exemplifying possible regional trend)

Littoral algal bioherms *	Fore-reef talus **	Basinal "turbidite" *
Paragenesis 1:	*Paragenesis 1:*	*Paragenesis 1:*
(a) detrital internal sediment	(a) calcite cement	(a) pressure-solution
(b) fibrous sparite		(b) granular sparite
(c) granular sparite		
Paragenesis 2:	*Paragenesis 2:*	*Paragenesis 2:*
(a) hematite replacement of framework concentrically around voids	(a) calcite cement	(a) minute fringe of micro-drusy sparite
(b) detrital internal sediment	(b) chalcedony open-space filling	(b) granular sparite
(c) fibrous sparite		
(d) granular sparite		
Paragenesis 3:	*Paragenesis 3:*	*Paragenesis 3:*
(a) fibrous sparite	(a) calcite cement	(a) granular sparite
(b) chemically precipitated, coarse dolomite, open-space filling	(b) chalcedony replacement	
(c) granular sparite		
Paragenesis 4:		*Paragenesis 4:*
(a) hematite open-space filling		(a) any of the above with:
(b) granular sparite		(b) pyritization and/or
		(c) silicification prior to or succeeding cementation
Paragenesis 5:		
(a) fibrous sparite		
(b) chemical and/or detrital internal sediment		
(c) granular sparite; a and b alternate to form up to six generations prior to c		

* Simplified after Wolf (1963a).

** After Newell et al. (1953).

bioherms and associated algal calcarenites. From one to six generations of fibrous sparite, sometimes separated by internal sediments such as minute pellets and iron oxide, fill open spaces in algal reefs and form the cement of the calcarenites (Plates 6-VIII, X, XIII, XIV, XIX). If any voids remained, they were subsequently occupied by clear granular sparite (Plates 6-V, VIII--XI, XIII, XIV, XIX). In numerous instances it can be demonstrated that fractures terminating in the voids permitted solutions to deposit the granular sparite (Plates 6-XI, XII, XIV). These distinct paragenetic relationships,

TABLE 6-V

Paragenetic model of limestones (after Wolf, 1963a)

(1) Pre-depositional stage, processes and products
(Based on limestone rock fragments, i.e., calclithite detritus, derived from an older carbonate source that underwent diagenesis before erosion)

(2) Syngenetic stage, processes and products
(a) syndepositional (e.g., framework accumulation)
(b) prediagenetic (e.g., purely physical reworking; mechanically deposited internal sediment)

(3) Diagenetic stage, processes and products *

(a) precementation ** (e.g., chemical internal sedimentation, replacement and corrosion of the framework)	(i) above high tide, i.e., subaerial ***
(b) syncementation (e.g., deposition of calcite in open cavities; chemical and mechanical internal sedimentation alternating with generations of cement)	(ii) intertidal ***
(c) postcementation (e.g., early fracturing permitting deposition of granular calcite)	(iii) below low tide ***

(4) Epigenetic stage, processes and products
(a) juxta-epigenetic ***,†
(b) apo-epigenetic ***,††

* In detailed studies of Recent and Pleistocene carbonates it may be possible to subdivide diagenesis further into i, ii, and iii.
** The suffix "cementation" can be replaced by "lithification" (Strakhov, 1958) depending on the product permitting subdivision. In the present case, the first generation of brown fibrous calcite was used.
*** Other subdivisions can be used depending on what type of sediments (i.e., marine or nonmarine, etc.) are under investigation (Wolf, 1965c).
† "juxta-" meaning "near" or "close-by".
†† "apo-" meaning "far", "remote".

which remain constant throughout the reefs, permit the subdivision shown in Table 6-V. The brown fibrous sparite forming the cement of the shallow-water limestones marks the syncementation stage and separates, therefore, the precementation from the postcementation stage. The clear granular calcite belongs to the postcementation period as it has not contributed to cementation of the shallow-water sediments, and was formed by different processes only after fracturing of the rock occurred. Obviously, there is a hiatus between the fibrous and granular sparite formation.

The deep-water Tolga calcarenite, on the other hand, was cemented by clear granular sparite at a much later stage than the equivalent Nubrigyn shelf deposits as indicated by considerable precementation pressure-solution. In other words, while the shelf deposits were in the syncementation stage, the basinal limestones were undergoing precementation diagenesis.

For regional paragenetic reconstructions of diagenetic and epigenetic alterations it may be important to find features that overlap in critical areas to permit a "time-correlation". For example, if hematite and pyrite geneses are confined to shallow and deep water facies, respectively, then one needs some criteria to prove that both did actually occur during the same paragenetic stage. Table 6-VI shows a simplified example where it is possible to demonstrate that hematite and pyrite geneses took place more or less penecontemporaneously, because both occurred before cementation and pressure-solution. If, for example, the pyrite in the basinal limestones had distinctly replaced both fossils and cement, the FeS_2 would have been of a later origin compared to the hematite. (Both precementation and postcementation pyrites are present in the Tolga calcarenite—the former is "sooty" and the latter occurs as cubes.)

TABLE 6-VI

Simplified example illustrating correlation of diagenetic products (after Wolf, 1963a)

Shelf sediment	Intermediate sediment	Basinal sediment
(1) Limestone accumulation	(1) limestone accumulation	(1) limestone accumulation
(2) Hematite debris	(2) hematite debris and "sooty" pyrite on fossils in places	(2) "sooty" pyrite on fossils and minute pyrite spheres
(3) Fibrous calcite cement	(3) occasional fibrous sparite, but mainly granular sparite	(3) pressure-solution
		(4) granular sparite cement

PALEOGEOGRAPHIC ENVIRONMENT-INDICATING DIAGENETIC PRODUCTS

Early diagenesis is controlled by surface and near-surface factors and its products are indicative of the environment. Some may reflect only very local conditions; others, however, may be useful criteria to interpret paleogeomorphologic and paleogeographic conditions. The petrographic discussion below is based on an Australian Devonian algal reef complex (Wolf, 1963a, 1965a) and serves as an example. It should be emphasized that although only the diagenetic products are used here to illustrate their usefulness in environmental interpretation, all other paleontologic, structural and stratigraphic criteria support the reconstructions made.

The following early diagenetic features were found to be indicative of a littoral environment for the algal bioherms: (1) internal open-space structures, i.e., incorporated former surface pits and surge channels, and the so-called stromatactis; (2) extensive internal detrital sedimentation; (3) certain internal chemical sediments, e.g., red iron oxide and dolomite; (4) fibrous calcite cement; (5) travertine; and (6) complex paragenesis.

The open-space structures, detrital sediments and chemical internal precipitates have been described earlier, and the latter are listed in Tables 6-II and 6-IV. The chemical composition of the interstitial fluids must have changed relatively quickly as indicated by the successive and alternating generations of iron oxide, calcium carbonate and dolomite precipitation; and bleached, leached or corroded host-rock walls. The paragenetic picture is very constant from bioherm to bioherm within the same unit. It seems very unlikely that such a complex paragenesis could occur either in a sublittoral environment, or under supralittoral conditions. First, extensive internal channel systems are not likely to form under sublittoral conditions. It is true that they can occur in subaerially-formed limestones such as eolianites, but these have mainly vertically oriented channels in contrast to the horizontal ones of the littoral carbonate sediments. Second, the diagenetically formed internal sediments, such as iron oxide and dolomite, are typical of littoral origin. If one admits the possibility that both iron oxide and dolomite internal cavity fillings could occur in limestones below low tide, then there is still a third factor, the complex cycles of fibrous calcite and internal sediments, to explain. For sediments to penetrate into a limestone framework, turbulent conditions and surging powerful currents, unlikely to occur in sublittoral environments, seem to be necessary. Admittedly, density currents and/or turbidity currents can transport sediment into and across the sublittoral zone. Under conditions below low tide, sediments would merely settle and, at the most, drift to and fro; but it seems unlikely that they could penetrate into a complexly channeled sediment framework.

Superficial observations made on Recent and Pleistocene limestones tentatively suggest that well-developed and extensive fibrous and, possibly, drusy calcite and aragonite development is confined to shallow-water and supralittoral environments. It is interesting to note that many beach-rocks have acicular, i.e., fibrous, carbonate cement, whereas subaerially cemented eolianites of Lord Howe Island, for example, show mainly drusy sparite (Plate 6-XVIII). These observations suggest that the Nubrigyn bioherms and associated calcarenites, which are characterized by fibrous calcite cement, were formed in a littoral environment.

In a number of thin-sections it has been observed that filamentous, unicellular algal mats and algal micrite colonies abut against overlying dense laminated travertine crusts which are composed of fibrous sparry calcite. Similar colonies in turn encrust the travertine layers. It appears that the travertine could have been formed only by exposure of the algal bioherms above sea water at low tide, whereas solution, evaporation and precipitation formed the sparry calcite crusts after dissolution of part of the algal colonies.

One can conclude from these discussions that, based on the diagenetic products alone, the Nubrigyn algal bioherms were formed most probably in a littoral environment (see Wolf, 1965c, for more details).

DIAGENESIS AND LIMESTONE CLASSIFICATION

The foregoing information on diagenetic alterations imposed on syngenetic limestone textures (Figs. 6-7—6-14) makes it clear that it is very difficult to follow one simple nomenclature and classification scheme for carbonate rocks, particularly for limestones. A scheme is necessary that suits both the practical and research geologists and is applicable in superficial and super-detailed studies, if a common meeting ground of ideas can be realized. Perhaps no such ideal classification is available. As shown in Table 6-VII (Wolf, 1963b), the descriptive and genetic stages can each be subdivided into two substages based on method and accuracy of the investigation carried out. During superficial studies only the size-nomenclature may be required. With the use of a binocular microscope it is possible to determine the grains or framework and the micrite matrix/cement ratio and textures, but it may not be possible to make a genetic interpretation of these components. This has to await detailed thin-section investigations. In the final analysis, when all depositional, stratigraphic and paleontologic information has been assembled, a paleogeomorphologic reconstruction is possible, and the sediments can be named accordingly, e.g., skeleton-orthosparite-beach calcarenite. For such a step-by-step build-up, the descriptive and genetic nomenclature must be kept separate, as repeatedly emphasized. Hence, terms that satisfy requirements in both stages are given in Tables 6-III and 6-VIII. The classification scheme

Fig. 6-7. Diagenetically altered organic-rich micrite, showing few centers of growth of slightly large microcrystalline calcite. Darker areas are possibly "dead oil". Loray Formation (Permian) from outcrop in Dead Horse Wash, White Pine County, Nevada; × 40.

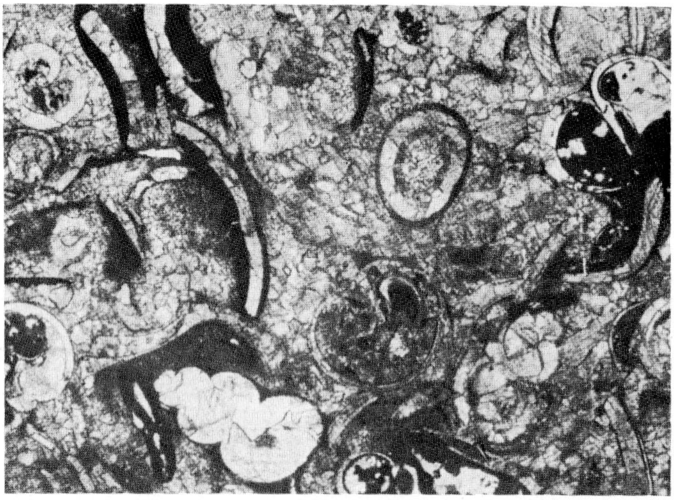

Fig. 6-8. Diagenetically altered skeletal limestone, showing formation of sparry calcite within brachiopod and gastropod shells, as well as in the matrix material. Loray Formation (Permian) from outcrop in Dead Horse Wash, White Pine County, Nevada; × 10.

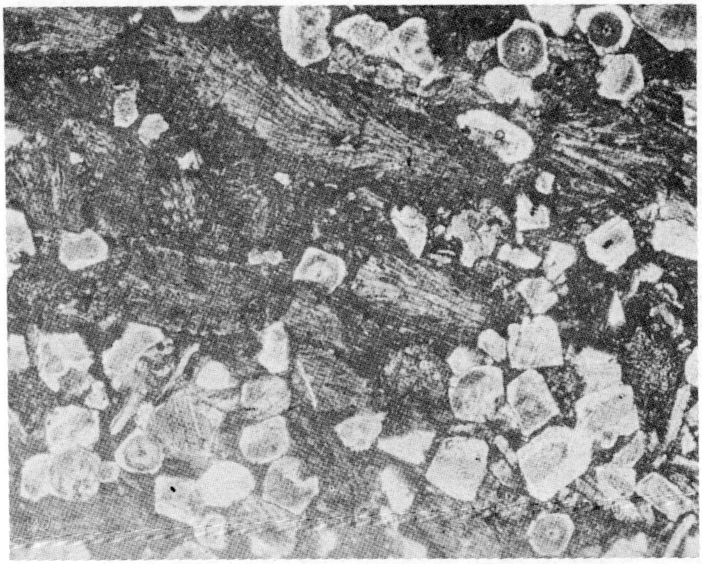

Fig. 6-9. Diagenetically altered bryozoan-encrinal limestone, illustrating selective diagenesis. Crinoid ossicles show authigenic overgrowth, but the lioclemid bryozoans and interstitial matrix are relatively unaffected. Gerster Formation (Permian) at type locality near Gerster Gulch, Tooele County, Utah; × 5.

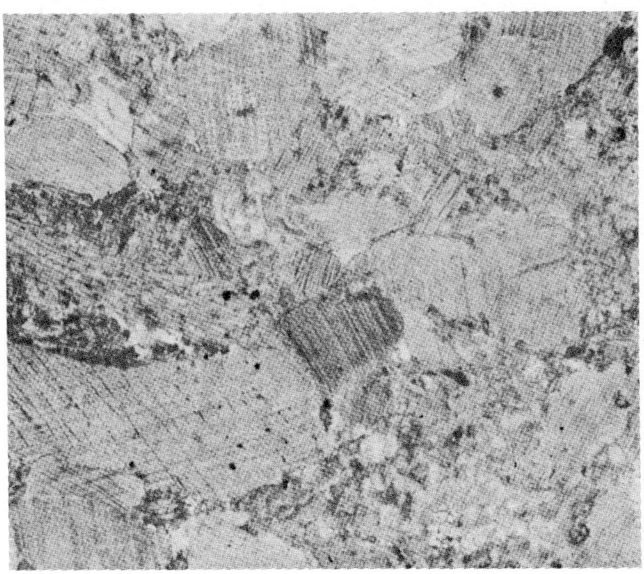

Fig. 6-10. Advanced stage of diagenesis of a criquinite, showing mostly relics of crinoid-stem fragments and development of calcite. Hall Canyon Member (Morrowan) of Oquirrh Formation in Oquirrh Mountains, Utah County, Utah; × 5.

Fig. 6-11. Early to medial stage diagenesis of calcarenite (criquinite variety), illustrating alteration of encrinal material and to a lesser degree the finer-grained, calcarenitic, interstitial matrix material. Hogan Formation (Desmoinesian) west of Wendover, water reservoir, Tooele County, Utah; × 5.

(Table 6-IX) is descriptive; the only difficulty lies in the recognition of a micrite biolithite in handspecimen, and it would be called in most cases "micrite limestone" until thin-section work furnishes more detail. Two examples on how the descriptive names can be easily changed into genetic terms by adding prefixes are presented in Table 6-VII.

Due to the diagenetic alterations of the matrix and cement in limestones, the micrite/sparite ratios may not be a true reflection of turbulence and washing ("winnowing" of some geologists) of the depositional environment, and Folk's (1959) concepts should be considered with care in this regard. His classification scheme, and the modified version given here, based on the syngenetic "grain or framework-matrix/cement ratio" is still useful and need not be discarded because of diagenetic alterations. In the scheme outlined above, i.e., the four substages leading from descriptive to genetic stages, the syngenetic diagenetic and epigenetic characters of limestones have been included (Table 6-IX). Hence, the lower end-member in Table 6-IX is either an unaltered micritic limestone or a crystalline limestone composed of pseudo-

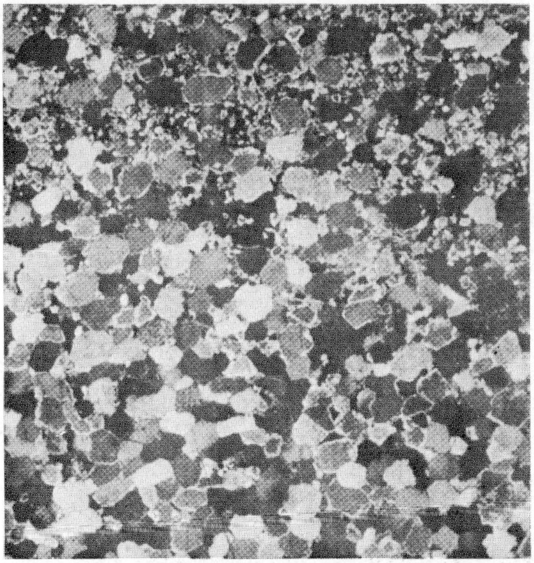

Fig. 6-12. Early diagenesis of a calcarenite, showing calcareous overgrowths on lime-pellet grains. Meadow Canyon Member (Derryan) of Oquirrh Formation, Cedar Mountains, Tooele County, Utah; × 20.

sparite or pseudomicrosparite. (Those interested in details of classification may wish to consult Ham and Pray, 1962, for example.)

Micrites

Limestones herein classed as micrites are those rocks originating from diagenesis of calcareous mud or ooze. Lime ooze may have a lower plasticity than clay, yet readily (and probably early in the diagenetic process) forms sets and systems of joints (diaclases), the development of which provides avenues for gas and liquid transfer. Before a discussion of diagenetic effects upon micrites can be expanded, it is necessary that certain concepts and nomenclature should be clarified. Folk (1959) termed micrite the lime-mud component (very fine-grained ooze or paste). Mud is very fine-grained (or crystalline) dense material which geologists have described as "lithographic", "cryptocrystalline", "cryptograined", "microcrystalline", "micrograined", etc. Leighton and Pendexter (1962) arbitrarily set the upper limit of the mud component at 0.031 mm, but some prefer this limit to be 0.005 mm (Table 6-III). An exact size limitation is not too critical (Baars, 1963). Numerous of the calcilutites and some calcisiltites fit into the category of micrites or micritic limestones, depending on the limit set for "micrite". A

Fig. 6-13. Coarsely-crystalline sparite, illustrating advanced stage of diagenesis of a calcarenitic criquinite. Ely Limestone (Derryan), Moorman Ranch area, White Pine County, Nevada; × 30.

prevalent tendency among some petrographers is to apply the textural term of "aphanitic" to the micrites, at least to those which have a micro- or cryptotexture. The term *aphanic* was proposed by DeFord (1946) as a textural term for carbonates, particularly limestones, which are crystalline (and/or grained), and the discrete particles of which are smaller than 0.1 mm. Microcrystalline (also micrograined) and cryptocrystalline (also cryptograined) are the two textural subdivisions. The term *aphanitic* is more loosely defined, and it is herein rejected as a textural term for carbonates, with the proposal that *aphanic* should be adopted because of its precise definition. Aphanic as a textural term has been applied successfully to petrographic studies of limestones (Mollazal, 1961) and of dolomites (Osmond, 1956). Many geologists, particularly sedimentary petrographers, use an upper limit within the medial silt-range to define micritic texture; as pointed out by Baars (1963), for most cases this is the practical limit for particle recognition.

The origin of lime-muds cannot be determined with accuracy in all cases, and it is probably that several mechanisms contribute to and operate during its formation (Table 6-III; Wolf, 1962, 1965b). Discrete particles may be chemi-

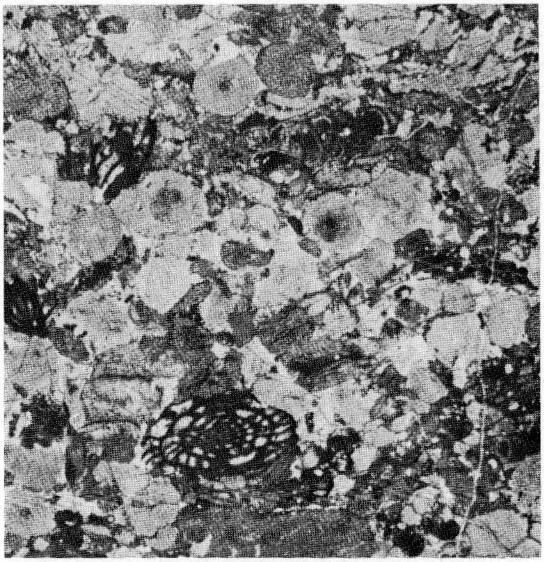

Fig. 6-14. Incipient dolomitization of a skeletal-detrital limestone, showing diagenesis of encrinal material and to a lesser degree other skeletal elements. Fusulinid test was silicified first, but later was partly replaced by dolomite. Ferguson Mountain Formation (Wolfcampian) in outcrop near top of the "Bear's Claw" north of Wendover, Tooele County, Utah; × 20.

cally or biochemically precipitated fine crystals, or finely comminuted "clastic" particulate material derived from originally larger particles and other sources. Lowenstam (1955) and Lowenstam and Epstein (1957) recognized the possibility that mud-sized needles of aragonite on parts of the Great Bahama Bank may be derived from calcareous Algae. Baars (1963) pointed out that several codiacean (green) Algae secrete clay-sized aragonite needles within their tissues; the genera *Penicillus*, *Rhipocephalus*, and *Udotea* were cited as examples. These aragonite needles disintegrate to produce lime-mud upon death of the Algae. It is because these Algae (and others) are important in modern lime-depositing seas and because of their short life cycle, that they may be extremely important sources of lime ooze. Furthermore, the small particle size and resultant small size of interparticle pores render the lime-muds particularly susceptible to diagenesis, especially pressure-solution and simple interparticle cementation. These characteristics also make the lime ooze amenable to diagenetic dolomitization early in the sedimentary history.

Lowenstam (1955) has stated that some calcilutites attributed to physicochemical precipitation have formed by breakdown of calcareous Algae, par-

TABLE 6-VII

Petrographic and petrologic stages of carbonate investigations (after Wolf, 1963b)

Descriptive		Genetic	
hand-lens investigation	binocular microscope investigation	petrographic microscope investigation	total petrographic summary including depositional structures, stratigraphy, and paleontology
Size nomenclature only, i.e., calcirudite, calcarenite, calcisiltite [1], calcilutite [1] (= micrite)	morphologic and size nomenclature	genetic, morphologic and size nomenclature	genetic, morphologic and size nomenclature with geomorphologic terminology
e.g., calcarenite	e.g., pellet—sparite—calcarenite	e.g., algal pellet—orthosparite—calcarenite	e.g., algal pellet—orthosparite—beach calcarenite
e.g., micrite	e.g., micrite	e.g., algal auto-micrite biolithite	e.g., algal auto-micrite knoll reef

[1] The term calcilutite is usually used for both clay- and silt-sized particles in the descriptive stage as they may not be distinguishable. In thin-section work, however, discrimination is possible.

ticularly the poorly calcified forms. In referring to the "white reef" in certain areas of Alberta, Belyea (1955) stated that much of the reef mass is fine-grained comminuted organic debris, and much of it is white dense limestone probably formed largely by lime-trapping Algae. Some lime ooze may precipitate on or near algal plants and form "algal slime", because the plants extract carbon dioxide from immediately adjacent sea water (Pray, 1958). Thomas and Glaister (1960) mentioned that in some Mississippian carbonate sequences, which they studied in the Western Canada Basin, microgranular carbonates graded vertically and laterally into chalky micrograined carbonates. They regarded part of the carbonates to be of chemical origin, but much of its represents "flour" formed by disintegration and abrasion of fossil debris and algal growths which developed in a shelf environment. Hambleton (1962) indicated that in the Missourian-age carbonate rocks in Socorro County, New Mexico, the dominant matrix material of back-reef facies is microcrystalline calcite ooze and "reef milk" (the latter being very fine-

TABLE 6-VIII

Components of allochthonous limestones (some are in situ products) (after Wolf, 1963b)

Descriptive-morphologic	Genetic-morphologic
pellets	faecal pellets bahamite pellets algal pellets
limeclasts	intraclasts extraclasts [1] (if rock is composed of more than 50% of extraclasts = calclithite)
oolites pisolites (= concentric fabrics) [1]	physicochemical algal weathering [2]
lumps	physicochemical oolites and pisolites algal oolites and pisolites weathering oolites and pisolites [2]
skeletons (floral and faunal)	e.g., coral, Bryozoa, Brachiopoda, and Algae
micrite	allomicrite } orthomicrite automicrite } pseudomicrite [3]
sparite and microsparite	orthosparite and orthomicrosparite pseudosparite and pseudomicrosparite [3]

[1] Of predepositional origin, i.e., from an older limestone source (Wolf, 1965b).
[2] Mostly a Recent or Pleistocene residual weathering product.
[3] Diagenetic to epigenetic product; in ancient rocks it is a penecontemporaneous product.
[4] Includes superficial oolites and circumcrusted particles.

grained, white and microcrystalline calcite), derived from abrasion of the reef core and reef flank. Edie (1958) recognized chalky (micritic) limestones in carbonates of Mississippian age in southeastern Saskatchewan, suggesting that their origin may be attributed to "flour" formed by disintegration and abrasion of fossil debris and algal growths under intense wave action in shoal areas. He stated (Edie, 1958, p. 105) that this flour "... might be expected to settle in the quiet-water environments of lagoons, intershoal areas of the shelf, and in the basinal areas." Some micrites possibly originated from "algal dust" at least in part. Wood (1941) first called attention to certain finer-grained varieties of Carboniferous limestones which he ascribed to an "algal dust" origin, thus coining the term. In applying this descriptive and genetic term, Carozzi (1960) noted that the grains themselves are angular with a diameter reaching 2–3 μ; he favored usage of the term "algal dust" when the fine-grained limestone contains associated algal tubes, or when it is

TABLE 6-IX

Limestone and dolomite classification scheme (modified after Folk, 1957, 1959; and Wolf, 1960)

Limestone						
micrite and/or sparite (%)	skeletons	limeclasts	pellets	oolites pisolites	lumps	organic in growths
	skeleton-limestone [1]	limeclast-limestone	pellet-limestone	oolite-(pisolite-) limestone	lump-limestone	coral-(algal-, etc. biolithite
10	skeleton-micrite-limestone [1] or skeleton-sparite-limestone [1]	limeclast-micrite limestone or limeclast-sparite-limestone	pellet-micrite-limestone or pellet-sparite-limestone	oolite-micrite-limestone or oolite-sparite-limestone	lump-micrite-limestone or lump-sparite-limestone	coral-micri biolithite, algal-micrit biolithite, etc. or coral-sparit biolithite etc.
50	micrite-[3] skeleton-limestone or sparite-skeleton-limestone	micrite-limeclast-limestone or sparite-limeclast-limestone	micrite-pellet-limestone or sparite-pellet-limestone	micrite-oolite-limestone or sparite-oolite-limestone	micrite-lump-limestone or sparite-lump-limestone	micrite [3]-c biolithite, micrite-alg biolithite, etc. or sparite-cor biolithite, etc.
90			micrite limestone [3]			micrite-biolithite [3]
			or sparite [3,4] (= crystalline) limestone			

[1] Use size-nomenclature, i.e., calcarenite, dolarenite, etc., instead of "limestone" wherever possible.

[2] State composition of impurities, i.e., quartzose, etc.

[3] Sparite/micrite ratios do not necessarily indicate degree of washing because the sparite may be pseudosparite. Also automicrite can form wave-resistant growths, e.g., algal micrite bioherms.

Dolomitized limestone and dolomite				
allochemical grains present			grains absent,	impurities
partially replaced by dolomite [5,6] (10—50%)	extensively replaced [5,6] (50—90%)	completely replaced [5] (>90%)	completely replaced (>90%)	present (10—50%)
dolomitic skeleton-limestone, pellet-limestone, etc.	calcareous skeleton-dolomite, pellet-dolomite, etc.	skeleton-dolomite, pellet-dolomite, etc.	aphanic, fine, medium, and/or coarse crystalline dolomite [8]	pebbly, gritty, sandy, silty, clayey, skeleton-dolomite, pellet-dolomite, etc.
dolomitic pellet-micrite-limestone [1] or dolomitic pellet-sparite-limestone [1]	calcareous pellet-micrite-dolomite [1] or calcareous pellet-sparite-dolomite [1]	pellet-micrite-dolomite [1] or pellet-sparite-dolomite [1]		e.g., sandy [2] skeleton-micrite-limestone, silty [2] dolomitic oolite-sparite, etc.
dolomitic micrite-pellet-limestone or dolomitic sparite-pellet-limestone	calcareous micrite-pellet-dolomite or calcareous sparite-pellet-dolomite	micrite-pellet dolomite or sparite-pellet-dolomite		e.g., sandy [2] dolomitic pellet-limestone, silty-sandy [2] skeleton-dolomite, etc.
dolomitic micrite [7]	calcareous dolomicrite [7]	dolomicrite	dolomicrite	e.g., clayey-micrite, dololutite, etc.
dolomitic sparite [7]		dolosparite [7]	"primary"?	

[4] Tufa, travertine, and caliche are often sparite limestones formed in situ.
[5] Note preferential dolomitization of matrix, etc.
[6] These columns are examples only. In fact any grain, colonial growth, matrix, and sparite can be replaced.
[7] Similar to limestones, dolomites range from dolomicrite to dolosparite.
[8] Dolomicrosparite and/or dolosparite may be used instead.

clearly derived from algal material. Later Carozzi and Soderman (1962) pointed out that petrographic studies of the Mississippian limestones in Indiana suggested that certain calcilutites developed from "algal dust" produced by phytoplankton. Algae are capable of precipitating micro- and cryptocrystalline calcite which, attendant upon attrition, abrasion and disintegration, yields aphanic-textured detrital lime particles which are in a sense "algal dust" (or algal allomicrite, Table 6-III).

The extent to which bacteria can precipitate directly lime ooze, which ultimately will result in micrite, is not fully understood. In his studies of bacterial precipitation of carbonates in sea water, Lalou (1957) emphasized that perhaps the role of bacteria is largely one of changing the physicochemical conditions of the medium, increasing its concentration of CO_2 up to saturation, enriching it in calcium and giving rise to an escape of H_2S by reducing sulfates. The effect of such reactions is to change the alkaline reserve of the medium, the pH, etc. It was his interpretation that the formation of carbonates by bacteria may be obtained if: (1) there is presence of assimilable organic matter in sufficient quantity, (2) the temperature is sufficiently high, (3) there is maximum light and sunshine, and (4) the waters are quiet and are seldom renewed. These conditions, he believed, are to be found in the lagoons and portions of the tropical sea water most isolated from the open seas.

During compaction of lime-mud, differential strain may result, which can vary from one depositional site to another, i.e., whether a lagoon, bank, miogeosyncline, etc. Nuclei of recrystallization will be set up giving rise to ultimate crystalline mosaic. It was pointed out by Wardlaw (1962) in his studies of diagenesis of the Irish Carboniferous limestones, that during recrystallization nuclei of strain-free grains originate at several points, the number of points increasing with time, and the strain-free grains may grow until they completely consume the matrix. Obviously, to produce a finely crystalline micrite or microsparite under such circumstances requires a large number of sites where new nuclei can develop.

Lime ooze has a high fluid content in the interparticle pore spaces. Thomas and Glaister (1960) studied porosity and facies relationships of some Mississippian carbonates in the Western Canada Basin and called attention to the fact that lime-mud, which formed in quiet-water environments of lagoons and shoal areas and which is chalky to clay-like, has a low oil-wetting ability and a high connate water saturation. Diagenetic dolomitization proceeds relatively fast in such ooze, and it would appear that dolomitization processes are strongly controlled by the presence of fluids in intergranular and intercrystalline pore spaces, particularly in those which have a high fluid content. It is noteworthy that calcium carbonate mud which precipitated as a colloidal gel, encrusting leaves of Algae, normally has a high fluid content;

during diagenesis a crypto- or micrograined limestone will form first and commonly "syneresis" cracks, joints, and primary contraction vugs will develop. Magnesium ions present in the original algal material may now be disseminated in the "algal dust" and will serve as nuclei for diagenetic dolomitization. With sufficient concentration of Mg^{2+} ions in the interparticle pore fluid, the transfer of Ca^{2+} ions out, and Mg^{2+} ions in, through the intergranular film is hastened and wholesale diagenetic dolomitization of the lime-mud can occur, particularly if additional magnesium ions are added at the interface or from the lime ooze beneath. Crypto- to microtextured chalky lime-mud that is rich in comminuted shell material, and/or cryptocrystalline or microcrystalline tests of calcareous composition, is also normally high in magnesium (from trace up to 12% and, exceptionally, more; cf. Correns, 1939). Percolating waters dissolve the calcium much faster than the magnesium (in accordance with the law of mass action) from a deposit of lime-mud composed of such detritus, and the relative amount of magnesium increases with progressive diagenesis. As noted by Sujkowski (1958), the Mg/Ca ratio approaches slowly a 1 : 1 value, with accompanying replacement giving rise to dolomite. He believed that such a diagenetic dolomite results in a much greater reduction of volume than takes place in the diagenesis of calcareous mud leading to limestone.

Numerous fine-textured limestones which petrologists may term calcilutites and calcisiltites in the field may, upon petrographic examination, be defined as micrites (Table 6-III). Originally, the sediment may have been crypto- or microcrystalline; such finely divided material (whether crystalline or grained, or both) can recrystallize by pressure-solution into a mosaic of larger crystals by the solution of the smallest, supersoluble grains and redeposition on the larger grains, or by grain growth (Bathurst, 1958). Pressure-solution is the transfer by solution of ions from a point of intergranular contact (where the crystal lattice is strained) by diffusion down the ion concentration gradient to a point of deposition on a crystal where there is no strain (Stauffer, 1962). Grain growth in limestones is defined by Bathurst (1958) as the ion transfer from one crystal lattice to another without any intervening solution. The process of ion migration in the solid state leads to the enlargement of the larger grains at the expense of the smaller. During diagenesis of a lime-mud to form micrite, particularly one containing particulate skeletal material (such as echinoderm ossicles), there will be transfer of ions with concomitant enlargement of the skeletal material. Inasmuch as crinoid and other echinoderm fragments consist of single large calcite crystals, they are commonly enlarged by the deposition of calcite in crystallographic continuity with the fragments (Stauffer, 1962). It should be pointed out here that if the lime-mud consists of finely comminuted material (by some geologists termed "matrix") in which there are embedded larger fragments, including skeletal material, diagenetic dolomitization (if such occurs) will affect the matrix material first; the particulate larger skeletal material is most resistant.

Siegel (1963) has noted that a factor that may influence diagenetic dolomitization of micrite is the polymorphic form of the calcium carbonate that is precipitated. Aragonite, because of its metastable state, should react more readily than calcite to magnesium-bearing waters to form dolomite. Vaterite is a more metastable form of calcium carbonate than aragonite and would, therefore, be even more likely to form dolomite. Zeller and Wray (1956) have demonstrated with laboratory studies that certain elements such as strontium and barium cause calcium carbonate to precipitate in the form of aragonite under conditions where the carbonate phase would normally be calcite. As pointed out in a preceding section, Siegel (1960) found that the alteration of aragonite to calcite in natural samples was inhibited by the presence of strontium, and that strontium might have to be removed before an alteration could take place. As he (Siegel, 1963) pointed out, a strontium-bearing aragonite might not react with sea waters to form dolomite, and the lithologic association observed in the geologic rock column would be limestone and gypsum, which is a relatively common pairing. "The role of the impurity ion must, then, be considered when speaking of the susceptibility of calcium carbonate to either early diagenetic or metasomatic alteration" (Siegel, 1963).

The mechanism of cementation of lime-mud during diagenesis to form micrite (as well as certain other limestones) presents numerous problems. In studying certain limestones in Indiana, Nitecki (1960) suggested these two possibilities: (1) dissolution at points of high compressive stress and reprecipitation at points of low stress; and (2) dissolution of the organically formed calcite because it is unstable for reasons other than stress, i.e., because there is an unstable amount of MgO present as impurities in the organic calcite, reprecipitation of stable calcite in pores will occur. In the first case, Nitecki noted that as long as the pore space is filled with water the grain-to-grain contact is limited; however, the existing pressures are hydrostatic except at the points of grain-to-grain contacts. The pores begin to fill gradually with precipitated calcite, giving rise to cement. As the process proceeds, the pressure becomes geostatic. Newly precipitated cement is nearer to the thermodynamic state of equilibrium than the organically precipitated, metastable calcite of the fossils. The result is a further growth of cement-like calcite in preference to the pre-existing organically precipitated crystals. Nitecki (1960) believed that, because the hydrostatic pressure is dependent upon the depth of the overburden, the solubility of limestone is higher at greater depths than at lesser depths (lower pressure). The $CaCO_3$ in solution will migrate to areas of lower pressure (lesser depths) where it will precipitate, fill the pores, and cement the sediments. The process of cementation will thus "proceed upward and will be generally accelerated because the pressure will be more geostatic in character" (Nitecki, 1960). These conclusions harmonize, in general, with statements presented herein, and add further credence to the suggestion that fluids highly charged with dissolved carbonate minerals can migrate toward the shelf area from the basin (greater depths and greater overburden) and give rise to diagenetic changes in the transition, hinge-line, or shelf lime-muds. Dolomitization does not occur because of lack of a copious supply of magnesium ions (and other factors as well), and the resultant diagenetic effect is cementation leading to lithification. As has been noted herein, dolomitization (if it does occur) does not necessarily occur in the same sediment, but the magnesium ions can migrate considerable distances through the interparticle fluids to cause diagenetic dolomitization in another realm. Perhaps this explains the presence of more areally extensive dolomites in the carbonates of Pennsylvanian and Permian age along and in the immediately adjacent shelfward portion of the Las Vegas Hinge Line in the three-corners area of Nevada—Arizona—Utah.

The occurrence of syngenetic pyrite and/or marcasite in micrites has been noted by many authors. Krumbein and Garrels (1952) pointed out that pyrite and calcite can form and be stable in an environment in which the pH is approximately 8.0 and in which the Eh is approximately −0.3. It is to be noted that the negative Eh value does not imply a stagnant environment. As emphasized by Moretti (1957), marine open-circulation conditions may exist down to the depositional interface, whereas below this level there may be a tendency toward a reducing environment due to depletion of oxygen. Thus, lime-muds beneath neritic, normal marine, open-circulation environments may have the property of the euxinic environment (Krumbein and Garrels, 1952). Moretti (1957) stated that if one assumes the depositional interface and the zero Eh level to be coincident, i.e., oxidizing conditions exist above the depositional interface whereas reducing conditions prevail below the depositional interface, then the decomposition of entombed organic matter would be anaerobic. Such decomposition would yield various products, including H_2S, and a reducing capacity would be rendered the environment and the S ion would be provided. Pyrite could form if sufficient amount of iron was introduced to the sea at the time of accumulation of the lime-muds. Syngenetic pyrite and/or marcasite would form under these conditions, and diagenetic iron sulfides could form at a later date. Iron monosulfides, such as hydrotroilite, could accumulate syngenetically, but under the effects of diagenesis would change to pyrite. It should be remembered that various strains of bacteria can cause iron to be taken into solution (such as at the provenance site) and be transferred to the depositional site where it is subsequently precipitated to react and form syngenetic products and possibly diagenetic minerals.

Diagenesis of pure lime ooze usually leads to fairly homogeneous micrite or micritic limestone as a result of compaction, with accompanying expulsion of water, and filling of pore spaces by micrite and by sparry cement. Presence of clay minerals retards the process of crystallization, and this is reflected in the texture of the indurated material. Perhaps influx into a sedimentary basin of pure lime-mud of micrograined texture for a prolonged period of time is an unusual circumstance. By the same token uninterrupted accumulation of microcrystalline lime ooze from supersaturated waters is an anomalous sedimentary feature of depocenters. Yet, limestones and "primary" dolomites (or dolomites of the restricted or evaporitic suite) of this category form thick and areally extensive members and formations in rocks of Precambrian to Pleistocene age in the Eastern Great Basin area of U.S.A. and elsewhere. These finely textured dolomites, dolosiltites, dolomicrites, micritic limestones, and micritic limestones with oolites are of particular significance in certain Permian units of the hinge-line area of southern Nevada. Thin-sections of some micrites, however, reveal presence of finely-divided organic matter (in some instances "dead oil") and micro-textured silica (not necessarily cement), indicating that other sediment was also introduced; the process of diagenesis did not obliterate the evidence.

Skeletal limestone

Rocks herein classified as skeletal limestones include those fragmental clastic and detrital rocks that have been given a number of names: bioclastic, fossiliferous—fragmental (e.g., criquinites), skeletal—detrital, and others. No single set pattern of diagenesis has been established for these sediments; grain growth, introduction of rim cement, and numerous processes collectively lumped under the catch-all term "recrystallization" occur with apparent rapidity in some skeletal limestones, and with variable speed and direction in others.

In his studies of the petrography and facies of some Upper Viséan (Mississippian) limestones in North Wales, Banerjee (1959) differentiated five limestone types, as follows: (1) shelly calcite-mudstone, (2) shelly calcite-siltstone, (3) coquina-lutite, (4) bioclastic calcarenite, and (5) crinoidal calcarenite. Petrographers will readily recognize that types (1) and (2) are transitional from the micrites on the one end to the skeletal limestones (3, 4 and 5) on the other end. His pure calcite-mudstone (grain size = 0.5—4.0 μ) is the micrite of some geologists. A limestone consisting dominantly of calcite-mudstone, but with some skeletal debris, is modified by the term "shelly". His "coquina-lutite" is a limestone, the dominant component of which is skeletal debris of sand and silt grade, which imparts to the rock a coarser texture than that of the shelly calcite-mudstone, though calcite-mudstone is present as matrix. In some respects this usage corresponds to certain nomenclature of Dunham (1962) who applied terminology of *grain-support* versus *mud-support*. Thus, if the skeletal limestone is grain-supported with minor amounts of lime-mud interstitial material, it will react to diagenesis differently than if it was mud-supported and skeletal particles were in the minority. Banerjee (1959), for example, divided his coquina-lutites into three subtypes on the basis of particle orientation and grain size, as follows: *Type 1*: without preferred planar shape orientation of skeletal particles; coarse-grained. *Type 2*: skeletal particles with preferred planar shape orientation parallel to the bedding plane, and having roughly the same grain size as *Type 1*. *Type 3:* the finest-grained of the three with more calcite-mudstone and with a preferred planar shape orientation of skeletal particles parallel to the bedding plane. As will be pointed out herein, some of these parameters of grain- and skeletal-orientation exert a significant influence on the processes of diagenesis of skeletal limestones.

Skeletal detritus is, of course, subject to abrasion and disintegration in high-energy environments which are typified by wave and current agitation and surf surge. Some of the skeletal particles may be reduced to sand-, silt-, and even clay-size grades in lower-energy environments, and the process may be to some degree syngenetic and to a degree diagenetic. For example, Dapples (1938) suggested that the continued size reduction of skeletal debris by scavengers might have produced the structureless calcilutites which are common in the Paleozoic. Ginsburg (1957) pointed out that boring blue-green Algae, although small, are extremely abundant in carbonates, and tiny filaments penetrate shell fragments. He stated that in the modern seas these organic destructive agents attack skeletal debris differentially; coral skeletons are most susceptible, whereas the dense skeletons of red Algae are most resistant. Detritus feeders such as holothurians, worms, crustaceans, echinoids, and others are instrumental in churning up sediment and in reducing coarse- and medium-textured skeletal detritus to fine-textured lime-mud.

Greensmith (1960) pointed out that in some Scottish limestones, a common feature of the fossiliferous carbonaceous varieties is the presence of early diagenetic microspheroidal and nonspheroidal pyrite which replaces the calcite shells and the carbonate of the matrix. He contended that their formation and the replacement reaction probably took place soon after burial because lenticular aggregates in the matrix sometimes show subsequent warping caused by compaction.

Reef-flank skeletal limestones are amenable to diagenetic changes; noteworthy among these are introduction of sparry calcite cement, skeletal grain growth, pressure-solution, and emplacement of rim cement. For example, Hambleton (1962) noted that in some Missourian age rocks of New Mexico the reef-flank deposits contain a profusion of gastropods, brachiopods, pelecypods, and cephalopods. A sparry calcite matrix cementing the fossil allochems suggested to him that strong local currents removed much of the microcrystalline calcite ooze. The dominant matrix material of the back-reef facies is microcrystalline calcite ooze and "reef milk" (very fine-grained, white and opaque microcrystalline calcite) derived from abrasion of the reef core and reef flank. It should be emphasized that in other occurrences of this "reef milk", the microcrystalline material can be preserved in the matrix rather than being washed out and is, therefore, subject to diagenetic changes, including dolomitization, in the proper environments.

Particulate material comprising newly deposited skeletal limestones does not react uniformly to diagenetic changes. As pointed out by La Porte (1962), the skeletons of many marine invertebrates consist of small masses of crystalline carbonate (calcite or aragonite) intimately intermixed with organic tissue. Details vary from one taxa to another. When the organic matter of skeletal material begins to decompose through oxidation or bacterial activity, the imbedded crystaline fraction is freed. Aragonite secreting corals, for example, produce upon total decomposition a type of sediment somewhat different than that produced by mollusks which may yield larger, hexagonal prisms. Diagenesis will affect one to a different degree than the other. Not to be overlooked in this assessment of diagenesis of skeletal limestones is the effect of Algae in secreting aragonite needles (Lowenstam, 1955). If a skeletal limestone consists in large measure of bioclastic algal detritus such as algal grains or lime clasts (not necessarily "algal dust") described by Wolf (1962, 1965a, b) and the debris contains an abundance of aragonite needles, it becomes obvious that the path of diagenetic alteraction will be different than in a brachiopod skeletal limestone, for example. Furthermore, presence of strontium in the aragonite may inhibit diagenesis in the algal bioclastic limestones. Another implication is "... that fossil calcilutites attributed to physicochemical precipitation or mechanically reduced skeletal carbonates may have been partially or largely derived from algally-secreted aragonite needles from ancestral Algae" (Lowenstam, 1955).

Many of the criquinites of the geologic rock record have resulted from the diagenesis of coquinas of echinoderm debris (= "criquinas"). The most obvious diagenetic process is the formation of optically continuous calcite overgrowth on crinoid or other echinoderm fragments, particularly the ossicles. Individual plates of modern echinoderm skeletons are made of optically oriented calcite crystals containing large interstices which become solid single crystals after death. The overgrowth is a continuation of this crystal as described in earlier sections. According to Bathurst (1958), this overgrowth can form by filling pore space or by replacing the lime-mud surrounding the crinoid fragments. He called the pore-filling overgrowth "rim cement" and the replacement overgrowth "syntaxial rims" (Fig. 6-2). Lucia

(1962) made a study of diagenetic effects in a crinoidal sediment in Devonian rocks of Texas and, in adhering to the usage of Bathurst, stated that the textural relationships between lime-mud and calcite overgrowth suggest that rim cementation is the dominant process in diagenesis. Of particular significance in Lucia's studies is a consideration of the effect of dolomitization of the crinoidal sediment. He noted that the original character of the sediment which was dolomitized can be reconstructed by noting how the crinoidal material was replaced; these two mechanisms were suggested: (1) a single crystal of dolomite in optical continuity with the single calcite crystal of the original crinoid fragment, a process referred to as pseudomorphic replacement, and (2) dolomite crystals not in optical continuity with the calcite of the original crinoid fragment, a process referred to as impingement. The most commonly observed mechanism in Lucia's studies is pseudomorphic replacement; he noted all stages from partial pseudomorphic replacement to complete pseudomorphic replacement with none of the original calcite left. Furthermore, he pointed out that the tendency for dolomite to replace single-crystal crinoid fragments with single dolomite crystals of the same crystallographic orientation suggests that it is difficult for dolomite to nucleate within a solid calcite crystal. The research of Lucia, which appears to be borne out by studies of many other petrographers, suggests that the sequence of dolomitization of crinoidal sediment (as observed in thin-sections) proceeds from dolomitization of the intercrinoid areas to dolomitization of the crinoid fragments. Lucia indicated that none of his thin-sections showed any dolomitization of the crinoid fragments unless the intercrinoid areas were entirely dolomite, with the exception of the small amount of impingement on their edges by the external dolomite crystals. Pseudomorphic replacement of crinoid fragments appears to take place mostly after the formation of internally impinging dolomite crystals. Lucia found no case in which the dolomite was composed solely of crinoid fragment pseudomorphs. In the rocks which he studied, the evidence proved that dolomitization occurred after rim cementation. If any of the dolomites had been composed solely of crinoid fragments and rim cement at the time of dolomitization, they would appear as dolomites composed essentially of crinoid pseudomorphs. Lucia (1962) stated that: "The presence of randomly oriented 0.1-mm dolomite rhombs between the crinoid fragments therefore discredits the argument that the intercrinoid areas had been filled with rim cement, and it implies that the intercrinoid areas were filled with a finer matrix." These arguments can be applied equally to other limestones of this category (= bioaccumulated skeletal detritus) in which bioclastic material consists of larger clasts in a matrix of smaller comminuted material. Non-dolomitization diagenetic effects will include crystallization (as well as recrystallization) of the matrix material first, followed by crystalline overgrowth (with or without optical continuity on the larger clasts). If lime-mud is also present, however, and the sediment is modally tripartite (i.e., consists of larger skeletal clasts such as crinoid ossicles, with a matrix of smaller skeletal debris, and impalpable lime-mud), the diagenesis may in some circumstances affect first the finest grade size material and then the larger particles. Certain clay minerals, and some clay-size particles, in the lime-mud may inhibit dolomitization there, but permit diagenesis to proceed directly to the matrix material and finally to the ossicles or other skeletal elements. Furthermore, leaching of the lime-mud may occur after rim cementation, thereby indicating that the interparticle lime-mud remained permeable to water. Lucia (1962) stated this thusly: "Where interparticle lime-mud was present, it inhibited the development of the calcite overgrowth and was available for selective leaching to form the visible porosity. The leaching process was not as effective where the lime-mud was supporting the load as where the crinoid fragments were supporting it." Consequently, skeletal grain-supported limes would differ in diagenetic effects from those that are mud-supported. The amount of porosity that develops during (or through) dolomitization may be related to the ratio of mud to crinoids or other echinodermal material, that is, the sediments containing the most echinoderm bioclastic material have the highest resultant porosity (see Lucia, 1962).

In his studies of the Mississippian carbonate deposits of the Ozarks, Moore (1957) pointed out that diagenetic effects were not limited merely to development of interlocking grains and to infiltration of fine calcareous mud, but were largely accomplished by precipitation of crystalline calcite out of solution. He demonstrated that none of the edges of crinoidal and other grains indicate the effect of solution, and thus concluded that most, if not all, of the cementing calcite was derived from the interstitial waters and not from the grains themselves. Secondary calcite was observed to occur as an approximately equigranular mosaic which lacks crystallographic continuity with adjacent crystalline echinoderm fragments. Moore added: "Lithification has been effected by compaction and calcite welding, not by recrystallization, although some secondary crystalline calcite is identifiable in various rock samples."

Lithoclastic (= detrital) limestone

Rocks not classified with skeletal types or with micrograined micrites can be termed lithoclastic (= "detrital" of Leighton and Pendexter, 1962; "intraclasts" and "calclithite" fragments of Folk, 1959; "limeclasts" of Wolf, 1963b, 1965b;) if the components are of calcareous composition and have been worn or reduced by attrition to yield a clastic texture. Some sediments that obviously formed by aggregation have been added to this category by Folk (1959), for example. Many calcarenites (particularly nonskeletal types) are to be classified in this group, and certainly many of the calcirudites and dolorudites are included here. During the various processes of diagenesis, newly deposited sediment of this large group is converted to pre-lithified and juxta-lithified equivalents by introduction, cementation and compaction of interstitial material, formation of crystalline material during authigenesis, development of coated grains and overgrowths of allogenic sediment, and possibly near-complete to complete recrystallization leaving only vestiges (= relics or "ghosts") of skeletal and/or litho-detrital material. Lithoclastic, detrital, or limeclast limestones have been termed mechanical limestones by some workers.

Andrichuk (1958) studied Late Devonian sedimentary carbonate rocks of central Alberta, Canada, and pointed out that the calcarenites, calcisilites, and calcilutites were formed primarily by two main processes as follows: (1) mechanical disintegration of organic skeletal material and redeposition as bioclastic limestone at, near, or a considerable distance from the original site of organic growth; and (2) chemical or biochemical precipitation of calcium carbonate in quiet or agitated waters in association with, or separate from, sites of active organic growth. He contrasted the two types, and compared the latter variety with the *bahamites* of Beales (1958), which are present-day deposits of the interior areas of the Bahama Banks (or ancient counterparts) and which are considered to consist predominantly of precipitated material that has aggregated into granules and composite grains (see also Illing, 1954). Andrichuk (1960) termed some of the calcarenites "pseudo-oolites", and indicated that: "pelletoid or pseudo-oolitic calcarenites and calcilutites may have formed by precipitation in a slightly supersaline environment in the interior of a bank as compared with the more normal salinity of waters in which bioclastic limestones were deposited" (cf. Illing, 1954; Beales, 1956). Among the diagenetic dolomites which Andrichuk (1960) recognized are those that have microsucrosic to coarse textures; he believed that diagenetic dolomitization of calcisiltites and calcarenites (whether of bioclastic or lithoclastic origin) accounts for these varieties. He stressed their significance by stating (Andrichuk, 1960): "The coarser dolomites with crystal size greater than 1/16 mm are considered to be of secondary origin where dolomitization occurred penecontemporaneously with deposition or at any time thereafter. These dolomites comprise the potential petroleum reservoirs." In discussing diagenetic effects of carbonate rocks of Mississippian age in the Lisbon area of the Paradox basin of the Four Corners area, Baars (1962) stated: "There, diagenesis has greatly increased the reservoir potential because of the solution of crinoid columnals. Diagenesis is of primary importance to petroleum geologists because of this close relationship with porosity."

Bathurst (1959b) studied diagenetic effects in Mississippian calcilutites and pseudobreccias in limestones of England and Wales, and indicated that in any limestone the *matrix* is an accumulation of one or more types of grain mosaic. He noted the presence of three dominant mosaics as follows: (1) granular cement and drusy mosaic, (2) rim-cemented single crystals, and (3) grain growth mosaic. He stated: "The calcilutites in their simplest form are rim-cemented carbonate muds or silts. Commonly, however, the original sediment was composed of aggregates of mud or silt, either faecal pellets or "grains" similar to Illing's (1954) Bahaman sands." He also noted that in some limestones grain growth mosaic is common and forms the pseudobreccias where the "fragments" are masses of grain growth mosaic which lie in a "matrix" of less altered limestone. Bathurst (1959b) also defined a mud aggregate ". . . as any aggregate of mud grains, usually having the size of a sand or silt particle, which has been mechanically deposited. Initially the aggregate may have been a faecal pellet (Eardley, 1938; Illing, 1954), or a rounded, subspherical aggregate of mud grains cemented originally by aragonite with no signs of organic control (as Illing's, Bahaman sands, 1954, et seq., which lithify to yield the Bahamites of Beales, 1958), or a fragment of algal precipitate (Wood, 1941; George, 1954, 1956; Lowenstam, 1955; Lowenstam and Epstein, 1957; Wolf, 1965a, b), or a spherical or ovoid growth form of a calcareous alga (Anderson 1950)."

Detrital, lithoclastic or limeclastic limestones, which contain larger clasts embedded in a matrix of finer detritus, commonly display variation in diagenetic effects, particularly those of dolomitization. Crystallinity of the matrix of calcarenites and calcirudites normally is coarser than that of the grains. Dolomitization appears to select the matrix in preference to the grains which may remain unaltered. Beales (1953) observed this effect in studying dolomitic mottling of Devonian limestones of Alberta, Canada, and considered that dolomitization took place at a time when the grains were still embedded in relatively porous mud. He stated: "The Palliser formation, laid down as limestone that was possibly magnesian, was subsequently altered to dolomitic limestone at an early stage in diagenesis. Secondary alteration and recrystallization produced the dolomitic mottling now so conspicuous in the rock." It was his contention that dolomitization began in the more susceptible centers, triggered further diffusion, and permitted dolomitizing solutions to spread. Dolomitization, he discovered, in the lower beds was localized along certain bedding laminae and spread irregularly from them; higher in the succession "worm burrows" and "Algae" were most affected.

Many calcilutites have apparently been diagenetically altered to a microcrystalline mosaic of interlocking anhedral crystals, from about 5 to 20 μ in average crystal size (= microsparite, Table 6-III). Calcarenites can alter to similarly-appearing rock, the crystalline mosaic of which is coarser textured (= sparite, Table 6-III) than that of altered calcilutites. In other words, detrital (lithoclastic) rocks can under the effects of diagenesis become finer textured but will have an interlocking anhedral mosaic as suggested earlier. These effects, however, are common (but not limited) to morè or less equigranular calcilutites and calcarenites. Diagenetic effects on these clastic carbonates, in which relatively larger grains are embedded in a "matrix" of finer texture, differ in that the matrix normally crystallizes to subhedral and euhedral forms that are not necessarily interlocked. Impingement and suturing may occur, however. Grain growth, authigenic overgrowth (including optical continuity with original clasts), rim cement, pressure-solution, and syntaxial rims are common diagenetic effects.

Chanda (1963) studied the effects of cementation and diagenesis of the Lameta Beds (Turonian) of Lametaghat, M.P., India, and noted that silicification starts as advancing fronts from the peripheries of the detrital grains and continues to grow at the expense of interstitial calcite, in the case of calcareous sandstones. In sandy limestones, however, silicification was not as extensive or as systematic. Chanda pointed out that the Lameta limestones are sandy microsparites, where the microsparites did not result from primary precipitation but have formed by aggrading recrystallization of micritic calcite. Diagenesis of these limestones, he noted, involved both selective and what he termed "perversive recrystallization". The microspars which developed on the floating clastic grains are always water-clear, whereas areas free of clastics are in places occupied by coarsely crystalline anhedral cloudy microspars.

Perhaps many lithoclastic limestones have experienced various degrees of diagenesis, not necessarily in uniform process, or as a continuum. Folk (1959) and Chanda (1963) offered certain criteria as evidences of recrystallization of microcrystalline calcite; because they are applicable to many lithoclastic limestones (as well as some micrites and skeletal limestones), they are repeated here: (1) the looseness of packing of clastic grains reguires aggrading recrystallization of microcrystalline calcite; (2) uniformity of size of the microspars of calcite; (3) patches of microspars grading by continual decrease of grain size into areas of normal microcrystalline ooze; (4) microspars have a radial fibrous form oriented perpendicular to the surface of clastic particles as an outwardly advancing aureole of recrystallization, and (5) relic patches of microcrystalline calcite and partially warping quartz grains, embedded in a mass of mosaic of microspars.

Pelletal and coated grain limestone

Limestones herein classed as pelletal and coated grain types include the faecal pellet and other pelletal limestones, and various oolitic and pisolitic types. Although many of these are intimately associated with reefal limestones on the one hand, and with detrital (lithoclastic) limestones on the other, they are treated separately here because diagenesis does not necessarily affect them as it may the other two. They may form in extremely shallow to moderately shallow waters, and normally develop best in agitated waters although some subtypes may form in quiet water environments. Many oolitic limestones develop in or adjacent to the reef complex where they are subject to movement by trans-reef currents as well as to surf-surge and wave activity. Diagenetic effects range from simple boring by Algae to complete obliteration of primary features during dolomitization, as well as complete silicification with faithful reproduction of internal details. Some of the particulate material in these limestones displays excellent concentric and/or radial features, whereas others are pseudo-oolites, sub-round to sub-spherical pellets, and superficial coated grains. Each reacts quite differently to diagenesis. Pelletal (= pelletoid) and pseudo-oolitic limestones may undergo a certain spectrum of diagenetic effects yielding a final product not too dissimilar to certain calcarenites; in fact, petrographic distinction may be difficult in some instances. Some may, in verity, resemble the "mud-aggregate" limestones of Bathurst (1959b).

One of the first effects of diagenesis on oolitic limestones is the development of water-clear to semi-transparent sparry calcite. Subsequently this orthosparite can be altered to an interlocking anhedral to subhedral mosaic of pseudosparite.

This mosaic can, by impingement, invade oolite envelopes (i.e., peripheral rings) and ultimately may take over all the rock. A "negative" relic of the oolite may remain, however, and display a dusty ring around the unaltered to slightly altered core or nucleus of the ovoid.

Petrographers are particularly concerned with dolomitization of pelletal and coated grain (= oolitic) limestones, if for no other reason than to evaluate reservoir potentialities. When carbonate sands composed of oolites and/or pisolites retain their original interparticle porosity, they are excellent reservoir rocks. If lime-mud, sparite, or other material binds and cements the particles, the resultant limestone may be devoid of effective porosity. Some oolite units have been subjected to leaching, and although the "rind" remains relatively unaffected, the nuclei are removed by solutions and a rock having considerable porosity results.

Beales (1958) made a careful study of various ancient carbonate rocks of Canada, and compared them to Bahaman type limestones. Inasmuch as aragonite is more susceptible to alteration than calcite, he pointed out that present-day Bahaman deposits, which are aragonitic, are subject to recrystallization. It was his contention that oolites show varying susceptibility to dolomitization; the matrix is most readily altered, followed by bahamite cores, oolitic envelopes, and coarsely crystalline skeletal cores, in that order. Regarding these bahamites and oolites, Beales (1958) stated: "Direct precipitation of calcium carbonate from sea water resulting in the formation of bahamites, or under more active water conditions of oolite, has probably formed very considerable thicknesses of limestone that occur throughout the geologic column from Late Precambrian to Recent time." Beales argued that a theory of aragonite needle agglutination for oolite growth is more satisfactory than one of direct precipitation. If true, dolomitization of such oolites may proceed with rapidity in some instances. It is to be remembered, however, that the conversion of aragonite to calcite may be a slow process under certain conditions. If protected by a covering of stable calcite, aragonite may be stable for a long period before inverting to calcite.

In areas of incipient dolomitization, oolites sometimes have dolomite rhombs concentrated in their nuclei (Brown, 1959). If oolites are encased in a calcarenitic matrix, then perhaps during recrystallization of this matrix material the periphery of the aragonitic oolites was converted to calcite. According to Brown (1959), when this protective layer was formed, the aragonite in the centers of the oolites remained unaltered until the oolites were fractured during compaction. Metastable aragonite material in the nuclei of the oolites then constituted natural foci for dolomitization. Brown's studies were concerned with diagenesis of a Late Cambrian oolitic limestone in Montana and Wyoming, but the principles are nonetheless worthy of consideration in petrographic investigations of other diagenetically altered oolitic limestones.

Edie (1958) made rather intensive studies of sedimentation of the Mississippian Mission Canyon and Charles formations in southeastern Saskatchewan, Canada, concluding that four environmental types are represented: (1) basin, (2) open marine shelf, (3) barrier bank, and (4) lagoon. Pisolitic, oolitic, and pseudo-oolitic (= pellet) limestones characterized the barrier banks. He observed that the pseudo-oolites are calcareous pellets composed of cryptocrystalline material and are similar in size to oolites but lack concentric layers. He believed that some of these pellets are chemical precipitates formed on the sea floor under moderately agitated water conditions similar to the calcareous sands of the Bahama Banks described by Illing (1954); but that some, if not most, of the pseudo-oolitic limestones may be largely of algal origin, and possibly represent both accretionary algal grains and "bioclastic" material formed by the fragmentation of algal colonies in areas of intense wave action. Wolf (1963a, 1965a, b) arrived at similar conclusions. If dolomitization affects sediments of the type described by Edie (1958), a rock having an earthy to sucrosic texture would likely result. It would have intercrystalline and interparticle porosity and commonly would contain dolomitized positive relics of fossils or fossil debris as well as relic oolites containing dolorhombs in the nuclei.

The petrographic studies of the Oil-Shale Group limestones of West Lothian and southern Fifeshire, Scotland, by Greensmith (1960) are quite informative. Oolitic texture is very common in all limestones of the group, and the carbonate of the ooliths appears to be an iron-rich dolomite (ω = 1.679—1.683), commonly set in a matrix of similar nature. Partial breakdown of this mineral to limonite during weathering gives many of the beds a distinctive light brown surface color. Coarse calcite euhedra are not common in the matrix, but internal pressure-solution effects have produced microstylolites. Evidence for dolomitization is almost neglibile and is expressed in the form of irregularly shaped, small transgressive vugs up to 1.2 × 0.10 mm in size. These sporadic cavities have a lining of coarse subhedral dolomite and often have a subsequent infill of a kaolinite-like mineral. In the words of Greensmith (1960): "In thin-section the coarse clear dolomite is seen to grade into the Fe-rich dolomite grains of the matrix which suggests that it represents a localized solution and reprecipitation effect hardly akin to the wholesale metasomatic changes associated with true dolomitization." It was noted by Greensmith that intimately intermingled with the ooliths in many of the limestones are similarly shaped and sized bodies to which the term "oolitoid" was applied. They lack the internal structure normally found in the oolites and consist of a fine-grained aggregate of iron-rich dolomite. Greensmith ruled out an origin due to recrystallization of oolites, as well as one associated with faecal pellets, but rather considered them to represent cross-sections of tubes that presumably resulted from activities of organisms such as worms. Seemingly, the presence of pelletal, oolitic, pisolitic, algal circumcrusted (Wolf, 1965b), and other noncoated, coated, and superficially coated grains in limestones, even to the point of comprising most of the rock, has been a point of dissension among some petrographers concerning diagenetic changes. Some regard one type as the alteration product of another. Many geologists regard each type as a distinctive sedimentary product, penecontemporaneous with sedimentation of the host rock. Terms like "spherulite", "axiolite", "ooloid", "oolith", etc. have been coined to define some of these coated grains. It is important, however, to make a distinction between a "superficial oolite" in which most of the particulate material consists of "superficial ooliths" with only thin external oolitic layers, and a true oolite which is dominantly composed of "ooliths" with well developed concentric structure. Most of the modern Bahaman oolitic sands are composed of superficial ooliths (Illing, 1954).

In the course of geological investigations of sedimentation in the Bimini, British West Indies region, Kornicker and Purdy (1957) discovered an area in the Bimini lagoon in which at least 90% of the sediment is composed of a single type of faecal pellets. The delicate faecal pellets were preserved because of extremely low agitation and current activity, scarcity of scavengers, and bacteriological precipitation of aragonite within the pellets. Of real significance, however, is the fact that during emergence at low spring tides desiccation results in permanent hardening of the pellets. The studies of Kornicker and Purdy, though suggestive, point up the importance of bacteriological precipitation of aragonite, and hardening through desiccation, in early diagenesis of various carbonate sediments. One can readily appreciate the importance of these early diagenetic changes leading to various types of limestones.

Newell and Rigby (1957) pointed out that faecal pellets, ooliths, ovoids, and a variety of grains termed "lumps" by Illing (1954) make up most of the bottom sands over great areas of the Bahama Banks. Some of the friable aggregates are bound together by algal mucus, whereas others are held together by calcium carbonate cement. Illing has identified these particles in all stages of cementation. As indicated by Newell and Rigby (1957): "They become firmer by precipitation of aragonite cement within the aggregate and as they are rolled about they lose their irregular shape, and the final grain, composed chiefly of cryptocrystalline aragonite, shows but little evidence of the original composite nature. Recrystallization of the fine detrital constituents takes place concurrently with cementation, quickly destroying the original texture." Newell and Rigby stated that Thorp (1936) probably was the first to record the large quantity of faecal pellets in the Bahamian sands and muds around the Andros Island. When fresh, the pellets are friable aggregates of fine detritus held together by mucus. They very soon become firmly bound together by aragonite cement, deposited perhaps through bacterial activity (Illing, 1954). They become finely crystalline, however, as a result of crystallization of the finest material.

Rusnak (1960) studied Recent oolites forming in the hypersaline environment of the Laguna Madre along the southern Texas coast and considered that the rate of carbonate precipitation and mechanical reorientation may very well be the controlling factors of primary crystalline orientation within oolites. He indicated that with rapid precipitation, individual needles may not assume a preferential orientation on the nucleus and thus will result in unoriented carbonate deposition. But with slower rate of precipitation they may become oriented radially, as in artificially precipitated oolites or spherulites (cf. Monaghan and Lytle, 1956; Lalou, 1957). It has been contended that where precipitation rate is very slow, crystallites become attached tangentially to the nucleus by rolling or agitation, or even become bent by mechanical rubbing. Rusnak (1960) stated: "... tangentially oriented oolite layers must be subjected to relatively high crystalline strain during the bending process. These strained crystallites may thus be more susceptible to recrystallization by diagenetic processes in response to a release of acquired strain."

Published reports on petrography of some limestones contain references to what is known as "granular" limestones. Many of these are not in the real sense of the word "grained" as pertains to lithoclastic or detrital limestones, but actually represent a stage of diagenesis of what were originally pelletal, oölitic, pisolitic or superficially coated granular limestones. Some such limestones are well sorted, and do contain a significant (but not dominant) proportion of crinoidal and algal material formed by attrition. Oolitic and related coated-grain material that formed in current-agitated waters is a dominant component. Perhaps floating, calcareous, planktonic Algae (Coccolithophoridae) contributed to some of the finely-divided, even microgranular, matrix material in which the oolites and ovoid bodies are embedded. If certain skeletal elements comprise the space between packed granules, porosity and permeability values may be high (Thomas and Glaister, 1960). Limestone of this category is particularly susceptible to diagenetic changes and, when dolomitized, gives rise to a rock having a crystalline-granular texture, sometimes with a reduced porosity and permeability.

The generalized term "recrystallization" has often been applied to diagenetically altered oölitic, pisolitic, and pelletal limestones, without specific reference to details of the alteration. Bathurst (1958) recognized two types of cement, for example, depending on whether the cementing material grew into void space or replaced the carbonate mud. Cement which develops into interparticle voids may be optically continuous on single crystal particles (e.g., crinoid ossicle) and thus be termed *rim cement.* This may be difficult to determine petrographically on some oolites and pisolites, however. The cement may consist of small crystals commonly oriented perpendicular to the void walls and give rise to fibrous and/or drusy cement (Fig. 6-2). This should be looked for in coated-grained limestones, particularly in those where the matrix has been dolomitized and created additional space in which the fibrous and/or drusy cement forms in the next step or phase of diagenesis. Furthermore, if two generations of dolomitization of the matrix material are represented, continued development of cement may preferentially form a coarse mosaic. Limestones of the type herein discussed may have lime-mud also present. If the cement occupies space previously taken up by the carbonate mud (but which has been washed out or leached away) and continues to enlarge pellets, granules, oolites, pisolites, etc., then *grain growth* may be instituted in some cases. Again, such phenomena are to be looked for in thin-sections.

Reefal limestone

Diagenesis of reef limestones, bioherms, biostromes, and comparable rocks built by wave-resisting organisms in the marine and lacustrine environments normally includes introduction of interstitial material, as well as many of those changes indicated for limestones discussed above. Because reefal limestones have already constructed a hard and relatively compact framework, diagenesis may be somewhat different in contrast to other limestone types. It is also true that some finely comminuted material (i.e., calcilutite, calcisiltite, calcarenite) may still remain in or near the reef framework after abrasion and disintegration of some of the reef rock, and this material is particularly susceptible to diagenetic changes. Lowenstam (1955) has stated that some of the calcilutites attributed to physicochemical precipitation may have formed by breakdown of calcareous Algae, particularly the poorly calcified forms. In writing about the "white reef" in certain Devonian reefs of Canada, Belyea (1955) stated that much of the reef mass consists of fine-grained comminuted organic debris, and that much of it is white dense limestone probably formed in large measure by lime-trapping Algae. Various references have been made to the "aphanitic" reef limestones (Hadding 1941, 1950; Henson, 1950; Wengerd, 1951; Newell et al., 1953; Wolf, 1962, 1965a, b, c). Possibly some of this fine-textured aphanic limestone within or near reef cores represents chemical or biochemical precipitates and products of recrystallization that brought about loss of the original texture as pointed out earlier.

Various workers have investigated dolomitized Devonian reefs in Alberta, Canada; Andrichuk (1958) stated: "... the threshold between a dolomitizing and non-dolomitizing environment appears to be very subtle and sensitive, and only a slight change in one of the factors affecting dolomitization may be sufficient to promote complete dolomitization in a limestone province ... less intense agitation and aeration and a less oxidizing environment would be more suitable for penecontemporaneous dolomitization ..."

A significant body of factual information is available concerning direct precipitation of calcite from marine and lacustrine waters, and the source of the calcium carbonate precipitated early in primary pores and interstices of reef limestone can be easily accounted for. Newell (1955) stated: "Surface waters, which are supersaturated with calcium carbonate, are warmed in the daytime over shallow reef flats several degrees above the waters of the open sea, and the solubility of the carbonate is further reduced by photosynthetic activity of reef plants. During ebb tides these reef-flat waters form a hydrostatic head a few inches above the surrounding sea ... Part of this water escapes seaward by sinking through the myriads of pores which riddle the reef flat, and calcium carbonate probably is deposited in transit." In the reefs which Newell studied there is an abundance (locally as much as one half of the rock mass) of fibrous calcite that is deposited over the surfaces of the frame builders. The prismatic structure of the calcite is radial with respect to the depositional surfaces. Newell noted that, in practically every example, deposition of the fibrous calcite clearly occurred in primary voids of the reef frame at a time when prevailing conditions prevented simultaneous accumulation of detrital sediment. Calcite was deposited directly from solution, and any remaining voids were filled by detritus. Identical features have been studied in detail by Wolf (1965c). It may be argued by some workers that the above-mentioned processes are not to be classed as diagenetic, but are syngenetic. Still others may favor a term such as "syndiagenetic", but this is to a certain degree only a play on words. Newell et al. (1953) stated: "Diagenesis, as illustraed by the Capitan reef complex, is chiefly the result of interactions between sediments and the fluids contained within them. Other factors, largely responsible for these reactions but also partly contributing independently to diagenesis, are biotic activity within the sediments, compaction, and the migration of ions and fluids." These workers indicated that the changes which take place below the temperature and pressure levels of metamorphism s. str. are considered to constitute the processes of diagenesis. Furthermore, because organic frame builders are in a sense lithified prior to most of the diagenetic processes, Newell et al. indicated that lithification is only one result of the processes which bring about postdepositional changes; it is too gradual a process, they pointed out, to restrict diagenesis (insofar as reefs are concerned, at least) to those changes which affect a sediment after deposition and up to, but not beyond, lithification. It is beyond the scope of this chapter to review all facets of diagenesis of the Captian reef complex and associated rocks, and so the interested reader is referred to Newell et al. (1953, chapter 6).

Petrologists studying the fabric of some reefs have called attention to the presence of "reef tufa", a particular variety of which is called "stromatactis" (Wolf, 1965c). Parkinson (1957) studied Lower Carboniferous reefs in northern England, and stated: "The most characteristic feature of the reefs, apart from the anomalous dips and the non-bedded nature of the calcite mudstone which comprises much of the rock, is the abundance of fibrous calcite, the "reef tufa". He made reference to occurrence of "reef tufa" in Permian reefs of western Texas, as reported by Newell (1955), but had this to say of the English reefs: "When it is recognized that algal remains are readily obliterated by recrystallization or disintegration, it seems possible that such organisms may have been of some importance in the English reefs. In this connection it is noteworthy that calcite muds, according to Wood (1941), might have originated from disintegrated algal deposits." Black (1954) referred to reef-like structures in various parts of the world where evidence of frame-building organisms is slight, but in which calcite mudstones are prominent. Furthermore, he suggested that recrystallization of algal skeletons could give rise to calcite mudstones of the knoll-reefs of England.

From the foregoing, it is apparent that petrologists and petrographers must exercise extreme caution in assessing diagenetic changes of reefal limestones. For example, one worker may term interstitial sparry calcite, that fills voids in frame-building organisms, recrystallized calcite, and it actually may be open-space sparite. Evidence for the latter conclusion is based largely on the work of Bathurst (1958) and was discussed in detail in earlier sections. Lime-mud, commonly detrital rather than directly precipitated crystalline cement, usually accompanies sparry calcite or granular cement development in voids of reefal limestones. Some of this lime-mud undoubtedly is the detrital infilling of remaining pore space as discussed in detail by Wolf (1965c), and some may actually represent aragonite needles secreted by Algae in an organic framework (Lowenstam, 1955); or it is an algal slime formed on or near algal plants: $CaCO_3$ precipitated as the plants extracted carbon dioxide from immediately adjacent sea water (Pray, 1958). The latter conclusion is worthy of further investigation, particularly for the bearing it may have on dolomitization of material within the reef *between* the originally formed organic framework (see Wolf, 1965c). Schlanger (1957) pointed out that during deep drilling operations on Eniwetok Atoll a dolomitized core was recovered from a depth of 4,078—4,100 ft, and subsequently was determined to be of Eocene age. Dolomite in much of the core is restricted to rod-shaped segments of articulate coralline Algae, identified as *Corallina.* Schlanger stated: "The dolomite crystals are definitely restricted to the Algae and their growth seems to be controlled by the shape of the rod. The area near the axis of the rod is occupied by a fine-grained mosaic of anhedral dolomite that grades outward into coarser, more euhedral crystals." He further noted presence of fine detrital filling (finely comminuted algal particles) in interseptal areas of unaffected corals in the core and termed it "paste" fill, which probably served as centers for dolomitization. Schlanger argued that there may be regions within the algal fragment in which the Mg-ion concentration is in excess of the "average" for the entire fragment. He further stressed the findings of Chave (1954) that considerable variation exists within a single algal colony, possibly due to seasonal temperature variation, or influence of metabolism of the Algae in inducing short-term high uptake of magnesium. Thus the Alga, as it grows, may contain disseminated dolomite nuclei whose size limits are too small to be detected even by X-rays. With time the dolomite nuclei, which evidently are more stable than the surrounding Ca—Mg solid solution under existing conditions, enlarge by diffusion of the originally adsorbed Mg ions in the structure and by addition of Mg ions from sea water (Schlanger, 1957).

Schwarzacher (1961) studied the petrology and structure of some Lower Carboniferous reefs in northwestern Ireland, and mapped "knoll-reefs" in detail. He noted that the reef limestone consists of clotted fine-grained calcareous mud (which he termed bahamite) that contains larger mud pebbles. Bryozoans were the only frame builders of significance. It was pointed out that during early diagenesis a cavity system was formed, probably due to sliding movements on the reef talus. This was soon filled with calcite and dolomite crystals, during a stage when the reef was still a mound on the sea floor. Lithification set in at a later stage and led to a preferred orientation of calcite grains. The relatively large volume of the cavities was filled first by calcite, but as pointed out by Schwarzacher: "Almost all cavities show some dolomite; if the cavities are lined with fibrous calcite then the dolomite crystallizes later; if fibrous calcite is missing then the dolomite may form the first lining on the calcite mudstone wall. Most dolomite occurs in well defined rhomb-shaped crystals whereby a definite growth relation of the crystals to the wall exists. . . . Most commonly, the 'c' axis is parallel with, and the longest diagonal of the rhomb is at right angles to, the wall." Wolf (1965a, c) has presented details on the Devonian algal reef-knolls of the Nubrigyn complex.

Not to be overlooked in any discussion of diagenesis of reefal limestones are the results of collapse attendant to removal of soluble materials such as evaporites in or adjacent to the reef. Some of these "evaporite-solution breccias" can be confused with "reef-edge breccias" (Greiner, 1956) and only through detailed petrologic and petrographic studies can the correct assessment be placed on diagenesis. Both types, it is true, are subject to dolomitizing solutions that migrate "updip" from the basin adjacent to the reef tract. The chaotic jumble of the blocks comprising the "solution breccia", however, do not resemble the rubble of frame-building organisms, which grade perceptibly into bioaccumulated calcarenitic material of the "reef-flank" or "reef-edge" breccias. As has been pointed out herein, the voids of these breccias can be filled partially or wholly by fibrous, drusy and/or granular sparry calcite. Should any space remain, "reef milk," "algal dust", or detrital, chemical, biochemical, or physicochemical carbonates may fill it. Complete to near-complete dolomitization of the entire rubble is not uncommon in some of the reef-talus material and collapse-breccias.

Sparry limestone

Petrographers have differences of opinion regarding origin of certain crystalline limestones and dolomites, particularly if lithologic association (e.g., reefs) does not contain suggestive data. To some workers, the formation of coarse sparry limestone is a function of metamorphism and, thus, outside the realm of diagenesis. When a limestone having such a texture is found interbedded in a sequence of limestones and other sedimentary rocks which are not metamorphosed, however, and some in verity are only slightly diagenetically altered, the evidence clearly suggests that diagenesis and epigenesis can create such a texture in limestones.

Sparry calcite cement, particularly in cores of reefs or adjacent thereto, and as infilling of mollusk and brachiopod shells, is quite common and was treated in large measure in preceding pages. Units adjacent to micrite reef cores commonly are composed of skeletal detritus (including reef-talus breccia) that is cemented by sparite. In some reefs, the coarsely crystalline material is composed of dolorhombs (dolosparite) and if of the open-space variety it may line cavities or void walls. Farther from the cores, however, an intermixing of sparry calcite-cemented skeletal detritus and sparite-in-calcarenite may occur, and still farther out micrite may be found (Cronoble and Mankin, 1963). The energy factor is an important one in development of sparry calcite: if one disregards the reef cores, the progression away from the cores through sparite (i.e., sparry cemented talus) to interdigitated talus, calcarenitic material, and finally micrite indicates in most instances a decrease in the energy in the depositional environments (Cronoble and Mankin, 1963).

Grain growth and recrystallization, pressure-solution, syntaxial rim cementation, and drusy and fibrous sparite cementation may work not only independently, but also one in harmony with at least one of the others, to form sparry limestone from rock which was not originally a sparite. One example may be the encrinal limestone (= criquinite) in which sparry calcite ultimately develops, particularly under slight differential load-stress, engulfing the crinoid fragments. The end result may be authigenic overgrowth only; or it may continue to near-complete recrystallization (i.e., combination of grain growth, recrystallization and cementation, all aided by pressure-solution), with obliteration of all but "negative" relics of the echinoderm fragments. A similar process possibly operates on some faecal pellet limestones and upon bahamites. Cements may completely engulf areas of fine carbonate mud, in particular the so-called "drewites" or algal-precipitated aragonite needles, with fine interparticle porosity. Coarse sparry calcite can result from this diagenetic process. If intercrinoidal voids are filled by sparry calcite which is enlarged by impingement or forms continuous overgrowths on monocrystalline carbonate fragments, such as crinoid ossicles, sparite will result. Should dolomitization be the process, complete obliteration of the original texture can ensue, or it can be "arrested" in some stage, giving a mottled or even coarse sucrosic appearance to the rock; fossils and/or other particulate material may be "positive" (recognizable as to organic remains or inorganic fragment), or they may be "negative" (= strongly suggestive of a vestige of a former fossil or other fragment, but proof lacking).

A most important consideration in the process of dolomitization to form dolosparite is that in most instances the rhombs grow by replacement as opposed to growing as a cement. Exceptions are to be noted, however. If, for example, a lime-mud contains particulate skeletal (i.e., crinoidal) or detrital (i.e., fragmental limestone clasts) material "floating" or embedded in the ooze, the mud is preferentially replaced first, and the particles may follow and be dolomitized in the order of their susceptibility to dolomitization. Cements occupying spaces previously occupied by carbonate mud may grow by replacement in optical continuity with a large host or by grain enlargement of the smaller particles to form a coarse mosaic of anhedral to euhedral crystals (Murray, 1960). Reduction in porosity by pressure-solution may be an important mechanism, yielding a limestone (or dolomitic rock or dolomite) that has an interlocking texture. If it is an interlocking mosaic of anhedra, in all likelihood it is a calcspar; whereas if it is composed of subhedral to euhedral grains, it may be a dolospar.

It is important to note that diagenesis does not always operate to form larger crystals and, thus create sparry limestones. Disregarding for a moment the process of recrystallization, one should be cognizant of the strong possibility of some calcarenites to be transformed during diagenesis into a rock with silt-sized calcite crystals (= microsparite) and realize that the calcisiltite is a product of alteration and not a primary detrital (that is, lithoclastic) limestone (Wardlaw, 1962). The petrographer should also be aware that recrystallization of large calcite crystals, whether infilling of voids or as interstitial sparry cement in reef framework, does not always result in fine-grained textures. Grain growth can readily produce sparite; some thin-sections show strained, twinned calcite crystals that have untwinned, unstrained rims of calcite in optical continuity with one of the sets of twin lamellae.

Certain bioclastic as well as lithoclastic limestones, particularly biocalcarenites and lithocalcarenites, are susceptible to pressure-solution that results in cementation by a sparry calcite or, ultimately, under proper conditions, by dolosparite. During the diagenetic processes, the larger fragments become etched and corroded around their boundaries and crystallized into subhedral or euhedral mosaic, or into sparite the discrete rhombs of which are in optical continuity with the host particles. Some grains invade other grains at contact points (Towse, 1957).

Thomas and Glaister (1960) in discussing facies and porosity relationships of certain Mississippian carbonate rocks of western Canada, stated: "With regard to the relation of dolomite development to textural features of original limestones, it has been observed that it preferentially occurs in open pores or in matrix (chalky, granular, and carbonate mud) material that surrounds the larger skeletal or non-skeletal grains. These larger grains are generally the last to show conversion to dolomite. Many skeletal fragments remain as calcite even when the remainder of the rock may be dolomite. The final type in this sequence is a dolomite with fossil casts." Among Pennsylvanian and Permian fusulinid-bearing limestones that have been dolomitized, one can commonly discover bank deposits and hinge-line accumulations of fusulinid coquinites in which interlocking subhedra of dolosparite are riddled with spindle-shaped cavities (molds of the Foraminifera). When silicification precedes dolomitization (as it often does), the fusulinids are faithfully replaced even to details of cell wall, whereas the matrix material has been converted to dolosparite.

Perkins (1963) made a detailed study of the petrology of a Middle Devonian limestone in southeastern Indiana, and mapped different carbonate facies. He indicated that the interstitial sparry calcite of the pelsparite facies is considered to be granular cement and not recrystallized micrite; the evidence supporting this conclusion was taken from the work of Bathurst (1958). Perkins considered the cement to be "granular" where the host particle, usually multigranular, lies in a mosaic of cement. Where the host is a single crystal, such as a crinoid fragment, and the cement forms a single rim in lattice continuity with it, the cement is called "rim cement", following the usage of Bathurst. Perkins also observed that sparry limestones commonly are not dolomitized, whereas micritic rocks were more susceptible to dolomitization due to their greater porosity and the greater surface area of the minute micrite grains. This again, bears out a well-established principle observed by many petrographers that calcspar resists dolomitization, whereas micrite and matrix (finely comminuted biocalcisiltite and biocalcarenite, also lithocalcisiltite and lithocalcarenite) may readily respond to dolomitizing solutions high in magnesium content. Possibly, some "catalysts" initiate nuclei of dolomitizing centers. Sparry calcite possibly can also form in various cavities, for example in animal burrows, gas-bubble pockets, worm-burrows, etc., yielding "eyes". Thus an ultimate lithification of the sediment can result in a "birdseye" limestone. The origin of "birdseyes" is discussed in detail by Wolf (1965c).

It is possible for dolorhombs to form in association with evaporite suites of sediments; with continued growth, a dolosparite can result. Miller (1961) noted clear, pale pink dolomite rhombohedra up to 0.14 mm long, disseminated with subhedral aragonite and calcite, in sludge dumps of salt extraction processing plants at Inagua, Bahamas. The rhombohedra are intricately associated with the other evaporite minerals; Miller suggested that in all probability the dolorhombs formed in minute voids where magnesium-rich brine accumulated or filtered through the waste sludge. Such a process conceivably could have operated in the geologic past within certain evaporite suites, and by so-called "filter-pressing" could have removed disseminated dolomite from one stratum only to concentrate it in another as dolosparite. The process, so it seems, would not of necessity be limited to dolomite but could have worked equally well with calcite.

Calcsparite and dolosparite possibly are more common in limestones and dolomitized limestones of the geologic record than published literature would suggest. Extensive cementation has been noted in young carbonate deposits which are sub-aerially exposed or are in the zone of meteoric waters along the Florida coast (Ginsburg, 1957). The Late Pleistocene Miami Oolite is thoroughly cemented by calcite at the exposed surface and below the ground-water level. But where it is still in the marine environment or above the ground-water table, it is friable and poorly cemented. Ginsburg (1957) noted that the oolite has a clear mosaic and partially recrystallized ooliths.

Sparry calcite as a term has been employed by Stauffer (1962) "... for the calcite which has been deposited from solution on a free surface" (orthosparite in Table 6-III). If the term is enlarged to embrace sparry dolomite as well, most of the criteria for recognition of sparite listed by Stauffer are applicable. For those dolosparites which have replaced other carbonate materials, criteria normally are readily available for recognition of the host rock. The following criteria, which are more or less directly indicative of sparry calcite, are taken from Stauffer (1960): (1) crystals in contact with a once free surface, such as oolites or inside of shell chambers; (2) crystals in the upper part of a former cavity which was partly filled with more or less flat-topped detrital sediment; (3) an increase in crystal size away from the wall of an allochem; (4) a decrease in the number of crystals away from the wall; (5) preferred orientation of the optic axes of crystals normal to the wall; (6) preferred orientation of the longest diameters normal to the wall; and (7) plane boundaries between crystals. In addition, Stauffer listed eight more criteria that he considered *suggestive* of open-space sparry calcite. He also presented criteria that are indicative and criteria that are suggestive of recrystallization in calcite. The interested reader is referred to the comprehensive and detailed article by Stauffer for additional information bearing on the subject of sparry calcite and recrystallized calcite.

FORMATION OF CARBONATE CONCRETIONS DURING DIAGENESIS

The escape of CO_2 during diagenesis appears to be one of the main driving forces for the formation of carbonate concretions. This can be seen on examining the following system of equilibriums presented by Strakhov (1954; see also Bissell and Chilingar, 1958):

$$\underset{(1)}{CO_2} \rightleftharpoons \underset{(2)}{H_2CO_3} \rightleftharpoons \underset{(3)}{(Ca, Mg, Fe, Mn)[HCO_3]_2} \rightleftharpoons \underset{(4)}{(Ca, Mg, Fe, Mn)\,CO_3} \qquad (6\text{-}1)$$

liquid phase — solid phase

During the first and second stages of diagenesis (as explained by Larsen and Chilingar, 1967), as a result of energetic bacterial activity, the amount of CO_2 in the interstitial waters is increasing. Thus, the above reaction goes to the right (from 1 → 2 → 3) and solid carbonates dissolve, resulting in higher alkalinity of interstitial waters. During the third stage of diagenesis, with decreasing amounts of CO_2, the reaction goes to the right (from 3 → 4) and especially close to the avenues (such as sandy layers) along which CO_2 can escape. As aragonite or calcite changes to dolomite, the avenues of increased porosity also enable CO_2 to escape and thus carbonates ($CaCO_3$, $FeCO_3$, $MgCO_3$, $MnCO_3$, or their mixtues) can precipitate. This also occurs in sandy layers inside clayey deposits. As a result of precipitation of carbonates and decrease in alkalinity of interstitial waters, the bicarbonates from adjoining clayey layers will move in to compensate for this created deficiency. As the new portions of CO_2 escape, additional carbonates are precipitated. This precipitation commonly occurs along certain horizons and around certain centers, giving rise to series of concretions. Uniform precipitation gives rise to sandstones cemented with various carbonates. The carbonate concretions are also commonly found inside clayey deposits close to the ventilation avenues along which escape of CO_2 can occur. Calcite concretions are found in highly calcareous sediments, whereas siderite concretions occur in sediments poor in calcareous material. The findings of Vital' (1959) indicate that many calcite and siderite concretions are found in sediments having low CO_2 content (<1%, and rarely 2--3%). Possibly this is due to secondary leaching-out of carbonates from rocks studied by Vital'.

Wide variations in physicochemical conditions within the sediment also account for considerable redistribution of substances during the third stage of diagenesis. For example, if a portion of sediment is characterized by higher pH (>8.0—8.5) whereas the other part by lower pH (~7), then $CaCO_3$ will move toward the area of high pH and the dissolved SiO_2 (from diatoms, sponge spicules, etc.) will move toward the zone of low pH where it will precipitate (Newell et al., 1953, p. 165; Strakhov et al., 1954, p. 593). Certain workers (Brodskaya and Timofeeva, both in Strakhov, 1959) studied carbonate concretions and their origin. During late diagenesis as a result of first-stage crystallization, or of crystallization of primary colloidal material with concomitant contraction, fractures form inside the concretions. These fractures do not reach the surface of concretions and end in V-pointed terminations. This observation led Vital' (1959, p. 236) to believe that crystallization and lithification start at the periphery, because as the loss in moisture and attendant volumetric contraction was reaching the central portions of concretions, the outer crust was already solid. The writers have observed that fractures are arranged parallel to the surface of the crust (concentric) or are diametrical, thus cutting the inner mass of concretions into sections.

TABLE 6-X

Relative effects of diagenesis on limestones (Wolf, 1963b; modified after Krumbein, 1942, table II)

	Compaction	Pressure-solution	Cementation	Inversion	Recrystallization	Solution
Particle size	—[1]	+ ++	—	—	+++	++
Shape and roundness	—	++ +++	—	— +	++	++
Surface texture	—	+++	+	—	++	+++
Particle orientation	+	— +	— + by crystallization	—	—	—
Mineral composition	—	—	++	+ expulsion of trace elements	+ expulsion of trace elements	+ ++
Porosity	+ sand ++ silt +++ mud	++	++ +++	?	+?	++
Permeability	+ sand ++ silt +++ mud	++	+++	?	+?	++
Color	+	—	—	—	+	— +
Paleoenvironmental indicator	only of indirect value	poor indicator	morphology of $CaCO_3$ cement; good indicator	?	too little information available	excellent to poor indicator

[1] Explanation of symbols: + = small to moderate effect; ++ = moderate to large effect; +++ = most strongly affected; — = negligible effect; ? = uncertain.

It is noteworthy that Vital' (1959, pp. 224—227) on studying the amounts of minor elements (Ni, Co, V, Cr) inside concretions and in surrounding sediments found much higher concentrations of these elements within the latter environment. He also observed that the amount of minor elements (with the exception of Cu) increases with increasing concentration of insoluble residue. This probably indicates that minôr elements did not migrate during diagenesis and are included inside concretions by mechanical means (together with captured particles of sediment). Sŭjkowski (1958,

Cavity formation	Internal sedimentation	Reworking	Dolomitization	Non-carbonate replacement	Authigenesis	
		++	—	—	—	Particle size
		+	—	—	—	Shape and roundness
		+	+	+	+	Surface texture
		++	—	—	+	Particle orientation
—	++	+	+++	+++	+++	Mineral composition
+++	+++	+	+ ++ +++	+?	+ ++	Porosity
+++	+++	+	+ ++ +++	+?	+ ++	Permeability
	++	+	+	+	+ ++	Color
may be excellent indicator	excellent	some are good indicators	fair to good; internal fillings excellent	good to excellent; silica, pyrite, hematite, etc.	good to excellent, e.g., glauconite	Paleoenvironmental indicator

p. 2704) pointed out that flint nodules in some beds of the English White Chalk contain a small amount of chalk, commonly in a state of loose aggregation, in their centers. The enclosed chalk is composed chiefly of shells of microorganisms. Some layers contain flints that are empty. This observation led Cayeaux (1897) to postulate that flints grew inward and not outward as was generally believed. Though the concretions are siliceous, the analogy to calcareous concretions should be pointed out. Probably most chert nodules, Liesegang concretions, and calcareous concretions in carbonate rocks originated during diagenesis.

SOME SIGNIFICANT RECENT DEVELOPMENTS [1]

The first edition of Bathurst's classical book *Carbonate Sediments and Their Diagenesis* appeared in 1971. In describing carbonate rocks, he made use of the terminology of Folk (1959), the scheme of Dunham (1962), and also certain terms of Grabau (1904). The first seven chapters of his book relate to such topics as: petrography of carbonate grains (skeletal structures, ooids, pisolites, and peloids); Recent carbonate environments (Great Bahama Bank, Florida, Gulf of Batabono, Persian Gulf, and British Honduras); Recent carbonate algal stromatolites; some chemical considerations; and growth of oöids, pisolites and grapestone. The remainder of the book concerns diagenesis of limestones, with the final chapter dealing with Recent dolomites.

In discussing diagenesis in a subaerial fresh-water environment and particularly the evolution of Recent and Pleistocene carbonates, Bathurst (1971, chapter 8) calls attention to modern realms of limestone accumulation. Such places as Eniwetok Atoll in the Marshall Islands of the tropical Pacific, Bermuda, Florida, Funafuti, and others were cited as localities where diagenesis proceeds rapidly in a subaerial environment to change unconsolidated lime-mud and skeletal, pelletal, and oolitic lime to indurated rock. It was noted that the micrite envelope (normally of algal origin) is of significance in diagenesis of oolitic and pelletal limestones. Calcitization of aragonite in skeletal lime-mud proceeds rather rapidly, particularly when exposed to subaerial conditions. Schlanger (1963, 1964) pointed this out for Guam and Eniwetok, noting that the in situ process of replacement of aragonite by calcite can yield a rock without significant porosity. This is a process involving growth of neomorphic spar. Crickmay (1945) similarly studied calcitization processes of young sediments of Lau, and stressed the wet polymorphic transformation requiring dissolution—precipitation. Bathurst (1971) assumed that the in situ wet polymorphic transformation process proceeds with the aid of a solution film thereby preserving a ghost of original wall structures in skeletal materials. Bissell and Chilingar (1958) stressed the significance of this intergranular film to expedite diagenesis of carbonate mud. Among skeletal material in which aragonite is calcitized, corals are more susceptible than mollusks or foraminiferids and the latter are more susceptible than echinoids to calcitization. Coralline algae have the highest content of $MgCO_3$, but are least susceptible of all to this process of changing aragonite to calcite.

In discussing cementation of lime-muds on Recent sea floors, Bathurst (1971, chapter 9) noted that these Recent carbonate sediments are being cemented to form hard layers at shallow depths of only a few meters or tens

[1] Since the publication of *Diagenesis in Sediments* (*Developments in Sedimentology, 8*, in 1967.)

of meters. It does not follow, however, that this cementation leads to consolidation. Studies made by various investigators in the Bahamas and around the southern tip of peninsular Florida demonstrate that cementation does occur in Holocene sediments at shallow marine depths. The present writers studied this facet of diagenesis along parts of the Florida Keys, observing that cementation occurs comparatively fast, particularly in the intertidal area. Subtidal cementation in restricted marine environments of Bermuda, Jamaica, and the Florida Keys normally proceeds rapidly in reef environments. This should be expected, because water in reefs is continually being replaced by new supplies of sea water that is supersatured with $CaCO_3$ and from which cement is precipiated. Pumping action of waves and tides would bathe reefs with new water. This is particularly true for the reef tract south of the Florida Keys.

Bathurst (1971, chapter 9) stressed *micritization* as a process of diagenesis of unconsolidated lime-mud. Emptied bores from such borers as arthropods, gastropods, or sponges (after death of the borer) becomc filled with micrite, and carbonate grains are gradually and centripetally replaced by micrite. Technically, this is not recrystallization; algal envelopes do form, however. Whereas Holocene envelopes are composed dominantly of micritic aragonite or high-magnesian calcite, ancient envelopes consist of a mosaic of equant crystals of low-magnesian micritic calcite.

One of the difficulties in studying thin-sections of limestones lies in recognizing cement, because not all sparry calcite in limestones is cement but, by contrast, is neomorphic in origin. Bathurst (1971) abandoned the terms "drusy" and "granular mosaic" in favor of the term "cement" which includes all passively precipitated, space-filling carbonate crystals which grow attached to a free surface. Bathurst (1971, chapter 9) assembled sixteen fabric criteria from various authors; they are briefly summarized as *Fabric Criteria for Cement:*

(1) Spar is interstitial (interparticle), with well-sorted and abraded particles that are in depositional contact with each other.

(2) There are two or more generations of spar.

(3) There are no relic structures such as are seen in neomorphic spar.

(4) Particles composed of micrite are not altered to spar.

(5) Micrite coats on particles are not altered to neomorphic spar.

(6) Mechanically deposited micrite is present but unaltered.

(7) Contacts between spar and particles are sharp.

(8) Margins of sparry mosaic coincide with surfaces that were once free, such as the surfaces of skeletal particles or of ooids, or molds of aragonitic shell fragments.

(9) Spar lines a cavity which it fills incompletely.

(10) Sparry mosaic occupies the upper part of a cavity whose lower part is

occupied by a more or less flat-topped internal sediment.

(11) The mass of sparry mosaic has the form to be expected of a pore filling or of an encrustation such as tufa.

(12) Intercrystalline boundaries in the mosaic are made up of plane interfaces.

(13) The size of the crystals increases away from the initial substrate of sparry mosaic.

(14) Crystal of the mosaic have a preferred orientation of optic axes normal to the initial substrate of the mosaic.

(15) Crystals of the sparry mosaic have a preferred shape orientation with longest axes normal to the initial substrate of the mosaic.

(16) Mosaics are characterized by a high percentage of enfacial junctions among triple junctions: percentages recorded so far range from 30 to 73%.

A word is in order concerning the enfacial junction: a triple junction is the meeting place of three intercrystalline boundaries. Most junctions betweeen interfaces are triple, and cannot be less in a mosaic; rarely there are four or more interfaces. Accordingly, an enfacial junction is a triple junction where one of the three angles is 180°. In case where three crystals grow together to give compromise boundaries, enfacial junctions cannot be formed. Where this does occur, a triple junction is formed with three angles none of which equals 180°. Furthermore, frequency of enfacial junctions in sparry mosaics is one of the least equivocal criteria for the distinction between cement and neomorphic spar.

Slices of limestones which contain coated grains (oolites, superficial oolites, etc.) commonly display a fibro-radial structure; in many insances this internal fabric is the result of calcitization of original aragonite fibrous structure. Oolites that form in present-day Great Salt Lake of Utah are now composed of aragonite; upon inversion to calcite (given adequate time) a fibro-radial structure is preserved.

There is, however, another structure which Bathurst (1971, chapter 10) described as a *radiaxial* fabric. He stressed the point that this is *not* to be confused with radial-fibrous (or, fibro-radial) fabric. Contrarily, it refers to the peculiar combination of curved twins, convergent optic axes and diverging subcrystals, with a cement crystal. Bathurst believes that radiaxial fabric developed in an entirely different manner from simple radial-fibrous crystals of aragonite or calcite cements. Crystals in radiaxial mosaic are elongate with length/breadth ratio (as observed in thin section) being as great as 7/1; they have a preferred orientation of longest axes normal to the cavity wall being filled. Furthermore, crystals do not extinguish uniformly under crossed polars; these crystals are generally composed of a number of subcrystals having slightly different positions of extinction. Subcrystals diverge away from the cavity wall; also, discrete crystals consist of a bundle of divergent subcrystals. Thus, radiaxial fabric characterizes features with a larger crystal. Possibly one explanation for development of radiaxial fabric is recrystallized aragonite cement. Regardless, it is evidence of one diagenetic overprinting in a limestone.

The present writers agree with Bathurst that growth of cement is controlled by four major factors, which are: (1) the form of the substrate to which crystals become attached, (2) the level of supersaturation in the solution, (3) composition of the solution, and (4) the rate of movement of the solute ions past the growing crystal faces. Thin-sections reveal that cement crystals grow syntaxially (lattice continuous) with host crystals in the substrate. This is common in criquinites (crinoidal limestones) where the cement overgrowth arises from one nucleus. The process of cementation by syntaxial (shared axes) overgrowth has been referred to by various authors as cementation by enlargement, secondary enlargement, rim cementation, and other terms. It is a truism that any surface composed of a crystal lattice with strong preferred orientation lends itself to oriented syntaxial overgrowth. This is true of prismatic layers in mollusks, brachiopods, and ostracods.

Bathurst (1971, chapter 11) also discussed the role of pressure-solution during postdepositional alteration of carbonate sediments. He emphatically stated that localized strain in a crystal immersed in its saturated solution must lead to localized dissolution; the process whereby grains undergo dissolution about their contacts is termed *pressure-solution*. Results of pressure-solution are clearly seen in coated-grain (oolitic) limestones, because one can observe in a slice where the original margin of an ooid has been affected and how much material was removed. This is remarkably well documented in a study of Lower Triassic carbonates of southern Nevada, U.S.A., made by Bissell (1970). Monomineralic isotropic grains (such as ooids, echinoderm plates, etc.) under even slight load are affected because the solubility product constant is higher near grain-to-grain contacts. A concentration gradient results, and solute ions are transferred away from the vicinity of the contacts into less concentrated solutions that occupy adjacent voids. Precipitation of $CaCO_3$ occurs on unstrained surfaces. The combined action of pressure-solution followed by precipitation is known as *solution transfer*. Bathurst pointed out that in an unconsolidated carbonate sediment under load, the distribution of strain at the surfaces of grains near their points of contact will be a function of grain orientation, size and shape, anisotropy of the crystal lacttice and (in polycrystalline grains) the fabric of the crystal mosaic. Bathurst believes that if pressure-solution acts after the precipitation of the second generation of cement, then it does so by the formation of stylolites. This may be true for some carbonate sediments, but studies made by the writers in 1970 on shoal-water Triassic carbonates (marine) revealed that stylolites form best in oolitic micrites during, and shortly following, precipitation of the first generation of cement. At this point in the present discussion, it is worthy of note that it was Kloden (1828) who described what he regarded as a fossil under the name *Stylolites sulcatus;* this, of course, is the pressure-solution zig-zag feature termed stylolite that develops during diagenesis in limestone. Three-dimensional views of these structures reveal that they are polygonal-shaped columns of mutually interpenetrating stylii. Amplitude of stylolites is a minimum measurement of the amount of soluble carbonate removed during their formation. Bathurst believes that the controlling factor in stylolitization is the orientation of the axis of linear stress (which is generally vertical), being a simple consequence of overburden. Furthermore, he indicates that the zig-zag form of the surface is presumed to be a consequence of lateral variations along the interface of solubility differences in the limestone. The writers agree with Bathurst that amplitude of finger-like stylolite columns gives the minimum thickness of lost material, and that in stylolites of large amplitude the sides of the columns become grooved comparable to piston-faults. The grooves, understandably, develop parallel to the axes of the columns. A point not stressed by Bathurst relates to the actual thickness of overburden; however, he stated his belief that stylolites are of postcementation origin, adding that they could begin to form in sediments which are only lightly-cemented. This is probably true, and the point is worth scoring that in a single bed (1 m thick, for example) numerous stylolites can form during the time the sediment is accumulating, shortly after deposition of a 1 m-thick unit. It is not necessary, the present writers contend, for a very thick pile of sediment to form before stylolitization begins. The process, one must add, can continue with additional accumulation of overburden. (See Coogan and Manus, 1975.)

Bathurst (1971) follows, to a certain degree, the usage of Folk (1965) and of Spry (1969) in discussing recrystallization; that is, recrystallization embraces any change in the fabric of a mineral or a monomineralic sediment, and the mineral is the same after as before the reaction. He noted, however, that three changs are possible, those of: (1) crystal volume, (2) crystal shape, and (3) crystal lattice orientation. Folk (1965) defined aggrading neomorphism as the process whereby finer crystal mosaics are replaced in situ by coarser crystal mosaics of the same mineral or its polymorph, without intermediate formation of apparent porosity. Included here would be diagenetic alteration of micrite or very fine skeletal detritus to form sparry calcite; it is an in situ wet process. Growth of neomorphic spar begins even in only slightly consolidated lime-mud; in this instance, it amounts to wet transformation of aragonite to calcite accompanied by a certain amount of passive dissolution—precipitation. Bathurst presented an excellent discussion of fabrics of sparry calcite which is not repeated here. The second enlarged edition of Bathurst's *Carbonate Sediments and Their Diagenesis* was published in 1975 and constitutes a classical reference work.

Many students of carbonate petrology and petrography have studied Cenozoic and Holocene carbonate rocks and sediments of Peninsular Florida and the Florida Keys. Within the past decade one of these studies covered the diagenesis of the Key Largo Limestone of a part of the Florida Keys (Stanley, 1966, pp. 1927—1947). The Key Largo Limestone is a Pleistocene reef limestone in the upper Florida Keys, consisting of an organic framework of coral colonies and an interstitial skeletal calcarenite. It was pointed out by Stanley that the constituent composition of the calcarenite facies indicates that aragonite and high-magnesium calcite were originally the chief mineralogic components. Fossils that initially were high-magnesium calcite have been diagenetically altered in situ to low-magnesium calcite through removal of excess magnesium by fresh water in the meteoric zone. Aragonite, by contrast, is currently being dissolved in the meteoric zone, and dissolved $CaCO_3$ is being deposited near the site of dissolution as low-magnesium calcite cement. Accordingly, Foraminifera and coralline Algae, which originally were composed of high-magnesium calcite, are now represented by stable, low-magnesium calcite. The dissolved aragonite was reprecipitated as calcite cement in the form of granular cementation and drusy mosaics.

In comparing carbonate deposition and postdepositional changes of Manlius Formation (Lower Devonian) of New York State with Recent analogs (south Florida and the Bahamas), La Porte (1967, pp. 73—101) noted that three facies can be recognized in the Manlius deposits: (1) supratidal, (2) intertidal, and (3) subtidal, all near mean sea-level. He noted that mudcracks and spar-filled vugs ("birdseye" structure) in limestones of the supratidal facies indicate frequent subaerial exposure. Dolomitization, where present, was penecontemporaneous and was inferred to have been similar to present-day supratidal dolomite formation in Florida and the Bahamas.

In a thought-provoking paper titled "Kinetics and diagenesis of carbonate sediments", Schmalz (1967, pp. 60—67) presented a simplified kinetic model of carbonate diagenesis. He indicated that although his model is based upon the assumption of first-order reactions, nonetheless the steady-state development of this model is independent of reaction order. His model, he stated, is compatible with results of laboratory investigations of reactions between both sea water and distilled water with mixed carbonate assemblages. Furthermore, consequences of the steady-state model are in close agreement with field observations in carbonate environments. He pointed out the need for detailed kinetic studies of the rates of solution and crystal growth of carbonate phases under a wide range of conditions. Such studies are under way, Schmalz noted, and hopefully should provide a more rigorous model, one that will define reaction mechanisms and will identify intermediate products of diagenetic reactions in great detail.

Robinson (1967, pp. 355—364) believes that diagenetic changes can best be observed in the early history of a homogeneous sediment. Accordingly, he carried out research on diagenesis and porosity development in Recent and Pleistocene oolites from southern Florida and the Bahamas. He observed that in the early stages of diagenesis, alteration of oolites is confined to lithification; calcium carbonate necessary for cementation was introduced from sources outside the sediment body, as well as from the dissolution of ooids. Cement formed in this manner is interparticle crystalline calcite. The mineral composition of the ooids remained essentially aragonite, with distinct structure. During this phase of diagenesis porosity and permeability were reduced due to cementation; however, during moderate cementation, rock rigidity was produced resulting in improved interconnection of pore spaces. Robinson noted that in advanced stages of diagenesis, ooid structure becomes less distinct and replacement of aragonite by calcite takes place. In such a replacement, ooids take on a soft friable texture, and when replacement is complete they assume a dense vitreous luster. Although porosity and permeability values vary as a result of this replacement, in general moderately good pore space connections prevail.

Because various processes of diagenesis operate early after deposition of lime-mud, it is of great importance that studies should be made on these carbonate muds in various realms. For example, Stockman et al. (1967, pp. 633—648) investigated some of the processes and results of production of lime-mud by Algae in south Florida. They concluded that one Algal genus, *Penicillus*, is a major contributor to fine aragonite mud. It was also their contention that since the flooding of parts of south Florida by rising sea level 4000—10,000 years ago, the present rate of production by *Penicillus* sp. could account for all the fine aragonite mud in the inner Florida Reef Tract and one third of the same material in the northeastern Florida Bay. They also observed that another source of fine lime-mud is the biological and mechanical breakdown of resistant skeletons, mollusks, Algae, corals, etc. These writers posed the following question: "To what extent can the breakdown of calcareous skeletons explain ancient lime-muds?" They partially answered this question by pointing out that a conservative estimate of the total production by fragile green Algae (ignoring contribution from outside sources) gives a range of 5—45 m/million years. They also stressed the point that there is every reason to believe that plants and animals with skeletons as fragile as *Penicillus* have existed since Cambrian times. Thus, diagenetic changes that Phanerozoic-age limestones have experienced were in large measure controlled by the composition of the particular taxa that produced the lime-mud.

Land (1967, pp. 914—930) made an in-depth study of diagenesis of skeletal carbonates, ranging in age from Miocene "chalky" aragonite pelecypods and corals from Maryland to Florida Pleistocene aragonite skeletal materials. He defined diagenesis as follows: ". . . diagenesis constitutes the alteration of sediments after the spatial configuration of constituent grains is no longer altered by currents, organisms, or desiccation; and until primary metastable phases have been eliminated." He pointed out that stabilization of metastable aragonite and Mg-calcites is an inevitable process and constitutes a large and important part of carbonate diagenesis. Land performed approximately seventy-five separate hydrothermal experiments at 285° C using three skeletal starting materials: a hydrocoral *Millepora alcicornis*, a scleractinian coral *Acropora cervicornis*, and a pelecypod *Argina pexata*. It was noted that aragonite inversion by this method produced coarse-grained equigranular mosaics, and the rate of inversion of seketal material seems to be only temperature-dependent near 300°. Non-skeletal aragonite is less reactive, and Mg-calcite inversion is very slow compared to aragonite inversion. Stabilization, Land pointed out, can occur in only a limited number of ways. "Solid-state" stabilization takes place by inversion or exsolution for aragonite and Mg-calcite, respectively. Stabilization can take place by total dissolution and reprecipitation. Lastly, stabilization can take place by replacement by either dolomite or a noncarbonate phase.

Roehl (1967, pp. 1979—2032) conducted a significant study of Early Paleozoic carbonates in which a comparison with Recent (Holocene) analogs was made. Roehl studied Ordovician and Silurian carbonate facies of the western Williston Basin in the western Dakotas and Montana, U.S.A., and compared their facies with Recent low-energy marine and subaerial carbonates of the Bahamas. He investigated intratidal, intertidal, and supratidal fabrics, and in his excellent paper discussed at great length distinctive reservoir fabrics represented. Despite replacement dolomitization, these fabrics originated under conditions similar to the sedimentary and diagenetic environments of western Andros Island, Bahamas, and the Trucial Coast, Persian Gulf. Both ancient and modern sediment examples, Roehl contended, reflect the importance of organic processes in fabric development. Furthermore, fabrics control the configuration, distribution, and quality of reservoir rock. Roehl discussed the subaerial diagenetic terrane in terms of protracted subaerial exposure toward the end of Silurian time in the western Williston Basin which he investigated. He observed that many of the sedimentological and diagenetic features pass imperceptibly from one zone to the other. Two kinds of diagenetic fabrics were distinguished: (1) "*epigenetic*" *fabrics* that result from diagenesis in the shallow subsurface, and (2) *karst fabric* which develops at the expense of exposed carbonate surfaces through erosion and soil-forming processes. As defined by him, "epigenesis" involved mostly supratidal dolomitization, either early or late in the history of the Ordovician and Silurian carbonates. Karsting is related to "epigenesis" because it is an early phase of vadose alterations.

Various investigators have studied carbonates in an effort to discover causes of grain breakdown and micrite formation. For example, Klement and Toomey (1967, pp. 1045—1051) attempted to determine the role of the blue-green Algae *Girvanella* in skeletal grain destruction and lime-mud formation in carbonate rocks in the Lower Ordovician of Texas. They concluded that marginal grain corrosion on calcareous grains present in these rocks (El Paso Group), particularly in mound carbonates, appears to be caused by the boring and perforating action of the primitive filamentous blue-green Algae *Girvanella.* Particle alteration and breakdown with the algal corrosion rim leads to micritization of the grain. The mechanisms of the boring and perforating process of the algal filaments, however, is not completely known. Their observations indicated that the micrite precipitation may be attributed to the algal life or decay process. Accordingly, they concluded that the work of grain-boring blue-green, green and red Algae appears to be a major agent in carbonate alteration, destruction, and ultimate micrite formation. These primitive Algae lived from the Precambrian time to the present, and are found in abundance in shallow-water environments (generally less than 50 m deep) and in all climatic zones. Thus, local blooms of these Algae may be an important factor in the early diagenesis of carbonate sediments.

Another study which relates to the significance of algal limestones and early diagenesis is that by Aitken (1967, pp. 1163—1178); this study concerns Cambrian and Ordovician carbonates of southwestern Alberta. Aitken proposed the new term *cryptalgal* (from the Greek word "kryptos", meaning hidden or secret) to identify carbonate rocks in which it is inferred (rather than observed) that Algae exercised considerable influence in their formation. He proposed the adjective "cryptalgalaminate" to identify planar-laminated carbonate rocks bearing evidence of algal-mat activity, and the term "thrombolite" for non-laminated cryptalgal bodies characterized by clotted fabric. Accordingly, stromatolitic and cryptalgalaminate carbonate sediments appear to be restricted to the intertidal zone, but thrombolites and oncolites do not, so Aitken contends. His classification of algal carbonates is as follows (p. 1164):

- algal carbonates
  - carbonates composed wholly or partly of skeletal calcareous algae (not further classified)
  - cryptalgal carbonates (noncalcareous, blue-green and red algae)
    - cryptalgal biolithites
      - oncolites (oncolitic carbonates)
      - stromatolites (stromatolitic carbonates)
      - thrombolites (strombolitic carbonates)
      - cryptalgalaminate carbonates
    - cryptal fragmental carbonates
      - cryptalgal biocalcirudite*
      - cryptalgal biocalcarenite*
      - cryptalgal biocalcisiltite*
      - cryptalgalaminate breccia

The significance of Aitken's paper lies in the fact that it is fundamental for researchers of petrology and petrography of limestones to make a clear distinction between carbonate rocks largely composed of skeletal calcareous Algae and the cryptalgal carbonate rocks. Diagenetic processes which operate on cryptalgal biolithites, oncolites, or stromatolites (for example) should proceed along somewhat contrasting avenues than along those of non-algal micrites, for example.

An epigenetic process occasionally overlooked by petrographers is that of dedolomitization. Evamy (1967, pp. 1204—1215) presented an excellent paper on this subject. He pointed out that dedolomitization, the reverse process of dolomitization, is brought about by solutions with a high Ca^{2+}/Mg^{2+} ratio reacting with dolomite to form calcium carbonate. Textural evidence of the process, Evamy pointed out, provides a useful indicator of near-surface diagenesis. He presented a schematic history of the diagenesis of certain dedolomitized limestones utilizing photomicrographs (from thin-sections) to document this process. It was also pointed out that clotted or "grumeleuse" texture of certain calcite rocks is believed to have resulted from dolomitization and subsequent dedolomitization. In studying rhombohedral pores in dedolomitized limestones, Evamy contended that such pores are developed through selective leaching of what he termed "dedolomite". He added that both high-magnesian calcite and argonite, if they are not leached, tend to alter to stable low-magnesian calcite during subsequent diagenesis. Ancient dedolomites are likely to be mineralogically stable. Accordingly, rhombohedral pores are only likely to be found where dedolomite has been leached shortly after dedolomitization. He also pointed out that much cal-

* (or derived dolomite)

cite-after-dolomite found at surface outcrop may be Recent and thus possibly grade into unaltered dolomite at depth. Calcite-after-dolomite in the subsurface, however, should be considered as fossil and indicative of former emergence. In pursuing this idea of dedolomitization further, K. de Groot (1967, pp. 1216—1220) performed experiments in the laboratory, noting that the following conditions are necessary for effective dedolomitization: (1) a high rate of water flow to remove Mg^{2+} formed, thus keeping the Ca^{2+}/Mg^{2+} ratio of the water constantly high; (2) a carbon dioxide partial pressure considerably lower than 0.5 atm; and (3) temperatures not higher than 50°C. The subject of dedolomitization is covered in greater detail in the next chapter.

Matthews (1967, pp. 1147—1153) studied diagenetic fabrics of biosparites from Pleistocene carbonates of Barbados, the West Indies. He stated: "If we are to truly understand the origin of carbonate rocks, it is equally important that we study diagenetic environments, for many carbonate rocks are as much the product of diagenesis, as they are the product of depositions." The present writers heartily agree with this statement. According to Matthews, the processes of diagenesis in Barbados carbonates took place in one or more of the following environments: (1) within the marine environment; (2) where marine water and fresh water intermix; (3) within a fresh-water phreatic zone; and (4) within a fresh-water vadose zone. His petrographic evidence tended to discount the importance of modification in the marine environment, but the relative quantitative importance of modification associated with the fresh-water table and fresh-water vadose zone was not determined. He did contend, however, that all the modifications he listed can be observed in sediments less than 150,000 years old.

Rucker (1968, pp. 68—72) studied carbonate mineralogy of cored sediments of the Exuma Sound, Bahamas, and found that aragonite, low Mg-calcite, and Mg-calcite were present in all of the 98 samples obtained. This carbonate mineral assemblage was interpreted as representing a mineralogically stable, recycled sediment derived from the extensive Bahama Banks, which were exposed during the latest Pleistocene sea level lowering. During this period of exposure, the mineralogically unstable, aragonite-rich bank sediments were subaerially altered to low-magnesium calcite. Accordingly, Rucker argued that diagenesis took place when sediment was subaerially exposed, rather than being a progressive alteration with depth in the cores, some of which were 134 cm long.

In an in-depth study of diagenetic patterns in the Wettersteinkalk (Middle Triassic), northern Limestone Alps, Bavaria and Tyrol, Germann (1968, pp.

490—500) summarized these patterns under (1) cementation (granular cement and fibrous cement), (2) authigenic mineral growth (dolomite poikilotopes), (3) crystal enlargement (granular mosaic, fibrous mosaic, and syntaxial replacement rims), (4) crystal diminution (granular mosaic), and (5) pressure-solution (microstylolites). Germann (p. 499) defined diagenesis as consolidation and loss of pore volume by means of compaction and cementation. He observed that the Wettersteinkalk almost completely has lost its porosity during early diagenetic cementation stages in which different cement types were deposited. No remarkable secondary porosity was caused by diagenesis. He concluded that different driving forces, with a tendency to produce diagenetic equilibrium by means of solution alteration, cause the original disequilibrium assemblage to more or less disappear. The result is a diagenetic change in crystal size from fine to coarse. Deep burial of the alpine Wettersteinkalk accompanied by tectonic stresses have intensified late diagenetic (epigenetic) alterations.

Numerous publications point out the importance of subaerial exposure of limestones to expedite early diagenesis. MacIntyre et al. (1968, pp. 660—664) called attention to subsea cementation of carbonate sediments at depths of about 50 ft on a submerged barrier reef off the west coast of Barbados, W.I. This cementation, which is taking place in an open marine environment, is attributed to the burrowing of the pelecypod *Gastrochaena hians* through carbonate sediment trapped and supported in a *Bubaris* sp. sponge mat. These investigators pointed out the following geological implications: (1) various biochemical processes are probably responsible for cementing carbonate sediments in open marine environments; and (2) burrowing organisms, such as *Gastrochaena hians*, can produce pelletoid limestones from sediments devoid of a mud-size fraction.

In his study of Recent (Holocene) sediments and carbonate diagenesis of South Bonaire, Netherlands Antilles, Lucia (1968, pp. 845—858) presented some very interesting facts and attendant interpretations; his photomicrographs deserve careful study by other petrographers. Lucia concluded that the Recent sediments of South Bonaire represent the filling of a depression in the Pleistocene surface. The sedimentation and diagenetic history proceeded as follows: (a) deposition of *Halimeda* grainstone; (b) deposition of a lithoclastic sediment with thin clay beds and cementation of this sediment to form a lacy carbonate crust; (c) deposition of an intertidal pelleted lime-mud; (d) completion of the coral rubble ridge; and (e) bedded gypsum deposition and incipient dolomitization of the pelleted lime-mud. According to Lucia, lithoclastic sediments with clasts composed largely of aragonite druse can be produced by multiple periods of cementation, fragmentation, and transportation. Textures of the Recent gypsum and anhydrite were found to be very similar to the textures of subsurface anhydrite.

Matthews (1968, pp. 1110–1119) studied carbonate diagenesis of a coral cap in the Barbados, West Indies. He stressed the fact that ancient carbonate rocks are composed of the stable minerals calcite and dolomite, whereas their Recent sediment analogs are predominantly composed of the unstable minerals argonite and high-magnesium calcite. This instability is, of course, of particular interest to students of carbonate diagenesis. Matthews concluded that there are certain geologically significant implication concerning interaction of meteoric water with aragonitic sediment: (1) With an aqueous phase saturated with respect to aragonite, the calcite growth step is much faster than the calcite nucleation. (2) Owing to an original homogeneous distribution of calcite nuclei, reef-associated carbonate muds and sands recrystallize more rapidly than the corals contained within them. This rapid recrystallization produces an intermediate stage in which a low-magnesium calcite matrix surrounds larger aragonitic skeletons. (3) The kinetics of water transport through the sediment may be regarded as a factor controlling the efficiency of the aragonite—calcite transformation near the surface of the sediment. If undersaturated water enters the aragonitic sediment and passes through it rapidly, only solution of aragonite may occur. (4) Under all climatic conditions represented on Barbados, the equilibration of aragonite to calcite requires a geologically significant length of time. Furthermore, lack of "evidence" of subaerial exposures does not necessarily imply that these surfaces have not been subaerially exposed.

Ferroan and non-ferroan calcite cements in Pleistocene—Holocene carbonates of the New Hebrides were studied by Colley and Davies (1969, pp. 554–558). Their petrographic study of stained biolithites from Erromango revealed an advanced state of diagenesis. Micrite is produced by grain diminution of Algal colonies and, probably, by fragmental abrasion. Other fabrics developed include granular mosaics and simple syntaxial rims. These geologists noted that a completely lithified, partially recrystallized rock has been produced in a very short geologic time period. Furthermore, they observed that the occurrence of ferroan calcite as a first stage is unusual, because it has hitherto been unreported in Holocene or Pleistocene carbonate sediments and, in addition, it is very rare as a first-stage cement in ancient carbonates.

More than one-half of the deep-sea carbonates in the Red Sea are believed to have precipitated inorganically (Milliman et al., 1969, pp. 724—736). Aragonite-cemented layers were formed during periods of glacially lowered sea level, when Red Sea waters were highly saline. These investigators assumed that the uppermost aragonite lithic layer is about 11,000 years old and, therefore, sedimentation rates in their cored samples ranged from 4 to 12 cm/1000 years. Inasmuch as carbonate content of the Red Sea sediments ranges from 50 to 80%, they calculated that 1–5 cm of carbonate is inorganically precipitated per thousand years. Thus, this is a new report of large-scale inorganic precipitation and lithification of Holocene deep-sea carbonates. Petrographers should be aware of this fact in their studies of ancient deep-sea carbonates.

Processes of diagenetic changes in oolitic limestones have intrigued petrographers over the years. Knewtson and Hubert (1969, pp. 954–968) investigated the diagenesis of oolitic calcarenites in the Ste. Genevieve Limestone (Mississippian) of Missouri. During early diagenesis of these carbonate sands, aragonite in fossils and in ooid laminae was dissolved by low-Mg water to form extensive moldic porosity within insoluble micrite envelopes. Precipitation of thin calcite crusts around the grains prevented collapse of most micrite envelopes. Intergranular and moldic pores were subsequently filled by mosaic calcite cement, Ferroan calcite spar of $(Ca_{0.990}\cdot Mg_{0.006}, Fe_{0.004})CO_3$ composition, based on electron microprobe analysis, was precipitated within the ooid molds in some beds. The ferroan calcite reflects concentration of iron derived from solution of iron-bearing ooid laminae in the water inside the molds.

In carbonate sediments, lithification resulting in the formation of a rock involves dissolution of unstable carbonate minerals, precipitation of stable minerals, and recrystallization. Gavish and Friedman (1969, pp. 980–1006) made an in-depth study of progressive diagenesis in Quaternary to Late Tertiary carbonate sediments in an effort to unravel some of the mysteries surrounding such diagenesis. They chose the Mediterranean coast of Israel as their area of study, and noted that calcite cement was introduced by meteoric water from outside the rocks which were being lithified. The rocks are loosely consolidated with early calcite cement which for the most part is restricted to the grain-to-grain boundaries. Mg-calcite changed to calcite without accompanying textural changes, but with a sharp reduction in the concentration of magnesium and manganese. Aragonite grains were not affected by dissolution. In the Ultimate Interglacial (probably 80,000–100,000 years ago) carbonate sediments, interparticle pore space was occluded by a drusy calcite mosaic and the rocks became well consolidated. Aragonitic skeletal fragments and tests were dissolved out, forming molds, and a drusy calcite mosaic was precipitated in the molds. Gavish and Friedman believe that the major diagenetic changes in subaerially exposed carbonates were completed in 80,000–100,000 years. In the intertidal marine environment, fibrous and cryptocrystalline cement of aragonite and Mg-calcite changes carbonate sediments into beachrock.

It has been pointed out (Neal, 1969, pp. 1040—1045) that carbonate petrologists generally regard cementation of limestones as a subaerial process. Neal (1969) studied diagenesis of the Blackjack Creek Formation (Pennsylvanian, Missouri). On staining the limestones with alizarin red-S and potassium ferricyanide, he observed that nonferroan calcite characterizes the subordinate dogtooth ("first generation") rim-cement, whereas ferroan calcite characterizes the dominant mosaic ("second generation") cement. He suggested that this compositional difference most likely records contrasting environments. It was his interpretation that the nonferroan first-generation cement is the product of conversion from aragonite that precipitated in the marine environment. The ferroan second-generation calcite, he contended, was precipitated as such from circulating ground water in primary pores lined with the earlier calcite and in pores resulting from solution of aragonitic bioclasts after partial lithification of the carbonate cement.

Schroeder (1969, pp. 1057—1073) set up a model of diagenesis after performing laboratory experiments at ordinary temperature and pressure for periods up to 240 days on Recent (Holocene) carbonate materials. He noted that fresh water and sea water dissolve aragonite and magnesium calcite prior to mineralogical changes. The observed rates of dissolution of calcium, magnesium, and strontium indicate that these elements are incorporated in aragonite and magnesium calcite in more than one way: in lattice positions, in lattice interstices, or in inclusions. According to Schroeder, this suggests the presence of more than one mineral phase in the skeletal materials which were studied. These phases differ in response to solution, in quantity and, probably, in chemical composition. Experimentally established sequence of preference in dissolution is as follows:

(1) In distilled water:
- a. Aragonite:
  - more soluble phase: Mg > Sr > Ca
  - less soluble phase: Mg > Ca > Sr
- b. Magnesium calcite: Mg > Ca > Sr

(2) In sea water:
- a. Aragonite: Mg > Ca > Sr
- b. Magnesium calcite: Mg > Ca > Sr

Schroeder concluded that diagenesis of sediments consisting of biogenic carbonates is an extremely complex process.

A thought-provoking paper titled "Diagenesis, chemical sediments, and the mixing of natural waters" by Runnells (1969, pp. 1188—1201) points out that the chemical effects of mixing of aqueous solutions in the laboratory include changes in the concentration and electrical properties of a solution, shifts in homogeneous equilibria, and precipitation and dissolution of a solid phase. Accordingly, similar chemical effects can be expected to occur in nature as a result of the mixing of natural waters and, in some instances, these may result in diagenesis. Runnells set up a simple classification that yielded 14 distinct categories of natural waters with 91 possible combinations of mixing of pairs of such waters. He pointed out that one obvious result of mixing of waters is the precipitation of the components of relatively insoluble minerals, such as barite and the carbonates. Amphoteric substances, such as aluminum and iron hydroxides, will also precipitate as a result of mixing. Less obvious are the possible diagenetic effects of the mixing of waters which differ only in salinity. The author of this paper did point out, however, that his hypothesis must be tested by geologic observations.

Although Choquette and Pray's (1970, pp. 207—250) classical paper on porosity in sedimentary carbonates concerns mainly nomenclature and classification, they point out that modifications in porosity represent a major and commonly the predominant diagenetic process in most sedimentary carbonates. The vast reduction in porosity, from that of the initial sediment to the negligible porosity of most ancient carbonates is accomplished largely by cementation. Also, the volume of cement filling former pores approaches or exceeds the volume of the framework in many carbonate rocks. Choquette and Pray (p. 238) pointed out that pore space, which reflects by its position and boundaries the depositional or diagenetic fabric elements of a sediment or rock, is termed "fabric selective". Porosity formed early in diagenesis is commonly fabric selective, in contrast to much of the porosity formed later when unstable carbonate minerals and most or all former pore space have been eliminated. Much carbonate porosity, they state, is fabric selective. (See Chilingarian and Wolf, 1976, pp. 737—751.)

Whitcombe (1970, pp. 334—338) discussed the diagenetic history of the Lower Limestone Shale Group (Carboniferous) of South Wales in terms of cementation, crystal enlargement, dolomitization, silica diagenesis and stylolitization. He noted that the initial high porosity was reduced practically to zero by cementation. Granular, drusy and rim cements were distinguished together with ferroan and nonferroan calcite cement. Cementation, he concluded, was an early diagenetic process. Crystal enlargement was the major neomorphic process. In the neomorphic mosaics, crystals and their boundaries show typical strained features. Stylolites occurred at different times during diagenesis, with maximum development in late diagenesis and epigenesis.

Submarine lithification of Holocene carbonates does occur, and in some areas it is of great significance in forming hard rocks. In their studies of submarine lithification of Jamaican reefs, Land and Goreau (1970, pp. 457—462) observed widespread formation of Mg-calcite cement in Recent reef-crest and forereef framework and interframework sediment. This cementation takes place immediately beneath the reef—water interface to depths of at least 70 m on the north coast of Jamaica. Chemically, the crystalline cement is most commonly a Mg-calcite containing 18.5 ± 1 mol% $MgCO_3$, as determined by both Debye—Scherrer powder diffraction method and by electron microprobe analysis. These investigators concluded that Mg-calcite micrites and cements appear to be at least as common as aragonite among Recent sediments. A primary Mg-calcite micrite is much more easily transformed into a calcite micrite by incongruent dissolution than is aragonite, which commonly undergoes grain-growth as it reacts.

Pingitore (1970, pp. 712—721) investigated diagenesis and porosity modification in uplifted coral reefs (*Acropora palmata*) in Pleistocene rocks of Barbados, West Indies. Earliest modifications of the coral took place in marine water, i.e., the entrapment of sedimentary void fill and chemical precipitation of aragonite needles. After uplift above sea level, exposure to fresh water resulted in solution of the coral skeleton. A major episode of diagenetic activity occurs, Pingitore noted, when the metastable aragonite skeleton inverts to calcite. This process involves a local volume for volume solution of aragonite and reprecipitation as calcite. Inasmuch as the new calcite is less dense than the aragonite, the excess material remains mobile in solution. The latter material then precipitates as calcite void fill in available open pore space, and is related in time and space (inches of water movement) to the region of active skeletal inversion.

Normally, petrologists do not think in terms of deep-water carbonate diagenesis, because studies of Pleistocene and Holocene carbonate sediments have largely been confined to shallow marine and subaerial environments. Scholle (1971, pp. 233—250), however, presented an interesting paper on diagenesis of deep-water carbonate turbidites of Upper Cretaceous Monte Antola flysch deposits, in the Northern Apennines, Italy. Deposition took place at oceanic depths, below the calcium carbonate compensation level. Relatively stable original mineralogy and absence of an early fresh-water flush in the Monte Antola sediments allowed considerable compaction to occur before cementation, yielding a close-packed texture with relatively small volumes of cement. Scholle (p. 249) suggested the following order or sequence of events for these deep-water carbonates: (1) A long interval of compaction beginning at the time of deposition and lasting about 2—15 million years; relatively little cementation took place, but consolidation was brought about by grain orientation, compaction, and dewatering. (2) Very early formation of pyrite in areas of concentrated organic matter; some calcite replacement took place. (3) When depth of burial exceeded about 500 m, pressure-solution and grain welding began to release significant amounts of calcium carbonate which produced nearly complete cementation of both the fine- and coarse-grained sediments. (4) Continued diagenesis of clay minerals produced a uniform chlorite—illite assemblage from an original suite of detrital clays.

Three generations of sparry calcite, defined by variations in fabric and their reaction with artificial stains, occupy primary and secondary voids in the Corallian Beds (Upper Oxfordian) of southern England (Talbot, 1971, pp. 261—273): (1) nonferroan calcite, (2) fibrous ferroan calcite, and (3) granular ferroan calcite. It was determined that cementation was accompanied by the replacement of skeletal aragonite, and that nonferroan calcite was precipitated from fresh ground waters. Fibrous ferroan calcite, which was precipitated soon after burial of the sediments, was probably derived from skeletal aragonite. Early burial cementation appears to have been largely confined to sediments containing appreciable quantities of skeletal aragonite. Subaerial and early burial diagenesis, Talbot noted, followed similar paths and in both processes calcite precipitation accompanied aragonite dissolution. Granular ferroan calcite was precipitated at depth, often after grain breakage and contact solution had taken place. All rocks that show these features have a granular ferroan calcite cement. Void filling often commenced in one environment (e.g., subaerial) and was completed in another, even after deep burial. Sediments containing abundant skeletal aragonite may be preferentially cemented due to the ready supply of calcium carbonate this material provides. Finally, Talbot noted that micrite envelopes are not essential as a cavity support during the replacement of skeletal aragonite via an intermediate void stage. A combination of mucilaginous films and crusts of calcite crystals can perform the same function.

Case-hardening by cementation of alluvium and colluvium, particularily in desert environments, is definitely a case of diagenesis of previously non-indurated clastic rocks in the subaerial realm. Lattman and Simonberg (1971, pp. 274—281) investigated case-hardening of carbonate alluvium and colluvium in the Spring Mountains, southern Nevada. They indicated that this process begins within one or two years after exposure and may proceed to the extent that samples break through enclosed clasts as readily as through the case-hardening matrix that encloses them. Cementation which causes this case-hardening appears to be due to solution of the sand-sized and finer carbonate fractions and subsequent deposition of calcite. Above present drainage channels, the case-hardening seems to be caused by surface runoff, and the rapidity and degree of cementation is controlled by the texture of layers and lenses within the colluvium and alluvium.

The ultrastructure of carbonate cements in a Holocene algal reef of Bermuda was studied in thin-sections and with S.E.M. micrographs by Ginsburg et al. (1971, pp. 472—482). Their studies demonstrate that the micrite cement crystallizes within voids and is not the recrystallization product of earlier carbonate cement. Their micrographs also show all stages from initial linings to complete fillings of voids as well as multiple generations of micrite. The crystallization of micrite within voids, they contend, offers an alternate explanation to the accepted view that fossil micrites result from recrystallization. These Bermuda algal reefs occur on the seaward edge of an oceanic platform where they are constantly flushed with sea water of normal composition. Evidently the cementation that occurs in this realm, as well as several other examples from the modern sea floor, requires no major modification of the composition of normal sea water. These investigators pointed out that the Bermuda reefs are, in geological terms, examples of penecontemporaneous cementation from normal marine waters.

Petrologists, who have investigated associations of carbonates and evaporites from the standpoint of oil and gas production, have discovered that in many areas diagenesis played a significant role in this process. Wardlaw and Reinson (1971, pp. 1759—1786) carried out such a study of carbonate and evaporite deposition and diagenesis of some Middle Devonian formations of south-central Saskatchewan, Canada. They contended that sulfate-reducing bacteria may have influenced the partial replacement of anhydrite by calcite. Under natural conditions in marine waters, sulfate reduction is probably limited only by the amount of organic matter which accumulates. Organic matter is oxidized by means of anaerobic bacterial processes, the product of which (ammonia) should accumulate during sulfate reduction. Such accumulation would favor the precipitation of carbonate. A distinctive pisolite cap on banks of the Winnepegosis Formation was interpreted by Wardlaw and Reinson as an inorganically formed vadose pisolite, which provided them with the evidence that there was subaerial exposure at the termination of bank development.

A study of marine diagenesis of carbonate sediment of Bonaire, Netherlands Antilles, was made by Sibley and Murray (1972, pp. 168—178). A discontinuous layer of lithified carbonate sand underlies a small part of the Lac, a large lagoon on the southeastern coast of Bonaire. This lithified layer consists of a grainstone cemented by acicular aragonite. Numerous lines of evidence presented by Sibley and Murray indicate that the cementation occurred in the marine environment. The lithified layer is at present continuously saturated with normal sea water. The four distinct types of micrite found in the beachrock and submarine cemented layer are: (1) structureless aragonite which fills or partially fills intraparticle pore space; (2) aragonite with a pellet-like fabric which fills or partially fills interparticle pore space; (3) aragonite coatings on single skeletal fragments with a gradational boundary between the fragment and coating; and (4) high- and low-magnesium calcite which coats single and multiple skeletal fragments with a sharp contact between grain and cement. Coralline algal encrustation was high-magnesium calcite, which results from micritization of encrusting coralline algae. Analyses of these coatings demonstrated that as micritization proceeds and the microstructure of the algal coating is destroyed, the mineralogy changes from high- to low-magnesium calcite.

Diagenetic implications of the distribution of high-magnesium calcite in lime-muds of the Great Bahama Bank were discussed by Husseini and Matthews (1972, pp. 179—182). Their studies of Pleistocene low-magnesium calcite analogs of Recent (Holocene) carbonate sediments indicate that the initial distribution of high-magnesium calcite in aragonite sediments may play a critical role in the preservation of sedimentary texture and fabric during fresh-water diagenesis. Accordingly, the distribution of high-magnesium calcite in Recent sediments provides a basis for speculation concerning potential diagenetic fabrics in analogous materials from older stratigraphic sequences. Abundance of calcite nuclei in ancient lime-muds is suggested to be a significant factor leading to the preservation of the texture of such sediments. The authors argued that the combination of abundant calcite nuclei and generally poor permeability of muddy sediments should lead to localized solution—reprecipitation during mineralogical stabilization.

Boyer (1972, pp. 205—210) made a study of grain accretion and related phenomena in unconsolidated surface sediments of the Florida Reef Tract, pointing out that these accretionary features include grain coatings, intragranular void fillings, and internal and external cements. He contended that all these are products of nonskeletal submarine carbonate precipitation and lithification. Such features are most abundant in sands along the platform edge near the outer margin of the reef tract, where flushing of sediment by ocean water is most intense, and generally become less abundant on the backreef toward the Keys. Boyer pointed out that a general progressive decline in abundance of accretionary features away from the Florida platform edge suggests that the abundance of such features could be a useful petrographic clue to paleogeography and water circulation during deposition of ancient calcarenites.

According to the investigations by Chafetz (1972, pp. 325--329), surface epigenesis of the Morgan Creek Limestone (Upper Cambrian, central Texas) resulted in dedolomitization, silicification, and aggradational recrystallization. Dedolomitized parts of the rock grade into "fresh" dolomite away from the surface. In intensely weathered samples, spar cement recrystallizes to large equant crystals. Intensity of recrystallization decreases inward from weathered surfaces. Chafetz concluded that identification of dedolomitization, a process which occurs most favorably under surface weathering conditions, aids in recognizing an unconformity between carbonate strata.

Non-skeletal aragonite and Mg-calcite are formed in the interior of hollow particles (chambers) in the West Indian and Mediterranean nearshore calcareous sediments according to studies made by Alexandersson (1972, pp. 441—460). Calcite has 15--17 mole percent $MgCO_3$ in solid solution. Intragranular fillings are best developed in sediments from turbulent environments; the carbonate material constitutes an addition to the particles and not a reorganized initial grain substance. Growth processes are supposed to represent slow precipitation under supersaturated conditions. During diagenesis Mg-calcite changes to aragonite, with changes in fabric ranging from ordered to random arrangement. Alexandersson discovered that biogenic aragonite induces nucleation and growth of aragonite, whereas biogenic calcite favors precipitation of Mg-calcite. He concluded that intragranular aragonite and Mg-calcite are characteristic constituents in modern shallow-water sediments from warm seas. They are particularly abundant in sediments from turbulent environments. Aragonite and Mg-calcite are both authigenic marine carbonates, which grow simultaneously under the same general conditions, and variation in mineralogy is not determined by the major physicochemical parameters of the sea water.

Al-Hashimi and Hemingway (1973, pp. 82--91) investigated Recent dedolomitization and origin of rusty crusts in dolomitic Carboniferous rocks of the Middle Limestone Group of the northeast coast of Northumberland, England. Both surface and near-surface forms of dedolomitization were found. The rusty crusts are caused by the presence of iron hydroxides in weathered crusts of ferroan (iron-rich) dolostones. The iron hydroxides and associated dedolomites are genetically related. The susceptibility of ferroan dolomites to dedolomitization is apparently caused by their metastability in the surface environment. Circulating sea or fresh water along permeable zones in the carbonates exposed at or near the present erosional surface is responsible for the oxidation and hydration of the ferrous iron content of the metastable ferroan dolomites, as well as the dedolomitization of these dolomites. Accordingly, the dedolomitization mechanism has been found to be a process of great potential in converting highly impermeable and impervious dolomitic rocks into porous and permeable carbonates. The detection of such a phenomenon in carbonate rocks at depths may indicate presence of an unconformity; it may also help in locating potential oil or gas reservoirs.

Syntaxial borders of calcite commonly form on calcite crystals; however, Zenger (1973, pp. 118—124) reported the occurrence of syntaxial calcite borders on dolomite crystals in carbonates of the Little Falls Formation (Upper Cambrian) in east-central New York. One type of border which he studied and termed "calcite rim" was interpreted to have formed by marginal dedolomitization. Another type of border consists of coarse, void-filling calcite also optically continuous with the dolomite. Zenger believes that apparently these "calcite envelopes" formed by passive precipitation of calcite, syntaxial with dolomite "nuclei". He pointed out that solutions with relatively high Ca^{2+}/Mg^{2+} ratios, which resulted from the passage of meteoric water through overlying Ordovician limestones may have acted as the dedolomitizing agent for the calcite rims.

Cores collected by drilling in the deep ocean are providing samples of deeply buried in situ ocean sediments in the process of becoming sedimentary rocks. Davies and Supko (1973, pp. 381—390) sampled oceanic sediments from deep-sea drilling and investigated diagenetic processes. They noted that carbonates have been dominant in deep-sea sediments since at least Jurassic time, except during Eocene time when siliceous sediments became dominant. Studies of long continuous sections of coccolith ooze have revealed a progressive sequence of diagenetic changes from break-up of individual coccoliths to solution and recementation. Cores obtained from the Central Pacific on the Magellan Rise, showed an almost continuous section of calcareous ooze over 1100 m in thickness. The top 200 m (Miocene to Recent in age) consists of loosely packed coccoliths with well-preserved sutures. Break-up of the coccoliths contributes prismatic calcite laths to the sediment, which appear as micrite under the microscope. From 200 to 600 m below the sediment—water interface (Late Eocene to Oligocene in age), the sediments consist of friable, pure white chalk. According to Davies and Supko, essentially all the pore water has been squeezed out at these depths and maximum grain contact was achieved short of actually crushing the grains. The section from 600 m to about 830 m consists of nanno-chalks, silicified limestone, and chert. In the nanno-chalks, the coccolith remains have been highly etched or coated with granular calcite. Foraminiferan tests have recrystallized and broken down to produce micrite. Below 825 m, well-lithified limestones and cherts are found. These are strongly cemented and the calcite cement crystals are of considerable size. Accordingly, a number of factors seem to interact to diagenetically alter biogenic oozes to micrites and pseudospar. These include overburden pressure, solution-welding, decay of cement-inhibiting organic films, and silicification.

Sarkisyan et al. (1973, pp. 1305—1313) studied the effects of postsedimentation processes on carbonate reservoir rocks in Paleozoic strata of the Volga-Urals Region, U.S.S.R. In pointing out that more than 45 percent [1] of the world's petroleum reserves are in carbonate formations, and that most of the oil in the Middle East (where more than half of the world's proved reserves are located) is produced from carbonate reservoirs, they stressed the importance of postdepositional changes that could alter porosity and permeability. Good carbonate reservoir rocks (high porosity and permeability) originate in the shallow parts of basins where a mobile water environment is present. The rocks they studied, are bioclastic limestones which were affected by various diagenetic and epigenetic processes, such as sulfatization, calcitization, silicification, and stylolitization. The latter improves the reservoir characteristics, whereas the former three adversely affect reservoir properties. It is important to note, however, that secondary mineral-formation processes indirectly improve the flow capacity of rocks (permeability) by creating heterogeneity, which favors the subsequent formation of fractures and solution cavities.

Rapid carbonate cementation of intertidal carbonate beach sands is a common phenomenon along the shoreline of Grand Cayman Island, West Indies, according to the studies of Moore (1973, pp. 591—602). Seemingly, beach rock cement precipitation has taken place under mixed meteoric—marine conditions. Although most cementation takes place on Grand Cayman in the intertidal zone, cementation also extends a considerable distance below mean low tide, indicating that beach-rock cementation probably merges seaward with submarine cemented hard grounds. Dominant cement fabrics and mineralogies comprise: (1) aragonite acicular crust, (2) clear, bladed-equant magnesian calcite crust, and (3) magnesian calcite micrite crust. The first two crusts were formed in intergranular pore spaces by either direct physical precipitation or secondarily as a byproduct of biologic activity. Ultrastructure of magnesian calcite micrite crusts show some evidence of direct biologic precipitation in the cementing process.

[1] This figure is probably closer to 60% at the present time (G.V.C.).

Land and Hoops (1973, pp. 614—617) made a study of sodium in carbonate sediments and rocks in an effort to determine a possible index to the salinity of diagenetic solutions. They pointed out that sodium is present in the Recent marine carbonate minerals (aragonite, calcite, Mg-calcite, and dolomite) greatly in excess of equimolar chloride concentration. The bulk sodium content of Recent carbonate sediments is greater than 1000 p.p.m. and decreases along with Sr, Mg, ^{18}O, and ^{13}C, as meteoric diagenesis takes place. According to Land and Hoops, the low sodium content of most ancient dolomites suggests that they are either primary products originating from solutions containing little sodium or have re-equilibrated with solutions low in sodium.

An investigation of diagenesis of Devonian styliolinid-rich pelagic carbonates from West Germany was made by Tucker and Kendall (1973, pp. 672—687), who ascertained that several stages of diagenesis were involved. The original sediment was believed to have been a sequence of calcilutites or calcisiltites, alternately rich and poor in *Styliolina*, interbedded with clay laminae or argillaceous limestone seams. On the sea floor, at a depth probably not exceeding a few 100 m, the *Styliolina*-rich layers became partially or wholly lithified by an epitaxial growth of acicular (aragonitic?) carbonate upon the *Styliolina* shells, which is comparable to the cementation of Pteropods in the Red Sea today. The acicular carbonate replaced the fine-grained carbonate matrix of the sediments. Two episodes of stylolite formation were separated by an episode of calcite vein formation, and are younger than the development of acicular carbonate replacement.

Gvirtzman et al. (1973, pp. 985—997) studied diagenetic control and distribution of uranium in coral reefs. They noted that the concentration of about 2 p.p.m. of uranium in the aragonitic skeletons of modern scleractinian corals is more-or-less constant, regardless of occurrence, anatomy, or taxonomy. Presence of aragonite or high-magnesian calcite cements usually raises the concentration of bulk samples to about 3 p.p.m. Organisms, such as corals and coralline algae, discriminate against the uptake of uranium while secreting their skeletons, whereas the uptake of uranium by mineral cements is less restrained. Aragonite cement contains about 3.6 p.p.m. and high-magnesian calcite cement has 2.6 p.p.m. uranium. During leaching by fresh water, aragonite of coral skeletons dissolves out. This creates hollow molds which are filled with drusy, low-magnesian calcite. In emergent reefs from the shores of the Red Sea, which display the effects of progressive diagenesis, this calcite is enriched in uranium (3.9 p.p.m.) beyond the contents found in marine cements. The level of concentration of uranium in low-magnesian calcite of diagenetically altered corals, these workers point out, is a function of the availability of uranium in meteoric waters. Furthermore, in aragonite as well as in high- and low-magnesian calcite uranium replaces calcium or occupies lattice vacancies in the crystal lattice.

The stratigraphy and mineralogy of a cored borehole on Barbados, W.I., was studied by Steinen and Matthews (1973, pp. 1012—1020) from the standpoint of phreatic versus vadose diagenesis. Their data supported a conclusion that meteoric water diagenesis may occur in the phreatic diagenetic environment. Carbonate sediments from the upper 1.5 m of the borehole have been in the vadose diagenetic zone since the initial emergence from the depositional environment. These carbonate sediments are for the most part mineralogically unstable and have not been affected by extensive dissolution. Sediments from borehole no. 17, inferred to have been exposed only to the vadose diagenetic environment for the past 100,000 + years, have experienced relatively slight diagenetic modification. On the other hand, a fresh-water phreatic lens has occupied parts of the borehole section for prolonged periods of time during the interglacial high stands of sea level. Total duration of the diagenesis in fresh phreatic environments ranges from around 5000 years in parts of the core to as long as 20,000 years. Portions of the core, which have been occupied by a fresh-water phreatic lens at least once, have experienced complete matrix mineralogic stabilization, significant precipitation of cement, and major development of moldic and solution-enlarged porosity.

Kahle (1974, pp. 30—39) studied ooids (oolite samples) from The Great Salt Lake of Utah, utilizing this area as an analogue for the genesis and diagenesis of ooids in marine limestones. He demonstrated that aragonite makes up more than 90% of the carbonate mineralogy of 13 samples studied and pointed out that the high degree of allochem recrystallization indicates that the aragonite in the allochems typically undergoes recrystallization to aragonite and not to calcite. Radial grain orientation is demonstrably not developed in many initially aragonitic Pleistocene marine ooids as a result of their partial to complete conversion to calcite. Kahle noted that many ooids in marine limestones probably formed initially with an entirely radial grain orientation in their rims, under a variety of environmental conditions. He was not certain that radial grain orientation in ooids is invariably a product of diagenesis.

Lindholm (1974, pp. 428—440) studied the fabric and chemistry of pore-filling calcite in septarian veins, as models for limestone cementation. It was suggested that the amount of magnesium in aqueous solution controls the crystal habit of crystallizing calcite which, in turn, affects the final fabric. The sequence of fibrous or bladed calcite followed by equant calcite suggests crystallization in waters where magnesium content decreased with time as a result of changing diagenetic environment with burial in marine lutites. Early-formed fibrous calcite may have crystallized as aragonite or as high-magnesium calcite (micron-sized or fibrous cement), but was later modified through neomorphism.

Various investigators have proven the occurrence of submarine cementation of reefs. An example from the Red Sea was the subject of a paper by Friedman et al. (1974, pp. 816—825). Their objective in this study was threefold: (1) to determine width of cementation zone (they showed that cements form a few centimeters below surfaces of reefs and eliminate pore space almost entirely within less than 60 cm below the surfaces); (2) to describe and illustrate the fabrics and kinds of minerals of marine cements in Red Sea reefs; and (3) to infer a possible mechanism of precipitation of carbonate cement. They noted that a pH level exceeding 9 (and even 10) appears to be necessary for the dissolution of quartz and the precipitation of carbonates. Furthermore, photosynthesis and respiration of the biomass in the reef cause a shift in the carbonate buffer system of sea water with the uptake of CO_2. Microlevels of pH of 10 and even 10.5 may likely be maintained in thin gel-like films or monomolecular layers ("skin effect") that cling to the surfaces of the framework builders of reefs. Such high pH levels may trigger the precipitation of carbonate cements in reefs.

Subsequent to a study of cores from a borehole on Barbados, West Indies, by Steinen and Matthews (1973, pp. 1012—1020), Steinen (1974, pp. 1008—1024) added details relating to phreatic and vadose diagenetic modification of Pleistocene limestones at Barbados. Data he obtained from a shallow cored borehole drilled through the 105,000 year old reef-tract sediments indicate that the carbonate sediments have been altered more rapidly and more extensively in fresh-water diagenetic environments than in the vadose or marine phreatic diagenetic environments. Total amounts of cement, mainly needle-fiber cement and dense micrite coatings, are high, whereas porosity is low. Packstones and grainstones in the lowest part of the borehole are largely unaltered and metastable carbonate mineralogy predominates.

According to Wachs and Hein (1974, pp. 1217—1231), lithification of Franciscan limestones did not occur immediately after deposition, as demonstrated by the preferred orientation of grains which is a result of compaction. Recrystallization of the micrite followed burial and compaction of the limestone. All degrees of recrystallization are found in the coccoliths that originally composed at least 50% of the micrite. During advanced stages of recrystallization, coccoliths and other grains were transformed to a continuous and uniform calcite mosaic. At this stage, the primary porosity of the sediment was destroyed and the sediment became completely lithified. Stylolites probably formed over a long period of time that started when the limestones were unlithified, and are common at the margins of chert bodies. Their remnants are also found within the chert, which formed only in micritic limestones containing Radiolaria. The eventual replacement of the original carbonate micrite between the radiolarian tests resulted in the transformation of a limestone bed to chert. Most stylolites are younger than veins, which, according to Wachs and Hein (1974, p. 1230), indicates that the veins also formed prior to the complete lithification of the sediment.

In a recent excellent book entitled *Marine Carbonates*, Milliman (1974) devoted Part IV to carbonate diagenesis: chapter 9 relates to carbonate degradation, chapter 10 concerns carbonate cementation, and chapter 11 is on dolomitization. The interested reader is encouraged to delve into a thorough analysis of Milliman's Part IV. In his chapter 9, he discussed biological erosion of carbonate substrate by borers and burrowers, grazers, browsers, and predators. Milliman stressed the point that biological erosion of carbonate substrates serves two important purposes in carbonate diagenesis. Firstly, it degrades and destroys carbonate components and, secondly, it creates secondary voids within the carbonate particles and substrate, which can be important in subsequent recrystallization.

Carbonate cementation, subject of chapter 10 of Milliman's book discusses petrographic terms as used by various investigators; table 70 (p. 271) of this chapter is presented here (see Table 6-XI).

Because of the many dangers and connotations presently inherent in the term "micrite", Milliman prefers the terms cryptocrystalline or submicrocrystalline. He divided carbonate cementation into two main groups: (1) intragranular cementation (cryptocrystallization), (2) intergranular cementation and lithification, and subdivided the latter into (a) intertidal (beachrock, mangrove reefs, and supratidal crusts) and (b) submarine lithification (shallow-water limestones and deep-water limestones).

First report of the occurrence of ooids from the Great Barrier Reef Province was that of Marshall and Davies (1975, pp. 285—291). They occur in a large area on the floor of the Capricorn Channel, in water depths of 100—120 m, and are believed to have formed in the Early Holocene. X-ray diffraction analysis showed that these ooids are composed of high-magnesium calcite. Electron probe studies indicated that magnesium occurs in a noncarbonate phase within the chambers of Foraminifera that acted as ooid nuclei. These ooids exhibit well defined concentric, radial, and granular fabrics. Diagenetic alteration of the ooids has resulted in (1) an overprint of

TABLE 6-XI

Terminologies for various carbonate cements and matrices (after Milliman, 1974, table 70, courtesy of Springer-Verlag)

Acicular	long, thin crystal, usually oriented normal to grain surface; usually aragonite
Blocky	massive, equant grains
Cryptocrystalline	Light brown to opaque tiny crystals, less than 4 microns in diameter (Purdy, 1963); similar to microcrystalline
Dentate	drusy cement with jagged, tooth-like edges
Drusy	a type of sparry cement; crystalline to subhedral crystals, coarser than about 10 microns, generally forming in voids (drusy mosaic) or as thin coatings (Bathurst, 1958); in drusy mosaic, cement crystals increase in size away from the component grains
Fibrous	similar to acicular
Granular	general term for cement crystals coarser than about 10 microns, but smaller than about 60 microns; overlaps with drusy
Micrite	microcrystalline calcite, less than 4 microns in diameter (Folk, 1959); less than 30 microns (Leighton and Pendexter, 1962)
Microcrystalline	subtranslucent crystals less than 4 microns in diameter (Folk, 1959); between 4 and 30 microns (Macintyre, 1967a)
Microspar	recrystallized matrix (Folk, 1965)
Pelletal	cement and matrix composed of numerous small pellets, 20 to 60 microns in diameter; the pellets are composed of cryptocrystalline grains
Submicrocrystalline	less than 4 microns in diameter (Macintyre, 1967a)
Sucrosic	sugary cement, common in dolomites
Sparry	clear, coarse crystals, generally coarser than 10 microns (Folk, 1959)

a secondary radial fabric on the primary concentric and radial ones, and (2) the progressive obliteration of these three fabrics by the growth of an equigranular mosaic, leading to the formation of a structureless ooid. During or after these changes, ooid nuclei have been replaced by high-magnesium calcite.

The writers applaud the address of the retiring President of the Society of Economic Paleontologists and Mineralogists Gerald M. Friedman in 1975 titled: "The making and unmaking of limestones or the downs and ups of porosity" (1975, pp. 379—398). As he pointed out, during lithification calcium carbonate is introduced from the outside during formation of a rock from the unconsolidated carbonate sediments. According to him, compaction accounts for only small reduction in pore space, which could have been quite large initially in an unconsolidated sediment. Waters responsible for cementation may be: (1) percolating fresh water on surface or in subsurface, (2) sea water; and (3) formation waters. In the phreatic zone (below the ground-water table), especially in moist climatic belts, the original pore space between particles as well as any secondary moldic pore space is filled with cements of various fabrics. As carbonate sediments in the phreatic zone are converted to limestones, they not only lose their pore space and become tight, but may also acquire postdepositional fabrics of recrystallization. During phreatic diagenesis, reefs can micritize beyond recognition. In the vadose zone (above the ground-water table), which in the arid regions may be thousands of feet thick, the pores remain largely open and secondary porosity may develop. Porous strata resulting from vadose diagenesis may occur interbedded, as cyclic sequences, with tight strata owing their origin to phreatic diagenesis. On the sea bottom, cementation may be related to bacterial activity. Bicarbonate ions, which may be generated here, may then combine with calcium to deposit a cemented calcium-carbonate deposit. The role of compaction fluids during diagenesis has been recently discussed by Chilingarian and Wolf (1976).

Kendall (1975, pp. 399—404) reported an instance of postcompactional calcitization of molluskan aragonite in a Jurassic limestone from Saskatchewan, Canada. Originally aragonitic bivalve shells from a thin limestone locally suffered severe compaction, with pressure solution and fracturing taking place prior to their calcitization. He believed that skeletal aragonite remained stable until after compaction, because the sediment was isolated from the ground-water system until relatively late in the sediment's diagenetic history. Inasmuch as this thin limestone, which contains the molluskan shells, is encased within argillaceous units, Kendall pointed out that possibly postcompactive calcitization always occurs in environments associated with surrounding impermeable sediments.

Abundant and delicately preserved hopper-shaped calcite pseudomorphs after halite occur in the supratidal—intertidal Joachim Dolomite (Middle Ordovician) of north Arkansas, according to a report by Handford and Moore (1976, pp. 387—392). Low Sr^{2+} concentration in the dolomicrite indicates that it has undergone extensive diagenesis by mixed meteoric—marine waters low in Sr^{2+} content. These investigators envisioned the following diagenetic sequence: Stage 1—dolomitization of tidal-flat carbonates and halite hopper growth. Stage 2—tidal-flat progradation and influx of mixed meteoric—marine water which dissolved the halite and precipitated both dolomite and calcite; early dolomite matrix began to convert to dolomicrite of present form and porosity was reduced. Stage 3—meteoric water completely dominated subsequent diagenesis by completing the dolomicrite conversion and eliminating the remaining porosity.

James et al. (1976, pp. 523—544) studied the facies and fabric specificity of early subsea cements in shallow Belize (British Honduras) reefs, noting that such early subsea cements and cemented sediments are restricted to marginal facies of shallow barrier and atoll reefs. In Belize, the reef-flat pavement, a subtidal bed of lithified coral conglomerate, 1 m thick and 10—100 m wide, occurs on the lee side of all marginal reefs. In seaward-facing spurs, cavities between in situ coral and *Millepora* are locally filled with cemented skeletal packstone. Cemented packstone and wackestone also partially fill intraskeletal pores and borings in coral fronds lying loose on the sea floor. These geologists reported that the predominant cement of internal sediments is Mg-calcite (14.5—18.6 mole % Mg) present as micrite and bladed spar. Aragonite, although common, is found only in coral pores. Crystal form of the cement and degree of lithification can be related to variations in the texture of the internal sediments.

Kobluk and Risk (1977, pp. 1069—1082) studied micritization and carbonate-grain binding by endolithic Algae. In their excellent paper, these authors pointed out that endolithic (boring) Algae are the direct or indirect agents of important erosive and early diagenetic processes in carbonate sediments. In addition to micrite envelopes produced by repeated boring and infilling of borings by precipitated micrite, the Algae also produce micrite envelopes outside grains by the calcification (cementation) of exposed dead endolithic filaments. The latter process reduces intergranular porosity. Algal filaments, which grow through the micrite envelopes into intergranular pores, live within the pores as chasmolithic algae, i.e., they live in holes not of their own creation. They may become calcified after death, producing an intertwined mesh of calcified filaments on which later micrite and microspar cements precipitate. As pointed out by Kobluk and Risk, the calcified intergranular filaments and associated cements further reduce intergranular porosity and serve to bind the grains. In the opinion of the writers of the present chapter, this will also reduce the permeability.

In addition, micrite envelopes may be produced beneath Algal-mucous coats through a process of etching and dissolution. This results in a highly microporous residue micrite (Kobluk and Risk).

According to Kobluk and Risk, *Girvanella* and similar Paleozoic and Mesozoic Algae may represent calcified Algae similar to those described previously and in which cement precipitates on dead algal thalli. *Girvanella* possibly represents calcified (cemented) filaments of several algal genera (a diagenetic taxon), rather than a single algal genus.

A model of development of intergranular and constructive-envelope calcified algal filament is presented in Fig. 6-15. Micropores left between calcified filaments are filled partly to completely by rhombohedral micrite and microspar cement. The latter precipitates on the palisade cement coating the algal filaments. Intergranular calcified filaments, which are filled with

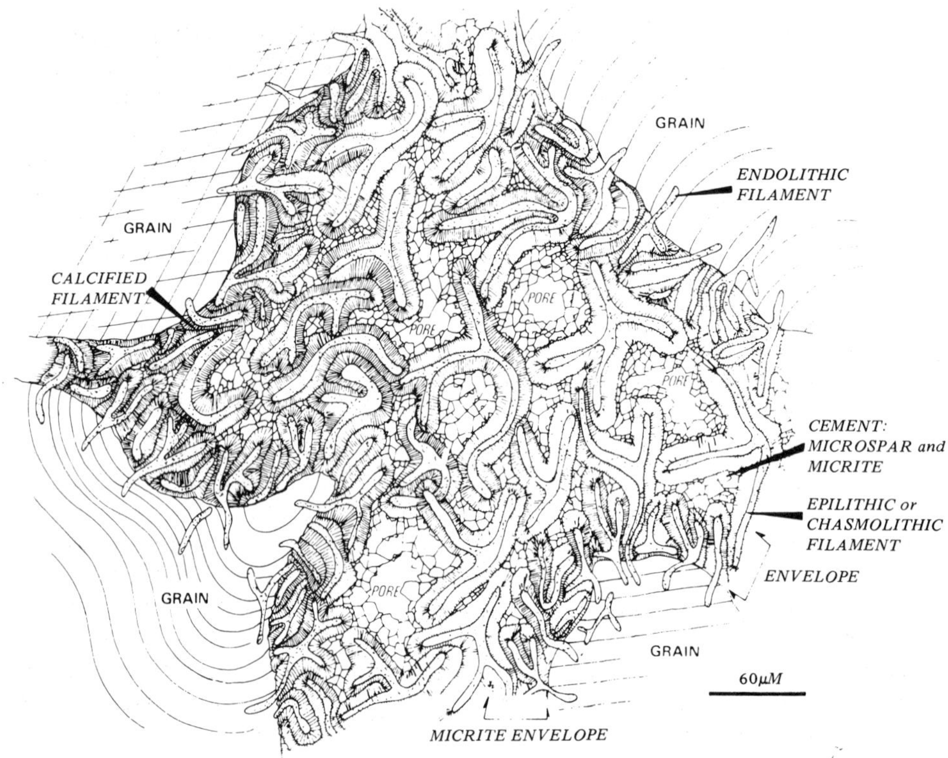

Fig. 6-15. Model of intergranular and constructive-envelope calcified algal filament development among four carbonate grains. (After Kobluk and Risk, 1977, fig. 8, p. 1079; courtesy Am. Assoc. Pet. Geol.)

micrite and encrusted in palisade cement, intertwine to produce a complex of interwoven calcified filaments.

On using scanning-electron microscopy, light microscopy, oxygen-isotopic analysis, and trace-element analysis, Scholle (1977, pp. 982—1009) made a very important contribution to our knowledge of chalk diagenesis. As pointed out by him, inasmuch as chalks consist largely of stable low-magnesium calcite, they undergo diagenetic alteration different from that of more widely studied aragonite and high-magnesium calcite-bearing, shallow-marine deposits.

According to Scholle (1977, p. 1004), chalks are subject to diagenetic loss of porosity either through early sea-floor cementation or through later burial-diagenesis. Fresh-water exposure or aging of sediments, alone, do not lead to significant cementation of chalks.

Although the rates of cementation as a function of burial depths are predictable, they can vary with the chemistry of interstitial fluids. The presence of fluids containing moderate amounts of magnesium in solution retards cement generation.

Usually, mechanical compaction predominates during the early diagenetic stage (dewatering), whereas pressure solution and reprecipitation are more important during later diagenesis. According to Scholle (1977, p. 1005), isotopic analysis can be used to confirm the patterns of chalk diagenesis. In some cases, it can be used as a measure of the maximum burial depth to which chalks have been subjected. Whereas carbon-isotopic analyses are not subject to strong diagenetic alteration and may reveal information about primary depositional conditions, bulk oxygen-isotopic analyses in chalks commonly are a better indication of diagenetic history.

Scholle (1977, p. 982) stated that with a few notable exceptions, the porosity and permeability of chalks decrease as a direct function of burial depth (see Figs. 6-16 and 6-17). The exceptions include the following: (1) overpressured chalk formations, i.e., effective (compactive, or grain-to-grain) stress, which is equal to the total overburden pressure minus the pore-fluid pressure, is less than normal; (2) formations in which carbonate reactions were reduced or terminated because of the oil entrance; and (3) zones with increased solution and cementation due to the tectonic stresses. A much steeper porosity versus depth gradient is observed in areas where fresh water entered the pores before major burial as compared to regions in which marine water remained in the pores.

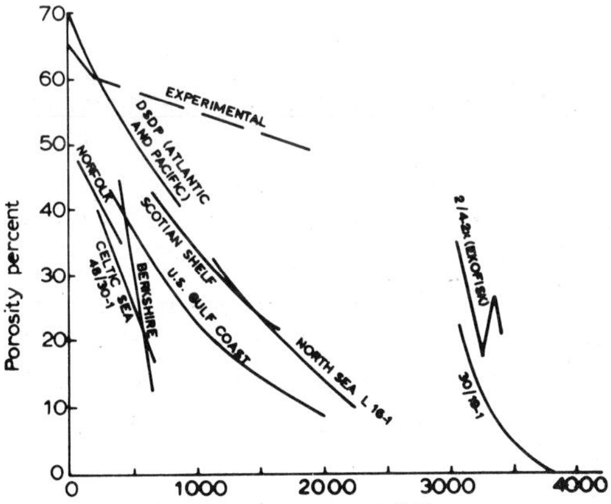

Fig. 6-16. Relationship between porosity and burial depth in chalks. (After Scholle, 1977, p. 992, fig. 5; courtesy Am. Assoc. Pet. Geol.) For additional data on compaction of carbonates, see Rieke and Chilingarian (1974).

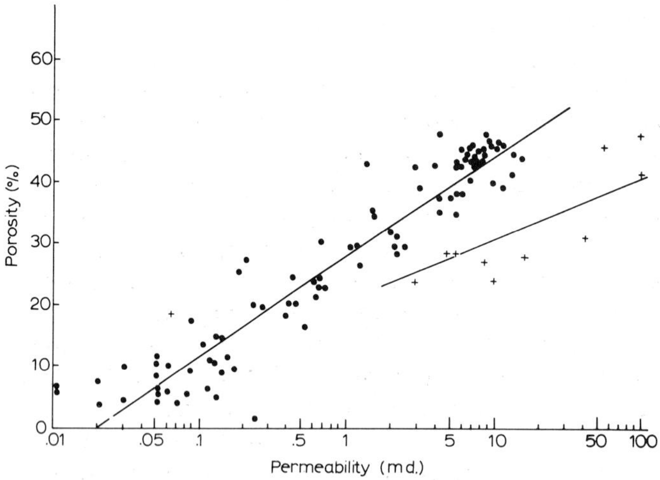

Fig. 6-17. Relationship between the porosity and permeability (air flow) for chalks and calcarenitic chalks. Solid circles represent pure chalks, whereas crosses designate coarser, calcarenitic chalks. Lines are least-square fits to data. (After Scholle, 1977, fig. 4, p. 989; courtesy Am. Assoc. Pet. Geol.)

According to Scholle (1977, p. 982), under normal conditions, a typical nannofossil chalk ooze has a 70% porosity at the water—sediment interface, which is reduced to about 35% at a depth of 1 km and to about 15% at a depth of 2 km. Porosity is essentially zero at a depth of 3 km; however, in areas such as the Ekofisk Field in the North Sea, large volumes of oil are produced from chalks having porosity (largely primary) as high as 40% at depths greater than 3 km. This is apparently due to the existence of overpressured conditions in this area (Central Graben).

Scholle (1977, p. 1005) concluded his excellent paper by the following profound statement: "Work on initially stable lithologies (such as chalks) may help to clarify the patterns of late diagenesis in other types of limestones as well."

In a recent book on *Compaction of Coarse-Grained Sediments, Vol. II*, Wolf and Chilingarian (1976) discussed compactional diagenesis of carbonate sediments and rocks. Compaction of coarse-grained carbonates (sands) was treated in detail in Vol. I of the same book by Coogan and Manus (1975).

In conclusion one can state that proliferation of literature on carbonate sediments and their diagenesis is so great that periodic (2—3 years) reviews in the form of books on the subject will constitute just as great a contribution to science as the original research work.

PRACTICAL APPLICABILITY OF DIAGENESIS

It is beyond dispute that a clear concept of the paleoenvironments is most important in the search for both non-metallic and metallic economic deposits in calcareous sediments. As many secondarily introduced accumulations are controlled by primary and secondary porosity and permeability, the investigation of diagenesis will, therefore, help in elucidating: (1) the locality and type of porosity and permeability; (2) permeability pinch-outs; (3) the factors that cause some non-reef limestones to be good reservoir rocks in one locality, whereas under seemingly similar conditions identical deposits are cap rocks; (4) the conditions that make shallow-water limestones either good or poor source rocks; (5) the likelihood of reef-flank and reef-core deposits occurring in one direction over others, and so forth. A useful approach for the reconstruction of paleoenvironments in oil exploration is the plotting of equal sparite, microsparite and micrite contents or ratios, among other parameters, as done by Stauffer (1962), for example; or plotting lines of equal Ca/Mg ratios as proposed by Chilingar (1953, 1956c). In one case, diagenetic and syngenetic features were used by Wolf (1965c) to prove a littoral environment of a reef complex. All this, of course, is based on a thorough understanding of diagenetic processes and products, for it is important to distinguish between cementation and recrystallization products, for instance.

Detailed work on carbonate diagenesis may also facilitate our understanding of certain metallic deposits in limestones as pointed out by Cloud et al. (1962), who mentioned that the location of particular lead, zinc and manganese deposits in carbonate rocks may well reflect some intrinsic chemical, biological, earlier diagenetic or textural property of the rock. Danchev and Ol'kha (1959), for example, studied uranium-bearing limestones to determine the parameters that controlled mineralization and the location thereof. They concluded that organic content controlled localization of the minerals, and that mineral dissemination predominates where the rock is least affected by recrystallization and leaching. The ore is of diagenetic origin and has been epigenetically redistributed.

GLOSSARY

Algal dust: angular to subangular medium- to dark-colored grains or crystals of carbonate, commonly 1—5 μ in diameter, derived from breakdown of algal felts, algally-precipitated aragonite needles, algal slime, and comminution of phytoplankton; associated with algal tubes, algal nodules, and other Algae or definite evidence thereof. Term proposed by Wood (1941), with certain details added by Carozzi (1960). "Algal dust" has also been called algal micrite (Wolf, 1965b) which occurs as allo- and automicrite types (Table 6-III).

Algal paste: dark gray to black finely-divided flecks, micrograined, microcrystalline, or cryptocrystalline in texture, forming a rather dense micritic limestone or dolomite, and associated with organic frame-builders such as corals, sponges, bryozoans, etc. Common, but not restricted, to the reef core. May actually represent compact, dense, diagenetically altered dust (Term used in a loose sense by Schlanger, 1957.)

Allo-: as used here, a prefix derived from allochthonous and indicating that the material has been transported before accumulation (Table 6-III).

Allochthonous: a term used here to designate sedimentary constituents which did not originate in situ; they were derived from outside of or within the area of depositional site and underwent transportation before final accumulation.

Allogenic: a term meaning "generated elsewhere," and applied to those constituents that came into existence outside of, and previously to, the rock of which they now constitute a part, e.g., extraclasts.

Aphanic: a term proposed by DeFord (1946) for the texture of carbonates, particularly limestones, in reference to crystalline (and/or grained) textures, the discrete particles of which are smaller than 0.005 mm (Table 6-III). Microcrystalline (also micrograined) and cryptocrystalline (also cryptograined) are the two textural subdivisions. *Aphanic* is used here to replace the term "aphanitic," which is loosely defined and is not utilitarian for carbonate rocks.

Apo-epigenesis: as used here, epigenesis affecting the sediments after diagenesis while they are far remote from the original environment of deposition under a relatively thick overburden. With an increase in temperature and pressure it grades into metamorphism (Wolf, 1963a).

Aragonite: a mineral, orthorhombic $CaCO_3$, dimorphous with calcite.

Authigenic: generated on the spot. Applied to those constituents that came into existence with or after the formation of the rock of which they constitute a part.

Auto-: as used here, a prefix derived from autochthonous and indicating that the material was formed in situ (Table 6-III).

Autochthonous: a term applied here to sedimentary rock components which originated and formed in situ, without undergoing prior transportation.

Bahamite: name proposed by Beales (1958) for the granular limestones that closely resemble the present deposits of the interior of the Bahama Banks, described by Illing (1954). The texture varies from calcisiltites to calcirudites, in which the grains are accretionary and commonly composite, consisting of smaller granules bound together by precipitated material into aggregate grains. Many misinterpretations of this rock type have been made (Wolf, 1965a, b).

Bank: a skeletal limestone deposit formed by organisms which do not have the ecologic potential to erect a rigid, wave-resistant structure. Contrasts with *reef,* which is a skeletal limestone deposit formed by organisms possessing the ecologic potential to erect a rigid, wave-resistant, topographic structure (Nelson et al., 1962).

Beach-rock: a friable to well-cemented beach sediment consisting of calcareous debris cemented by calcium carbonate (Ginsburg, 1953).

Biolithite: a term applied to faunal and/or floral organisms that grew and remained in situ (Folk, 1959).

Birdseye: spots or tubes of sparry calcite in limestones (Hall, 1847). Perkins (1963) pointed out that these "calcite eyes" are common to pelsparites, and may have resulted from one of the following (or certain combinations thereof): (1) precipitation of sparry calcite in animal burrows, or in worm tubes; (2) soft-sediment slumping or mud-cracking; (3) precipitation of sparry calcite in tubules resulting from escaping gas bubbles; (4) reworking and rapid redeposition of soft sediment to produce a rock with very vaguely defined proto-intraclasts, semicoherent clouds of calcareous mud, and irregular patches of spar; and (5) recrystallization of calcareous mud in patches. For alternative interpretation see Wolf (1965c).

Blady calcite: see *Cement.*

Boundstone: applies to most reef rock, stromatolites, and some biohermal and biostromal rocks in which the original components were bound together during deposition, and remain substantially in position of growth (Dunham, 1962).

Breccia: a rock made up of angular rock fragments, most of which are larger than 2.0 mm in diameter. Carozzi (1960) mentioned a recrystallization breccia that results from the differentiation in place of a homogeneous calcilutite. Recrystallization began at numerous points scattered throughout the rock but was incomplete, and as a result the recrystallized patches appear as fragments in a groundmass that was spared by the process. It is here recommended that the definition of Carozzi be expanded to include other carbonate rocks such as calcisiltites, calcarenites, etc.

Bryalgal: term proposed by Bissell (1964) for organic frame-building combination of bryozoans and Algae which create a rigid, wave-resistant limestone mass that forms banks, and is reefal or at least is intimately associated with reefs. The deposit is in situ; in some occurrences, one organism encrusts the other. Algae, for example, may encrust bryozoans; they may also encrust corals, stromatoporoids, sponges, and other framebuilders.

Calcareous: as used here, referring to calcitic and aragonitic material.

Calcilutite to calcirudite: a range of terms suggested by Grabau (1904; 1913) for limestones indicating the size of the calcareous components as given below:
calcilutite = clay-sized calcareous particles,
calcisiltite = silt-sized calcareous particles,
calcarenite = sand-sized calcareous particles,
calcirudite = gravel-sized calcareous particles.
(Compare with *Dololutite to Dolorudite.*)

Calcite: a mineral, calcium carbonate, $CaCO_3$, hexagonal—rhombohedral, dimorphous with aragonite.

Calclithite: a limestone containing 50% or more of fragments of older limestone eroded and redeposited (Folk, 1959). The individual fragments are called extraclasts (Wolf, 1963b, 1965b). (Compare with *Intraclast.*)

Caliche: it is a lime-rich deposit found in soils and formed by capillary action drawing the lime-bearing waters to the surface where, by evaporation, the lime is precipitated (Pettijohn, 1957). In bajadas, intermonts, alluvial fans and colluvium of parts of the Great Basin of Western United States some of the caliche deposits are dolomitic due to presence of extensive dolomite rubble. Caliche, whether calcareous and/or dolomitic, also cements alluvial fans to form *fanglomerate.*

Calcsparite: see *Sparite.*

Carbonate rock: a sedimentary rock composed of more than 50% calcite, aragonite, and/or dolomite.

Catagenesis: see *Epigenesis.*

Cement: chemically precipitated material into voids and in situ onto the surfaces of the host-framework. The calcareous cement in limestones may be of different crystal size-grades: micrite (often mistaken for detrital matrix), microsparite, and sparite. The morphological and textural types are granular, fibrous, blady, and drusy. Carbonate cement often resembles products formed by recrystallization and grain growth. The cryptocrystalline carbonate is too small to be resolved by an ordinary petrographic microscope and appears as a dense mass. The granular sparry cement consists of more or less equidimensional crystals. The fibrous sparite occurs as very thin elongate fibres, whereas the blady type has somewhat wider elongate crystals. The term drusy sparite does not refer to a single crystal but to the textural relation of aggregates composed of crystals that increase in size and elongation with increasing distance from the host. Drusy calcium carbonate usually changes into a granular type. (See Table 6-III and Fig. 6-2.)

Cementation: the process of chemical precipitation of material into voids and in situ onto the surfaces of the host-framework. (Compare with internal chemical sedimentation.)

Clast: an individual constituent of detrital sediment or sedimentary rock produced by the physical disintegration of a larger mass either within or outside the basin of accumulation. (See also *Extraclast* and *Intraclast.*)

Coated grains: grains possessing concentric or enclosing layers of calcium carbonate; for example, oolites, pisolites, superficial oolites, algal-encrusted skeletal grains (Leighton and Pendexter, 1962), and circumcrusts (Wolf, 1962, 1965b).

Cone-in-cone: a concretionary structure occurring in marls, etc., characterized by the development of a succession of cones one within another (Holmes, 1928).

Contemporaneous: existing together or at the same time in contrast to penecontemporaneous.

Coquina: carbonates consisting wholly, or nearly so, of mechanically sorted fossil debris. Most commonly applied to the more or less cemented coarse shell debris. For the finer shell detritus of sand size or less, the term *microcoquina* is more appropriate (Pettijohn, 1957).

Coquinite: indurated equivalent of coquina.

Criquina: coquina of crinoidal debris.

Criquinite: indurated equivalent of criquina.

Cryptocrystalline: crystalline material that is so fine that it cannot be resolved by a petrographic microscope (Williams et al., 1955). Electron microscope studies, however, show distinct crystalline features. Cryptocrystalline carbonate forms the finest part of the micrite (Table 6-III).

Dense: compact; having its parts crowded together. Not necessarily restricted to fine-textured carbonate rocks, although commonly applied in this sense by many petrographers.

Depocenter: contraction of depositional center; refers to an environment of sedimentary deposition, without particular restriction as to whether it is a basin, bank, shelf, trough, etc. (Murray, 1952). Geosynclines, particularly of the miogeosynclinal type, consist of basins, troughs, swells, banks, welts, incipient-to-prominent highs, accessways, thresholds, reefs, barriers, lagoons, hinge-lines, and various other repositories of sediment accumulation. Centers of carbonate deposition, i.e., depocenters, make up these repositories of geosynclines, intra-cratonic basins, platforms, etc. (Bissell, 1962).

Detrital limestone: limestone composed of fragments that have been transported before accumulating. (Detrital is synonymous with "allochthonous".)

Detritus: transported material not formed in situ. (Detritus is synonymous with allochthonous material, allochthonous fragments, and debris.)

Diagenesis: it includes all physicochemical, biochemical and physical processes modifying sediments between deposition and lithification, or cementation, at low temperatures and pressures characteristic of surface and near-surface environments. In general, diagenesis, is divisible into pre-, syn-, and post-cementation or lithification processes. Diagenesis, as defined in this chapter, takes an intermediate position between syngenesis and epigenesis, the former grading into diagenesis by syndiagenesis, and the latter grading into metamorphism. Under unusual conditions, however, diagenesis as defined here may grade directly into metamorphism (see epigenesis). Because reef limestones, and other limestones which are constructed in situ by organic frame-builders, are largely in and of themselves lithified to a degree, the definition must be expanded for this particular group of limestones to include the interactions between sediments and the fluids contained within them below the temperature and pressure levels of metamorphism sensu stricto, and in a similar sense between fluids and framework, infilled detritus framework, and combinations thereof.

Dololutite to dolorudite: a range of terms applied to sedimentary dolomites composed of constituents ranging in size from clay to gravel, similar to those in limestones, as follows:
dololutite = clay-sized dolomite particles,
dolosiltite = silt-sized dolomite particles,
dolarenite = sand-sized dolomite particles,
dolorudite = gravel-sized dolomite particles.

Dolomite: (1) a mineral, $CaMg(CO_3)_2$, hexagonal rhombohedral. (2) A carbonate rock composed predominantly of the mineral dolomite; in normal routine petrographic work, dolomite (or dolostone of some geologists) is a carbonate rock composed of more than 50% by weight of the mineral dolomite. More practically, areal percentages are used instead of weight percentages.

Dolomitic: where used in a rock name, "dolomitic" refers to those rocks that contain 5—50% of the mineral dolomite, as cement and/or grains or crystals. Dolomitic can be applied to the large spectrum of sedimentary rocks that are dolomite-bearing, and also to limestones which have been dolomitized to a degree but not completely.

Dolomitic mottling: incipient or arrested dolomitization, or arrested (or incomplete) dedolomitization. Common to limestones that have large particulate skeletal or non-skeletal material embedded in finer-textured matrix. Under the effects of dolomitization there is a preferential replacement or alteration of the matrix but not of the large particles. Also common to more or less homogeneous textured limestones that have been incompletely dolomitized, leaving patches, blotches, laminae, or other structures unaffected.

Dolomitized: refers to rocks or portions of rocks in which limestone textures are discernible, but which have been changed to dolomite.

Dolosparite: see *Sparite.*

Drusy: see *Cement.*

Earthy: refers to a variety of slightly argillaceous carbonate with earthy texture generally closely associated with chalky deposits and commonly showing similar porosity values. Microtextured (0.01 mm and slightly less) (Thomas, 1962).

Endogenic: as used here, referring to components derived from within the sedimentary formation.

Epigenesis (or *catagenesis*) as used here, it includes all processes at low temperature and pressure that affect sedimentary rocks after diagenesis and up to metamorphism. Epigenesis has been subdivided into juxta- and apo-epigenesis (Wolf, 1963b, 1965c). It is possible that under unusual conditions syngenesis and diagenesis grade directly into metamorphism. For example, unconsolidated sediments may be exposed to volcanic high temperatures and metasomatic material and undergo metamorphism before diagenesis. Also, sediments partly undergoing cementation may be metamorphosed by shallow intrusions causing an increase of temperature and possibly pressure before epigenesis could occur:

$$\left.\begin{array}{l}\text{syngenesis}\\ \text{diagenesis}\end{array}\right\} \rightarrow \text{metamorphism}$$

epigenesis
↓
metamorphism

Evaporite-solution breccia: solution breccias are created when intervening soluble evaporites—salt, anhydrite, gypsum, etc.—are dissolved away, letting the carbonate beds crush under the weight of overlying sediments. An extremely angular collapse breccia results, in which the matrix is of essentially the same material as the rock fragments (Sloss and Laird, 1947; Greiner, 1956). These chaotic breccias normally are associated with evaporites, and may also be adjacent to reef limestones which, attendant to removal of the evaporites, collapse and may be "healed" or cemented by calcareous and/or dolomitic material.

Exogenic: as used here, referring to components derived from outside, i.e., from either above or below, the sedimentary formation.

Extraclast: fragment of calcareous sedimentary material produced by erosion of an older rock outside the depocenter in which it accumulated (Wolf, 1963b, 1965a, b). (Compare with *Intraclast* and *Calclithite.*)

Fibrous: see *Cement.*

Flour: chalky-appearing, finely comminuted material in limestones or dolomites, generally formed by disintegration and abrasion of fossil debris and algal growths under intense wave action, surf-surge, and current action in shoal areas. It may represent clay-sized particulate carbonate mud formed through attrition, or may result from chemical flocculation, biochemical activity, or through other means. These micro-grained, chalky carbonates may be due to disintegration and abrasion of fossil detritus on banks and shelves that are subject to the high energies of waves and currents.

Grain growth: this process acts in monomineralic rocks of low porosity. The intergranular boundaries migrate causing some grains to grow at the expense of their neighbors. The reaction takes place in the solid state, ions being transferred from one lattice to another without solution. Larger grains tend to replace smaller ones, and a fine mosaic is gradually replaced by a coarser. As grain growth proceeds, many of the enlarged grains are themselves replaced by their more successful neighbors (Bathurst, 1958). In limestones grain growth appears to affect only the very fine mosaics with grain diameters ranging from 0.5 to 4.0 μ. These include calcite-mudstones, the walls of Foraminifera, algal frameworks, bahamite particles, and ooliths (Bathurst, 1959b).

Grainstone: mud-free carbonate rocks, which are necessarily grain-supported, are termed grainstone; some are current laid, whereas others form as a result of mud being bypassed while locally produced grains accumulate, or of mud being washed (= winnowed) out (Dunham, 1962).

Grain-supported: carbonate sedimentary rock in which grains are so abundant as to support one another, just as they do in mud-free rocks (Dunham, 1962).

Granular: see *Cement.*

Grumous: a term signifying clotted, aggregated, flocculated. As applied to sedimentary carbonate rocks it refers to micro- and macroscopic aggregation of lime-mud particles and other flocculated or otherwise clotted and aggregated, irregularly-shaped material. In a sense comparable to bahamite, but commonly of smaller dimension.

Halmyrolysis: the chemical rearrangements and replacements that occur while the sediment is still on the sea floor (Pettijohn, 1957). It is sometimes called submarine weathering.

Hypogenic: a term applied to material that is derived from within the earth interior in contrast to supergenic components. (See also *Supergenic*.)

Impingement: a mechanism or process in dolomitization in which dolomite crystals replace limestone, commonly skeletal particles such as crinoid ossicles and plates, but not in optical continuity with the calcite of the original particle (Lucia, 1962).

Intergranular porosity: void space between grains, whether bioclastic or lithoclastic. In sedimentary carbonate rocks the term granular commonly refers to the grains, whether skeletal or nonskeletal.

Internal filling: a collective term including both internal sediments and cement that fill cavities within a sedimentary formation (Wolf, 1963a, 1965c).

Internal chemical sediment: allochthonous chemically precipitated sediment both formed and deposited intraformationally in cavities (Wolf, 1963a, 1965c). (Compare with *Internal mechanical sediment* and *Internal filling.*)

Internal mechanical sediment: allochthonous clastic sediment brought in from the surface, or derived by intraformational abrasion, and deposited in cavities within the sedimentary formation (Wolf, 1963a, 1965c).

Internal sedimentation: allochthonous sediment derived from the surface or from within the rock framework and accumulated in cavities within the sedimentary rock formation. It is a collective term including both mechanical and chemical internal sediments (Wolf, 1963a, 1965c).

Interstitial: of, pertaining to, existing in, or forming an interstice or interstices.

Intraclast: fragment of more or less consolidated calcareous sedimentary material produced by erosion within a basin of deposition and redeposited there (Folk, 1959). (Compare with *Extraclast.*)

Intraformational: formed by, existing in, or characterizing the interior of a geological formation.

Intragranular porosity: pore space or voids within individual particles, particularly skeletal material. Of significance in leached ostracodal, Foraminiferal, algal, and oolitic limestones, but, like intergranular porosity is sometimes adversely affected by diagenetic processes.

Inversion: the process by which unstable minerals change to a more stable form of the same chemical composition (except for a possible change in contents of trace elements and/or isotopes) but with a different lattice structure.

Juxta-epigenesis: epigenesis affecting the sediment after diagenesis while it is near the original environment of deposition either under a relatively thin overburden or, if regression occurred, while exposed above sea level (Wolf, 1963a). (Compare with *Epigenesis* and *Apo-epigenesis.*)

Limestone: a sedimentary rock composed of at least 50% calcium carbonate material. For practical microscopic work, it is a carbonate consisting of 50% or more, by areal percentage, of calcite or calcareous material.

Lithification: that complex of processes that converts a newly deposited sediment into an indurated rock. It may be contemporaneous with, or occurs shortly or long after deposition.

Lithoclastic: autochthonous and allochthonous carbonate detritus: mechanically formed and deposited carbonate clasts, derived from previously formed limestones and/or dolomites, within, adjacent to, or outside the depositional site. (See also *Detrital, Limeclast, Intraclast, Extraclast,* and *Calclithite.*)

Lithographic: pertaining to a compact carbonate rock having about the same particle size and textural appearance as the stone used in lithography (DeFord, 1946). Characterized by conchoidal fracture. Numerous micro- and crypto-textured micritic limestones and dolomites are lithographic.

Littoral: belonging to, inhabiting or taking place on or near the shore between low-tide and high-tide level. (See *Sublittoral* and *Supralittoral.*)

Lump: a descriptive term applied to an aggregate grain composed of two or more pellets, oölites, skeletons, etc., or fragments thereof. The aggregation or accretion can form by physicochemical, algal, and weathering processes. Calcareous grains such as pellets lying in contact with each other on the sea bed tend to become cemented or welded together and form lumps (Illing, 1954; Wolf, 1965b).

Matrix: if the particles in the calcareous rock are of different orders of size grades, the term matrix is used for the material that fills the interstices between the larger grains. Matrix is thus the material in which any sedimentary particle is embedded. The matrix may be either microtextured or granular. With an increase in matrix percentage, a limestone grades into a deposit composed solely of micrite, of calcisiltite, or of calcarenite. Granular matrices tend to become more poorly sorted as particle size increases. Some prefer to restrict "matrix" to clay-sized or micritic components surrounding coarser material.

Metamorphism: this term refers to the mineralogic, textural and structural adjustment of solid rocks to physical or chemical conditions at higher temperatures and pressures than those under which the rock in question originated.

Micrite: a descriptive term for calcareous crystalline and/or grained materials less than 0.005 mm in diameter (Table 6-III) as used here. (Folk, 1959, used 0.004 mm, whereas Leighton and Pendexter, 1962, drew the limit at 0.031 mm). Micrite that is so finely crystalline that it cannot be resolved by a petrographic microscope is called "cryptocrystalline". It is consolidated or unconsolidated ooze or lime-mud of either chemical or mechanical origin, and possibly of biologic, biochemical, and physico-chemical origin. It is used by some geologists as synonymous with calcilutite (clay-sized particles). The exact range of both micrite and calcilutite, however, has been differently placed by other workers. (See *Orthomicrite* and *Pseudomicrite*, and Table 6-III).

Micrograined: a grain-size term pertaining to carbonate particles smaller than 0.02 mm and larger than 0.005 mm in diameter; *microclastic* is more or less synonymous.

Microsparite: see *Sparite*.

Mud aggregate: any aggregate of mud grains, usually having the size of a sand or silt particle, which has been mechanically deposited. Initially, the aggregate may have been a faecal pellet, or a rounded, sub-spherical aggregate of mud grains cemented originally by aragonite with no signs of organic control, or a fragment of algal precipitate, or a spherical or ovoid growth form of a calcareous alga (Bathurst, 1959b).

Mudstone: muddy carbonate rocks containing less than 10% grains (10% of bulk); the name is synonymous with calcilutite, except that it does not specify mineralogic composition, and does not specify that the mud is of clastic origin (Dunham, 1962).

Mud-supported: muddy carbonate rock which contains more than 10% grains, but not in sufficient amount to be able to support one another; such grains are "floating," and thus they are mud-supported.

Oolite or oolith: a spherical to ellipsoidal body up to 2 mm in diameter which may or may not have a nucleus, and has concentric or radial structure or both. It is accretionary. The term is descriptive. At least three genetic possibilities exist identical to those mentioned for pisolites. If particles lack concentric or radial features, one should refrain from calling them false or pseudo-oolites, but name them pellets. (See *Pisolite* and *Pellet.*)

Open-space structures: they are structures in carbonate rocks which formed by the partial or complete occupation with internal fillings composed of internal sediments and/or cement of one to several generations (Wolf, 1963a, 1965c).

Organic lattice: reef-building framework, and some bank deposits constructed by organic frame-builders, in situ.

Orthomicrite: it is a genetic term applied to micrite that has not undergone secondary changes such as recrystallization and grain growth. Two types are recognizable: allochthonous and autochthonous micrite named allomicrite and automicrite, respectively (Wolf, 1963b). (See *Micrite* and *Pseudomicrite*, and Table 6-III.)

Orthosparite: see *Sparite.*

Paragenesis: a general term for the order of formation of associated minerals, textures, and structures in time succession, one after another.

Pelagosite: it is a deposit generally white, gray to brownish with a pearly luster, composed of $CaCO_3$ with higher $MgCO_3$, $SrCO_3$, $CaSO_4$, H_2O and SiO_2 contents than in normal limy sediments (Revelle and Fairbridge, 1957).

Pellet: a spherical, sub-spherical, ovoid, to irregular-shaped small particle composed of clay-sized to fine silt-sized material and devoid of any internal structure. Micrite pellets have been called pseudo-oolites, false oolites, etc. Three genetic types appear to be of significance: faecal, bahamite, and algal pellets.

Penecontemporaneous: a term used in connection with the formation of sedimentary rocks, and implies "formed at almost the same time". (Compare with *Contemporaneous.*)

Pressure-solution: a preferential solution takes place on the higher stressed parts of a grain and deposition of matter on surfaces with lower potential energies. The pressure is supplied by the overburden and should result in a recognizable grain fabric, with the grains flattened at right angles to the pressure. Regarded as perhaps the most important process in closing the original pore space of sediment (Bathurst, 1958, 1959b). Microcrystalline calcite can recrystallize by pressure-solution into a mosaic of larger crystals by the solution of the smallest, supersoluble grains and redeposition on the larger grains (Stauffer, 1962).

Primary: characteristic of or existing in a rock at the time of its formation. This definition is too all-inclusive and vague in detailed studies and its use should be discouraged. It can be used unambiguously as a very general colloquial term in connection with genetic discussions only if the context leaves absolutely no doubt. (See *Secondary.*)

Pseudobreccia: masses of grain-growth mosaic which lie in a "matrix" of less altered limestone; most of these are visible to the naked eye. The "fragments" are irregularly shaped patches of coarse calcite mosaic usually between 1 and 20 mm in diameter, and are dark gray in handspecimen. They lie in the finer, pale-gray "groundmass" of calcite-mudstone. In thin-section the "fragments" appear light and the "groundmass" dark (Dixon and Vaughan, 1911; Bathurst, 1959b).

Pseudomicrite: it is a genetic term applied to micrite formed by secondary changes such as "grain diminution" or "degenerative recrystallization" of faunal and floral material (Wolf, 1963b). The causes of this process are still poorly understood. (See *Micrite* and *Orthomicrite*, and Table 6-III.)

Pseudomorphic replacement: a diagenetic process whereby the original character of a limestone is altered during dolomitization; skeletal material, and specifically crinoidal material for example, is replaced in such a manner that single crystals of dolomite are in optical continuity with the calcite of the original crinoid fragment. The term contrasts with the process termed *impingement,* which does not give rise to the optical continuity of dolomite crystals with the original crinoid fragment (Lucia, 1962).

Pseudo-oolites: calcareous pellets which have cryptocrystalline and/or microcrystalline internal texture, and are of similar size and shape to oolites but lack concentric structure. These particles can form as faecal, bahamite, and algal pellets, whereas others are formed by the abrasion of micritic limestones. In general, "pseudo-oolite" is a synonym of "pellet".

Pseudosparite: see *Sparite.*

Recrystallization: this term is usually used loosely for a number of processes that include inversion, recrystallization sensu stricto, and grain growth, all of which may result in textural and crystal-size changes. Recrystallization proper occurs when nuclei of new, unstrained grains or crystals appear in or near the boundaries of the old, strained ones. These nuclei grow until the old mosaic has been wholly replaced by a new, relatively strain-free mosaic with a nearly uniform grain size. Its coarseness depends on the density of the initial nucleation. Where the nuclei are widely spaced there is an intermediate porphyroblastic stage (Bathurst, 1958). As used by Folk (1959), recrystallization is a process wherein original crystal units of a particular size and morphology become converted into crystal units with different grain size or morphology, but the mineral species remains identical before and after the process occurs. Bathurst (1958, 1959b) presented criteria for recognition of various diagenetic fabrics and made a plea for the elimination of the term "recrystallization" in favor of specific recognition of the individual process.

Aggradation recrystallization results in the enlargement of the crystals, whereas degradation recrystallization gives rise to a relative decrease in size of crystals or grains. The latter process has also been termed "grain diminution" and "degenerative recrystallization" (see text).

Reef: a structure erected by frame-building or sediment-binding organisms. At the time of deposition, the structure was a wave-resistant or potentially wave-resistant topographic feature. A reef is thus a skeletal deposit. By contrast, a *bank* is a skeletal limestone deposit formed by organisms which do not have the ecologic potential to erect a rigid, wave-resistant structure. Reef and bank deposits, therefore, denote origin, whereas the terms *bioherm* and *biostrome* denote shape (Lowenstam, 1950; Cloud, 1952; Nelson et al., 1962).

Reef complex: the aggregate of reef, fore-reef, back-reef, and inter-reef deposits which are bounded on the seaward side by the basin sediments and on the landward side by the lagoonal sediments. (See Nelson et al., 1962, for an exhaustive treatment of skeletal limestones, including reef terminology.)

Reef milk: matrix material of the back-reef facies, consisting of microcrystalline white and opaque calcite ooze, and derived from abrasion of the reef core and reef flank (Hambleton, 1962).

Reef tufa: fibrous calcite which forms thin to thick deposits, layered or unlayered, in the myriads of voids in reef and other organic frame-builders; the fibrous calcite is prismatic in structure and is radial in respect to the depositional surfaces. The fibrous calcite, or reef "tufa" is deposited directly upon the framework of the reef and within the various voids and interstices, from supersaturated waters. The mechanism may be largely physicochemical, or, aided by profuse algal growth to extract CO_2 from the water, may also be biological to biochemical. Development of reef tufa follows and/or accompanies growth of organic frame-builders, and precedes infilling of detritus such as lime-mud, calcarenite, etc. (Newell, 1955; Parkinson, 1957; Wolf, 1965c). Many types of "reef tufa" have been called "stromatactis".

Reefal: as used herein, a purely descriptive and non-genetic term having reference to carbonate deposits in and adjacent to any of the numerous varieties of reefs and to any or all of their intergral parts.

Rim cement: cement which grows into interparticle voids and is optically continuous on single crystal particles such as crinoid fragments. Thus, the host is a single crystal and the cement forms a single rim in lattice continuity with it. The overgrowth is a continuation of this crystal, and the overgrowth can form by filling the pore space (Bathurst, 1958).

Saccharoidal: a descriptive term which, in general, means "sugary" texture. More specifically it is a product or result of dolomitization in which crystallization or recrystallization creates a new texture. It may be first-stage crystallization, but more commonly is recrystallization that occurs early in the newly-deposited lime-mud. It does not alter gross primary structures of the sediment such as ripple mark, thin bedding, etc., but does tend to destroy minor structures such as shells of organisms. Saccharoidal texture is recognized by the well-developed rhombs of dolomite of approximately uniform size resting one against the other with point contact and commonly separated by exceptionally large as well as small pore openings. The fabric displays loose packing, and suggests that dolomitization occurred when the grains were loose and before compaction altered the original texture (i.e., a packing typical of loose beach and shoreline sands). Recrystallization of the original calcite grains destroys the original particle-size distribution and substitutes a new, highly restricted crystal-size distribution ranging from medium- to coarse-sand dimensions (Dapples, 1962).

Secondary: a general term applied to rocks and minerals formed as a consequence of alteration. This term is too all-inclusive and ambiguous in detailed studies and should be used only as a very general colloquial term when misinterpretation is absolutely impossible. (See also *Primary*.)

Skeletal: pertaining to debris derived from organisms that secrete hard material around or within organic tissue. The term *bioclastic* is considered to be synonymous with skeletal. (Nelson et al., 1962, use "skeletal" in a somewhat different sense. See also Leighton and Pendexter, 1962, for discussion of terms skeletal and skeletal limestone.)

Solution transfer: this is a translation of the German *Lösungsumsatz*. It refers to the solution of detrital particles around their points of contact where elastic strain and solubility are enhanced (pressure-solution), followed by redeposition on less strained particle surfaces (Bathurst, 1959b).

Sparite: it is an abbreviation of, and is therefore synonymous with, sparry calcite. Sparite, as used here, is a descriptive term applied to any transparent or translucent crystalline calcite and aragonite. It can occur in numerous morphologic forms, namely, granular, drusy, fibrous, and blady. Three possible origins are recognized: (1) precipitation into open voids, (2) recrystallization, and (3) grain growth. The first is distinguished by adding the genetic prefix ortho-, and the latter two by pseudo-. Microsparite ranges in diameter from 0.005 mm to 0.02 mm, whereas sparite is larger than 0.02 mm (Table 6-III).

The prefix dolo- is used to indicate sparry dolomite crystals, i.e., dolomicrosparite and dolosparite. Some workers prefer the prefix calc- to distinguish calcsparite from the dolomitic variety, but to some sparite is automatically understood to mean the calcareous variety.

Sparry: see *Sparite.*

Spherulite: as used here, a small spherical or spheroidal particle composed of a thin dense calcareous outer layer with a sparry calcite core. It can originate by recrystallization or grain growth and the central sparite is then a typical pseudosparite. On the other hand, spherulites can be formed as minute bodies by biological processes and the open space is then filled by orthosparite, e.g., *Calcisphaera* of algal origin. As defined by Pettijohn (1957), however, spherulites are minute bodies of oölitic nature in which only a radial structure is visible. The surfaces of such bodies, unlike those of oölites, are somewhat irregular.

Stromatactis: these are open-space structures with horizontal flat to nearly flat bottoms, and are filled by internal sediments and/or cement. They have been termed "reef tufa" by some. Their genesis has been variously interpreted as being caused by the burial of soft organisms which upon decomposition left an open space. More recent studies, however, show that they are most likely syngenetic voids in calcareous sediments, which are or are not changed by subsequent corrosion and corrasion. Algae are only indirectly responsible by overgrowing surface pits and channels, and thus form an internal cavity system (Wolf, 1963a, 1965c). It seem that *Stromatactis* are most common in micritic limestones formed by calcareous Algae, that left little or no evidence in most occurrences in Great Britain, North America, etc., but are well preserved in one Australian locality (Plates 6-I—XXIV).

Schwarzacher (1961) described the fabric of some Lower Carboniferous reefs of northwestern Ireland, and noted that in some places calcite grows into what at one time have been hollow spaces; this was interpreted to represent either recrystallization phenomena or remains of frame-building organisms, i.e., *Stromatactis.* Schwarzacher referred to Bathurst (1950) who recognized the cavity nature of structures he described under the name of *Stromatactis,* and tentatively interpreted them as hollow molds of organisms which presumably disappeared at an early stage in diagenesis. Lowenstam (1950) regarded *Stromatactis* as a rigid frame-building organism. (See *Reef tufa.*)

Stromatolite: laminated sediment formed by calcareous Algae, which bind fine detritus and/or calcium carbonate precipitated biochemically. The deposit may form irregular accumulations or structures that may remain fairly constant in shape, e.g., *Collenia.*

Subaerial: formed, existing, or taking place on the land, in contrast to subaqueous.

Sublittoral: belonging to, inhabiting or taking place in the bottom environment extending from low-tide level to approximately 100—150 ft below low-tide level.

Sucrosic: contraction of saccharoidal, thus meaning "sugary" texture.

Supergenic: a term applied to those processes and products caused by material derived from descending fluids and gases. (See *Hypogenic.*)

Supralittoral: belonging to, inhabiting or taking place in the near-shore region above high-tide level.

Syndeposition: see *Syngenetic.*

Syneresis cracks or vugs: cracks or vugs formed by a spontaneous throwing off of water by a gel during aging. Thomas and Glaister (1960) pointed out that in some Mississippian carbonates of the Western Canada Basin calcium carbonate evidently was precipitated as a colloidal gel encrusting leaves of sea plants (photochemical removal of carbon dioxide from sea water by the plants, causing precipitation). The end-result was the production of cryptograined limestone which contains "syneresis" cracks and associated primary contraction vugs. When these vugs are filled by sparite, they resemble "birdseyes".

Syngenesis: as used here, the processes by which sedimentary rock components are formed simultaneously and penecontemporaneously. Syngenesis has been subdivided into syndeposition and pre-diagenesis. The former comprises processes responsible for the formation of the sedimentary framework, whereas the latter is responsible for those parts that were introduced subsequently but before the principal processes of diagenesis began, i.e., internal mechanical sedimentation. The latter does not constitute part of diagenesis because its products are formed by ordinary sedimentary deposition and do not alter the sedimentary framework as such.

Syntaxial rim: a mechanism of replacement overgrowth, which develops during diagenesis as a syntaxial extension of a detrital single crystal (e.g., a crinoid fragment). During recrystallization or grain growth some of the newly formed crystals become optically oriented with a detrital grain, commonly a crinoid ossicle, and form the so-called syntaxial rim. It is not to be confused with similar optically oriented overgrowth formed by chemical precipitation of calcium carbonate in voids (Bathurst, 1958, 1959b). In studying diagenetic effects of a crinoidal sediment, Lucia (1962) observed that the textural relationship between lime-mud and calcite overgrowth suggest that rim cementation is the dominant process; furthermore, dolomitization occurred after rim cementation.

Terrigenous: land-derived; refers particularly to sediments resulting from erosion of the land.

Travertine: calcium carbonate, $CaCO_3$, usually of light color and commonly concretionary and compact, deposited from solution in ground and surface waters. It is a more dense and often banded variety, in contrast to tufa. (See also *Tufa* and *Caliche.*)

Tufa: a chemical, spongy, porous sedimentary rock composed of calcium carbonate, deposited from solution in the water of a spring or of a lake, or from percolating ground-water. (See also *Reef tufa, Travertine, Caliche.*)

Vaterite: it is a metastable hexagonal form of calcium carbonate, $CaCO_3$. It is doubtful if it occurs in the geologic column, but if so, such occurrences are rare.

Welding: term used in reference to crystal welding, in which discrete crystals and/or grains become attached one to another during compaction and in large measure through diagenesis. Pressure-solution, and solution transfer are likely the operative processes. Welding can continue beyond normal diagenesis to epigenesis.

Winnow: the Old English word is *windwian,* and has reference to exposure to the wind such that lighter particles are blown away, thus winnowing grain, and the word winnow is a contrivance of winnowing grain. In this chapter winnowing can apply only to eolianites, and not to water-moved limestones. The term *washed* is preferred in the latter case.

REFERENCES

Aitken, J.D., 1967. Classification and environmental significance of cryptalgal limestones and dolomites, with illustrations from the Cambrian and Ordovician of Southwestern Alberta. *J. Sediment. Petrol.,* 37: 1163—1178.

Alderman, A.R. and Skinner, H.C.W., 1957. Dolomite sedimentation in the southeast of South Australia. *Am. J. Sci.,* 225: 561—567.

Alderman, A.R. and Von der Borch, C.C., 1963. A dolomite reaction series. *Nature,* 198: 465—466.

Alexandersson, E.T., 1976. Actual and anticipated petrographic effects of carbonate undersaturation in shallow sea water. *Nature,* 262 (5570): 653—657.

Alexandersson, E.T., 1972. Intragranular growth of marine aragonite and Mg-calcite: evidence of precipitation from supersaturated seawater. *J. Sediment. Petrol.,* 42: 441—460.

Al-Hashimi, W.S. and Hemingway, J.E., 1973. Recent dedolomitization and the origin of the rusty crusts of Northumberland. *J. Sediment. Petrol.,* 43: 82—91.

Amstutz, G.C., 1959. Syngenese und Epigenese in Petrographie und Lagerstättenkunde. *Schweiz. Mineral. Petrogr. Mitt.,* 39: 2—84.

Amstutz, G.C. (Editor), 1964. *Sedimentology and Ore Genesis.* Elsevier, Amsterdam, 184 pp.

Anderson, F.W., 1950. Some reef-building Algae from the Carboniferous rocks of northern England and southern Scotland. *Proc. Yorkshire Geol.* Soc., 28: 5—28.

Andrichuk, J.M., 1958. Stratigraphy and facies analysis of Upper Devonian reefs in Leduc, Stettler and Redwater areas, Alberta. *Bull. Am. Assoc. Pet. Geol.,* 42: 1—93.

Andrichuk, J.M., 1960. Facies analysis of Upper Devonian Wabamun Group in west-central Alberta, Canada. *Bull. Am. Assoc. Pet. Geol.,* 44: 1651—1681.

Baars, D.L., 1962. Permian System of Colorado Plateau. *Bull. Am. Assoc. Pet. Geol.,* 46: 149—218.

Baars, D.L., 1963. Petrology of carbonate rocks. In: *Shelf Carbonates of the Paradox Basin — Four Corners Geol. Soc., Field Conf., 4th,* pp. 101—129.

Banerjee, A., 1959. Petrography and facies of some Upper Viséan (Mississippian) limestones in North Wales. *J. Sediment. Petrol.,* 29: 377—390.

Barghoorn, E.S., Meinschein, W.G. and Schopf, J.W., 1965. Paleobiology of a Precambrian shale. *Science,* 148: 461—472.

Barnes, I., 1965. Geochemistry of Birch Creek, Inyo County, California, a travertine depositing creek in an arid climate. *Geochim. Cosmochim. Acta*, 29: 85—112.
Barnes, I. and Back, W., 1964. Dolomite solubility in ground water. *U.S. Geol. Surv., Prof. Pap.*, 475-D: D179—D180.
Bathurst, R.G.C., 1958. Diagenetic fabrics in some British Dinantian limestones. *Liverpool Manchester Geol. J.*, 2: 11—36.
Bathurst, R.G.C., 1959a. The cavernous structure of some Mississippian stromatactis reefs in Lancashire, England. *J. Geol.*, 67: 506—521.
Bathurst, R.G.C., 1959b. Diagenesis in Mississippian calcilutites and pseudobreccias. *J. Sediment. Petrol.*, 29: 365—376.
Bathurst, R.G.C., 1971. *Carbonate Sediments and their Diagenesis.* Elsevier, Amsterdam-London-New York, 620 pp.
Bathurst, R.G.C., 1975. *Carbonate Sediments and their Diagenesis* (2nd ed.). Elsevier, Amsterdam-London-New York, 658 pp.
Bausch, W.M., 1965. Dedolomitisierung und Recalcitisierung in fränkischen Malmkalken. *Neues Jahrb. Mineral. Monatsh.*, 1965 (3): 75—82.
Bausch, W.M. and Wiontzek, H., 1961. Petrografische Untersuchungen am Hauptdolomit von Rehden. *Erdöl, Kohle, Erdgas, Petrochem.*, 14: 686—692.
Baxter, J.W., 1963. Experimental replacement of oölitic limestone by fluorite. *Trans. Ill. State Acad. Sci.*, 56: 53—58.
Beales, F.W., 1953. Dolomitic mottling in Palliser (Devonian) Limestone, Banff and Jasper National Parks, Alberta. *Bull. Am. Assoc. Pet. Geol.*, 37: 2281—2293.
Beales, F.W., 1956. Conditions of deposition of Palliser (Devonian) Limestone of southwestern Alberta. *Bull. Am. Assoc. Pet. Geol.*, 40: 848—870.
Beales, F.W., 1958. Ancient sediments of Bahaman type. *Bull. Am. Assoc. Pet. Geol.*, 42: 1845—1880.
Bein, A., 1977. Shelf basin sedimentation: mixing and diagenesis of pelagic and clastic Turonian carbonates, Israel. *J. Sediment. Petrol.*, 47 (1): 382—391.
Belyea, H.R., 1955. Cross-sections through the Devonian Systems of the Alberta Plains. *Geol. Surv. Can., Pap.*, 55-3.
Bergenback, R.E. and Terriere, R.T., 1953. Petrography of Scurry Reef, Scurry County, Texas. *Bull. Am. Assoc. Pet. Geol.*, 37: 1014—1029.
Berger, W.H., 1972. Deep-sea carbonates: dissolution facies and age-depth constancy. *Nature*, 236 (5347): 392—395.
Birse, D.J., 1928. Dolomitization processes in the Paleozoic horizons of Manitoba. *Trans. R. Soc. Can., Sect. IV*, 22: 215—222.
Bischoff, J.L., 1958. Kinetics of calcite nucleation: magnesium ion inhibition and ionic strength catalysis. *J. Geophys. Res.*, 73 (10): 3315—3321.
Bischoff, J.L., 1969. Temperature controls on aragonite-calcite transformation in aqueous solution. *Am. Mineral.*, 54: 149—155.
Bissell, H.J., 1959. Silica in sediments of the Upper Paleozoic of the Cordilleran area. In: *Silica in Sediments — Soc. Econ. Paleontol. Mineral., Spec. Publ.*, 7: 150—185.
Bissell, H.J., 1962. Pennsylvanian and Permian rocks of Cordilleran area. In: *Pennsylvanian System in the United States.* Am. Assoc. Pet. Geol., Tulsa, Okla., pp. 188—263.
Bissell, H.J., 1963. Lake Bonneville: geology of Southern Utah Valley, Utah. *U.S., Geol. Surv., Prof. Pap.*, 257-B: 101—128.
Bissell, H.J., 1964. Ely, Arcturus and Park City Groups (Pennsylvanian-Permian) in eastern Nevada and western Utah. *Bull. Am. Assoc. Pet. Geol.*, 48: 565—636.
Bissell, H.J. and Chilingar, G.V., 1958. Notes on diagenetic dolomitization. *J. Sediment. Petrol.*, 28: 490—497.

Bissell, H.J. and Chilingar, G.V., 1962. Evaporite type dolomite in salt flats of western Utah. *Sedimentology*, 1: 200—210.
Black, W.W., 1952. The origin of the supposed tufa bands in Carboniferous reef limestones. *Geol. Mag.*, 89: 195—200.
Black, W.W., 1954. Diagnostic characters of the Lower Carboniferous knoll-reefs in the north of England. *Trans. Leeds Geol. Assoc.*, 6: 262—297.
Borchert, H., 1962. Der Wassergehalt bei der Metamorphose der Kalisalze. *Ber. Geol. Ges.*, 1: 145—194.
Borchert, H., 1964. Über Faziestypen von marinen Eisenerzlagerstätten. *Ber. Geol. Ges.*, 2: 161—193.
Bosellini, A. and Ginzburg, R.N., 1971. Form and internal structure of Recent algal nodules (rhodolites) from Bermuda. *J. Geol.*, 79 (6): 669—682.
Botvinkina, L.N., 1960. Diagenetic stratification. *Dokl. Akad. Nauk S.S.S.R., Earth Sci. Sect.*, 125: 213—215.
Boyer, B.W., 1972. Grain accretion and related phenomena in unconsolidated surface sediments of the Florida reef tract. *J. Sediment. Petrol.*, 42: 205—210.
Bramkamp, R.A. and Powers, R.W., 1958. Classification of Arabian carbonate rocks. *Bull. Geol. Soc. Am.*, 69: 1305—1318.
Braun, H., 1961. Die Entstehung der marin-sedimentaren Eisenerzlagerstätten Norddeutschlands. *Z. Erzbergbau Metallhuettenges.*, 14: 466—484.
Brown, C.W., 1959. Diagenesis of Late Cambrian oölitic limestone, Maurice Formation, Montana and Wyoming. *J. Sediment. Petrol.*, 29: 260—266.
Bruevich, S.V. and Vinogradova, E.G., 1947. Chemical composition of interstitial solutions of Caspian Sea. *Gidrokhim. Mater.*, 13 (1/2).
Bryner, L.C., Beck, J.V., Davis, D.B. and Wilson, D.G., 1954. Micro-organisms in leaching sulfide minerals. *Ind. Eng. Chem.*, 46: 2587—2592.
Buchta, H., Leutner, R. and Wieseneder, H., 1963. The extractable organic matter of pelite and carbonate sediments of the Vienna Basin. In: *Organic Geochemistry*, P.D.I., Vienna, pp. 335—354.
Campbell, C.V., 1962. Depositional environments of Phosphoria Formation (Permian) in southeastern Bighorn Basin, Wyo. *Bull. Am. Assoc. Pet. Geol.*, 46: 478—503.
Canal, P., 1947. Observations sur les caractères pétrographiques des calcaires dolomitiques et des dolomies. *C. R. Soc. Géol. Fr.*, 1947: 161—162.
Carozzi, A., 1949. Sur une particularité des calcaires pseudoolithiques de l'Urgonien (Nappe de Morcles). *Arch. Sci. (Geneva)*, 2: 348—350.
Carozzi, A., 1960. *Microscopic Sedimentary Petrography*. Wiley, New York, N.Y., 485 pp.
Carozzi, A. and Soderman, J.G.W., 1962. Petrography of Mississippian (Borden) crinoidal limestone at Stobo, Indiana. *J. Sediment. Petrol.*, 32: 397—414.
Carroll, J.J. and Greenfield, L.J., 1963. Mechanism of calcium and magnesium uptake from the sea by marine bacteria. *Geol. Soc. Am., Progr. Annu. Meet., 1963, Abstr.* p. 30A.
Cayeaux, L., 1897. Contribution à l'étude micrographique de terrains sédimentaires. *Soc. Géol. Nord, Mém.*, 4 (2): 589 pp.
Cayeaux, L., 1935. *Les Roches sédimentaires de France — Roches carbonatées*. Masson, Paris, 447 pp.
Chafetz, H.S., 1972. Surface diagenesis of limestone. *J. Sediment. Petrol.*, 42: 325—329.
Chanda, S.K., 1963. Cementation and diagenesis of the Lameta Beds, Lametaghat, M.P., India. *J. Sediment. Petrol.*, 33: 728—738.
Chave, K.E., 1954. Aspects of the biochemistry of magnesium, 1. Calcareous marine organisms; 2. Calcareous sediments and rocks. *J. Geol.*, 62: 266—283; 587—599.

Chave, K.E., Deffeyes, K.S., Weyl, P.K., Garrels, R.M. and Thompson, M.E., 1962. Observations on the solubility of skeletal carbonates in aqueous solution. *Science*, 137: 33—34.
Chilingar, G.V., 1953. Use of Ca/Mg ratio in limestones as a geologic tool. *Compass*, 30: 29—34.
Chilingar, G.V., 1955a. Physical-chemical characteristics of basins and sediments of Taman peninsula. *Bull. Am. Assoc. Pet. Geol.*, 39: 1668—1669.
Chilingar, G.V., 1955b. Review of Soviet literature on petroleum source rocks. *Bull. Am. Assoc. Pet. Geol.*, 39: 764—768.
Chilingar, G.V., 1956a. Note on direct precipitation of dolomite out of sea water. *Compass*, 34: 29—34.
Chilingar, G.V., 1956b. Relationship between Ca/Mg ratio and geologic age. *Bull. Am. Assoc. Pet. Geol.*, 40: 2256—2266.
Chilingar, G.V., 1956c. Use of Ca/Mg ratio in porosity studies. *Bull. Am. Assoc. Pet. Geol.*, 40: 2489—2493.
Chilingar, G.V., 1956d. Joint occurrence of glauconite and chlorite in sedimentary rocks: A review. *Bull. Am. Assoc. Pet. Geol.*, 40: 394—398.
Chilingar, G.V., 1958a. A short note on authigenic minerals. *Compass*, 35: 294—297.
Chilingar, G.V., 1958b. Some data on diagenesis obtained from Soviet literature. *Geochim. Cosmochim. Acta*, 13: 213—217.
Chilingar, G.V., 1959a. The Caspian Sea and its sediments. A review. *Int. Geol. Rev.*, 1 (1): 105—111.
Chilingar, G.V., 1959b. Formation of sediments in recent basins. A review. *Int. Geol. Rev.*, 1 (3): 74—81.
Chilingar, G.V., 1962a. Dependence on temperature of Ca/Mg ratio of skeletal structures of organisms and direct chemical precipitates out of sea water. *South. Calif. Acad. Sci., Bull.*, 61 (1): 45—60.
Chilingar, G.V., 1962b. Possible loss of magnesium from fossils to the surrounding environment. *J. Sediment. Petrol.*, 32: 136—139.
Chilingar, G.V. and Bissell, H.J., 1961. Dolomitization by seepage refluxion. *Bull. Am. Assoc. Pet. Geol.*, 45: 679—681.
Chilingar, G.V. and Bissell, H.J., 1963a. Is dolomite formation favored by high or low pH? *Sedimentology*, 2: 171—172.
Chilingar, G.V. and Bissell, H.J., 1963b. Formation of dolomite in sulfate and chloride solutions. *J. Sediment. Petrol.*, 33: 801—803.
Chilingar, G.V. and Terry, R.D., 1954. Relationship between porosity and chemical composition of carbonate rocks. *Pet. Eng.*, 26: 341—342.
Chilingarian, G.V. and Wolf, K.H., 1975. *Compaction of Coarse-Grained Sediments, 1.* Elsevier, Amsterdam, 552 pp.
Chilingarian, G.V. and Wolf, K.H., 1976. *Compaction of Coarse-Grained Sediments, 2.* Elsevier, Amsterdam, 808 pp.
Choquette, P.W. and Pray, L.C., 1970. Geologic nomenclature and classification of porosity in sedimentary carbonates. *Bull. Am. Assoc. Pet. Geol.*, 54: 207—250.
Choquette, P.W. and Traut, J.D., 1963. Pennsylvanian carbonate reservoirs, Ismay Field, Utah and Colorado. In: *Shelf Carbonates of the Paradox Basin — Four Corners Geol. Soc., Field Conf., 4th.*, pp. 157—184.
Clayton, R.N. and Degens, E.T., 1959. Use of carbon-isotope analyses for differentiating fresh-water and marine sediments. *Bull. Am. Assoc. Pet. Geol.*, 43: 890—897.
Clayton, R.N. and Epstein, S., 1958. The relationship between $^{18}O/^{16}O$ ratios in coexisting quartz, carbonate and iron oxides from various geological deposits. *J. Geol.*, 66: 352—373.

Cloud, P.E., 1942. Notes on stromatolites. *Am. J. Sci.*, 240: 363—379.
Cloud, P.E., 1952. Facies relationships of organic reefs. *Bull. Am. Assoc. Pet. Geol.*, 36: 2125—2149.
Cloud, P.E., 1962. Environment of calcium carbonate deposition west of Andros Island, Bahamas. *U.S. Geol. Surv., Prof. Pap.*, 350: 138 pp.
Colley, H. and Davies, P.J., 1969. Ferroan and non-ferroan calcite cements in Pleistocene—Recent carbonates from the New Hebrides. *J. Sediment. Petrol.*, 39: 554—558.
Coogan, A.H. and Manus, R.W., 1975. Compaction and diagenesis of carbonate sands. In: G.V. Chilingarian and K.H. Wolf (Editors), *Compaction of Coarse-grained Sediments*, 1, pp. 79—166.
Correns, C.W., 1939. Die Sedimentgesteine. In: T.F.W. Barth, C.W. Correns and P. Eskola (Editors), *Die Entstehung der Gesteine*. Springer, Berlin, pp. 116—262.
Correns, C.W., 1950. Zur Geochemie der Diagenese. Das Verhalten von $CaCO_3$ und SiO_2. *Geochim. Cosmochim. Acta*, 1: 49—54.
Craig, H., 1957. Isotopic standards for carbon and oxygen and correction factors for mass-spectrometric analysis of carbon dioxide. *Geochim. Cosmochim. Acta*, 12: 133—149.
Crickmay, G.W., 1945. Petrography of limestones; geology of Lau, Fiji. *Bernice P. Bishop Mus. Bull.*, 181: 211—258.
Cronoble, W.R. and Mankin, C.J., 1963. Genetic significance in the limestones of the Coffeyville and Hogshooter Formations (Missourian), northeastern Oklahoma. *J. Sediment. Petrol.*, 33: 73—86.
Curtis, C.D. and Krinsley, D., 1965. The detection of minor diagenetic alteration in shell material. *Geochim. Cosmochim. Acta*, 29: 71—84.
Daetwyler, C.C. and Kidwell, A.L., 1959. The Gulf of Batabano, a modern carbonate basin. *World Pet. Congr., Proc., 5th, N.Y., 1959, Sect. 1*, pp. 1—21.
Danchev, V.I. and Ol'kha, V.V., 1959. Some problems on the origin of uranium mineralization in connection with the study of effective porosity of ore-bearing carbonate rocks. *Izv. Akad. Nauk S.S.S.R., Ser. Geol.*, 7: 15—26.
Dapples, E.C., 1938. The sedimentational effects of the work of marine scavengers. *Am. J. Sci., Ser. 5*, 36: 54—65.
Dapples, E.C., 1942. The effect of macro-organisms upon near-shore marine sediments. *J. Sediment. Petrol.*, 12: 118—126.
Dapples, E.C., 1962. Stages of diagenesis in the development of sandstones. *Bull. Geol. Soc. Am.*, 73: 913—934.
Davies, T.A. and Supko, P.R., 1973. Oceanic sediments and their diagenesis: some examples from deep-sea drilling. *J. Sediment. Petrol.*, 43: 381—390.
Deelman, J.C., 1975. "Pressure solution" re-examined. *Neues Jahrb. Geol. Paläontol. Abh.*, 149 (3): 384—397.
DeFord, R.K., 1946. Grain size in carbonate rock. *Bull. Am. Assoc. Pet. Geol.*, 30: 1921—1928.
Degens, E.T., 1959. Die Diagenese und ihre Auswirkungen auf den Chemismus von Sedimenten. *Neues Jahrb. Geol. Paläontol., Monatsh.*, 1: 73—84.
Degens, E.T., 1964. Über biochemische Umsetzung im Frühstadium der Diagenese. In: L.M.J.U. van Straaten (Editor), *Deltaic and Shallow Marine Deposits*. Elsevier, Amsterdam, pp. 83—92.
Degens, E.T., 1965. *Geochemistry of Sediments*. Prentice-Hall, Englewood Cliffs, N.J., 342 pp.
Degens, E.T. and Epstein, S., 1962. Relationship between $^{18}O/^{16}O$ ratios in coexisting carbonates, cherts and diatomites. *Bull. Am. Assoc. Pet. Geol.*, 46: 534—542.

Degens, E.T. and Epstein, S., 1964. Oxygen and carbon isotope ratios in coexisting calcites and dolomites from recent and ancient sediments. *Geochim. Cosmochim. Acta*, 28: 23—44.

Dietrich, R.V., Hobbs Jr., C.R.R. and Lowry, W.D., 1963. Dolomitization interrupted by silicification. *J. Sediment. Petrol.*, 33: 646—663.

Dixon, E.E.L. and Vaughan, A., 1911. The Carboniferous succession in Gower. *Q. J. Geol. Soc. London*, 67: 447—571.

Donahue, J., 1965. Laboratory growth of pisolite grains. *J. Sediment. Petrol.*, 35: 251—256.

Dunham, R.J., 1962. Classification of carbonate rocks according to depositional texture. In: *Classification of Carbonate Rocks — Am. Assoc. Pet. Geol., Mem.*, 1: 108—121.

Eardley, A.J., 1938. Sediments of Great Salt Lake, Utah. *Bull Am. Assoc. Pet. Geol.*, 22: 1305—1411.

Edie, R.W., 1958. Mississippian sedimentation and oil fields in southeastern Saskatchewan. *Bull. Am. Assoc. Pet. Geol.*, 42: 94—126.

Elias, G.K., 1963. Habitat of Pennsylvanian algal bioherms. In: *Shelf Carbonates of the Paradox Basin — Four Corners Geol. Soc., Field Conf., 4th*, pp. 185—203.

Emery, K.O., 1956. Sediments and water of Persian Gulf. *Bull. Am. Assoc. Pet. Geol.*, 40: 2354—2383.

Emery, K.O. and Rittenberg, S.C., 1952. Early diagenesis of California Basin sediments in relation to origin of oil. *Bull. Am. Assoc. Pet. Geol.*, 36: 735—806.

Emery, K.O., Tracey Jr., J.I. and Ladd, M.S., 1954. Geology of Bikini and nearby atolls. *U.S. Geol. Surv., Prof. Pap.*, 260-A: 263 pp.

Epstein, S., Graf, D.L. and Degens, E.T., 1964. Oxygen isotope studies on the origin of dolomites. In: H. Craig, S.L. Miller and G.J. Wasserburg (Editors), *Isotopic and Cosmic Chemistry* (dedicated to H.C. Urey on the occasion of his 70th birthday). North Holland, Amsterdam, pp. 169—180.

Erickson, A.J., 1965. Temperatures of calcite deposition in the Upper Mississippi Valley lead—zinc deposits. *Econ. Geol.*, 60: 506—528.

Evamy, B.D., 1967. Dedolomitization and the development of rhombohedral pores in limestones. *J. Sediment. Petrol.*, 37: 1204—1215.

Evans, J.W., 1900. Mechanically formed limestones from Junagarh (Kathiawar) and other localities. *Q. J. Geol. Soc. London*, 56: 559—583.

Fairbridge, R.W., 1957. The dolomite question. In: Regional Aspects of Carbonate Deposition. *Soc. Econ. Paleontol. Mineral., Spec. Publ.*, 5: 125—178.

Feray, D.E., Heuer, E. and Hewatt, W.G., 1962. Biological, genetic and utilitarian aspects of limestone classification. In: *Classification of Carbonate Rocks — Am. Assoc. Pet. Geol., Mem.*, 1: 20—32.

Flügel, E. and Flügel-Kahler, E., 1962. Mikrofazielle und geochemische Gliederung eines obertriadischen Riffes der nördlichen Kalkalpen. *Mitt. Mus. Bergbau, Geol. Tech. Landesmus. "Joanneum", Graz*, 24: 1—128.

Folk, R.L., 1956. *Some Informal Thoughts on Recrystallization in Limestones.* Univ. Texas, Austin, Texas, 5 pp. (unpublished).

Folk, R.L., 1957. *Petrology of Sedimentary Rocks.* Hemphill's, Austin, Texas, 120 pp.

Folk, R.L., 1959. Practical petrographic classification of limestones. *Bull. Am. Assoc. Pet. Geol.*, 43: 1—38.

Folk, R.L., 1962a. Petrography and origin of the Silurian Rochester and McKenzie shales, Morgan County, West Virginia. *J. Sediment. Petrol.*, 32: 539—578.

Folk, R.L., 1962b. Spectral subdivision of limestone types. In: *Classification of Carbonate Rocks — Am. Assoc. Pet. Geol., Mem.*, 1: 62—84.

Franke, W., 1971. Structure and development of the Iberg/Winterberg reef. In: *Sedimentology of Parts of Central Europe, Guidebook VIII, Int. Sediment. Congr., 1971*: 83—89.
Friedman, G.M., 1959. Identification of carbonate minerals by staining methods. *J. Sediment. Petrol.*, 29: 87—97.
Friedman, G.M., 1965. Early diagenesis and lithification of carbonate sediments. *J. Sediment. Petrol.*, 34: 777—813.
Friedman, G.M., 1975. The making and unmaking of limestones or the downs and ups of porosity. *J. Sediment. Petrol.*, 45: 379—398.
Friedman, G.M., Amiel, A.J. and Schneidermann, N., 1974. Submarine cementation in reefs: example from the Red Sea. *J. Sediment. Petrol.*, 44: 816—825.
Füchtbauer, H., 1948. Einige Beobachtungen an authigenen Albiten. *Schweiz. Mineral. Petrogr. Mitt.*, 28: 709—716.
Fyfe, W.S. and Bischoff, J.L., 1965. The calcite-aragonite problem. In: L.C. Pray and R.C. Murray (Editors), *Dolomitization and Limestone Diagenesis — Soc. Econ. Paleontol. Mineral., Spec. Publ.*, 13: 3—13.
Garrels, R.M. and Dreyer, R.M., 1952. Mechanism of limestone replacement at low temperatures and pressures. *Bull. Geol. Soc. Am.*, 63: 325—380.
Garrels, R.M., Dreyer, R.M. and Howland, A.L., 1949. Diffusion of ions through intergranular spaces in water-saturated rocks. *Bull. Geol. Soc. Am.*, 60: 1809—1828.
Gavish, E. and Friedman, G.M., 1969. Progressive diagenesis in Quaternary to Late Tertiary carbonate sediments: sequence and time scale. *J. Sediment. Petrol.*, 39: 980—1006.
George, T.N., 1954. Pre-Seminulan Main Limestone of the Avonian series in Breconshire. *Q. J. Geol. Soc. London*, 110: 283—322.
George, T.N., 1956. Carboniferous Main Limestone of the East Crop in South Wales. *Q. J. Geol. Soc. London*, 111: 309—322.
Germann, K., 1968. Diagenetic patterns in the Wettersteinkalk (Ladinian, Middle Trias), northern Limestone Alps, Bavaria and Tyrol. *J. Sediment. Petrol.*, 38: 490—500.
Ginsburg, R.N., 1953a. Beach rock in South Florida. *J. Sediment. Petrol.*, 23: 85—92.
Ginsburg, R.N., 1953b. Early diagenesis and lithification of shallow-water carbonate sediments in South Florida. In: *Regional Aspects of Carbonate Deposition — Soc. Econ. Paleontol. Mineral., Spec. Publ.*, 5: 80—100.
Ginsburg, R.N. and James, N.P., 1976. Submarine botryoidal aragonite in Holocene reef limestones, Belize. *Geology*, 4: 431—436.
Ginsburg, R.N. and Schroeder, J.H., 1973. Growth and submarine fossilization of algal cup reefs, Bermuda. *Sedimentology*, 20: 575—614.
Ginsburg, R.N., Marszalek, D.S. and Schneidermann, N., 1971. Ultrastructure of carbonate cements in a Holocene algal reef of Bermuda. *J. Sediment. Petrol.*, 41: 472—482.
Glover, J.E., 1955. Petrology and petrography of limestones from the Fitzroy Basin, Western Australia. *Comm. Aust. Bur. Miner. Res., Geol. Geophys., Rep.*, 18: 22 pp.
Glover, J.E., 1965. Studies in the diagenesis of some Western Australian sedimentary rocks. *J. R. Soc. W. Aust.*, 46: 33—56.
Gorsline, D.S., 1963. Environments of carbonate deposition, Florida Bay and the Florida Straits. In: *Shelf Carbonates of the Paradox Basin — Four Corners Geol. Soc., Field Conf., 4th*, pp. 130—143.
Goto, M., 1961. Some mineralo-chemical problems concerning calcite and aragonite, with special reference to the genesis of aragonite. *Fac. Sci., Hokkaido Univ.*, 825: 573—640.
Grabau, A.W., 1904. On the classification of sedimentary rocks. *Am. Geol.*, 33: 228—247.

Grabau, A.W., 1913. *Principles of Stratigraphy*. Seiler, New York, N.Y., 1185 pp.

Graf, D.L., 1960. Geochemistry of carbonate sediments and sedimentary carbonate rocks. *Ill. State Geol. Surv., Circ.*, 297, 298, 301, 308.

Graf, D.L. and Goldsmith, J.R., 1955. Dolomite—magnesian relations at elevated temperatures and CO_2 pressures. *Geochim. Cosmochim. Acta*, 7: 109—128.

Graf, D.L. and Lamar, J.E., 1950. Petrology of Fredonia oölite in Southern Illinois. *Bull. Am. Assoc. Pet. Geol.*, 34: 2318—2336.

Greensmith, J.T., 1960. Introduction to the petrology of the Oil-Shale Group limestones of west Lothian and southern Fifeshire, Scotland. *J. Sediment. Petrol.*, 30: 553—560.

Greiner, H.R., 1956. Methy dolomite of Northeastern Alberta: Middle Devonian reef formation. *Bull. Am. Assoc. Pet. Geol.*, 40: 2057—2080.

Griffin, R.H., 1942. Dolomitic mottling in the Platteville Limestone. *J. Sediment. Petrol.*, 12: 67—76.

Gundu Rao, C., 1972. On aggradational recrystallisation in argillaceous micritic limestones. *Oil Nat. Gas Comm. Bull.*, 9 (1).

Gundu Rao, C., 1975. Pseudopellets from Karaikal Well No. 6, Cauvery Basin and their significance. *Oil Nat. Gas Comm., Bull.*, 11 (2): 47—50.

Gundu Rao, C. and Siddiqui, Z.R., 1974. Garudamangalam Limestone, an Upper Cretaceous analog of a modern sandy beach rock from the Trichinopoly Upper Cretaceous, Cauvery Basin, S. India. *Oil. Nat. Gas Comm. Bull.*, 11 (1): 43—60.

Gvirtzman, G., Friedman, G.M. and Miller, D.S., 1973. Control and distribution of uranium in coral reefs during diagenesis. *J. Sediment. Petrol.*, 43: 985—997.

Hadding, A., 1941. The pre-Quaternary sedimentary rocks of Sweden, 6, Reef limestones. *Lunds Univ. Årsskr., Avd. 2 (N.F.)*, 37 (10): 137 pp.

Hadding, A., 1950. Silurian reefs of Gotland. *J. Geol.*, 58: 402—409.

Hadding, A., 1956. The lithological character of marine shallow water limestones. *K. Fysiogr. Sällskap. Lund, Förh.*, 26: 1—18.

Hadding, A., 1958. Origin of the lithographic limestones. *K. Fysiogr. Sällskap. Lund, Förh.*, 28: 21—32.

Halbach, P., Der Einfluss der Mineralführung auf die Aufbereitbarkeit des Erzes bei der Starkfeldscheidung der Eisenerzlagerstätte von Pegnitz (Oberfranken). *Bergbauwissenschaften*, 12: 3—12.

Halla, F., Chilingar, G.V. and Bissell, H.J., 1962. Thermodynamic studies on dolomite formation and their geologic implications: an interim report. *Sedimentology*, 1: 296—303.

Ham, W.E., 1952. Algal origin of the "Birdseye" Limestone in the McLish Formation. *Proc. Okl. Acad. Sci.*, 33: 200—203.

Ham, W.E. and Pray, L.C., 1962. Modern concepts and classifications of carbonate rocks. In: *Classification of Carbonate Rocks — Am. Assoc. Pet. Geol., Mem.*, 1: 2—19.

Hambleton, A.W., 1962. Carbonate-rock fabrics of three Missourian stratigraphic sections in Socorro County, N. Mex. *J. Sediment. Petrol.*, 32: 579—601.

Handford, C.R. and Moore Jr., C.H., 1976. Diagenetic implications of calcite pseudomorphs after halite from the Joachim Dolomite (Middle Ordovician), Arkansas. *J. Sediment. Petrol.*, 46: 387—392.

Harbaugh, J.W., 1961. Relative ages of visible crystalline calcite in Late Paleozoic limestones. *Geol. Surv. Kansas, Bull.*, 152 (4): 91—126.

Henderson, D.M. and Rhodes, F.H.T., 1956. Dolomitization of the Platteville Limestone. *Trans. Ill. State Acad. Sci.*, 48: 166—172.

Henson, F.R.S., 1950. Cretaceous and Tertiary reef formations and associated sediments in Middle East. *Bull. Am. Assoc. Pet. Geol.*, 34: 215—238.

Hohlt, R.B., 1948. The nature and origin of limestone porosity. *Q. Colo. Sch. Mines*, 43 (4): 51 pp.
Holmes, A., 1921. *Petrographic Methods and Calculations.* Murby, London, 515 pp.
Honess, A.P. and Jeffries, C.D., 1940. Authigenic albite from the Lowville Limestone at Bellefonte, Pennsylvania. *J. Sediment. Petrol.*, 10: 12—18.
Honjo, S., Fischer, A. and Garrison, R., 1965. Geopetal pyrite in fine-grained limestones. *J. Sediment. Petrol.*, 35: 480—488.
Husseini, S.I. and Matthews, R.K., 1972. Distribution of high-magnesium calcite in lime muds of the Great Bahama Bank: diagenetic implications. *J. Sediment. Petrol.*, 42: 179—182.
Illing, L.V., 1954. Bahaman calcareous sands. *Bull. Am. Assoc. Pet. Geol.*, 38: 1—95.
Illing, L.V., 1959. Deposition and diagenesis of some Upper Paleozoic carbonate sediments in Western Canada. *World Pet. Congr., Proc., 5th, N.Y., 1959, Sect. 1*, pp. 23—52.
Ingerson, E., 1962. Problems of the geochemistry of sedimentary carbonate rocks. *Geochim. Cosmochim. Acta*, 26: 815—847.
Irwin, M.L., 1965. General theory of epeiric clear water sedimentation. *Bull. Am. Assoc. Pet. Geol.*, 49: 445—449.
Jaanusson, V., 1961. Discontinuity surfaces in limestones. *Bull. Geol. Inst. Univ. Uppsala*, 40 (35): 221—241.
James, N.P., Ginsburg, R.N., Marszalek, D.S. and Choquette, P.W., 1976. Facies and fabric specificity of early subsea cements in shallow Belize (British Honduras) reefs. *J. Sediment. Petrol.*, 46: 523—544.
Jankowski, G.J. and ZoBell, C., 1944. Hydrocarbon production by sulfate-reducing bacteria. *J. Bacteriol.*, 47: 447.
Johnson, J.H., 1951. An introduction to the study of organic limestones. *Q. Colo. Sch. Mines*, 46 (2): 185 pp.
Johnson, J.H., 1957. Petrography of limestones, geology of Saipan, Mariana Islands. *U.S. Geol. Surv., Prof. Pap.*, 280-B-D: 177—187.
Julivert, de Zamarreño, 1963. Estudio petrográfico de las Calizas de la formación Rosablanca de la región de la Mesa de los Santos. *Bol. Geol., Univ. Ind. Santander*, 15: 5—34.
Kahle, C.F., 1965. Possible roles of clay minerals in the formation of dolomite. *J. Sediment. Petrol.*, 35: 448—453.
Kahle, C.F., 1974. Ooids from Great Salt Lake, Utah, as an analogue for the genesis and diagenesis of ooids in marine limestones. *J. Sediment. Petrol.*, 44: 30—39.
Karcz, I., 1964. Grain-growth fabrics in the Cambrian dolomites of Skye. *Nature*, 204 (4963): 1080—1081.
Kaye, C.A., 1959. Shoreline features and Quaternary shoreline changes, Puerto Rico. *U.S., Geol. Surv., Prof. Pap.*, 317-B: 140.
Kendall, A.C., 1975. Post-compactional calcitization of molluscan aragonite in a Jurassic limestone from Saskatchewan, Canada. *J. Sediment. Petrol.*, 45: 399—404.
Khvorova, I.V., 1958. *Atlas of Carbonate Rocks of the Middle and Upper Carboniferous of the Russian Platform.* Izv. Akad. Nauk S.S.S.R., Moscow, 170 pp.
Kitano, Y. and Hood, D.W., 1965. The influence of organic material on the polymorphic crystallization of calcium carbonate. *Geochim. Cosmochim. Acta*, 39: 29—41.
Kleber, W., 1964. Zum Problem der topotaktischen Umwandlung von Aragonit in Calcit. *Neues Jahrb. Mineral., Monatsh.*, 1964: 368—369.
Klement, K.W. and Toomey, D.F., 1967. Role of blue-green alga *Girvanella* in skeletal grain destruction and lime-mud formation in the Lower Ordovician of West Texas. *J. Sediment. Petrol.*, 37: 1045—1051.

Knewtson, S.L. and Hubert, J.F., 1969. Dispersal patterns and diagenesis of oolitic calcarenites in the Ste. Genevieve Limestone (Mississippian), Missouri. *J. Sediment. Petrol.*, 39: 954—968.
Kobluk, D.R. and Risk, M.J., 1977. Micritization and carbonate-grain binding by endolithic algae. *Bull. Am. Assoc. Pet. Geol.*, 61 (7): 1069—1082.
Kornicker, L.S. and Purdy, E.G., 1957. A Bahaman faecal-pellet sediment. *J. Sediment. Petrol.*, 27: 126—128.
Krumbein, W.C., 1942. Physical and chemical changes in sediments after deposition. *J. Sediment. Petrol.*, 12: 111—117.
Krumbein, W.C. and Garrels, R.M., 1952. Origin and classification of chemical sediments in terms of pH and Eh potentials. *J. Geol.*, 60: 1—33.
Kukal, Z., 1964. Lithology of carbonate formations. *Sb. Geol. Véd.*, 6: 123—165.
Lalou, C., 1957. Studies on bacterial precipitation of carbonates in sea water. *J. Sediment. Petrol.*, 27: 190—195.
Land, L.S., 1967. Diagenesis of skeletal carbonates. *J. Sediment. Petrol.*, 37: 914—930.
Land, L.S. and Goreau, T.F., 1970. Submarine lithification of Jamaican reefs. *J. Sediment. Petrol.*, 40: 457—462.
Land, L.S. and Hoops, G.K., 1973. Sodium in carbonate sediments and rocks: a possible index to the salinity of diagenetic solutions. *J. Sediment. Petrol.*, 43: 614—617.
Landes, K.K., 1946. Porosity through dolomitization. *Bull. Am. Assoc. Pet. Geol.*, 30: 305—318.
La Porte, L.F., 1962. Paleoecology of the Cottonwood Limestone (Permian), northern Mid-Continent. *Bull. Geol. Soc. Am.*, 73: 521—544.
La Porte, L.F., 1967. Carbonate deposition near mean sea-level and resultant facies mosaic: Manlius Formation (Lower Devonian) of New York State. *Bull. Am. Assoc. Pet. Geol.*, 51: 73—101.
Larsen, G. and Chilingar, G.V., 1967. Introduction. In: G. Larsen and G.V. Chilingar (Editors), *Diagenesis of Sediments*, Elsevier, Amsterdam, pp. 1—17.
Lattman, L.H. and Simonberg, E.M., 1971. Case-hardening of carbonate alluvium and colluvium, Spring Mountains, Nevada. *J. Sediment. Petrol.*, 41: 274—281.
Lees, A., 1964. The structure and origin of the Waulsortian (Lower Carboniferous) "Reefs" of West-Central Eire. *Philos. Trans. R. Soc. London, Ser. B*, 247: 483—531.
Lees, A., 1973. Platform carbonate deposits. *Bull. Cent. Rech. Pau-S.N.P.A.*, 7 (1): 177—192.
Leighton, M.W. and Pendexter, C., 1962. Carbonate rock types. In: *Classification of Carbonate Rocks. Am. Assoc. Pet. Geol., Mem.*, 1: 33—61.
Lindholm, R.C., 1974. Fabric and chemistry of pore filling calcite in septarian veins: models for limestone cementation. *J. Sediment. Petrol.*, 44: 428—440.
Lindström, M., 1963. Sedimentary folds and the development of limestone in an Early Ordovician sea. *Sedimentology*, 2: 243—276.
Linscott, R.O., 1958. Petrography and petrology of Ismay and Desert Creek zones, Four Corners Region. In: *Geology of the Paradox Basin — Intermount. Assoc. Pet. Geol., Annu. Field Conf., Guidebook*, pp. 146—152.
Logan, B.W., 1961. Cryptozoön and associated stromatolites from the Recent, Shark Bay, Western Australia. *J. Geol.*, 69: 517—533.
Lowenstam, H.A., 1950. Niagaran reefs of the Great Lakes area. *J. Geol.*, 58: 430—487.
Lowenstam, H.A., 1954. Factors affecting the aragonite/calcite ratios in carbonate-secreting marine organisms. *J. Geol.*, 62: 284—322.
Lowenstam, H.A., 1955. Aragonite needles secreted by algae and some sedimentary implications. *J. Sediment. Petrol.*, 25: 270—272.

Lowenstam, H.A., 1961. Mineralogy, $^{18}O/^{16}O$ ratios and magnesium contents of recent and fossil brachiopods and their bearing on the history of the oceans. *J. Geol.*, 69: 241—260.
Lowenstam, H.A., 1964. Sr/Ca ratio of skeletal aragonites from the recent marine biota at Palau and from fossil Gastropoda. In: H. Craig, S.L. Miller and G.J. Wasserburg (Editors), *Isotopic and Cosmic Chemistry*. North-Holland, Amsterdam, pp. 114—132.
Lowenstam, H.A. and Epstein, S., 1957. On the origin of sedimentary aragonite needles of the Great Bahama Bank. *J. Geol.*, 65: 364—375.
Lucia, F.J., 1961. Dedolomitization in the Tansill (Permian) formation. *Bull. Geol. Soc. Am.*, 72: 1107—1110.
Lucia, F.J., 1962. Diagenesis of a crinoidal sediment. *J. Sediment. Petrol.*, 32: 848—865.
Lucia, F.J., 1968. Recent sediments and diagenesis of South Bonaire, Netherlands Antilles. *J. Sediment. Petrol.*, 38: 845—858.
MacIntyre, I.G., Mountjoy, E.W. and d'Anglejan, B.F., 1968. An occurrence of submarine cementation of carbonate sediments off the west coast of Barbados, W.I. *J. Sediment. Petrol.*, 38: 660—664.
Marcher, M.V., 1962. Petrography of Mississippian limestones and cherts from the northwestern Highland Rim, Tenn. *J. Sediment. Petrol.*, 32: 819—832.
Marfil, R., Alonso, J.J. and Garcia, M.C., 1971. Estudio del material cementante del Trias inferior de la Cordillera Ibérica. *Estud. Geol.* (Spain), 27: 427—439.
Marshall, J.F. and Davies, P.J., 1975. High-magnesium calcite ooids from the Great Barrier Reef. *J. Sediment. Petrol.*, 45: 285—291.
Matter, A., 1974. Burial diagenesis of pelitic and carbonate deep-sea sediments from the Arabian Sea. In: R.B. Whitmarsh, O.E. Weser, D.A. Ross et al. (Editors), *Initial Reports of the Deep Sea Drilling Project*, 23: 421—469.
Matthews, R.K., 1967. Diagenetic fabrics in biosparites from the Pleistocene of Barbados, West Indies. *J. Sediment. Petrol.*, 37: 1147—1153.
Matthews, R.K., 1968. Carbonate diagenesis: equilibration of sedimentary mineralogy to the subaerial environment; coral cap of Barbados, West Indies. *J. Sediment. Petrol.*, 38: 1110—1119.
Maxwell, W.G.H., Day, R.W. and Fleming, P.J.G., 1961. Carbonate sedimentation on the Heron Island Reef, Great Barrier Reef. *J. Sediment. Petrol.*, 31: 215—230.
Maxwell, W.G.H., Jell, J.S. and McKellar, R.G., 1963. A preliminary note on the mechanical and organic factors influencing carbonate differentiation, Heron Island Reef, Australia. *J. Sediment. Petrol.*, 33: 962—963.
Mayer, F.K., 1932. Ueber die Modifikation des Kalzium-karbonats in Schalen und Skeletten rezenter und fossiler Organismen. *Chem. Erde*, 7: 346—350.
McKee, E.D., Chronic, J. and Leopold, E.B., 1959. Sedimentary belts in lagoon of Kapingamarangi Atoll. *Bull. Am. Assoc. Pet. Geol.*, 43: 501—562.
McLain, J.F. and Schlanger, S.O., 1957. Tertiary reef and associated limestone facies from Louisiana and Guam. *J. Geol.*, 65: 611—627.
Middleton, G.V., 1961. Evaporite solution breccias from the Mississippian of southwest Montana. *J. Sediment. Petrol.*, 31: 189—195.
Miller Jr., D.N., 1961. Early diagenetic dolomite associated with salt extraction process, Inagua, Bahamas. *J. Sediment. Petrol.*, 31: 473—476.
Milliman, J.D., 1974. *Marine Carbonates, 1. Recent Sedimentary Carbonates.* Springer, New York-Heidelberg-Berlin, 375 pp.
Milliman, J.D., Ross, D.A. and Ku, Teh-Lung, 1969. Precipitation and lithification of deep-sea carbonates in the Red Sea. *J. Sediment. Petrol.*, 39: 724—736.
Mišik, M., 1970. Verwischung der ursprünglichen Merkmale in kalkigen Sedimenten und Kalken bei der Dolomitisierung. *Verh. Geol. Bundesanst. (Austria)*, 4: 673—688.

Mollazal, Y., 1961. Petrology and petrography of Ely Limestone in part of eastern Great Basin. *Brigham Young Univ. Stud., Geol. Ser.*, 8: 3—35.
Monaghan, P.H. and Lytle, M.L., 1956. The origin of calcareous oöliths. *J. Sediment. Petrol.*, 26: 111—118.
Moore Jr., C.H., 1973. Intertidal carbonate cementation, Grand Cayman, West Indies. *J. Sediment. Petrol.*, 43: 591—602.
Moore, R.C., 1957. Mississippian carbonate deposits of the Ozark region. In: *Regional Aspects of Carbonate Deposition — Soc. Econ. Paleontol. Mineral., Spec. Publ.*, 5: 101—124.
Moret, L., 1940. Rôle probable des Holothuries dans la genèse de certains sédiments calcaires. *C. R. Soc. Géol. Fr.*, 1940: 11—12.
Moretti, F.J., 1957. Observations on limestones. *J. Sediment. Petrol.*, 27: 282—292.
Morris, R.C. and Dickey, P.A., 1957. Modern evaporite deposition in Peru. *Bull. Am. Assoc. Pet. Geol.*, 41: 2467—2474.
Morteani, G., 1965. Strontiumgehalte und Spurenelemente in Schwerspatvererzungen des Belierophonkalkes und des Bozener Quarzporphyrs nordöstlich von Larvis (Provinz Trento, Nord-italien). *Bergbauwissenschaften*, 12: 1—6.
Müller, G., Schöll, W.U. and Tietz, G., 1973. Diagenetic development of a Precambrian limestone as interpreted from a modern analogue. *Sedimentology*, 20: 529—537.
Müller, W., 1964. Unterschiede in den chemischen und physikalischen Eigenschaften von fluviatilen, brachischen und marinen Sedimenten. In: L.M.J.U. van Straaten (Editor), *Deltaic and Shallow Marine Deposits.* Elsevier, Amsterdam, pp. 293—300.
Murray, G.E., 1952. Stratigraphy. In: *Geology of Beauregard and Allen Parishes — La. Dep. Conserv., Geol. Bull.*, 27: 224 pp.
Murray, R.C., 1960. Origin of porosity in carbonate rocks. *J. Sediment. Petrol.*, 30: 59—84.
Neal, W.J., 1969. Diagenesis and dolomitization of a limestone (Pennsylvanian of Missouri) as revealed by staining. *J. Sediment. Petrol.*, 39: 1040—1045.
Nelson, H.F., Brown, C.W. and Brineman, J.H., 1962. Skeletal limestone classification. In: *Classification of Carbonate Rocks — Am. Assoc. Pet. Geol., Mem.*, 1: 224—252.
Neuman, A.C., Gebelein, C.D. and Scoffin, T.P., 1970. The composition, structure, and erodability of subtidal mats, Abaco, Bahamas. *J. Sediment. Petrol.*, 40 (1): 274—297.
Newell, N.D., 1955. Depositional fabric in Permian reef limestones. *J. Geol.*, 63: 301—309.
Newell, N.D. and Rigby, J.K., 1957. Geological studies on the Great Bahama Bank. In: *Regional Aspects of Carbonate Deposition — Soc. Econ. Paleontol. Mineral., Spec. Publ.*, 5: 15—72.
Newell, N.D., Rigby, J.K., Whiteman, A.J. and Bradley, J.S., 1951. Shoal water geology and environments, eastern Andros Island, Bahamas. *Bull. Am. Mus. Nat. Hist.*, 97: 29 pp.
Newell, N.D., Rigby, J.K., Fischer, A.G., Whiteman, A.J., Hickox, J.E. and Bradley, J.S., 1953. *The Permian Reef Complex of the Guadalupe Mountains Region, Texas and New Mexico.* Freeman, San Francisco, Calif., 236 pp.
Niggli, P. and Niggli, E., 1952. *Gesteine und Minerallagerstätten, 2. Die Exogenen Gesteine und Minerallagerstätten.* Birkhäuser, Basel, 554 pp.
Nitecki, M.H., 1960. A carbonate vein in limestone. *J. Sediment. Petrol.*, 30: 624—625.
Ogniben, L., 1957. Petrografia della serie solfifera siciliana e considerazioni geologiche relative. *Serv. Geol. Ital., Mem. Descr. Carta Geol. Ital.*, 33: 275 pp.
Orme, G.R. and Brown, W.W.M., 1963. Diagenetic fabrics in the Avonian limestones of Derbyshire and North Wales. *Proc. Yorkshire Geol. Soc.*, 34: 51—66.

Osmond, J.C., 1956. Mottled carbonate rocks in the Middle Devonian of eastern Nevada. *J. Sediment. Petrol.*, 26: 32—41.
Otte, C. and Parks, J.M., 1963. Fabric studies in Virgil and Wolfcamp bioherms, New Mexico. *J. Geol.*, 71: 380—396.
Pantin, H.M., 1958. Rate of formation of a diagenetic calcareous concretion. *J. Sediment. Petrol.*, 28: 366—371.
Parkinson, D., 1957. Lower Carboniferous reefs of northern England. *Bull. Am. Assoc. Pet. Geol.*, 41: 511—537.
Peña, J.A. de la and Marfil, R., 1975. Estudio petrológico del Pérmico de la Cordillera Iberica: zona de Torre La Hija (N.E. de Molina de Aragón, Guadalajara). *Estud. Geol. (Spain)*, 31: 513—530.
Perkins, R.D., 1963. Petrology of the Jeffersonville Limestone (Middle Devonian) of South-eastern Indiana. *Bull. Geol. Soc. Am.*, 74: 1335—1354.
Peterson, J.A. and Ohlen, H.R., 1963. Pennsylvanian shelf carbonates, Paradox Basin. In: *Shelf Carbonates of the Paradox Basin — Four Corners Geol. Soc., Field Conf., 4th*, pp. 65—79.
Pettijohn, F.J., 1957. *Sedimentary Rocks.* Harper and Row, New York, N.Y., 2nd ed., 718 pp.
Philcox, M.E., 1963. Banded calcite mudstone in the Lower Carboniferous "Reef" knolls of the Dublin Basin, Ireland. *J. Sediment. Petrol.*, 33: 904—913.
Pia, J., 1933. Die rezenten Kalksteine. *Z. Kristallogr. Mineral. Petrogr., Abt. B, Mineral. Petrog. Mitt.* (Ergänzungsband), pp. 1—420.
Pingitore Jr., N.E., 1970. Diagenesis and porosity modification in *Acropora palmata*, Pleistocene of Barbados, West Indies. *J. Sediment. Petrol.*, 40: 712—721.
Powers, R.W., 1962. Arabian Upper Jurassic carbonate reservoir rocks. In: *Classification of Carbonate Rocks. Am. Assoc. Pet. Geol., Mem.*, 1: 122—192.
Pray, L.C., 1958. Fenestrate bryozoan core facies, Mississippian bioherms, southwestern United States. *J. Sediment. Petrol.*, 28: 261—273.
Pray, L.C. and Wray, J.L., 1963. Porous algal facies (Pennsylvanian) Honaker Trail, San Juan Canyon, Utah. In: *Shelf Carbonates of the Paradox Basin — Four Corners Geol. Soc., Field Conf., 4th*, pp. 204—234.
Purdy, E.G., 1963. Recent calcium carbonate facies of the Great Bahama Bank. *J. Geol.*, 71: 334—355; 472—497.
Purdy, E.G., Pusey III, W.C. and Wantland, K.F., 1975. Continental shelf of Belize — regional shelf attributes. In: *Studies in Geology, 2. Belize Shelf — Carbonate Sediments, Clastic Sediments and Ecology*. Am.Assoc. Pet. Geol., Tulsa, Okla., pp. 1—52.
Reijers, T.J.A., 1974. Diagenesis in the reefal facies of the Middle to Upper Devonian Portilla Limestone Formation of N.W. Spain. *Breviora Geol. Asturica*, 18 (3): 33—48.
Revelle, R., 1934. Physico-chemical factors affecting the solubility of calcium carbonate in sea water. *J. Sediment. Petrol.*, 4: 103—110.
Revelle, R. and Fairbridge, R., 1957. Carbonates and carbon dioxide. In: J.W. Hedgpeth (Editor), *Treatise on Marine Ecology and Paleoecology, 1. Ecology — Geol. Soc. Am., Mem.*, 67: 239—296.
Revelle, R. and Fleming, R.H., 1934. The solubility product constant of calcium carbonate in seawater. *Pac. Sci. Congr. Proc., Pac. 5th, 1934*, 3: 2089—2092.
Rich, M., 1963. Petrographic analysis of Bird Spring Group (Carboniferous-Permian) near Lee Canyon, Clark County, Nevada. *Bull. Am. Assoc. Pet. Geol.*, 47: 1657—1681.
Rieke, H.H. III and Chilingarian. G.V., 1974. *Compaction of Argillaceous Sediments.* Elsevier, Amsterdam, 424 pp.
Rittenberg, S.C., Emery, K.O., Hülsemann, J., Degens, E.T., Fay, R.C., Reuter, J.H., Grady, J.R., Richardson, S.H. and Bray, E.E., 1963. Biogeochemistry of sediments

in Experimental Mohole. *J. Sediment. Petrol.*, 33: 140—172.
Rivière, A. and Vernhet, S., 1964. Contribution à l'étude de la sédimentologie des sédiments carbonatées. In: L.M.J.U. van Straaten (Editor), *Deltaic and Shallow Marine Deposits*. Elsevier, Amsterdam, pp. 356—361.
Robinson, R.B., 1967. Diagenesis and porosity development in Recent and Pleistocene oolites from southern Florida and the Bahamas. *J. Sediment. Petrol.*, 37: 355—364.
Roehl, P.O., 1967. Stony Mountain (Ordovician) and Interlake (Silurian) facies analogs of Recent low-energy marine and subaerial carbonates, Bahamas. *Bull. Am. Assoc. Pet. Geol.*, 51: 1979—2032.
Romm, I.I., 1950. Geochemical characteristics of recent deposits of Taman peninsula. In: *Recent Analogues of Petroliferous Facies*. Gostoptekhizdat, Moscow.
Ross, C. and Oana, S., 1961. Late Pennsylvanian and Early Permian limestone petrology and carbon isotope distribution, Glass Mountains, Texas. *J. Sediment. Petrol.*, 31: 231—244.
Rucker, J.B., 1968. Carbonate mineralogy of sediments of Exuma Sound, Bahamas. *J. Sediment. Petrol.*, 38: 68—72.
Rukhin, L.B., 1958. *Grundzüge der Lithologie*. Akademie Verlag, Berlin, 806 pp.
Rukhin, L.B., 1961. *Principles of Lithology*. Gostoptekhizdat, Leningrad, 2nd ed., 779 pp.
Runnells, D.D., 1969. Diagenesis, chemical sediments, and the mixing of natural waters. *J. Sediment. Petrol.*, 39: 1188—1201.
Rusnak, G.A., 1960. Some observations of Recent oölites. *J. Sediment. Petrol.*, 30: 471—480.
Russell, R.J., 1962. Origin of beach rock. *J. Geomorphol.*, 6: 1—16.
Sabins Jr., F.F., 1962. Grains of detrital, secondary, and primary dolomite from Cretaceous strata of the Western Interior. *Bull. Geol. Soc. Am.*, 73: 1183—1196.
Sackett, W.M., 1964. The depositional history and isotopic organic carbon composition of marine sediments. *Mar. Geol.*, 2: 173—185.
Sander, B., 1951. *Contributions to the Study of Depositional Fabrics: Rhythmically Deposited Triassic Limestones and Dolomites*. Am. Assoc. Pet. Geol., Tulsa, Okla., 207 pp.
Sarin, D.D., 1962. Cyclic sedimentation of primary dolomite and limestone. *J. Sediment. Petrol.*, 32: 451—471.
Sarkisyan, M.A., Politykina, M.A. and Chilingarian, G.V., 1973. Effect of postsedimentation processes on carbonate reservoir rocks in Volga-Urals Region, U.S.S.R. *Bull. Am. Assoc. Pet. Geol.*, 57: 1305—1313.
Schlanger, S.O., 1957. Dolomite growth in coralline algae. *J. Sediment. Petrol.*, 27: 181—186.
Schmalz, R.F., 1956. The mineralogy of the Funafuti drill cores and its bearing on the physico-chemistry of dolomite. (Abstract.) *J. Paleontol.*, 30: 1004—1005.
Schmalz, R.F., 1967. Kinetics and diagenesis of carbonate sediments. *J. Sediment. Petrol.*, 37: 60—67.
Scholle, P.A., 1971. Diagenesis of deep-water carbonate turbidites, Upper Cretaceous Monte Antola Flysch, northern Apennines, Italy. *J. Sediment. Petrol.*, 41: 233—250.
Scholle, P.A., 1977. Chalk diagenesis and its relation to petroleum exploration: oil from chalks, a modern miracle? *Bull. Am. Assoc. Pet. Geol.*, 61 (7) 982—1009.
Schroeder, J.H., 1969. Experimental dissolution of calcium, magnesium and strontium from Recent biogenic carbonates: a model of diagenesis. *J. Sediment. Petrol.*, 39: 1057—1073.
Schroeder, J.H., 1972. Fabrics and sequences of submarine carbonate cements in Holocene Bermuda cup reefs. *Geol. Rundsch.*, 61 (2) 708—730.

Schroeder, J.H., 1973. Submarine and vadose cements in Pleistocene Bermuda reef rock. *Sediment. Geol.*, 10: 179—204.
Schroeder, J.H., 1974. Carbonate cements in Recent reefs of the Bermudas and Bahamas keys to the past? *Ann. Soc. Géol. Belg.*, 97: 153—158.
Schroeder, J.H. and Zankl, H., 1974. Dynamic reef formation: a sedimentological concept based on studies of Recent Bermuda and Bahama reefs. In: *Proceedings Second International Coral Reef Symposium, 2.* Great Barrier Reef Committee, Brisbane, pp. 413—428.
Schwarzacher, W., 1961. Petrology and structure of some Lower Carboniferous reefs in northwestern Ireland. *Bull. Am. Assoc. Pet. Geol.*, 45: 1481—1503.
Seibold, E., 1962. Kalk-Konkretionen und karbonatisch gebundenes Magnesium. *Geochim. Cosmochim. Acta*, 26: 899—909.
Sibley, D.F. and Murray, R.C., 1972. Marine diagenesis of carbonate sediment, Bonaire, Netherlands Antilles. *J. Sediment. Petrol.*, 42: 168—178.
Sidwell, R. and Warn, G.F., 1954. Diagenesis of Pennsylvanian limestones, Upper Pecos Valley, New Mexico. *J. Sediment. Petrol.*, 24: 255—262.
Siegel, F.R., 1960. The effect of strontium on the aragonite-calcite ratios of Pleistocene corals. *J. Sediment. Petrol.*, 30: 297—304.
Siegel, F.R., 1961. Variations of Sr/Ca ratios and Mg contents in Recent carbonate sediments of the Northern Florida Keys area. *J. Sediment. Petrol.*, 31: 336—342.
Siegel, F.R., 1963. Artificially induced thermoluminescence of sedimentary dolomites. *J. Sediment. Petrol.*, 33: 64—72.
Siever, R., Beck, K.C. and Berner, R.A., 1965. Composition of interstitial waters of modern sediments. *J. Geol.*, 73: 39—73.
Singlewald, J.T. and Milton, C., 1929. Authigenic feldspar in limestone at Glens Falls, New York. *Bull. Geol. Soc. Am.*, 40: 463—468.
Skeats, E.W., 1918. The formation of dolomite and its bearing on the coral reef problem. *Am. J. Sci.*, 45: 185—200.
Skinner, H.C.W., Skinner, B.J. and Rubin, M., 1963. Age and accumulation rate of dolomite-bearing carbonate sediments in South Australia. *Science*, 139: 335—336.
Sloss, L.L. and Laird, W.M., 1947. Devonian System in northwestern Montana. *Bull. Am. Assoc. Pet. Geol.*, 31: 1404—1430.
Stanley, S.M., 1966. Paleoecology and diagenesis of Key Largo Limestone, Florida. *Bull. Am. Assoc. Pet. Geol.*, 50: 1927—1947.
Stauffer, K.W., 1962. Quantitative petrographic study of Paleozoic carbonate rocks, Caballo Mountains, New Mexico. *J. Sediment. Petrol.*, 32: 357—396.
Stehli, F.G. and Hower, J., 1961. Mineralogy and early diagenesis of carbonate sediments. *J. Sediment. Petrol.*, 31: 358—371.
Steinen, R.P., 1974. Phreatic and vadose diagenetic modification of Pleistocene limestone: petrographic observations from subsurface of Barbados, West Indies. *Bull. Am. Assoc. Pet. Geol.*, 58: 1008—1024.
Steinen, R.P. and Matthews, R.K., 1973. Phreatic vs. vadose diagenesis: stratigraphy and mineralogy of a cored borehole on Barbados, W.I. *J. Sediment. Petrol.*, 43: 1012—1020.
Stockman, K.W., Ginsburg, R.N. and Shinn, E.A., 1967. The production of lime mud by algae in South Florida. *J. Sediment. Petrol.*, 37: 633—648.
Strakhov, N.M., 1953. Diagenesis of sediments and its significance for sedimentary ore formation. *Izv. Akad. Nauk S.S.S.R., Ser. Geol.*, 5: 12—49.
Strakhov, N.M., 1956a. Types and genesis of dolomite rocks. In: *Types of Dolomite Rocks and their Genesis.* Akad. Nauk S.S.S.R., Moscow, pp. 5—27.

Strakhov, N.M., 1956b. Distribution and genesis of Upper Carboniferous dolomite rocks of Samarskaya Luka. In: *Types of Dolomite Rocks and their Genesis.* Akad. Nauk S.S.S.R., Moscow, pp. 185—208.

Strakhov, N.M., 1958. *Méthodes d'Etudes des Roches sédimentaires.* Bur. Rech. Géol. Géophys. Minières, Paris, pp. 535—542.

Strakhov, N.M., 1959 (Editor). *Toward Knowledge of Diagenesis of Sediments.* Akad. Nauk S.S.S.R., Moscow, 295 pp.

Strakhov, N.M., 1960. *Principles of Theory of Lithogenesis, 1. Types of Lithogenesis and their Distribution on the Earth's Surface.* Akad. Nauk S.S.S.R., Moscow, 212 pp.

Strakhov, N.M., 1976. *Problems of Geochemistry of Recent Oceanic Lithogenesis.* Akad. Nauk S.S.S.R., Moscow, 299 pp.

Strakhov, N.M., Brodskaya, N.G., Knyazeva, L.M., Razzhivina, A.N., Rateev, M.A., Sapozhnikov, D.G. and Shishova, E.S., 1954. *Formation of Sediments in Recent Basins.* Akad. Nauk S.S.S.R., Moscow, 791 pp.

Sugden, W., 1963. Some aspects of sedimentation in the Persian Gulf. *J. Sediment. Petrol.*, 33: 355—364.

Süjkowski, Zb.L., 1958. Diagenesis. *Bull. Am. Assoc. Pet. Geol.*, 42: 2692—2717.

Swinchatt, J.P., 1965. Significance of constituent composition, texture, and skeletal breakdown in some recent carbonate sediments. *J. Sediment. Petrol.*, 35: 71—90.

Taft, W.H., 1961. Authigenic dolomite in modern carbonate sediments along the southern coast of Florida. *Science,* 134 (3478): 561—562.

Taft, W.H., 1962. Cation influence on the recrystallization of metastable carbonates, aragonite and high-Mg calcite. *Geol. Soc. Am., Progr. Annu. Meet., 1963*: 162A.

Taft, W.H., 1963. Cation influence on diagenesis of carbonate sediments. *Geol. Soc. Am., Progr. Annu. Meet., 1963*: 162A.

Talbot, M.R., 1971. Calcite cements in the corallian beds (Upper Oxfordian) of southern England. *J. Sediment. Petrol.*, 41: 261—273.

Tarr, W.A., 1925. Is the chalk a chemical deposit? *Geol. Mag.*, 62: 252—264.

Teodorovich, G.I., 1950. *Lithology of Carbonate Rocks — Paleozoic of Ural-Volga Region.* Akad. Nauk S.S.S.R., Moscow, 213 pp.

Teodorovich, G.I., 1955. Towards the question of formation of sedimentary calcareous-dolomite rocks. *Tr. Inst. Nefti, Akad. Nauk S.S.S.R.*, 5: 75—107.

Teodorovich, G.I., 1958. *Study of Sedimentary Rocks.* Gostoptekhizdat, Leningrad, 572 pp.

Teodorovich, G.I., 1961. *Authigenic Minerals in Sedimentary Rocks.* Consultants Bureau, New York, N.Y., 120 pp.

Termier, H. and Termier, G., 1963. *Erosion and Sedimentation.* Van Nostrand, London, 434 pp.

Terzaghi, R.D., 1940. Compaction of limemud as a cause of secondary structure. *J. Sediment. Petrol.*, 10: 78—90.

Tester, A.C. and Atwater, G.I., 1934. The occurrence of authigenic feldspars in sediments. *J. Sediment. Petrol.*, 4: 23—31.

Thomas, G.E., 1962. Grouping of carbonate rocks into textural and porosity units for mapping purposes. In: *Classification of Carbonate Rocks — Am. Assoc. Pet. Geol., Mem.*, 1: 193—223.

Thomas, G.E. and Glaister, R.P., 1960. Facies and porosity relationships in some Mississippian carbonate cycles of Western Canada Basin. *Bull. Am. Assoc. Pet. Geol.*, 44: 569—588.

Thorp, E.M., 1936. Calcareous shallow-water marine deposits of Florida and the Bahamas. *Carnegie Inst. Wash. Pap., Tortugas Lab.*, 29 (452): 37—119.

Towse, D., 1957. Petrology of Beaver Lodge Madison Limestone reservoir, North Dakota. *Bull. Am. Assoc. Pet. Geol.*, 41: 2493—2507.

Tucker, M.E. and Kendall, A.C., 1973. The diagenesis and low-grade metamorphism of Devonian styliolinid-rich pelagic carbonates from West Germany: possible analogues of Recent pteropod oozes. *J. Sediment. Petrol.*, 43: 672—687.

Turmel, R.J. and Swanson, R.G., 1976. The development of Rodriguez Bank, a Holocene mudbank in the Florida reef tract. *J. Sediment. Petrol.*, 46 (3): 497—518.

Twenhofel, W.H., 1942. The rate of deposition of sediments: a major factor connected with alteration of sediments after deposition. *J. Sediment. Petrol.*, 12: 99—110.

Usdowski, H.E., 1962. Die Entstehung der kalkoölithischen Fazies des norddeutschen unteren Buntsandsteins. *Beitr. Mineral. Petrogr.*, 8: 141—179.

Usdowski, H.E., 1963. Die Genese der Tutenmergel oder Nagelkalke (cone-in-cone). *Beitr. Mineral. Petrogr.*, 9: 95—110.

Valyashko, M.G., 1962. *Geochemical Regularities in Formation of Potassium Salt Deposits.* Akad. Nauk S.S.S.R., Moscow, 397 pp.

Van Straaten, L.M.J.U., 1948. Note on the occurrence of authigenic feldspar in non-metamorphic sediments. *Am. J. Sci.*, 246: 569—572.

Van Straaten, L.M.J.U. (Editor), 1964. *Deltaic and Shallow Marine Deposits.* Elsevier, Amsterdam, 464 pp.

Van Tuyl, F.M., 1916. The origin of dolomite. *Iowa Geol. Surv., Bull.*, 25: 251—422.

Vinogradov, A.P., 1953. The elementary chemical composition of marine organisms. *Sears Found. Mar. Res., Yale Univ., Mem.*, 2.

Vishnyakov, S.G., 1951. Genetic types of dolomite rocks. *Dokl. Akad. Nauk S.S.S.R.*, 76 (1): 112—113.

Vishnyakov, S.G., 1951. Genetische Typen der Dolomitgesteine. *Dokl. Akad. Nauk S.S.S.R.*, 76 (2): 111—114.

Vital', D.A., 1959. Carbonate concretions in Mesozoic deposits of Russian platform. In: N.M. Strakhov (Editor), *Towards Knowledge of Diagenesis of Sediments.* Akad. Nauk S.S.S.R., Moscow, pp. 196—237.

Von Engelhardt, W., 1960. *Der Porenraum der Sedimente.* Springer, Berlin-Göttingen-Heidelberg, 207 pp.

Von Engelhardt, W., 1961. Zum Chemismus der Porenlösung der Sedimente. *Bull. Geol. Inst. Univ. Uppsala*, 40: 189—204.

Wachs, D. and Hein, J.R., 1974. Petrography and diagenesis of Franciscan limestones. *J. Sediment. Petrol.*, 44: 1217—1231.

Walker, T.R., 1957. Frosting of quartz grains by carbonate replacement. *Bull. Geol. Soc. Am.*, 68: 267—268.

Walker, T.R., 1962. Reversible nature of chert—carbonate replacement in sedimentary rocks. *Bull. Geol. Soc. Am.*, 73: 237—242.

Wallace, R.C., 1913. Pseudobrecciation in Ordovician limestones in Manitoba. *J. Geol.*, 21: 402—421.

Wallace, R.C., 1927. Recent work in dolomitization. *Natl. Acad. Sci., Natl. Res. Counc., Comm. Sedimentation, Rep.*, pp. 64—70.

Wardlaw, N.C., 1962. Aspects of diagenesis in some Irish Carboniferous limestones. *J. Sediment. Petrol.*, 32: 776—780.

Wardlaw, N.C., 1976. Pore geometry of carbonate rocks as revealed by pore casts and capillary pressure. *Bull. Am. Assoc. Pet. Geol.*, 60 (2): 245—257.

Wardlaw, N.C. and Reinson, G.E., 1971. Carbonate and evaporite deposition and diagenesis, Middle Devonian Winnipegosis and Prairie Evaporite formations of South-Central Saskatchewan. *Bull. Am. Assoc. Pet. Geol.*, 55: 1759—1786.

Weber, J.N., Bergenback, R.E., Williams, E.G. and Keith, M.L., 1965. Reconstruction of depositional environments in the Pennsylvanian Vanport Basin by carbon isotope ratios. *J. Sediment. Petrol.*, 35: 36—48.

Weller, J.M., 1959. Compaction of sediments. *Bull. Am. Assoc. Pet. Geol.*, 43: 273—310.

Wells, A.J., 1962. Recent dolomite in the Persian Gulf. *Nature*, 194 (4825): 274—275.

Wells, J.W., 1957. Coral reefs. In: H.S. Ladd (Editor), *Treatise on Marine Ecology and Paleoecology, 1 — Geol. Soc. Am., Mem.*, 67: 609—631.

Wengerd, S.A., 1951. Reef limestones of Hermosa Formation, San Juan County, Utah. *Bull. Am. Assoc. Pet. Geol.*, 35: 1039—1051.

Werner, F., 1961. Zu Verkittungsvorgängen an Psammiten. *Geol. Rundsch.*, 51: 507—517.

Weyl, P.K., 1960. Porosity through dolomitization: conservation-of-mass requirements. *J. Sediment. Petrol.*, 30: 85—90.

Weynschenk, R., 1951. The problem of dolomite formation considered in the light of research on dolomites in the Sonnwend Mountains (Tirol). *J. Sediment. Petrol.*, 21: 28—31.

Whitcombe, P.J., 1970. Diagenesis of the carbonate rocks of the Lower Limestone Shale Group (Carboniferous) of South Wales. *J. Sediment. Petrol.*, 40: 334—338.

Wickman, F.E., Blix, R. and Von Ubisch, H., 1951. On the variations in the relative abundance of the carbon isotopes in carbonate minerals. *J. Geol.*, 59: 142—150.

Williams, H., Turner, F.J. and Gilbert, C.M., 1954. *Petrography*. Freeman, San Francisco, Calif., 406 pp.

Wilson, J.L., 1969. Microfacies and sedimentary structures in "deeper-water" lime mudstones. In: G.M. Friedman (Editor), *Depositional Environments in Carbonate Rocks. A Symposium — Soc. Econ. Paleontol. Mineral., Spec. Publ.*, 14: 4—19.

Wilson, J.L., 1974. Characteristics of carbonate-platform margins. *Bull. Am. Assoc. Pet. Geol.*, 58 (4): 810—824.

Wolf, K.H., 1960. Simplified limestone classification. *Bull. Am. Assoc. Pet. Geol.*, 44: 1414—1416.

Wolf, K.H., 1962. The importance of calcareous algae in limestone genesis and sedimentation. *Neues Jahrb. Geol. Paläontol., Monatsh.*, 1962 (5): 245—261.

Wolf, K.H., 1963a. *Syngenetic to Epigenetic Processes, Paleoecology, and Classification of Limestones, in particular with Reference to Devonian Algal Limestones of Central New South Wales.* Thesis, University of Sydney, Sydney, N.S.W. (unpublished).

Wolf, K.H., 1963b. Limestones. Australian National Univ., Canberra, A.C.T. (unpublished).

Wolf, K.H., 1965a. Petrogenesis and paleoenvironment of Devonian algal limestones of New South Wales. *Sedimentology*, 4: 113—178.

Wolf, K.H., 1965b. Gradational sedimentary products of calcareous Algae. *Sedimentology*, 5: 1—37.

Wolf, K.H., 1965c. Littoral environment indicated by open-space structures in algal reefs. *Palaeogeogr., Palaeoclimatol., Palaeoecol.*, 1: 183—223.

Wolf, K.H., 1965d. "Grain-diminution" of algal colonies to micrite. *J. Sediment. Petrol.*, 35: 420—427.

Wolf, K.H. and Chilingarian, G.V., 1976. Compactional diagenesis of carbonate sediments. In: G.V. Chilingarian and K.H. Wolf (Editors), *Compaction of Coarse-Grained Sediments, II.* Elsevier, Amsterdam, pp. 719—768.

Wolf, K.H. and Conolly, J., 1965. Petrogenesis and paleoenvironment of limestone lenses in Upper Devonian red beds of New South Wales. *Palaeogeogr., Palaeoclimatol., Palaeoecol.*, 1: 69—111.

Wolf, K.H., Chilingar, G.V. and Beales, F.W., 1967a. Elemental composition of carbonate skeletons, minerals, and sediments. In: G.V. Chilingar, H.J. Bissell and R.W. Fairbridge (Editors), *Carbonate Rocks*, Elsevier, Amsterdam, pp. 23—149.

Wolf, K.H., Easton, J.A. and Warne, S., 1967b. Techniques of examining and analyzing carbonate skeletons, minerals and rocks. In: G.V. Chilingar, H.J. Bissell and R.W. Fairbridge (Editors), *Carbonate Rocks*, Elsevier, Amsterdam, pp. 253—341.

Wood, A., 1941. "Algal dust" and the finer-grained varieties of Carboniferous limestone. *Geol. Mag.*, 78: 192—200.

Wray, J.L., 1962. Pennsylvanian algal banks, Sacramento Mountains, New Mexico. In: *Geoeconomics of the Pennsylvanian Marine Banks in Southeast Kansas — Kans. Geol. Soc., Annu. Field Conf., 27th, Guidebook*, pp. 124—128.

Young, R.B., 1935. A comparison of certain stromatolitic rocks in the dolomite series of South Africa, with modern algal sediments in the Bahamas. *Trans. Geol. Soc. S. Afr.*, 37: 153—162.

Zankl, H. and Schroeder, J.H., 1972. Interaction of genetic processes in Holocene reefs off North Eleuthera Island, Bahamas. *Geol. Rundsch.*, 61 (2): 520—541.

Zaytseva, E.D., 1954. Vertical distribution of biogenous elements in interstitial solutions of Bering Sea. *Dokl. Akad. Nauk S.S.S.R.*, 99 (2): 289—291.

Zeller, E.J. and Wray, J., 1956. Factors influencing precipitation of calcium carbonate. *Bull. Am. Assoc. Pet. Geol.*, 40: 140—152.

Zenger, D.H., 1973. Syntaxial calcite borders on dolomite crystals, Little Falls Formation (Upper Cambrian), New York. *J. Sediment. Petrol.*, 43: 118—124.

ZoBell, C.E., 1942. Changes produced by micro-organisms in sediments after deposition. *J. Sediment. Petrol.*, 12: 127—136.

ZoBell, C.E., 1946. Studies on redox potential of marine sediments. *Bull. Am. Assoc. Pet. Geol.*, 40: 477—513.

References added in proof

Hall, J., 1847. Cryptozoön, N.G., cryptozoön proliferum, n. sp. *Annu. Rep. New York State.*

Klöden, K.F., 1828. *Beiträge zur Mineralogischen und Geognostischen Kenntniss der Mark Brandenburg*, 1. Dieterici, Berlin.

MacIntyre, I.G., 1967. *Recent Sediments off the West Coast of Barbados, W.I.* Unpubl. Ph. D. Thesis, McGill University, Montreal.

Maxwell, W.G.H., 1962. Lithification of carbonate sediments in the Heron Island Reef, Great Barrier Reef. *J. Geol. Soc. Aust.*, 8: 217—238.

Spry, A., 1969. *Metamorphic Textures.* Pergamon, Oxford, 350 pp.

Chapter 7

DOLOMITES AND DOLOMITIZATION

GEORGE V. CHILINGAR, D.H. ZENGER, H.J. BISSELL and K.H. WOLF

INTRODUCTION

Definitions and classification

Although there are numerous kinds of dolomite, one statement that may safely be made is that the significant processes of dolomitization are post-depositional or diagenetic. According to numerous investigators, primary precipitation of dolomite crystals, particularly in marine environments, although probably represented in some Holocene carbonate sediments, is of little significance when considering the great bulk of ancient dolomite. The economic significance of dolomite as a host for oil and mineral deposits results from porosity changes which occurred during diagenesis.

Clarification of terminology, although a relatively unexciting task, must be faced considering the profusion of different usages of terms such as syngenesis, diagenesis, and epigenesis. The matter is further complicated, even given an unequivocal vocabulary, by the difficulty of determining the relative time and location of dolomitization in many circumstances. For example, diagenesis has been used by most workers in this country to cover the processes modifying sediments and rocks, excluding metamorphism and, commonly, weathering. Chilingar et al. (1967, pp. 313—314), however, used the term only up to lithification of the sediment. The same workers (p. 322), as well as many others, considered *syngenesis* as separate from diagenesis, whereas, according to the *AGI Glossary of Geology* (1973 edition), some would apply the term syngenesis to initial diagenesis, thus overlapping with the "early diagenesis" of others. At the other end, *epigenesis* may refer to changes occurring after lithification of a sediment, although it has often been used to indicate a relationship to secondary structures, that is, in a postorogenic sense. Consequently, many geologists, especially Soviet, prefer the term *catagenesis* in referring to processes occurring after lithification.

Fairbridge (1957) classified dolomites as early diagenetic (penecontemporaneous), late diagenetic, and postorogenic. In his later paper (1967), he employed a similar scheme including the categories *syndiagenesis* (with initial and early burial stages), *anadiagenesis* (the compaction and maturation stage during which lithification occurs), and *epidiagenesis* (the emergent or postorogenic phase of diagenesis). Friedman and Sanders (1967, p. 308) classed dolomites as *syngenetic* (formed penecontemporaneously in its environment of deposition as a micrite or as fine-grained crystals), *detrital*, *diagenetic* (formed by stratigraphic types of replacement of calcium carbonate sediments during or following consolidation of the sediments), and *epigenetic* (formed by replacement with localization of the dolomite by structural control). It is interesting to note that Friedman and Sanders considered that although diagenetic dolomite could form penecontemporaneously, it would be difficult to distinguish it from the syngenetic (essentially primary) type. This was also pointed out by Chilingar and Bissell (1963a).

Badiozamani (1973) presented a genetic classification of dolomites and his basic framework is presented in Fig. 7-1. Wilson's recent text (1975, p. 311) presents a more elaborate scheme based on that of Badiozamini; the reader is referred to Wilson for more details.

Freeman (1972, pp. 445—446) suggested that a reasonable basis for a genetic classification would be a geochemical framework, admitting that this could be elusive. As can be seen in Fig. 7-2, Freeman's use of "syngenetic" and "epigenetic", although having a connotation of time, referred primarily to the source of the dolomitizing solutions. Syngenetic dolomitization, which passes beyond the depositional stage, results from solutions that have been generated predominantly at the surface, whereas epigenetic dolomitization is that resulting from dolomitizing waters that evolved mainly in the subsurface.

In their thorough analysis of porosity types, Choquette and Pray (1970,

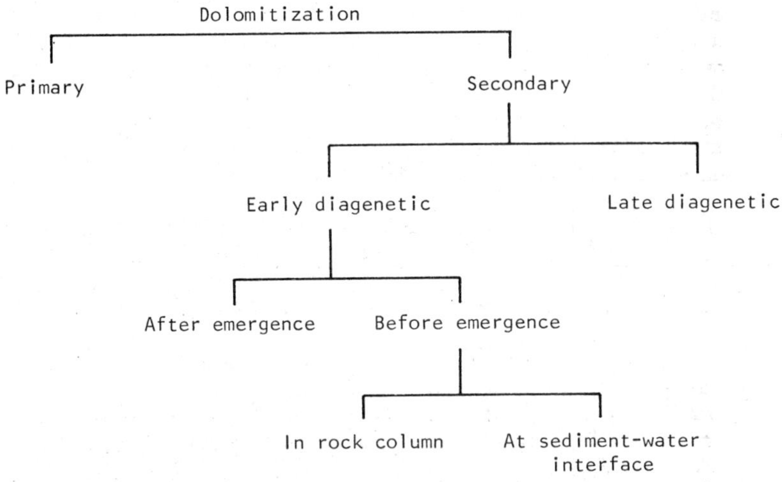

Fig. 7-1. Framework of Badiozamani's genetic classification of dolomites. (Modified from Badiozamani, 1973.)

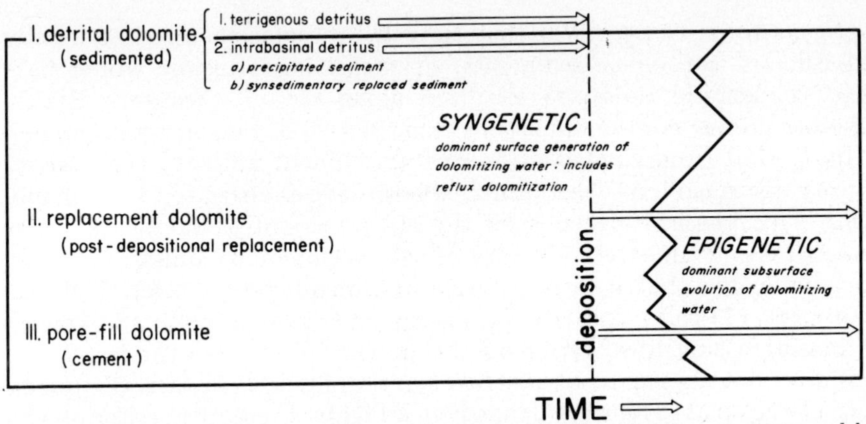

Fig. 7-2. Classification of products and genesis of dolomitization as portrayed by Freeman (1971, p. 447; courtesy Am. Assoc. Pet. Geol.)

pp. 219—220) discussed three major stages in the history of a sedimentary carbonate body subsequent to deposition (theoretically the classification could be applied to dolomites):

(1) *Eogenetic stage*: the time interval between deposition and burial of the newly deposited sediments or rock below the depth of significant influence by processes that operate from the surface or depend for their effectiveness on proximity to the surface.

(2) *Mesogenetic stage*: the time interval during which the sediments or rocks are buried at depth below the major influence of processes operating directly from or closely related to the surface. It constitutes the entire time between the geologically brief, early stage of burial and a final phase of imminent erosion.

(3) *Telogenetic stage*: the time interval during which long buried carbonate rocks are influenced by processes associated with the formation of an unconformity.

In view of the conflicting use of terms and with the realization that, very likely, there will never be a scheme satisfactory to all, the present authors consider three categories of dolomite, excluding a metamorphic variety, as follows:

(1) *Primary*: that formed essentially at the sediment—water interface by direct precipitation from solution.

(2) *Diagenetic*: that formed by replacement of calcium carbonate sediments in the postdepositional but prelithification phase. "Penecontemporaneous" generally refers to the earliest stages of diagenesis, but could be followed directly by epigenesis.

(3) *Epigenetic or catagenetic*: that formed by replacement of essentially lithified calcium carbonate sediments, or rocks. Postorogenic would be a variety of epigenetic dolomite localized along tectonic features. Hydrothermal dolomite is considered here as another type. Epigenesis, then, embraces the period from consolidation of the sediment until the rock arrives at the very "surface" of weathering, where dedolomitization is probably dominant. This phase, especially for the older dolomites, represents by far the greatest bulk of time in the history of most sedimentary units.

It is appreciated that it is not always an unequivocal matter to distinguish between: (1) early and late diagenetic, and (2) diagenetic and epigenetic dolomites. Using lithification as the boundary between the latter subdivisions does not appear to be ideal to some geologists. In fact, Choquette and Pray (1970, p. 218) claimed that it is a highly impractical criterion and they posed the thought-provoking question "What constitutes lithification?" The present writers, however, contend that lithification serves as well as any factor. Much of the evidence for epigenetic (or catagenetic) dolomitization (which is discussed in detail later), both in the field and under the microscope, shows that dolomitization occurred after the sediment was sufficiently well cemented to be coherent, even though the remaining porosity may be quite high. Generally, diagenetic dolomitization is accomplished by fluids generated at the surface (as is the case with Freeman's (1972) syngenetic dolomites). One may also state that fluids evolving in the subsurface (including compaction fluids) are the agents affecting mostly late diagenetic and epigenetic dolomites. Dolomitizing waters from the surface, however, may promote dolomitization in the early stages of epigenesis or later during postorogenic phases when the rocks may be subjected to reactions with descending meteoric waters. In any event, the determination of the source of dolomitizing solutions in ancient carbonates is in many cases most difficult and one is commonly forced to turn to carbon and oxygen isotope analyses for such distinctions. Early diagenetic dolomitization is that of shallow burial, whereas late diagenetic and epigenetic dolomites may form under shallow or deep burial conditions.

Inclusion of hydrothermal dolomitization as a kind of epigenetic dolomite necessitates a concern for the ambiguous and subjective boundary between hydrothermal and metamorphic processes. Muffler and White (1969) pointed out that mineralogical and chemical interactions between rocks and hot water are generally considered as hydrothermal, particularly when the interactions are related to fractures, veins, or disseminated ores. They realized, however, that in the Salton Sea Geothermal Field, hydrothermal alteration can be treated in the same manner as metamorphism, the difference between the two being essentially semantic in that case.

Classifications involving these temporal aspects are basically genetic in philosophy. Dolomites have also been classed descriptively, examples being those of Folk (1959), Leighton and Pendexter (1962), Powers (1962), Thomas (1962), Chilingar et al. (1967, table 9), and several schemes compiled by Bissell and Chilingar (1967). Although many of these descriptive classifications are based on such objective features as texture, fabric, and composition, they may not completely avoid genetic implications and, in fact, some attempt to achieve this (Chilingar, 1957); an example is the seemingly straightforward type "dolomitized limestone" (Leighton and Pendexter, 1962, p. 57). The present writers appreciate the practical working nature of an objective descriptive classification, but believe that a genetic one (i.e., primary, diagenetic, and epigenetic) is more appropriate for this particular discussion.

The dolomite problem

Difficulties associated with the classification of dolomites are in large part related to the subject of the formation of dolomite. The authors hasten to point out, however, that although the purpose of the chapter is to treat the diagenesis of dolomitization, it does not pretend to cover the history of thought on such matters as the origin, classification, and occurrence of dolomites. Many reviews in English are available, in particular the older classics by Steidtmann (1911) and Van Tuyl (1916), and the more recent works by Fairbridge (1957), Ingerson (1962), Sonnenfeld (1964), Friedman and Sanders (1967), and Zenger (1972b). Friedman and Sanders (1967, pp. 280—289) and Bathurst (1971, pp. 517—543) presented good summaries of Holocene marine occurrences of dolomite.

It was not long after the discovery and description of dolomite by Dolomieu (1791 — original reference not seen; see Dunbar and Rodgers, 1957, p. 219), that a major anomaly was recognized: although dolomites formed a great portion of late Precambrian and Paleozoic carbonates, they subsequently decreased markedly in significance (Chilingar, 1956a; Zenger, 1972b, p. 107). For some time, and until some two decades ago, it was believed that modern marine dolomite existed primarily as isolated rhombs in Holocene sediments (Fairbridge, 1957, p. 164). In comparison with the earlier abundance of dolomite, this practical absence of the mineral in Holocene carbonates, and the attendant restriction of the application of uniformitarianism, promoted the formulation of the "dolomite problem" (Zenger, 1972b, p. 107). However, with the numerous discoveries of penecontemporaneous, mainly peritidal dolomite since the mid 1950's (for example, Alderman and Skinner, 1957; Curtis et al., 1963; Deffeyes et al., 1965; Illing et al., 1965; Shinn et al., 1965; Müller and Tietz, 1966; Shinn, 1968; see also Fig. 7-3), many felt that a solution to the problem was imminent. Such opti-

Fig. 7-3. Holocene, mud-cracked dolomitic supratidal crust, west side Andros Island, Bahama Banks. (After Zenger, 1972, fig. 1; courtesy, Geol. Soc. Am.)

mism was given much momentum by the published accounts of ancient analogs of peritidal dolomite (for example, Laporte, 1967; Matter, 1967; Roehl, 1967; Schenk, 1967; West et al., 1968; Textoris, 1969; Thompson, 1970). But the extent and bulk of Holocene dolomite are unimpressive when compared to those of ancient dolomites. Also, as Zenger (1972a) has shown, there has been an overextrapolation between the modern occurrences of penecontemporaneous peritidal dolomite and dolomites in the geologic record; as will be discussed briefly later, many ancient carbonates, dolomitized penecontemporaneously or later during diagenesis, either did not accumulate in a peritidal environment or were not replaced there.

To account for this "discrepancy" between the significance of modern and ancient dolomites remains one of the intriguing questions in earth history. Part of the difference could relate to the vast epeiric seas of the past which provided situations conducive to more widespread penecontemporaneous or early diagenetic dolomitization. Another contributing factor is the epigenetic dolomitization which has not as yet had the opportunity to modify Holocene carbonates. Furthermore, the discrepancy might be partially attributable to some change in crustal and/or oceanic conditions, such as possible changes in Ca/Mg ratio of sea water with time (Chilingar, 1956a).

Strakhov (1953, p. 24) believed that most of the diagenetic dolomites formed in the past, accumulated in Late Paleozoic time. The probability of high CO_2 content in the atmosphere and high Mg/Ca ratio of sea water during the Precambrian and Early Paleozoic times convinced some writers (Strakhov, 1953; Chilingar, 1956b) that dolomite precipitated directly out of sea water, in certain realms at least, during various stages of sedimentation. Süjkowski (1958, pp. 2715—2716), however, made this statement: "Since Precambrian time, sediments have been more or less of the same type as recent ones, and, as temperature, air composition, and other geographical factors were not far removed from those known today, it is only logical to assume that diagenesis followed the same lines as today. This is shown by identical rock types throughout the geological column."

Recently, a unique proposal has been made by Mottl and Holland (1975) that is based on evidence, in large part experimental, showing that at high temperatures (between 70°C and 500°C) Mg^{2+} is removed from sea water and enters basalt. They proposed that the increase in temperature to account for the Mg^{2+} flux from sea water to basalt could be due to the increase in sea-floor spreading, and accompanying activity at mid-oceanic rises during the Mesozoic. This roughly coincides in time with a notable increase in limestone/dolomite ratios (Chilingar, 1956a); Mottl and Holland suggested that this depletion of Mg^{2+} cation from sea water is responsible for the drastic decrease in dolomite during the last 200 million years.

Inasmuch as it is not the purpose of this chapter to explore this problem further, the reader is referred to summaries of these aspects by Ingerson (1962, pp. 834—836) and Zenger (1972b, pp. 120—122).

PRIMARY DOLOMITE

There is a continuing controversy over the significance of primary dolomite, as defined here. The following statement of Ingerson (1962, p. 830) is often quoted, ". . . it has never been demonstrated unequivocally that dolomite has ever been precipitated in nature directly from solution . . ." On the other hand, Chilingar (1956a), Deelman (1975; 1976, written communication) and Davies and Ferguson (1975) claim to have directly precipitated dolomite (see section "Synthesis of dolomite"). The findings of Beloborodova et al. (1972) also indicate the possibility of direct precipitation of dolomite out of Precambrian sea water. Favorable factors include: (1) Mg^{2+}/Ca^{2+} ratio close to 1, and (2) high alkalinity and high pH, which insures high concentration of CO_3^{2-} anions in solution. Field and petrographic evidence indicate that at least the great portion of dolomite, modern and ancient, is of the replacement type. Some indicators of a replacement origin include dolomitized reefs (Fig. 7-4), fossils (Figs. 7-5, 7-6), ooids (Fig. 7-7), calcispheres, peloids (Fig. 7-8), calcite inclusions in poikilotopic dolomite crystals (Figs. 7-8, 7-9), and other relationships characteristic of late diagenetic and epigenetic dolomitization to be discussed in detail later. (See also Figs. 7-10, 7-11, 7-12, 7-13, 7-14, and 7-15).

Despite this widespread evidence and the minimal success in directly precipitating dolomite (or even protodolomite, as contrasted with magnesium-rich calcite) at earth surface temperatures and pressures, there have been numerous suggestions of a primary origin for the extremely fine-grained unfossiliferous dolomites associated with evaporite sequences. Even for some of these dolomites, however, a metasomatic origin has been proposed (Newell et al., 1953, p. 121). The replacement theory has dominated (Dunbar and Rodgers, 1957, p. 242) and many geologists have not seriously considered primary dolomite formation as being widely applicable. The serious chemical

Fig. 7-4. Portion of completely dolomitized favositid biostrome, with vugs, Lockport Formation (Middle Silurian) immediately east of Niagara Falls, New York. The unit represents original subtidal site of formation. (After Zenger, 1972, fig. 7; courtesy Geol. Soc. Am.)

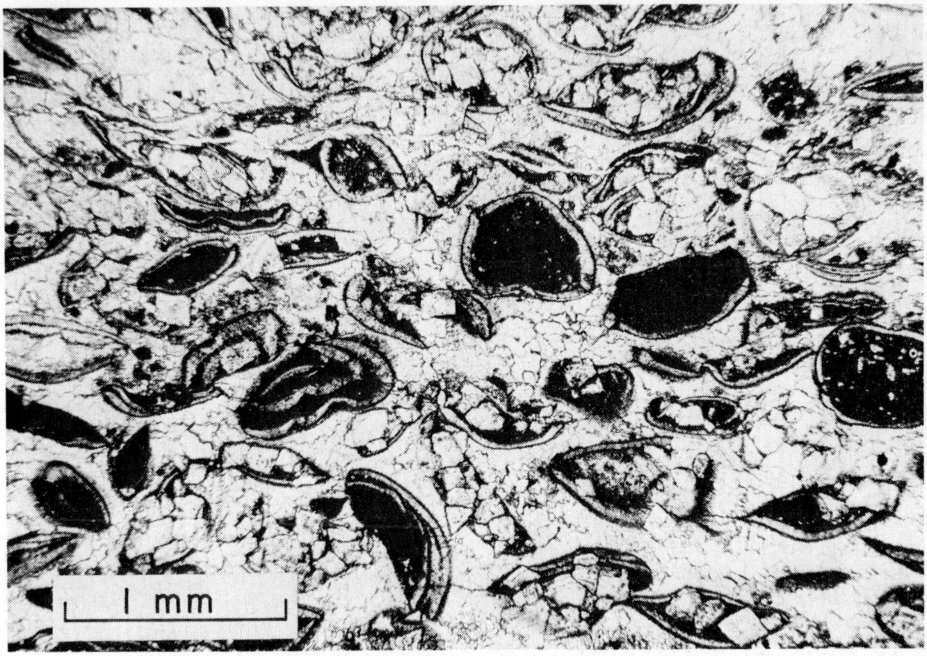

Fig. 7-5. Photomicrograph (plane polarized light) of thin-section of dolomitic (approximately 15% dolomite) limestone with tests of ostracode *Eukloedenella* partially replaced by dolomite (large, light crystals, commonly rhombic, within and transecting tests). Lockport Formation (Middle Silurian), Sherrill, New York. (Photograph by D.H. Zenger.)

and kinetic problems for dolomitization in general are all the more critical for a primary nucleation. Degens and Epstein (1964), in observing a lack of significant difference between δO^{18} of calcite and dolomite in coexisting pairs of these minerals, concluded that dolomite does not directly precipitate from aqueous solution but must form by solid-state replacement with no exchange of CO_3^{2-}. On the other hand, Clayton et al. (1968) claimed to have found distinctions between both the carbon and oxygen isotope ratios of coexisting pairs of calcite and dolomite in Deep Spring Lake, California. Consequently, they concluded that dolomite formation might best be explained by precipitation rather than by cationic replacement. Fritz and Smith (1970) acknowledged the confusion in the interpretation of stable-isotope data and suggested that small differences in $\delta 0^{18}$ between replacement dolomite and associated limestones might be accounted for by an initial precipitation of protodolomite which has an oxygen-isotope composition approximating that of calcite. The isotopic composition of the subsequent

Fig. 7-6. Photomicrograph (crossed polarizers) of thin-section of completely dolomitized fossil-fragmental sample of Joslin Hill Member (Herkimer Formation, Middle Silurian, Utica, New York). Note well-preserved nature of bryozoan zoaria and pelmatozoan fragments in quartzose (white grains are quartz) dolomite "matrix". Deposition interpreted as subtidal and dolomitization as diagnetic. (Photograph by D.H. Zenger.)

replacement dolomite is then controlled by the protodolomite—water fractionation.

In most Holocene dolomite occurrences there is evidence for a replacement origin and these penecontemporaneous types will be classed under the diagenetic category. For example, protodolomite is a replacement product of chiefly aragonitic sediments on Bonaire, Netherlands Antilles (Deffeyes et al., 1965), on the Florida Keys (Shinn, 1968), in the Bahamas (Shinn et al., 1965), and in the Persian Gulf (Illing et al., 1965). At some localities, however, the precipitation of dolomite may be occurring. There is a suggestion of primary nucleation of dolomite crystals in the Coorong Lagoon and associated ephemeral lakes in southern Australia (Alderman and Skinner, 1957; Skinner, 1963). It is extremely difficult to demonstrate the act of precipitation in a natural environment (Bathurst, 1971, p. 523). The whitings observed there, and elsewhere, could be the result of wave or current agitation. Neither is it possible to determine whether suspended sedi-

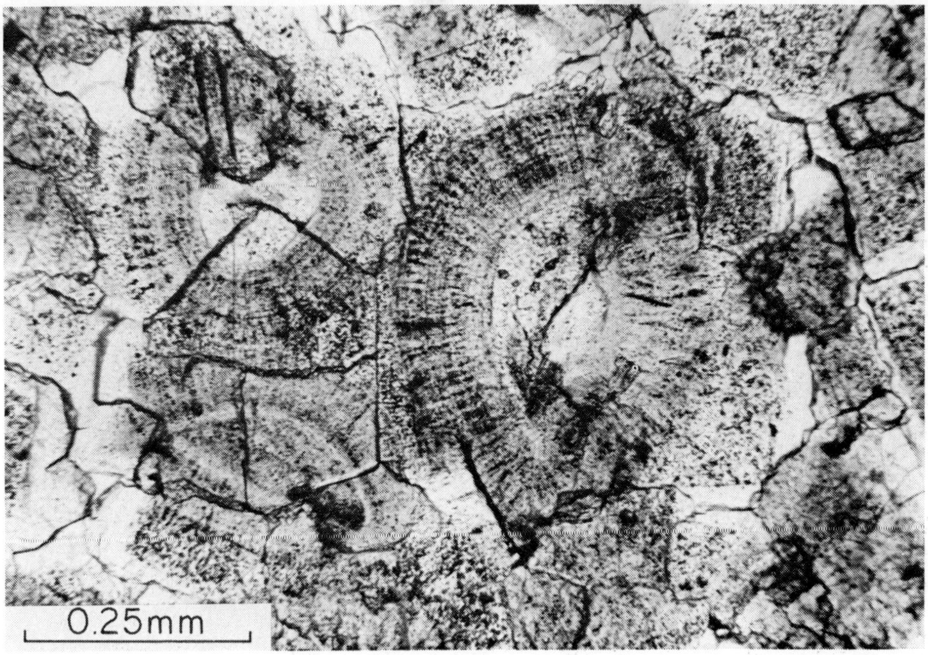

Fig. 7-7. Photomicrograph (plane polarized light) of thin-section of oolitic dolomite from Lockport Formation (Middle Silurian), central New York. Dolomitization is complete (note dolomite crystal boundaries transecting relect ooids). (Photograph by D.H. Zenger.)

ments are freshly nucleated crystals. Some of the several carbonate minerals forming there are undoubtedly primary, but others are probably the result of reactions between crystals and the saline water (Bathurst, 1971, pp. 523—524).

The dolomite forming in Deep Spring Lake (a playa in east-central California), although not in a marine environment, is believed by most workers familiar with its occurrence to be a direct precipitate. Jones (1961, p. 201) first proposed a primary origin based on the highest percentages of dolomite within areas of more frequent flooding. The euhedral form of the dolomite rhombs, which bear no consistent orientation to the host calcite crystals, suggested a similar origin to Peterson et al. (1966, p. 258). Clayton et al. (1968, p. 431) agreed, their conclusions being based on the difference between carbon and oxygen isotope ratios of coexisting calcite and dolomite (mentioned previously) and a dolomite—interstitial water oxygen isotope fractionation factor of about 1.035 (which is outside the range of fractionation between water and calcite). Barnes and O'Neil (1969) reported both

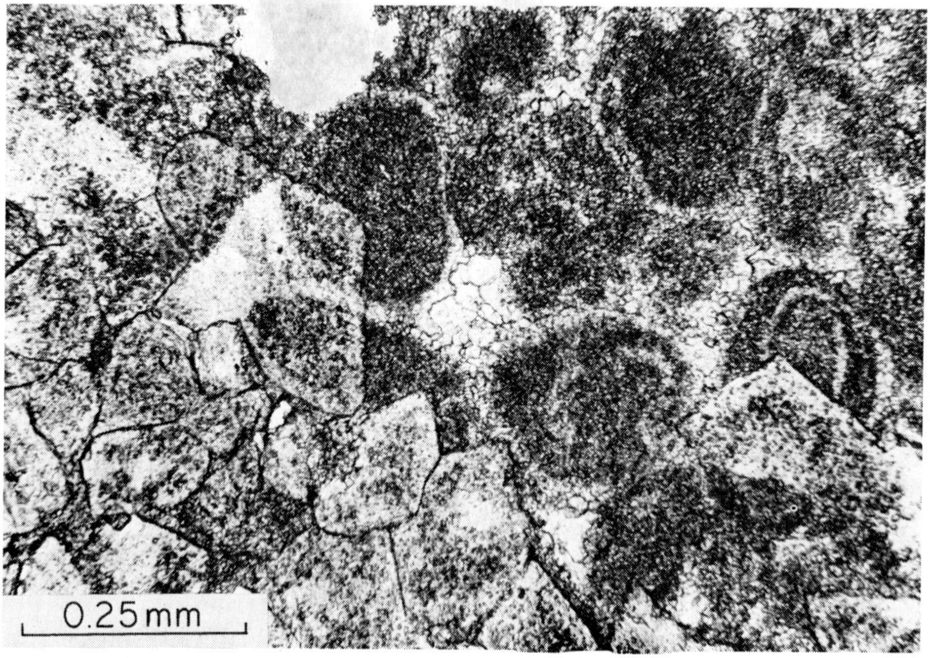

Fig. 7-8. Photomicrograph (plane polarized light) of thin-section of calcareous dolomite, Lost Burro Formation (Middle and Upper Devonian, east-central California) showing dolomite crystals replacing calcitic peloids and matrix; note calcitic inclusions in poikilotopic dolomite. (Photograph by D.H. Zenger.)

calcium-rich and stoichiometric dolomite forming presently along streams draining magnesium-rich serpentinite terrain in northern California, and later (1971, p. 713) they proposed that these dolomites might be directly precipitated from these alkaline waters.

Behrens and Land (1972) concluded that laterally extensive dolomite beds of Holocene age in Baffin Bay, Texas, formed by direct precipitation from marine waters. Their evidence consists of a lack of replacement textures, an enrichment of dolomite of some 3‰ in δO^{18} relative to coexisting aragonite and magnesium-calcite, and a much higher Sr^{2+} content (about 800 p.p.m.) than that of ancient "replacement" dolomites.

Miller (1961) described the lack of dolomite formation in a salt extraction process by evaporation of sea water, but he did observe clear dolomite rhombs up to 0.14 mm long disseminated with subhedral aragonite and calcite in the sludge dump which had been above the water level about 8 years. There is no evidence that the dolomite formed by replacement, leaving the possibility that the rhombs formed as a "direct precipitate" in minute voids

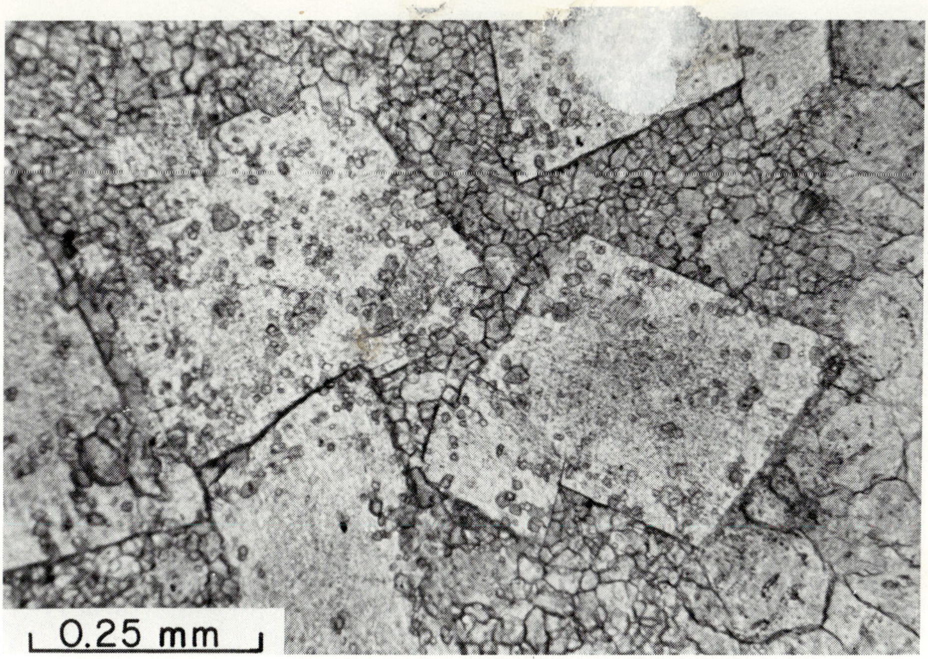

Fig. 7-9. Photomicrograph (ordinary light) of thin-section of poikilotopic dolomite (large, lighter crystals) replacing calcisiltite (darker, owing to staining with Alizarin Red S). Note inclusions of calcite. Lost Burro Formation (Middle and Upper Devonian), Funeral Mountains, east-central California. (Photograph by D.H. Zenger.)

in which magnesium-rich brine accumulated or filtered through the waste sludge.

Petrographically, it may be nearly impossible to distinguish between primary and penecontemporaneous dolomites in Holocene sediments owing to their very small crystal size. Such a distinction becomes even more complicated when dealing with the ancient rocks. The possibility of precipitation of more widespread dolomite must be kept in mind. Possibly, studies of isotopic and trace-elemental composition may prove helpful in this area although, typically, there are differences of opinion in the interpretation of such data. Weber (1964) claimed that "primary" dolomites, as contrasted with "secondary" dolomites, contained significantly higher amounts of Al, Ba, Fe, K, Li, Zn, and Na, whereas Sr is significantly concentrated in the "secondary" dolomite crystals. Weber's "primary", however, would include the penecontemporaneous replacement type, which in ancient rocks would appear petrographically similar to directly precipitated ones. In the work referred to above, Behrens and Land (1972, p. 159), on the other hand,

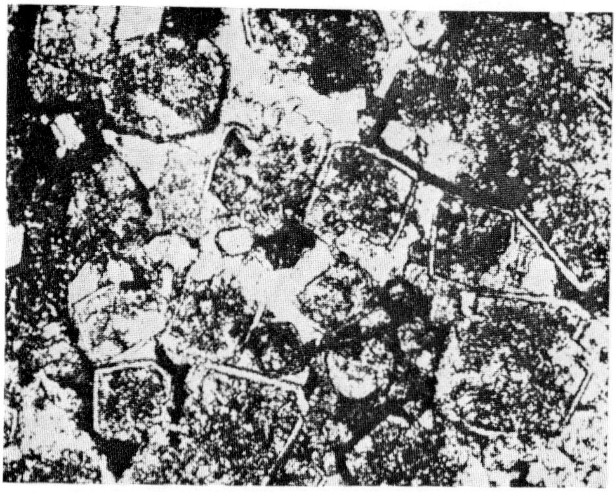

Fig. 7-10. Dolomitic limestone with complex texture. The dolomite rhombohedrons are composed of core and thin rim. The cores are composed of very fine-grained calcite with dolomite background, whereas the rims are pure dolomite. Areas between rhombohedrons are filled with medium-grained calcite. Insoluble residue: 1.70%, $CaCO_3$: 75.37%, $CaMg(CO_3)_2$: 22.88%. Upper Carboniferous, Samarskaya Luka. (After Khvorova, 1958, fig. 175, p. 127.) Diagenetic dolomite, later altered during epigenesis; ×45.

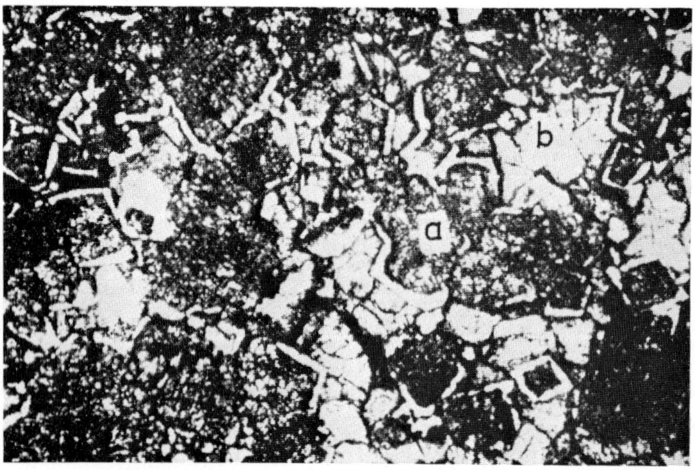

Fig. 7-11. Dolomitic limestone having complex texture. Dolomite rhombohedrons have cores with very fine-grained calcite and light dolomite rims. In places where dolomite rhombohedrons are in close contact, the rim disappears and dolomite crust is created which borders several adjoining rhombohedrons (a). The areas between aggregates of rhombohedrons, which apparently previously constituted pore spaces, are filled with medium-grained calcite (b). Thin-section was colored with K_2CrCO_4. Upper Carboniferous, Samarskaya Luka. (After Khrovora, 1958, fig. 177, p. 127.) Diagenetic dolomite, later altered during epigenesis; ×45.

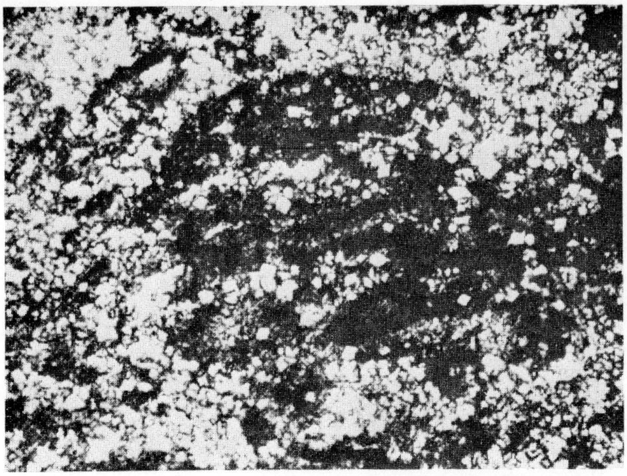

Fig. 7-12. Strongly dolomitized algal (*Dvinella*) limestone. In the mass of very fine-grained dolomite, one can observe numerous poorly preserved algal tubes (Family *Beresel-leae*). Podol'skiy horizon, Onega. (After Khvorova, 1958, fig. 181, p. 128.) ×45.

showed that presumed primary precipitates in Baffin Bay contained higher amounts of strontium than ancient dolomites, which could have lost much of this element through diagenesis as has been theorized for ancient lime-stones (Kinsman, 1969).

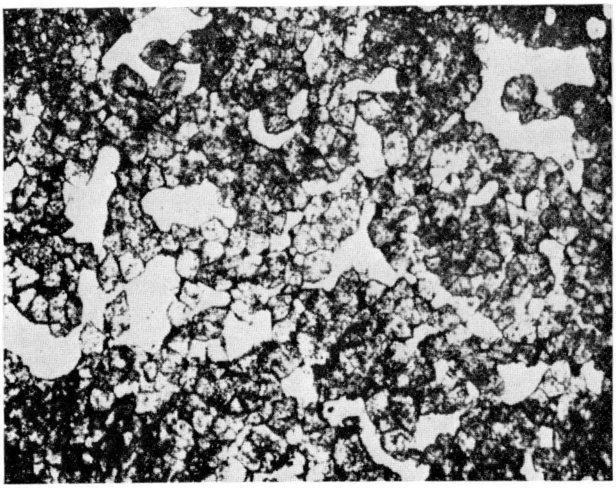

Fig. 7-13. Fine-grained, very porous dolomite. The pores are intercommunicated. Upper Carboniferous, R. Pinega. (After Khvorova, 1958, fig. 244, p. 139.) ×20.

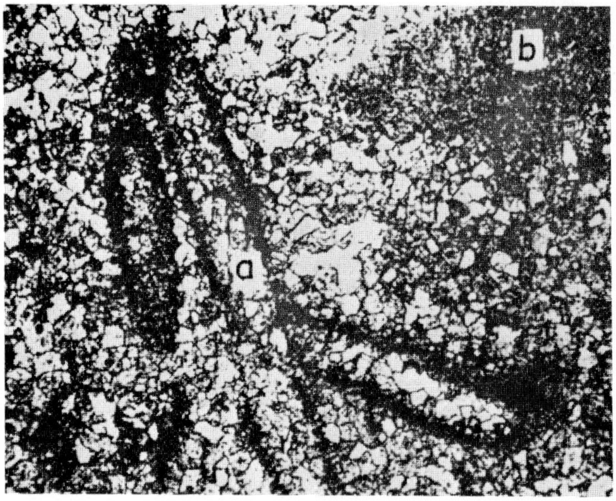

Fig. 7-14. Fine-grained dolomite with the green Algae *Siphoneae* (*a*). The walls of the latter are replaced by finer crystals than those of the inner portions and the main rock mass. Non-dolomitized remains of crinoid (*b*). Upper carboniferous, Samarskaya Luka. (After Khvorova, 1958, fig. 255, p. 141.) ×45.

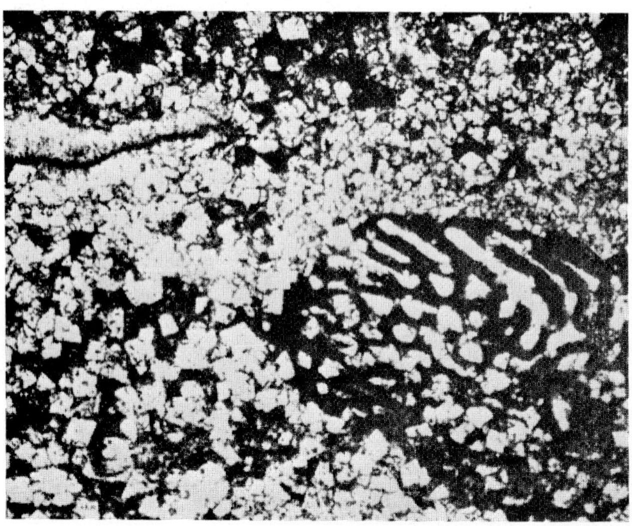

Fig. 7-15. Fine-grained dolomite with remains of micro-grained calcite and non-dolomitized organic material. Upper Carboniferous, Samarskaya Luka. (After Khvorova, 1958, fig. 186, p. 129.) ×20.

According to many geologists[1], primary dolomite, if having formed in the past, represents an insignificant contribution toward the total mass of ancient dolomites, which often are of the coarse-grained replacement types.

CHEMISTRY OF DOLOMITIZATION

General

Still incomplete is the knowledge of the chemistry involved in the formation of dolomite. Still relevant is the statement by Hsü (1967, p. 188): "Under the circumstances of our present imperfect knowledge, postulates of the conditions that are favorable for the precipitation of dolomite or for the replacement of calcite by dolomite have been speculative. Much has been written that dolomite genesis is related to pH, Eh, or P_{co_2} of the sedimentary or diagenetic environment. Preceding analysis of our present knowledge has furnished very little basis for such postulates."

Dolomite synthesis—experimental and theoretical

Numerous attempts have been made to synthesize dolomite in the laboratory, not only by direct precipitation but also by replacement. Generally, it may be said that these attempts have met with little success, especially when the experimental conditions, such as temperature, pressure, and partial pressure of CO_2 (P_{co_2}), have been maintained near those existing at the earth's surface. Chilingar (1956c), Fairbridge (1957), and Siegel (1961) provided summaries of the numerous earlier attempts to synthesize dolomite.

Success has been achieved relatively easily under higher-than-normal temperatures, particularly in the hydrothermal environment. In a very important contribution, Graf and Goldsmith (1956) treated certain magnesium-rich calcitic starting materials in the presence of H_2O and $H_2O + CO_2$ over a temperature range of 25°C—450°C. At temperatures over 200°C, and depending on the starting materials, some reactions produced stable dolomite of 1 : 1 composition, ordered with respect to Ca and Mg. They concluded (p. 174) "... at elevated temperatures ionic mobility is great enough for cation ordering to take place ...", thus permitting the development of stable, ordered dolomite. In fact, synthetic dolomite, produced under dry conditions between 500°C and 800°C (Graf and Goldsmith, 1955) appeared to be identical to natural dolomite. These results suggested that at about 500°C ionic mobility is sufficiently great for dolomite to form even by dry, solid-state

[1] Including Dr. D.H. Zenger.

cation diffusion. On the other hand, runs at lower temperatures (about 70°C) did not result in the production of ordered, stoichiometric dolomite but rather, calcium-rich and imperfectly ordered (i.e., lacking the superstructure X-ray reflections of true dolomite) material designated protodolomite (Graf and Goldsmith, 1956, p. 184). Land (1967, pp. 923—927) obtained replacement dolomite, in the form of reaction rims, by reacting aragonitic and magnesium-rich skeletal material with solutions of known calcium and magnesium concentration at 300°C and 100 atm pressure. Such syntheses of dolomite at these high temperatures facilitated by hydrothermal conditions is in keeping with the phase diagram of the system $CaCO_3$—$MgCO_3$ described by Goldsmith et al. (1961, p. 52).

Despite the success of synthesizing dolomite at high temperatures (and pressures), most attempts at room temperature were unsuccessful. [1] Berner (1971, p. 149) stated: "The chemistry of the formation of dolomite is very complex. For one thing, it has never been synthesized at sedimentary temperatures. The many reports in the literature of dolomite synthesis at low temperatures have incorrectly used the term dolomite for a ~50 mole% magnesian calcite whose major X-ray diffraction peaks match those of dolomite. Unless long-range Ca—Mg ordering can be demonstrated by the presence of at least faint traces of the principal-ordering X-ray reflections, then the material should not be called dolomite . . ."

Glover and Sippel (1967) acknowledged the general failure to synthesize either a range of magnesian calcites or of dolomite having any degree of ordering at temperatures of 0—40°C at atmospheric pressure. They, as well as McCauley and Roy (1966), succeeded in producing crystallized magnesian calcite, but in the range of dolomite composition these showed no evidence of ordering. Graf and Goldsmith (1956, p. 184) proposed that the crystallochemical differences between Ca^{2+} and Mg^{2+} are too subtle at earth surface conditions to permit the available energy there to carry the ordering beyond the protodolomite stage. They concluded that ancient, typically ordered dolomite may have evolved from a protodolomite precursor. Glover and Sippel (1967, p. 612) also emphasized the critical nature of kinetics in dolomite formation at low temperatures. Time available in the natural environment might make the difference between order and disorder.

Recently, there have been numerous claims of dolomite syntheses, some of which may represent significant breakthroughs. Gaines and Heffner (1973) produced a cation-ordered "protodolomite" at 100°C and atmospheric pressure by reacting aqueous solutions of magnesium and calcium chlorides with aragonite. They envisioned a solution—precipitation mechanism controlled by desolvation of Mg^{2+} ions at the growing crystal surface.

[1] "Successful" attempts, however, still remain to be confirmed.

The resulting "protodolomite" is practically indistinguishable from natural Holocene dolomite formed under marine hypersaline conditions.

Chazen and Ehrlich (1973) reported the synthesis of "probably" ordered dolomite from aragonite at atmospheric P_{co_2} and room temperature (25°C). These workers placed ground dolomite in artificial sea water analogs of various strengths and found that it dissolved incongruently to yield aragonite with detectable amounts being formed in less than one day. The aragonite was partially reconverted to dolomite by allowing the solution with aragonitized dolomite to slowly evaporate to dryness for a period of about seven weeks. The resulting solids were analyzed by electron-probe microanalysis which showed a zonation from the center outward as follows: original dolomite → aragonite → new dolomite; X-ray diffraction analysis suggests ordering.

Davies and Ferguson (1975) claimed to have synthesized dolomite in a tank experiment simulating a shallow water, organically rich, evaporitic carbonate environment which they allowed to evolve over a period of eight to nine months. An important ingredient was the layer of decaying filamentous algae placed between layers of crushed calcite and nesquehonite. These workers reported that dolomite had precipitated within this organic material. The exact nature of their "dolomite" is not stipulated and is documented by a very small peak in the position of the main dolomite reflection, i.e., (211). Based on their experiment, as well as on their finding of disordered dolomite within organic material in a small bay at One Tree Reef, Great Barrier Reef, Australia, they believed that alkalinity was the most significant factor in their syntheses.

One of the more interesting pieces of recent work regarding dolomite synthesis is that of Deelman (1975; 1976, written communication) as it reportedly involves precipitation of dolomite under sedimentary conditions and also describes a theoretical model of crystal growth. Deelman concluded that the mixed-layer structure could only form as a sedimentary mineral through layer-by-layer growth, as contrasted with the mechanism of spiral growth in which quicker development of carbonate crystals leads to a statistical distribution of Ca^{2+} and Mg^{2+}. He mentioned magnesium-calcite and protodolomite which show the random distribution of the two cations and which lack the ordered layered structure of dolomite. According to Deelman (1975, p. 472): "The change of a layer of $CaCO_3$ into the next sheet of $MgCO_3$ requires a constant change in energy content, that is, in the thermodynamics of the medium. This change might be related to the concentrations of the two different cations, to the concentration of the bicarbonate ion (Mg and Ca having different solubilities), or to temperature variations. Changes in thermodynamic conditions can only lead to layer-by-layer growth if a difference in adsorption behavior between the two cations exist. And, in fact, a considerable difference in physicochemical properties of Mg^{2+} and Ca^{2+} ions does exist (the valency is the same but the diameters of the two ions are quite distinct, leading to different electrostatic properties)." Deelman recognized a periodicity in some natural occurrences of dolomite such as the "dynamic" lakes reported by Müller et al. (1972), playa-type lakes, and intratidal and supratidal environments. It should be pointed out, however, that these natural "dolomites" are poorly ordered and are best classed as protodolomites.

Following this reasoning, Deelman set up an experimental procedure. A series of preliminary experiments (Deelman, 1976, written communication) led to the particular run reported in 1975 (p. 472): "Bicarbonate solutions on the basis of aqua demin with relatively low Mg/Ca ratios (smaller than 5) yielded, after slow, periodically interrupted desiccation, . . . aragonite and a dolomite-like mineral. Although the main peak of the latter mineral is located exactly at 31° (Cu-K_{α} radiation) in X-ray diffractograms, the signal is still too weak to allow unequivocal identification." Fig. 7-16 was provided by Deelman to demonstrate layer-by-layer growth of his experimentally produced, dolomite-like mineral. Subsequently, Deelman (1976, written communication) apparently has had more success precipitating dolomite but the judgment of his documentation awaits his forthcoming publication.

Lippmann (1973, pp. 150—151) has summarized equations for the dolomitization of calcareous sediments by magnesium-bearing solutions:

$$2\ CaCO_3 + MgSO_4 \rightarrow CaMg(CO_3)_2 + CaSO_4 \qquad (7\text{-}1)$$

and:

$$2\ CaCO_3 + MgCl_2 \rightarrow CaMg(CO_3)_2 + CaCl_2 \qquad (7\text{-}2)$$

The low solubility of $CaSO_4$ coupled with the common natural occurrence of dolomite and gypsum or anhydrite led Lippmann to believe that eq. 7-1 would represent a more probable reaction producing dolomite. But, he pointed out that sea water and most other brines, which have commonly been considered as the necessary type of dolomitizing solution, contain both sulfate and chloride ions and it is impossible to assign given anions to specific cations. Consequently, it has become common practice to generalize eqs. 7-1 and 7-2 to give:

$$2\ CaCO_{3\,(solid)} + Mg^{2+} \rightarrow CaMg(CO_3)_{2\,(solid)} + Ca^{2+} \qquad (7\text{-}3)$$

Although he accepted this as the overall reaction governing dolomitization, Lippmann claimed that the mechanism of the formation of ionic crystals of dolomite from ions in solution should be expressed by:

$$Ca^{2+} + Mg^{2+} + 2\ CO_3^{2-} \rightarrow CaMg(CO_3)_{2\,(solid)} \qquad (7\text{-}4)$$

Basically there is little difference between reactions 7-3 and 7-4 in that

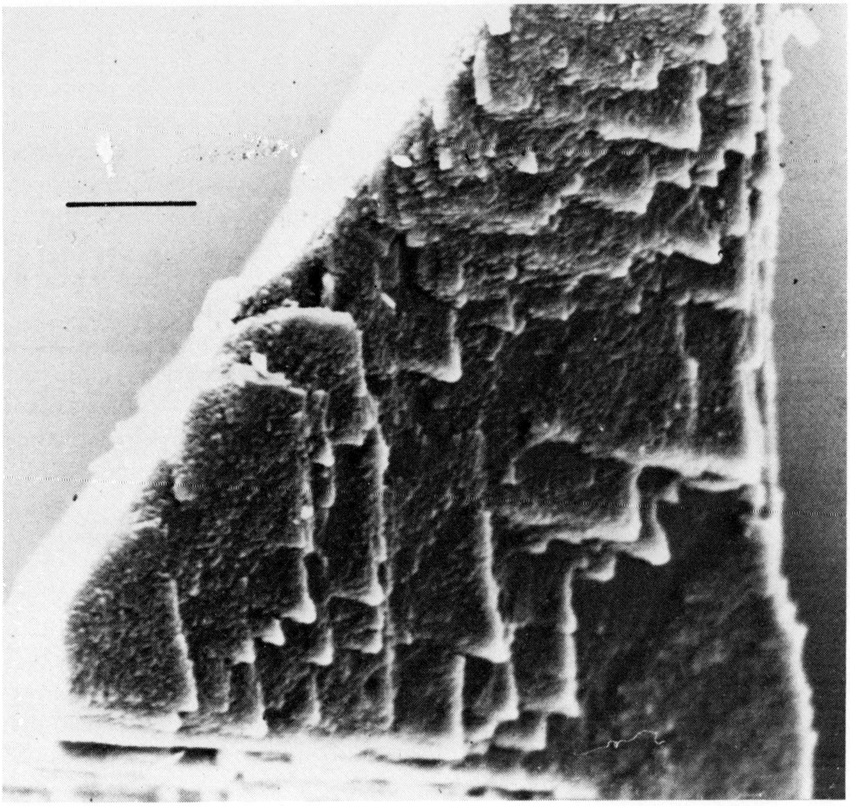

Fig. 7-16. Scanning electron photomicrograph of layer-by-layer growth on experimentally produced dolomite-like material. Bar = 10 μm. (After Deelman, 1975, fig. 3; reproduced by permission of Geol. Soc. Am.)

the starting material, $CaCO_3$, in 7-3 must first undergo dissolution in order to produce Ca^{2+} and CO_3^{2-} ions. Whereas eq. 7-4 expresses no assumption regarding which relative concentrations lead to the formation of dolomite, eq. 7-3 implies low starting concentrations of Ca^{2+} and CO_3^{2-}, owing to the relatively low solubility of $CaCO_3$ and high Mg^{2+} concentrations required to maintain equilibrium conditions.

Two important points are involved in these lines of reasoning: (1) the assumption of a "wet" process of dolomitization, and (2) the emphasis that a high Mg^{2+}/Ca^{2+} ratio in the solution is necessary for dolomitization as represented by eq. 7-3. These points and their relation to the chemistry and kinetics of dolomitization warrant additional remarks.

A wet or dry process?

The matter of a dry (i.e., solid-state) versus wet process of dolomitization has been widely debated in recent years. The search for a mechanism of dolomite crystal growth has proceeded in two distinct but related approaches: the study of the bulk-thermodynamic properties of entire crystals and their solutions and that of the surface layers of crystals (Bathurst, 1971, p. 536).

Liebermann (1967) attempted to precipitate dolomite under conditions which he judged to be characteristic of coastal areas with restricted circulation. He bubbled CO_2 through artificial sea water at temperatures of 5—10°C and with salinities 3—6 times that of natural sea water; the pH was adjusted by adding Na_2CO_3 and NH_3. Liebermann believed that the elimination of $CaSO_4$ was important in obtaining the precipitate which formed in the pH range of 8—9. This precipitate included magnesian calcite, disordered dolomite, and an ordered dolomite. One may question the degree of similarity between Liebermann's system and that in nature, as well as the presence of true dolomite (Zenger, 1972b, p. 111). One of his assumptions, however, is significant (Liebermann, 1967, p. 241): the precipitation of a crystalline compound from an aqueous solution implies that the compound must be in true ionic solution at the moment of precipitation. He criticized the commonly quoted equation for dolomitization, i.e., eq. 7-3. Assuming the magnesium ions are supplied by sea water, Liebermann contended that the dissolved magnesium sulfate would be used up first, followed by the magnesium chloride as indicated by the following equations:

$$2\ Ca(HCO_3)_2 + MgSO_4 \rightarrow CaCO_3 \cdot MgCO_{3\,(solid)}$$

$$+ CaSO_{4\,(solid)} + 2\ CO_{2\,(gas)} + 2\ H_2O \qquad (7\text{-}5)$$

and:

$$2\ Ca(HCO_3) + MgCl_2 \rightarrow CaCO_3 \cdot MgCO_{3\,(solid)}$$

$$+ CaCl_2 + 2\ CO_{2\,(gas)} + 2\ H_2O \qquad (7\text{-}6)$$

These reactions are essentially similar to eqs. 7-1 and 7-2 above. Liebermann's hypothesis (p. 242), then, is that dolomitization is a two-step process consisting of the solution of calcium carbonate by hypersaline sea water and the subsequent precipitation of calcium carbonate and magnesium carbonate as the double salt.

It was mentioned earlier that on the basis of similar oxygen isotopes in coexisting calcite and dolomite, Degens and Epstein (1964) assumed that sedimentary dolomite formed by solid-state replacement of Ca^{2+} ions in $CaCO_3$ by Mg^{2+} ions from aqueous solution without exchange of CO_3^{2-} ions. In their detailed study of the growth of dolomite crystals in Deep Spring Lake, Peterson et al. (1966) concluded that the precipitated surface layer of the crystals consisted of a magnesian calcite and that ordering to a slightly calcium-rich dolomite occurred within 100 Å of the surface by diffusion of Mg^{2+} ions from the surface inward. Clayton et al. (1968) suggested that these calcium-rich surface layers resulted from contamination by calcite in the lake sediments in which the dolomite crystals are growing. They concluded that dolomite can form in the lake either by direct precipitation or by alteration of a calcite precursor by a solution—precipitation mechanism, which would be a wet process.

There is an essential "... absence of dry diffusion in carbonates at earth-surface temperatures as shown by the presence of metastable magnesian calcites in fossils as old as Eocene (Chave, 1954) that were embedded in relatively impervious shales" (Graf and Goldsmith, 1956, p. 185). On the other hand, there are examples of well-crystallized Pleistocene dolomite. Consequently, the amount of dolomite formed by solid-state ionic diffusion at the earth's surface or under shallow burial conditions probably is minor compared to that produced by the "wet" replacement of a calcareous precursor. Bathurst (1971, p. 542) concluded that "... dolomitization is always a wet process, involving exchange of oxygen and carbon isotopes between precursor, solution, and precipitate". Nucleation within a carbonate crystal by diffusion at metamorphic temperatures, however, may be a significant process.

Thermodynamics, kinetics and the low-temperature relationship of dolomitization to concentrated solutions with high Mg^{2+}/Ca^{2+} ratios

The experimental work of Liebermann (1967), Chazen and Ehrlich (1973), Gaines and Heffner (1973) and others intimate that hypersalinity is a prerequisite for dolomitization. As a result of the ancient and modern association of dolomites and evaporites and the conclusions of numerous geochemists studying the chemistry and kinetics of dolomite synthesis, there developed the generalization that dolomite is essentially an evaporite mineral owing its origin to the effect of brines characterized by a high Mg^{2+}/Ca^{2+} ratio. Kinsman (1973), for example, stated that one of the requisites for dolomite formation within the sabkhas of the Persian Gulf is a solution with a high Mg^{2+}/Ca^{2+} ratio; the apparent dolomitizing fluids in that area are of marine origin and their Mg^{2+}/Ca^{2+} ratios range from 7 to 22. Chilingar (1956a) was the first investigator to suggest that the high Mg/Ca ratio (≈ 7.8) of water is a prerequisite for the precipitation of dolomite. He also stressed the need for high CO_2 pressure, which would increase alkalinity ($[CO_3^{2-}]$ + $[HCO_3]$).

At this point the present writers would like to re-emphasize the significance of the relationship between thermodynamics and kinetics in the study of dolomitization. Krauskopf (1967, p. 87) referred to the discrepancy among geochemists of the value of the solubility product (K_D), reported as 10^{-17}—10^{-19}. Hsü (1967) favored the values nearer 10^{-17}, as did Barnes and Back (1964), based on the composition of ground waters in dolomites. He admitted, however, that this interpretation is not accepted by those who question whether equilibrium has been attained, or even approximated, between a ground water and the carbonate minerals of its host rocks. In spite of the difficulty of determining the exact value of K_D owing to such questions as incongruent versus congruent dissolution, the effects of associated ions on the stability of carbonate minerals (Winland, 1969), and the very matter of attainment of equilibrium, the present writers accept the value of 10^{-17} as at least an approximation (see Blatt et al., 1972, p. 483; Lippmann, 1973, p. 157). Because no precipitation of dolomite was occurring in cave waters with an activity product for dolomite higher than 10^{-15} (Holland et al., 1964), Krauskopf (1967, p. 87) concluded that supersaturation with respect to dolomite can persist for long periods of time without formation of the mineral. Using the data for the activity products of various ions in sea water as presented by Garrels and Thompson (1962), Lippmann (1973, pp. 157—158) calculated the activity product $[Ca^{2+}][Mg^{2+}][CO_3^{2-}]^2$ as $10^{-15.01}$ which further suggests that sea water is considerably supersaturated with regard to dolomite. Such findings are supported by Blatt et al. (1972, p. 485) who showed that the Mg^{2+}/Ca^{2+} ratio in sea water is significantly above the equilibrium value (slightly less than unity) and reaction 7-3 should proceed in the direction of dolomite formation at the expense of pre-existing calcite. Thermodynamically, the Mg^{2+}/Ca^{2+} ratio of sea water lies within the stability field of dolomite (Lippmann, 1973, p. 158), which is less soluble than calcite and aragonite in that medium.[1] Blatt et al. (1972, pp. 485—486) stated that "... no known dolomitization is taking place today in sea water ..." and that "... the requirements for dolomitization are two-fold. Water must have a sufficiently high aMg/aCa ratio. In a marine-like water, the activity ratio must be above 8.4 for efficient dolomitization. In addition, a driving force must exist that will move thousands of pore volumes of this fluid through the calcium carbonate sediment or rock."

As mentioned previously, the slow rate of cation ordering accounts for the

[1] It might be noted that in an experimental and theoretical study, Sass (1965) contended that sea water in contact with calcite is undersaturated with respect to dolomite.

essential lack of stoichiometric, well-ordered dolomite in Holocene carbonates and in the products of laboratory attempts to synthesize dolomite at low temperatures. A related kinetic factor appears to be the "dehydration barrier" (Lippmann, 1973, p. 76) that must be overcome for anhydrous magnesium compounds (i.e., magnesite and dolomite) to form from sea water instead of the hydrated metastable phases hydromagnesite and nesquehonite. Energy must be supplied to overcome the H_2O-Mg^{2+} bond that inhibits the entry of magnesium into the magnesite or dolomite lattice. Gaines (1974) considered that desolvation of the hydrated ions might be a rate-controlling factor in dolomitization. This kinetic obstacle favors the growth of calcium carbonate crystals (particularly aragonite), because Ca^{2+} is less hydrophilic than Mg^{2+}. Other studies of the kinetics of dolomitization suggest that at higher Mg^{2+} ion concentration this inhibitional effect can be reversed and the crystallization of dolomite enhanced. Consequently, there evolved the concept of a requirement of a high Mg^{2+}/Ca^{2+} ratio in the potential dolomitizing fluid. Because such high ratios are common in surface and near-surface hypersaline solutions, dolomite was quite widely accepted as an evaporitic mineral (Friedman and Sanders, 1967, pp. 334, 338). Some authors seriously doubt the validity of this generalization (Zenger, 1972a, pp. 8—9) and some believe that possibly too much emphasis has been placed on the necessity for a high Mg^{2+}/Ca^{2+} ratio for dolomitization to proceed. Many ancient dolomites appear to have formed in a normal marine environment. Some evaporites are associated with limestone, rather than with dolomite, in the geologic record (Sonnenfeld, 1964, p. 126). In Holocene peritidal crusts of Shark Bay, Australia, evaporitic conditions prevail and gypsum is produced; however, the carbonate material is primarily aragonite (Davies, 1970, p. 179). There is practically an absence of dolomite in the evaporitic Purbeckian limestones of the Dorset coast, England (I.N. West, 1970, oral communication), and Hudson (1970) described calcite pseudomorphs after gypsum in the Middle Jurassic Great Estuarine Series of Scotland with no dolomite being observed.

There is disagreement over the need of a high Mg^{2+}/Ca^{2+} ratio for the formation of Holocene dolomite-bearing sediments. Shinn (1968, p. 615) recorded high concentrations of ions and high Mg^{2+}/Ca^{2+} ratios in the interstitial solutions of the supratidal crust on Sugarloaf Key, Florida. Atwood and Bubb (1970), on the other hand, claimed that their analyses of the waters from the same crust showed near-normal concentrations of ions and Mg^{2+}/Ca^{2+} ratios. Interstitial waters of dolomite crusts on northern Andros Island, Bahamas, also showed essentially normal chlorinities and Mg^{2+}/Ca^{2+} ratios (O.P. Bricker, 1970, written communication; see Zenger, 1972a, p. 8). Such studies must be interpreted very carefully. There is a possibility that there has been a change in the chemistry of the interstitial fluids since dolomitization of the crusts, some of which are hundreds of years old. Chilingar (1956a) stated that the Ca/Mg ratio is decreasing with time and the present-day sea water reached a value necessary for dolomite formation. This suggested the existence of present-day formation of dolomite, which was not reported at that time (1956).

Apparently, dolomite is forming in the Coorong and associated ephemeral lakes over a wide range of salinities and Mg^{2+}/Ca^{2+} ratios including values below those of normal sea water. Rather, it has been suggested that dolomite formation is related to high pH owing to luxuriant plant growth (Skinner, 1963, p. 459). Lippmann (1968; 1973, pp. 84—85, 159—161, 167—190) stressed the positive influence of high alkalinity (i.e., CO_3^{2-} concentration) on dolomite formation. If the CO_3^{2-} activity is sufficiently high, these anions possess the capability of dislodging water dipoles from surface cations of a growing crystal and then becoming bonded into them (cations). Because of the extreme slowness of the reactions leading from metastable forms to true dolomite, Lippmann studied the synthesis of structurally analogous norsethite — $BaMg(CO_3)_2$ — with its similar magnesite-like layers. Inasmuch as norsethite formation is enhanced by a high CO_3^{2-} concentration, Lippmann proposed that high alkalinity promotes dolomitization rather than the often-invoked high Mg^{2+}/Ca^{2+} ratios produced by evaporation. Low activity of CO_3^{2-} in sea water will enhance the inhibition caused by the dehydration barrier even though the dearth of CO_3^{2-} may be accompanied by high Mg^{2+} and Ca^{2+} activities (in order for thermodynamic equilibrium to be maintained). Furthermore, Lippmann claimed, based on the analogy with the norsethite synthesis, that if the reaction:

$$CaCO_{3\,(solid)} + Mg^{2+} + CO_3^{2-} \rightarrow CaMg(CO_3)_{2\,(solid)} \qquad (7\text{-}7)$$

is considered the most likely path to dolomite formation, there is no need for concern over the removal of Ca^{2+}, which many, including Bissell and Chilingar (1958, p. 494), have recognized as a problem if reaction 7-3 is followed instead of eq. 7-7. Lippmann (1973, pp. 160—161) expressed serious doubt that supersaturation with regard to dolomite is greatly increased by evaporation. Although Mg^{2+} activity is increased, that of CO_3^{2-} is reduced because this anion is partially lost from the evaporating sea water and is partially captured by the increased number of Mg^{2+} and Na^+ ions to form complexes. He did point out, however, that elevated Mg^{2+}/Ca^{2+} ratios may promote dolomitization if the increase is caused not only by evaporation, but by the presence of high $[CO_3^{2-}]$ in solution, which decreases the Ca^{2+} concentration in accordance with the constant ionic product $[Ca^{2+}]$ $[CO_3^{2-}]$.

Davies and Ferguson (1975), whose experimental work has been described (see section "synthesis of dolomite"), believed that high alkalinity, produced by photosynthesis or by decaying organic matter, is as critical in dolomitization, or more so, than the often invoked Mg^{2+}/Ca^{2+} ratio.

Barnes and O'Neil (1969, 1971) have demonstrated that alkalinity is a controlling factor in dolomite formation along nonsaline streams draining serpentinite terrain in the Coast Ranges of Oregon and California. According to Chilingar et al. (1967), in all probability diagenetic dolomitization occurs in strongly reducing to weakly reducing environments having high alkalinity ($[A] = [HCO_3^-] + 2[CO_3^{2-}]$) and a pH > 8 (up to 9 and higher). Teodorovich (1958, pp. 93—103) suggested that replacement dolomites (with some glauconite and oxides and hydroxides of iron) also form in weakly oxidizing to oxidizing environments having pH of 8 (7.8)—9. He believed that many replacement dolomites with relict organic and oolitic textures formed in oxidizing environments that have a pH of 7.2—7.8. Furthermore, Chilingar and Bissell (1963a) assigned a pH value of ⩽8 to the environment necessary, in their evaluation, for the formation of "primary" sedimentary dolomites.

Teodorovich (1958, p. 305) believed that Haidinger's reaction could account for some dolomitization in waters saturated with $CaSO_4$, according to this reaction:

$$CaCO_3 + MgSO_4 \rightarrow MgCO_3 + CaSO_4 \quad \text{(in solution)}$$
$$CaCO_3 + MgCO_3 = CaCO_3 \cdot MgCO_3 \downarrow \qquad (7\text{-}8)$$

The second part of this reaction, however, does not occur at high concentrations of $MgSO_4$. Thermodynamics of this reaction were discussed by Halla et al. (1962). On the basis of experimental data, Valyashko (1962, p. 47) presented the main reaction which occurs between calcium bicarbonate and sulfate solutions:

$$Ca(HCO_3)_2 + MgSO_4 \rightleftharpoons CaSO_4 + Mg(HCO_3)_2$$
$$CaSO_4 \xrightarrow{\text{hydration}} \downarrow CaSO_4 \cdot 2\,H_2O \qquad (7\text{-}9)$$
$$Mg(HCO_3)_2 \xrightarrow{\text{hydrolysis}} x\,Mg(OH)_2 \cdot y\,MgCO_3 \cdot z\,H_2O \downarrow$$

With small concentration of $MgSO_4$ in solution, the above reaction practically does not occur, but instead there is decomposition of calcium bicarbonate with formation of calcite, as follows:

$$Ca(HCO_3)_2 \rightarrow CaCO_3 + H_2O + CO_2 \qquad (7\text{-}10)$$

As the $MgSO_4$ increases, first gypsum starts to form, followed by basic carbonates of magnesium. At this time, $CaCO_3$ practically disappears from the bottom phase (see also Chilingar and Bissell, 1963b).

In addition to reaction 7-9, reaction 7-11 also occurs very slowly:

$$2\ Ca(HCO_3)_2 + MgSO_4 \rightarrow CaMg(CO_3)_2 + CaSO_4 + 2\ H_2O + 2\ CO_2 \quad (7\text{-}11)$$

Valyashko (1962, p. 55) obtained individual, rhombohedral crystals of dolomite (identified by crystallo-optical analysis) in the laboratory at atmospheric conditions (low CO_2 pressure). Possibly this reaction could account, in a measure at least, for certain extensive dolomite formation at higher CO_2 pressures.

In the past few years, geochemical work has indicated that even at earth surface conditions, high salinities and high Mg^{2+}/Ca^{2+} ratios are not required for dolomitization. One of the pioneer studies in this regard is that of the geochemistry and hydrology of the Tertiary limestone aquifer of central Florida by Hanshaw et al. (1971). They recognized that the three-phase system calcite—dolomite—water occurs at an Mg^{2+}/Ca^{2+} ratio of unity in ground-water systems. It is important to note that this is also the figure given by Lovering (1969, p. 749) and about that recorded by Blatt et al. (1972, p. 485). Dolomitization is occurring in the Tertiary aquifer by solutions having Mg^{2+}/Ca^{2+} ratios just above 1. They concluded that, despite low Mg^{2+}/Ca^{2+} ratios as compared with that of sea water, the circulation of water in such a dynamic ground-water system can provide a large Mg^{2+} ion supply for dolomitization. In this particular case, the magnesium is provided by a mixing of the ground water and underlying hypersaline water. Badiozamani (1973) and Folk and Land (1975) have presented models which are not dependent on the Mg^{2+}/Ca^{2+} ratio in solution. Both models also involve the mixing of sea water and ground water to produce solutions capable of dolomitization at relatively low concentrations and Mg^{2+}/Ca^{2+} ratios (these models are designated "dorag" [1] and "schizohaline" by Badiozamani and by Folk and Land, respectively). According to Baziozamani, theoretical calculations indicate that mixing meteoric ground waters with up to 30% sea water results in undersaturation with respect to calcite, whereas dolomite saturation increases. In the case of solutions containing between 5 and 30% sea water, calcite can be replaced by dolomite. Folk and Land (1975) dealt with both thermodynamic and kinetic aspects of dolomite formation. Under evaporitic conditions, precipitation is rapid and, consequently, it is difficult for true dolomite to form because there is insufficient time (and energy) for the exact Ca—Mg ordering; thus, aragonite and magnesian calcite form preferentially. Only if Mg^{2+}/Ca^{2+} ratios are very high can dolomite form in such a situation and then it is aphanitic and imperfectly ordered. At reduced salinities, however, dolomite can nucleate at lower Mg^{2+}/Ca^{2+} ratios, approaching 1 : 1 in ground water. The absence of foreign ions and the slow rate of crys-

[1] Persian word for "mixed blood".

tallization typical of phreatic environments are conducive to the growth of clear (limpid), euhedral dolomite rhombs. They proposed that an ideal environment would be a mixing zone of fresh waters and saline waters, where the salinity would be lowered but a high Mg^{2+}/Ca^{2+} ratio maintained. This would reduce the effect of foreign ions and provide conditions favorable for slow crystallization of dolomite. Recently, however, Carpenter (1976) has argued that such mixing of sea water with ground water actually reduces the thermodynamic drive toward the dolomitization of calcite, which he feels depends solely on the Ca/Mg activity ratio in the aqueous solution. Badiozamani (1976) presented further arguments that his dorag model is sound both thermodynamically and kinetically and the disagreement persists, which future research work will probably resolve.

There are, then, various conditions under which dolomite can nucleate. Kinetics and a lack of suitable environments seem to be major constraining factors in the near absence of Holocene ordered dolomite. Most observations are limited to the very uppermost few meters of modern sediments, however, and a large portion of the diagenetic environment of *today* (such as that described by Hanshaw et al., 1971) has not been examined. Will the protodolomite forming in the peritidal zone become ordered, stoichiometric dolomite? Although many would favor this idea, Lippmann (1973, p. 188) questioned whether protodolomites are precursors of dolomite. It has been shown that protodolomite can be persistent, having been reported from rocks as old as Paleozoic (Goldsmith and Graf, 1958; Füchtbauer and Goldschmidt, 1965).

Geochemistry of deeper-burial dolomitization

There appear to be fewer geochemical problems when considering deeper-burial and hydrothermal dolomitization. Whatever may be the significance of a high Mg^{2+}/Ca^{2+} ratio in surface on near-surface environments, this factor definitely becomes less important with higher temperatures. The number of non- or partially-hydrated Mg^{2+} ions becomes greater at higher temperatures, thus decreasing this inhibitory effect (Murata et al., 1972, p. 4). Also, kinetics is not significant; the conditions necessary to obtain equilibrium are stable over longer periods of time than on the earth's surface. Consequently, time and temperature will permit dolomitization of limestone precursors providing that the reacting solutions have a higher magnesium content than those solutions that are at equilibrium with calcite and dolomite. Most pore waters contain more Mg^{2+} than corresponds to this equilibrium (Usdowski, 1968, pp. 28—31; see Fig. 7—17). Lovering (1969, p. 743) showed the influence of temperature on the equilibrium Ca^{2+}/Mg^{2+} ratio which ranges from about 0.15 at 20°C to 35.0 at 420°C. Such general findings were supported

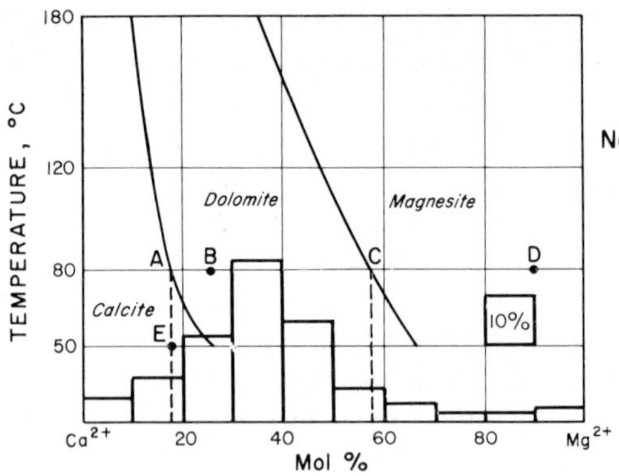

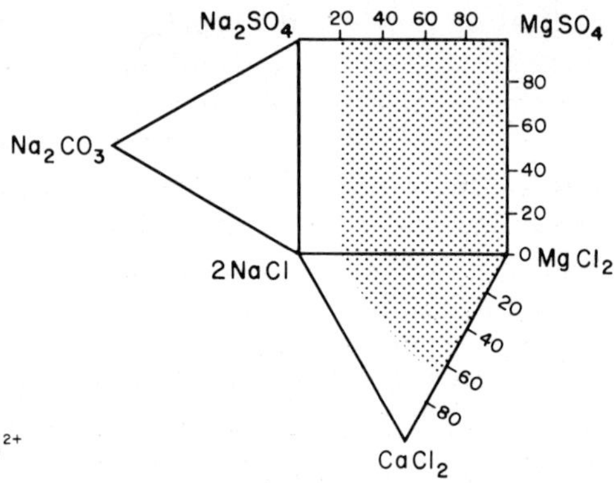

Fig. 7-17. The equilibrium curves (intermediate values of the equilibrium data for points 8, 10, 22 and 16 of Fig. 7-20) for calcite—dolomite and dolomite—magnesite and the frequency of Ca^{2+} and Mg^{2+} cation occurrence in pore solutions. (After Usdowski, 1968, p. 30; courtesy Springer-Verlag.)

Fig. 7-18. Region of dolomite formation in saturated chloride and sulfate solutions. (After Valyashko, 1962, p. 57, fig. 14.)

by Murata et al. (1972) who reported that at a temperature of 60°C, solutions with Mg^{2+}/Ca^{2+} ratios of 0.33—2.00 are at equilibrium with dolomite, whereas those with lower and higher ratios are at equilibrium with calcite and magnesite, respectively.

Figure 7-18 shows the region of dolomite formation in saturated chloride and sulfate solutions, whereas Fig. 7-19 indicates the solubility of the $CaCO_3$—$MgCO_3$—H_2O system at $P_{CO_2} \approx 1$ atm and temperatures ranging from 0 to 70°C. In the latter figure, the points of intersection between the bisectrix and dolomite saturation curves show the composition of solutions saturated with respect to pure dolomite, whereas the solubilities of pure $CaCO_3$ and $MgCO_3$ are shown on the ordinate and abscissa, respectively. The solubilities of mixtures of dolomite + calcite and dolomite + magnesite are shown by the junction (nodal) points. The curve connecting these junction points to the left of the bisectrix represents the solubility of mixtures of dolomite and calcite, whereas the one at the right of the bisectrix represents mixtures of dolomite and magnesite.

Schematic representation of the phase-relationships in the 5-component system (Ca^{2+}—Mg^{2+}—CO_3^{2-}—SO_4^{2-}—Cl_2^{2-}—H_2O) by an equilateral prism is presented in Fig. 7-20. The composition of the solutions are given in mole percent and each side of the prism represents 100 mole percent. The stability fields of Ca—Mg—chloride (close to the edge 1 to 2) and of $MgSO_4$-hydrate are enlarged. Fig. 7-20 also shows the phase relationships of a part of the 6-com-

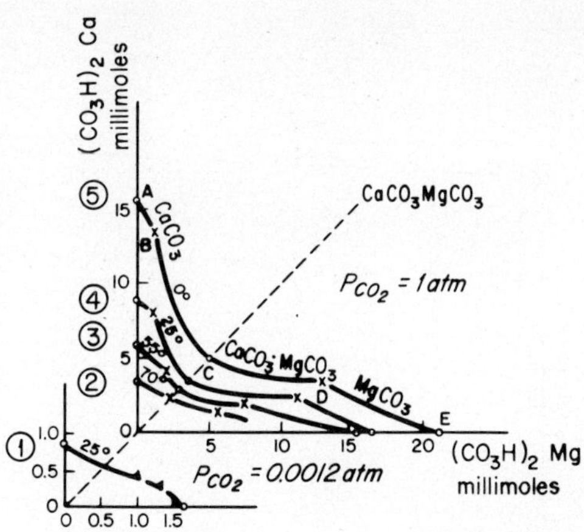

Fig. 7-19. Solubility of $CaCO_3$—$MgCO_3$—H_2O system at $P_{CO_2} \approx 1$ atm and $P_{CO_2} \approx 0.0012$ atm and temperatures ranging from 0° to 70°C. Points between ordinate and 45°-line represent solubility of calcite—dolomite mixtures, whereas those between 45°-line and the abscissa represent dolomite—magnesite mixtures. The amounts of $Mg(HCO_3)_2$ and $Ca(HCO_3)_2$ are expressed in mmole/1000 G solution. (After Yanat'eva, 1950, 1954; also consult Chilingar, 1956g, and Baron and Favre, 1958.)

ponent system (Na_2^{2+}—Ca^{2+}—Mg^{2+}—CO_3^{2-}—SO_4^{2-}—Cl_2^{2-}—H_2O). The difference between the two systems is that in the 6-component system halite must be added as additional phase to each solid component. In addition, NaCl is present in the solutions up to saturation concentration (Usdowski, 1968, p. 23). The phase relationships of the 5-component and 6-component systems, which are based on experimental studies made at 50°, 80°, 120°, and 180°C below the vapor pressures of the solutions, are presented in Table 7-I. For both systems, in the gaps between the solid phases, solution and vapor must be added to all solid phases.

According to Usdowski (1968, p. 24), the most important equilibria involved in the formation of dolomite in sediments are those between calcite and dolomite (surface 8-10-22-16, Fig. 7-20) and between magnesite and dolomite (surface 7-9-23-17, Fig. 7-20). The following two equations have been presented by Usdowski:

$$2\,CaCO_3 + Mg^{2+} + (CO_3^{2-}, SO_4^{2-}, Cl_2^{2-}) \rightleftharpoons CaMg(CO_3)_2 + Ca^{2+} + (CO_3^{2-}, SO_4^{2-}, Cl_2^{2-}) \quad (7\text{-}12)$$

$$2\,MgCO_3 + Ca^{2+} + (CO_3^{2-}, SO_4^{2-}, Cl_2^{2-}) \rightleftharpoons CaMg(CO_3)_2 + Mg^{2+} + (CO_3^{2-}, SO_4^{2-}, Cl_2^{2-}) \quad (7\text{-}13)$$

Fig. 7-17 shows the fields of stability of dolomite, calcite, and magnesite.

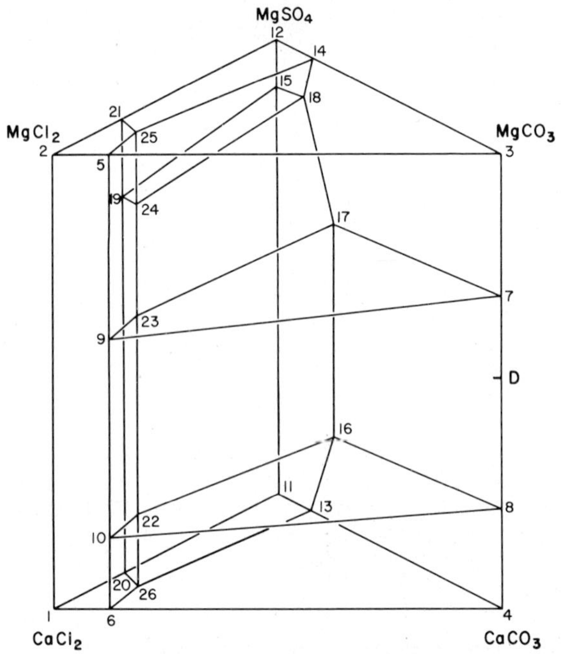

Fig. 7-20. Schematic diagram of the phase relationships in the 5-component system (Ca^{2+}—Mg^{2+}—CO_3^{2-}—SO_4^{2-}—Cl_2^{2-}—H_2O) by an equilateral prism. *D* = dolomite; point 11 = $CaSO_4$. (After Usdowski, 1968, p. 22, fig. 1.)

Calcite, dolomite, and solutions are stable along the left curve of the diagram, whereas the right curve represents the equilibrium between magnesite, dolomite and solutions. For example, if a solution having composition *B* (dolomite field) comes in contact with limestone at 80°C, dolomite forms and solution composition changes from *B* to *A*, where reaction ceases. If the composition of solution coming in contact with limestone is represented by point *D* (magnesite field), magnesite forms first and composition of solution changes from *D* to *C*. At point *C*, magnesite reacts with solution to form dolomite. The equilibrium *A* is attained after all magnesite is used up. Thus, the only prerequisite for the formation of dolomite is the presence of reaction solutions having higher Mg content than the solutions that are at equilibrium with calcite and dolomite, which is the case in most instances as shown by the frequency of Ca^{2+} and Mg^{2+} in pore solutions (Fig. 7-17).

Point *E* in Fig. 7-17 represents pore solution having a Mg^{2+} content of 18 mole % at 50°C. This solution is stable with respect to calcite; however, if temperature is raised above 80°C (calcite—dolomite boundary), the solution enters the dolomite field and dolomitization of limestone should occur.

TABLE 7-I

The phase relationships of the 5-component system (Ca^{2+}—Mg^{2+}—CO_3^{2-}—SO_4^{2-}—Cl_2^{2-}—H_2O) and the 6-component system (Na_2^{2+}—Ca^{2+}—Mg^{2+}—SO_4^{2-}—Cl_2^{2-}—H_2O). In the case of the 6-component system, halite must be included as an additional solid (after Usdowski, 1968, p. 23)

Points in Fig. 7-20	Solid phases
6—26—13—4—10—22—16—8	calcite
10—22—16—8—9—23—17—7	dolomite
9—23—17—7—24—18—5—25—14—3	magnesite
26—13—22—16—23—17—24—18—19—15—20—11	anhydrite
24—18—25—14—21—12—19—15	$MgSO_4$-hydrate
1—20—26—6—2—21—25—5	Ca-Mg chloride
26—22—16—13	calcite + anhydrite
10—22—16—8	calcite + dolomite
22—23—17—16	dolomite + anhydrite
9—23—17—7	dolomite + magnesite
23—24—18—17	magnesite + anhydrite
24—25—14—18	magnesite + $MgSO_4$-hydrate
24—19—15—18	anhydrite + $MgSO_4$-hydrate
22—16	calcite + dolomite + anhydrite
23—17	magnesite + dolomite + anhydrite
24—18	magnesite + anhydrite + $MgSO_4$-hydrate
22	calcite + dolomite + anhydrite + Ca—Mg chloride
23	magnesite + dolomite + anhydrite + Ca—Mg chloride
24	magnesite + anhydrite + $MgSO_4$-hydrate + Ca—Mg chloride

SEDIMENTOLOGY AND PETROLOGY OF DIAGENETIC DOLOMITIZATION

Introduction

In the early diagenetic environment, the sediments are in closest proximity to a major potential source of Mg^{2+}, that is, sea water. The uppermost carbonate sediments, or those affected primarily by penecontemporaneous processes to start with, commonly have considerable porosities and can be relatively freely bathed by interstitial waters with chemistry similar to that of sea water. In this "initial stage" (Fairbridge, 1967, pp. 33—34) oxidizing conditions prevail, assuming free water circulation above the sediment—water interface. Organic matter buried with the carbonate sediments provide nutrients for burrowing organisms that disturb the surface layers of the fresh material keeping them well oxidized. The activity of certain heterotrophic bacteria, by their production of organic acids or by their decomposition of organic matter, can cause a rise in the P_{CO_2} and an accompanying sharp drop in pH immediately below the sediment surface (Fairbridge, 1967, pp. 33—34; ZoBell, 1942, p. 131; 1946, p. 492). Although the ensuing effect on carbonates could be but slight, Fairbridge mentioned the possible dissolutional results. In his "early burial" stage of syndiagenesis, the conditions are generally anaerobic owing to the exhaustion of oxygen by aerobic bacteria in the sediments above. Here the pH has resumed a higher value similar to or perhaps slightly above that at the surface. According to Fairbridge (1967, p. 36): "The reducing zone generally leaves a characteristic mark on the syndiagenetic phase because it is inevitably the last environment of a sedimentational stage and thus leaves its imprint on sediments for all time. Although its products may subsequently be modified, the evidence is never totally effaced." This imprint of the reduction zone is all the more impressive if the sediment first passed through the zone of oxidation. Coarse-grained carbonate sediments provide a greater opportunity for oxidation than lime muds which may become calcilutite containing marcasite or pyrite.

Porosity in dolomites is commonly discussed in terms of the void space that results from dolomitization or that which develops secondarily (e.g., by solution and fracturing) in dolomitic rocks. For example, Murray and Lucia (1967, p. 29) theorized that if the sediments being dolomitized are the only source of carbonate (local source), porosity can be formed and maintained. If, on the other hand, carbonate is brought in with the dolomitizing water (distant source), porosity will be destroyed. Assuming that either Mg^{2+} or CO_3^{2-} is brought to the site of dolomitization, then predolomitization porosity needs to be considered. For most sediments there is a negative correlation between compaction (which in turn is related to the amount of overburden) and porosity. Most experimental work in this regard has been done on terrigenous sediments, particularly muds. For example, Athy (1930) reported porosities of 45–50% for surface clays but 3–4% for shales at a depth of 6000 ft.[1] The decrease in porosity with increased depth is greatest in the first few hundred feet of burial (Brown, 1969, p. 488). (For details on compaction of fine-grained and coarse-grained sediments, see Rieke and Chilingarian, 1974, and Chilingarian and Wolf, 1975, 1976).

Primary porosity in carbonate sediments ranges from 40 to 70% (Choquette and Pray, 1970, p. 211), depending on the type, with most carbonate rocks possessing extremely low porosities. In the case of the more porous reservoir rocks, a good deal of the porosity is secondary (i.e., due to later solution, late diagenetic dolomitization, etc.). Fruth et al. (1966) experimentally compacted Bahamian sediments and produced compactional features observed in natural limestones. Most of the sediment types tested showed a high initial compaction below about 30 bars pressure, but above 350 bars the compaction curves are nearly parallel. Using the density of packing of oölitic grainstones as an index, Coogan (1970) showed compaction ranging from zero in unburied Pleistocene Bahamian oölite to nearly 90% for parts of the Smackover Limestone (Jurassic, Louisiana) at about 10,000 ft. Ginsburg (1957, p. 91) demonstrated that although medium sand-sized carbonate sediments undergo slow continuous compaction with burial, finer grained sediments experience the greatest amount of compaction in the first foot. Because such a shallow compaction can hardly be due to the effect of overburden, Ginsburg attributed the repacking to factors such as mixing by organisms, the gel-like character of such sediments, and escape of bacterial gases. Brown (1969, p. 492), however, referred to a lack of appreciable compaction in cores of modern carbonate mud taken in the Bahamas and Florida Bay area. Pray (1960) has suggested that there is commonly little evidence for compaction of carbonates. Yet there is the ubiquitous occurrence of the "merged peloidal fabric" in lithified equivalents of carbonate muds, although they may not all be the result of compactional effects.[2] Whether due to compaction and/or cementation, the primary porosity in carbonate sediments decreases downward in the diagenetic environment. The porosity generally becomes very low as lithification occurs and the rock becomes part of the mesogenetic zone. Secondary porosity such as vugs and orogenic-induced fracture systems may produce new void space through which solutions may move. Generally, however, the movement of solutions, especially compaction fluids, is much more efficient during diagenesis than in epigenesis. But again the matter of time must be considered. Although it is impossible to delimit "diagenetic" in years, this stage is much shorter than

---

[1] 1 ft = 0.30481 m.

[2] Experimental work of Shinn et al. (1977) on lime-muds showed the existence of appreciable compaction with no noticeable crushing of shells.

the epigenetic phase, and the older the rock, the greater the difference. It should be realized, however, that some sediments may never become truly lithified prior to the telogenetic phase.

Diagenetic dolomitization and porosity

As pointed out by Bissell and Chilingar (1958, p. 493), a study of the processes of diagenetic dolomitization should involve consideration of porosity changes. The replacement of calcite by dolomite involves a contraction (increase in porosity) of about 12—13% (Chilingar and Terry, 1954) if the reaction proceeds as follows:

$$2\ CaCO_3 + Mg^{2+} \rightarrow CaMg(CO_3)_2 + Ca^{2+} \qquad (7\text{-}14)$$

Obviously, this contraction will occur only if there is no additional precipitation of carbonates in the pores and there is no subsequent compaction. The majority of carbonate petrologists agree that dolomitization gives rise to porosity providing a solid framework is available which will minimize the effects of subsequent compaction. For example, the findings of Lucia (1962) indicate that porosity in the dolomite facies in Devonian crinoidal rocks in the Andrews South Devonian Field (Andrews County, Texas) was formed during diagenetic dolomitization. He found that porosity is generally related to the ratio of the lime-mud to crinoid fragments. Chilingar and Terry (1954) also showed that a definite relationship exists between porosity and degree of dolomitization. (See Chilingar et al., 1972, and Sarkisyan et al., 1973.)

T.F. Gaskell of British Petroleum Co., Ltd. (personal communication, 1963) determined the porosity and density of carbonate reservoir rocks in southwestern Iran. The average density values for the different oil fields, grouped in ranges of porosity of 0—4.0, 4.1—8.0, 8.1—12.0, and >12.1% are presented in Table 7-II and Fig. 7-21. The mean values were weighted according to the number of observations for each oil field. A certain amount of the density scatter may be due to impurities in the limestones, variation in the amount of initial, primary porosity, variation in the degree of secondary cementation subsequent to dolomitization, etc. The gradual trend of density from 2.70 g/cm^3 at the low porosity to 2.80 g/cm^3 for the high-porosity group indicates that dolomitization gives rise to porosity. Inasmuch as at 20°C the density of calcite is 2.71 and that of dolomite is equal to 2.87, the average values given in Table 7-III correspond to the percents of dolomitization given in Table 7-II. These results (Table 7-II) are in close accord with those obtained by Chilingar and Terry (1954).

Diagenetic dolomitization may include two major stages, viz, (1) early diagenetic and (2) late diagenetic. In the case of early diagenetic dolomitization, due to subsequent compaction the porosity resulting from dolomitiza-

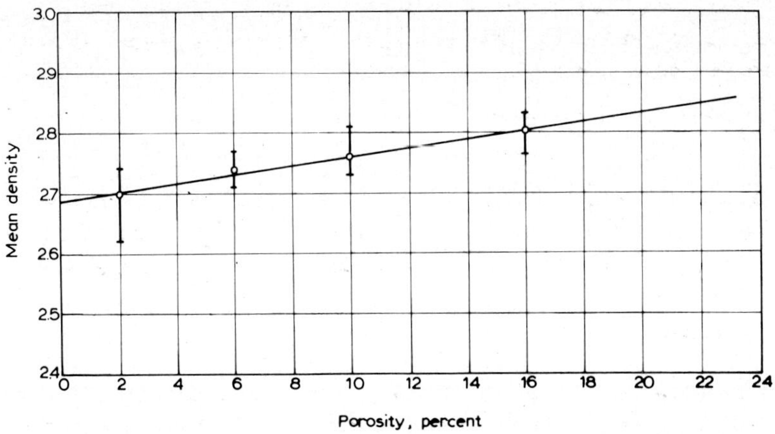

Fig. 7-21. Relationship between density and porosity of Iranian carbonate rocks. (After T.F. Gaskell, British Petroleum Co. Ltd., personal communication, 1963.)

tion is appreciably lowered (or disappears completely). Chilingar and Bissell (1961) stated that the only time dolomitization does not give rise to porosity is possibly during the very early stages of diagenesis. Subsequent cementation, however, commonly destroys the porosity created by dolomitization. Dedolomitization, which also destroys porosity, commonly is not recognized. Harbaugh (1967) presented an excellent and thorough discussion on the effect of dolomitization on porosity.

TABLE 7-II

Relationship between porosity and density of Iranian carbonate rocks

Name of oil field	Porosity range [1] (%)			
	0—4.1	4.1—8.0	8.1—12.0	⩾12.1
Haft Gel	2.68 ± 0.04 (14)	2.73 ± 0.09 (9)	2.75 ± 0.14 (8)	2.78 ± 0.12 (20)
Naft Khaneh	2.62 ±— (1)	2.77 ± 0.08 (3)	2.81 ± 0.05 (9)	2.83 ± 0.05 (24)
Gach Saran	2.71 ± 0.09 (7)	2.77 ± 0.08 (10)	2.78 ± 0.08 (11)	2.79 ± 0.08 (8)
Agha Jari	2.74 ± 0.09 (7)	2.74 ± 0.09 (10)	2.73 ± 0.06 (5)	2.81 ± 0.09 (9)
Naft Sefid	2.67 ± 0.04 (3)	2.73 ± 0.04 (3)	2.76 ± 0.06 (9)	2.76 ± 0.08 (12)
Lali	2.74 ± 0.08 (6)	2.74 ± 0.03 (6)	2.79 ± 0.04 (2)	2.79 ± 0.03 (4)
M-i-S	2.67 ± 0.12 (11)	2.71 ± 0.11 (8)	2.73 ± 0.12 (9)	2.80 ± 0.06 (12)
Mean	2.70 (49)	2.74 (49)	2.76 (53)	2.80 (89)

[1] The ± figures are mean square errors of the average values, and the figures in parentheses are the numbers of observations.

TABLE 7-III

Relationship between porosity and density

Porosity (%)	Density (g/cm^3)	Dolomitization (%)
0–4.1	2.70	0
4.1–8.0	2.74	20
8.1–12.0	2.76	32
⩾12.1	2.80	58
	2.84	82

In the following section are presented some of the models for dolomitization that have evoked a good deal of interest. The reader is encouraged to refer to the classical work of Friedman and Sanders (1967) for an excellent summary of various types of dolomites, modern and ancient.

Some models and examples of diagenetic dolomitization

Shallow evaporative processes. The process of seepage refluxion was popularized by Adams and Rhodes (1960) well before the appearance of most publications on penecontemporaneous Holocene dolomite. According to their model, dense, hypersaline brines enriched in Mg^{2+}, as a result of prior precipitation of gypsum in an evaporitic lagoon, would sink and flow seaward along the bottom until possibly reaching a permeable zone consisting of metastable carbonates (Fig. 7-22). Seeping through these porous sediments and rocks by displacing connate water, these brines could provide both a source of Mg^{2+} ions and a vehicle to remove the Ca^{2+} ions in eq. 7-3.

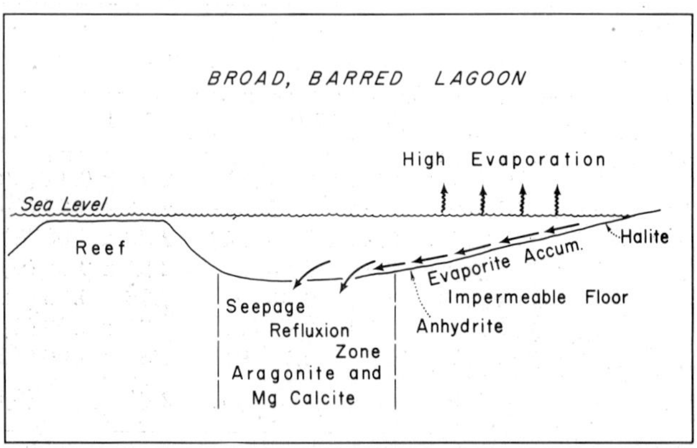

Fig. 7-22. Diagram of model for dolomitization by seepage refluxion from an evaporitic lagoon. (Modified after Adams and Rhodes, 1960, p. 1915.)

Their model was applied to the Permian Basin of West Texas and, in this particular case, it was inferred that reflux extended to depths of hundreds of meters and that the fluids passed through rocks with differing degrees of permeability. More porous zones were favored and impervious limestone lenses were bypassed (Adams and Rhodes, 1960, p. 1917). In view of the depth of penetration and the rocks involved, in the case of this model both diagenetic and epigenetic dolomitization would occur, although the dolomitizing solutions are generated at the surface; the process would seem to be most effective under shallow burial conditions.

Subsequently, the seepage refluxion model has been invoked to explain relatively recent dolomitization, particularly at Bonaire, Netherlands Antilles. Deffeyes et al. (1965) proposed that dense, Mg^{2+}-rich brines resulted from evaporation in a supratidal lake (Pekelmeer) to the extent that gypsum was precipitated and the Mg^{2+}/Ca^{2+} ratio consequently raised. Inasmuch as surficial sediments in contact with these brines are dolomitized, it was envisioned that such fluids could seep into the underlying carbonates and dolomitize them. As a likely nearby "fossil" example, they pointed to the Plio-Pleistocene beds on northern Bonaire in which dolomite passes downdip into limestone. Later, Lucia (1968) reported on a borehole drilled through the floor of the Pekelmeer in which no dolomite was found and the interstitial water was normal marine where brines had been expected. A thin clay layer of generally low permeability was discovered which separates sediments with normal sea water (below) from those with hypersaline water (above). Springs of normal sea water represent local sites where the ash bed is broken or missing. These springs represent the upper ends of permeability channels in the underlying Pleistocene (Murray, 1969). Most of the year sea water is drawn onto the land through these springs because of the hydrostatic head resulting from sea level being higher than the level of the lake. During a short period in summer, however, the springs are inactive and pressure on the landward side is greater than on the seaward side. Murray concluded that return flow, or reflux, should occur at this time.

Even before these restricted possibilities for reflux were appreciated, there was the problem of demonstrating that the rate of this process is appropriate for the time available (Bathurst, 1971, p. 532). There is, then, some question as to the significance of seepage refluxion not only at Bonaire, but more generally in the geologic record, especially where only a small topographic head was available along a featureless shoreline. Furthermore, the slight difference in density between normal sea water and hypersaline waters may be a serious deterrent to the efficiency of the model (Hsü and Siegenthaler, 1969). On the other hand, Murray (1969, p. 1012) contended that earmarks of seepage refluxion are common in ancient rocks—such as tongues of dolomite extending beneath dolomitized supratidal sediments and projections of dolomite beneath stratigraphically controlled dolomite. Lucia (1972, p. 160) stated: "Extensive reflux dolomitization is characteristic of evaporite shoreline carbonate deposits."

Many instances of dolomitization have been attributed to this process. For example, Berner (1965, p. 1298) suggested reflux action to explain some of the dolomitization of transported and buried reef skeletal material beneath the atolls at Funafuti, Kita-daito-jima, and Eniwetok: "This reaction could have taken place within a restricted backreef lagoon or tidal flat, or in the shallow subsurface by the downward and seaward migration of the overlying brine."

There is little doubt that seepage of evaporitic fluids has occurred in the Lockport Formation (Middle Silurian) of New York (Zenger, 1965b, pp. 126—127). Overlain by the well-known evaporitic Salina Group, the Lockport Formation contains considerable amounts of gypsum and anhydrite primarily in epigenetic vugs and fractures. If hypersaline solutions passed through the lithified rock, however, it is conceivable that they also did so much earlier and this is suggested by some pseudomorphs after celestite (?) observed in the matrix of some of the dolomites. In central New York, although the overlying Salina Group deposits maintain their evaporitic characteristics, there is little evidence of brines having passed through the formation. At its easternmost extent, dolomitic beds of the Ilion Member are sandwiched between thicker intervals of impervious shale suggesting an unlikely situation for dolomitization by seepage refluxion.

Friedman and Sanders (1967, pp. 287—289) discussed the mechanism of "capillary concentration", as originally applied by Mueller (1960) to Chilean nitrate deposits. According to this model, interstitial waters consist initially of normal sea water, that transpires upward through sea marginal sediments and becomes concentrated at the surface by evaporation. Again the Mg^{2+}/Ca^{2+} ratios would increase owing to the precipitation of gypsum. Friedman and Sanders concluded that many of the reported "supratidal crusts" formed in this manner (Fig. 7-23). Replenishment of water could be provided by sea water moving laterally and upward through the sediments from the adjacent lagoon. The model has been applied to dolomitization of buried intertidal sediments within the sabkhas of the Persian Gulf (Illing et al., 1965) and to the formation of dolomite crusts in the Florida—Bahama region (Shinn and Ginsburg, 1964). In both of these situations, however, periodic floodings into the supratidal zone might be more significant than the capillary movement. Friedman and Sanders (1967, p. 289) even suggested that the Holocene dolomitic crusts on Bonaire may have formed by capillary concentration of waters moving in through the coral rubble ridge separating the Pekelmeer and the crusts from the ocean. Thompson (1970, p. 1284) considered this process as the most likely one to account for dolo-

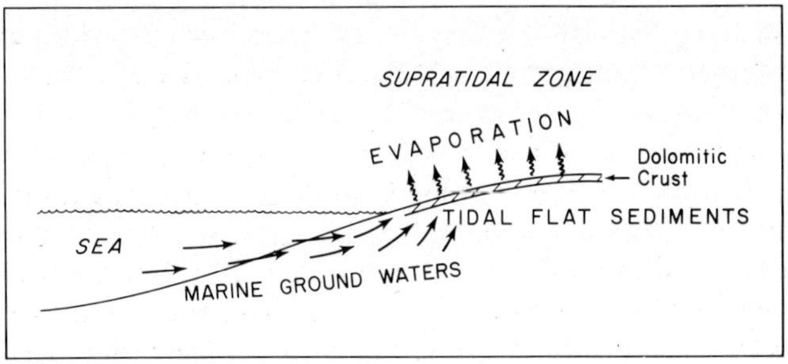

Fig. 7-23. Diagram of model for dolomitization by capillary concentration or by evaporative pumping. (After Zenger, 1972b, p. 119; courtesy J. Geol. Educ.)

mitization of supratidal sediments in the Ordovician Juniata Formation of the central Appalachians.

Hsü and Siegenthaler (1969), strongly opposed to seepage refluxion as a model accounting for any significant dolomitization, proposed a mechanism, based on laboratory experiments, somewhat similar in some respects to "capillary concentration", for the movement of dolomitizing solutions (Fig. 7-23). In their model, interstitial waters move in response to a vertical hydraulic gradient induced by evaporative loss at the surface, rather than to capillary movement caused directly by the action of surface tension; they applied the term "evaporative pumping" to this mechanism.

The foregoing models both include hypersaline solutions as requisites for dolomitization. Whereas many dolomite units may owe their origin to evaporitic conditions and the presence of brines, these are not universal requirements, as has been discussed under "Chemistry". To consider *all* dolomites as evaporitic (Friedman and Sanders, 1967, pp. 334, 338) is an overgeneralization (Zenger, 1972a, p. 9).

The terms "penecontemporaneous", "supratidal", and "evaporitic" have been associated with dolomitization since the discoveries of the several Holocene occurrences in the 1950's and 1960's. These discoveries prompted the recognition of numerous ancient analogs such as: certain facies of the Manlius Formation (Lower Devonian) of New York (Laporte, 1967); Middle Ordovician rocks of western Maryland (Matter, 1967); thin dolomites in the Macumber Formation (Mississippian) of the Canadian Maritime Provinces (Schenk, 1967); Carboniferous rocks of Ireland (West et al., 1968); some facies within a terrigenous suite in the Juniata Formation (Upper Ordovician) of the central Appalachians (Thompson, 1970); Upper Pennsylvanian shoreline deposits of Iowa and Nebraska (Fagerstrom and Burchett, 1972); subsurface Ordovician and Silurian rocks of the Williston Basin (Roehl, 1967); and portions of the El Paso Group (Lower Ordovician) of west Texas and the Jeffersonville Limestone (Middle Devonian) as interpreted by Lucia (1972). Textoris (1969) believed that much of the dolomite in the Appalachian Basin is of a penecontemporaneous, supratidal origin. In his paper, he summarized evidence that he considered diagnostic of such dolomitization: (1) limestone in interbedded limestone—dolomite sequences containing features of intertidal and shallow subtidal deposits; (2) occurrence of dolomite closer to terrigenous sources and limestone farther away; (3) desiccation cracks; (4) birdseye structures; (5) intraclasts; (6) evaporitic crystal molds; (7) some types of burrows (vertically oriented); (8) algal stromatolites forming low, laterally linked hemispheroids (LLH of Logan et al., 1964); and (9) micritic microfacies replaced to various degrees by euhedral dolomite crystals. To these could be added thinly laminated dolomite, which suggests algal-mat activity (Freeman, 1966; Shinn et al., 1969). For the most part, such evidence, especially taken collectively, appears to support the interpretations in these particular cases. Zenger (1972a), however, has pointed out that overgeneralization has resulted and that a considerable portion of ancient dolomitized carbonates does not represent initial supratidal accumulation and/or supratidal dolomitization. Some of the several pieces of evidence listed by Textoris (1969) and used by workers drawing these analogies between modern and ancient deposits are equivocal. For example, laterally linked Holocene stromatolites, other than in the Bahama—Florida Bay area, are most typical of the intertidal zone. Furthermore, subtidal stromatolites do form today (Gebelein, 1969), although rarely. Possibly, they were much more abundant in the Precambrian and Early Paleozoic (Garrett, 1970) when grazing invertebrates, similar to those feeding on algal mats today, were uncommon or absent.

There is considerable doubt about the often-quoted generalization that dolomites are characterized by higher insoluble residue content than associated limestones (Hatfield and Rohrbacher, 1966; Zenger, 1965a; Lumsden, 1973, 1974) which would be taken as indicative of a more landward depositional site of dolomites. Berry and Boucot (1971, p. 49), in their regional synthesis of the Silurian of North America, showed that, with few exceptions, the "... Limestone Suite rims that of the dolomites and separates it from a terrigenous suite around the continent ...".

Taken individually, such features as lamination of sediment, association with evaporites (as crystals, molds, or pseudomorphs), birdseye structures, and, particularly, the presence of dolomite itself, may be suggestive but certainly not indicative of supratidal conditions; however, the collective evidence can be impressive. To distinguish between the intertidal and supratidal paleoenvironments might be very difficult and the use of the term peritidal, to include the entire tidal flat, as introduced by Folk (1970), often is more appropriate. Also difficult is the determination of whether the dolomitization process itself was supratidal and penecontemporaneous or not (Zenger, 1972b, p. 116). The most persuasive evidence for this origin consists of a strong correlation between dolomite content and beds interpreted as supratidal. Certainly the determination of the timing of dolomitization becomes even more complicated when dealing with a completely dolomitized sequence. An example illustrating possible problems is that of Freeman's (1966) description of the Plattin and Joachim formations (Ordovician) of northern Arkansas. Although the units contain such peritidal characteristics as algal stromatolites, desiccation cracks, burrows, and halite molds, Freeman interpreted the dolomitization as postlithification, based on localization of dolomite crystals along stylolites and floating in spar.

Zenger has undertaken field and petrographic studies of various Paleozoic dolomite units as follows: Little Falls Formation, Upper Cambrian, east-central New York (Zenger, 1969, 1972c); Herkimer Formation, Middle Silurian, central New York (Zenger, 1971); Lockport Formation, Middle Silurian, New York (1965b); Lost Burro Formation, Middle and Upper Devonian, east-central California (Zenger and Pearson, 1969). There is abundant evidence that the great bulk of dolomite in these units resulted from metasomatism of calcium-carbonate sediments representing various environments of deposition (Zenger, 1972a). Details of the stratigraphy of these units can be found in the above references. Table 7-IV summarizes their general depositional environments as gauged by criteria permitting a comparison with modern and reported ancient supratidal dolomites. Portions of the Little Falls and upper Lockport Formations contain evidence (algal stromatolites — see Figs. 7-24 and 7-25, desiccation cracks, intraclasts) of peritidal deposition. Most of the other units (Herkimer, Lost Burro, lower Lockport) show evidence (bioherms, abundant neritic fossils) of subtidal accumulation. Dolomitization is primarily postpenecontemporaneous and is not related to secondary structures. Some dolomite is of diagenetic origin, but dolomitization commonly was not completed until after lithification (epigenetic, as defined here). Even in those units, which are most likely candidates for peritidal, penecontemporaneous dolomites (i.e., Little Falls and upper, main mass of the Lockport), the dolomite is ordered (Fig. 7-26), unlike that in Holocene occurrences. Also, the grain size is much coarser than in the possible modern analogs. It is possible that time and the subjection of penecontemporaneous peritidal protodolomite to the diagenetic environment could account for both the ordering and the coarse grain size observed. But, there are numerous examples of both disordered and extremely fine-grained dolomite in the geologic record. These probable peritidal dolomites have maintained their characteristics; thus, subsequent diagenesis cannot masquerade all such protodolomites. Many of the extensive sheet-like dolomites from the record show no definitive evidence of either peritidal accumulation or peri-

TABLE 7-IV

Summary of characteristics and suggested environments of deposition of studied dolomite units

Selected points for comparison with modern and reported ancient supratidal dolomites	Lockport Formation		Herkimer Formation Joslin Hill Member
	Gasport Member	"Upper" Lockport	
Algal stromatolites	none observed	common at certain horizons; mainly LLH [1]	none observed
Evaporites	relatively common; mainly sulfates in vugs	common, mainly sulfates in vugs; pseudomorphs may represent syngenetic origin	none observed
Sedimentary structures	minor cross-laminations	intraclasts; vuggy and porous zones; birdseyes rare; ripple marks; oöids; peloids; minor desiccation cracks	cross-laminations; ripple marks (para-ripples common); intraclasts minor
Fossils	bioherms; abundant marine fauna: brachiopods, corals, pelmatozoans, stromatoporoids, etc.	fossiliferous zones, some biostromal; corals and stromatoporoids most common; brachiopods, ostracodes, trilobites	abundant marine fauna: brachiopods, mollusks, pelmatozoans, ostracodes, trilobites, etc.
Insoluble content versus dolomite/calcite ratio	untested; apparently no significant correlation	no significant correlation	not applicable; essentially no calcite present
Nature of dolomite	crystal size mainly 20—100 μ; well ordered; apparently stoichiometric	crystal size mostly 40—200 μ; well ordered; apparently stoichiometric	crystal size mainly 40—250 μ; well ordered; dolomite is ferroan
Nature of dolomitization	mainly diagenetic replacement; no conclusive evidence of any penecontemporaneous dolomitization	mainly diagenetic replacement; local occurrences of rhombs in micrite may indicate some initial penecontemporaneous dolomitization	mainly early diagenetic replacement; no conclusive evidence for any penecontemporaneous dolomitization
General environment of deposition	shallow subtidal shelf; normal salinity	shallow subtidal, at times partly restricted to peritidal and in part evaporitic	shallow subtidal shelf; normal salinity

[1] LLH = laterally linked hemispheroids (Collenia type); SH = discrete, vertically stacked hemispheroids (cryptozoon type).

Little Falls Formation	Lost Burro Formation	
	"Lower"	"Upper"
relatively common; LLH, SH[1], possible flat algal laminations	possible minor flat algal laminations	none observed
none observed	none observed	none observed
intraclasts; cross-laminations; desiccation cracks; oöids; peloids; grapestone; vuggy; birdseyes rare; vertical burrows	desiccation cracks, cross-laminations in lower quartzitic portion; peloids	peloids
extremely rare; sporadic zones with lingulid brachiopods	rare; few zones with brachiopods, corals, and stromatoporoids	stromatoporoid biostromes; less common brachiopods and corals; calcispheres
not applicable; calcite very minor and late diagenetic	not applicable; calcite very minor	no correlation
crystal size mainly 40—180 μ; well ordered; apparently stoichiometric	crystal size mainly 25—300 μ; well ordered; apparently stoichiometric	crystal size mainly 40—100 μ; well ordered; apparently stoichiometric
mainly diagenetic replacement; complete dolomitization precludes determination of any penecontemporaneous dolomitization	mainly diagenetic replacement; complete dolomitization precludes determination of any initial penecontemporaneous dolomitization	mainly diagenetic replacement
shallow subtidal to mainly peritidal (chiefly intertidal); no evidence of hypersaline conditions	shallow subtidal to perhaps partly peritidal; normal or near-normal salinity	shallow subtidal; normal or near-normal salinity

Fig. 7-24. Algal stromatolite (LLH) in Little Falls Formation, Little Falls, New York (note pocket knife, upper right corner, for scale). Environment of deposition of original carbonate sediments, now completely dolomitized, interpreted as peritidal. (After Zenger, 1972, fig. 3; courtesy Geological Society of America.)

tidal dolomitization. In addition, there have been numerous reports of now dolomitized, initially subtidal calcium-carbonate sediments (Fadeyev, 1962; Schmidt, 1965; Asquith, 1967; Berry and Boucot, 1971; Mossler, 1971; and others).

Sea water—meteoric water mixing. In connection with the concept of schizohaline dolomitization discussed in the section on chemistry, Land (1973a, b) has proposed a model based on his studies of diagenesis of Pleistocene carbonates in Jamaica. There, as elsewhere in the Caribbean, Middle Pleistocene reefs are lithified and commonly dolomitized. The contact between the Middle Pleistocene Yarmouth(?) Hope Gate Formation and the overlying "Sangamon" Falmouth Formation is erosional, indicating that diagenesis and dolomitization of the Hope Gate deposits occurred before Falmouth time. The Hope Gate dolomite occurs as two different textural types: (1) a micrite-replacement product with grain size ranging from identifiable silt-sized particles down to less than 0.1 μm in diameter, and (2) a void-

Fig. 7-25. Photomicrograph (ordinary light) of thin-section of marginal portion of completely dolomitized algal stromatolite (left half of illustration — relict laminations trending steeply near center) and detrital filling (whitish quartz grains, relict peloids) on right. Note coarseness of dolomite crystals. Little Falls Formation, Middleville, New York. (Photograph by D.H. Zenger.)

filling dolomite consisting of large sparry crystals. The spar fills voids created by the solution of aragonite and the coarseness of the spar suggests precipitation from a relatively dilute solution (Folk and Land, 1975). No marine sediments unrelated to reef deposition that could correspond to prograding supratidal environments have been found. The outcrop belt of the Hope Gate Formation is relatively narrow (a few hundred meters) and Land (1973a, p. 71) doubted that a Pekelmeer-type environment or a sabkha could have ever existed. Luxuriant reef growth, as well as a lack of supporting evidence, argue against hypersaline conditions that would have accompanied seepage refluxion. The Hope Gate dolomites have low contents of Sr^{2+} (200 p.p.m.) and Na^{+} (350 p.p.m.), precluding their precipitation from normal marine waters. If normal or moderately hypersaline solutions were responsible for Hope Gate carbonates, then the Sr^{2+} to Na^{+} ratio of the minerals should be typical of marine precipitates. Kinsman's (1969) work on strontium partitioning shows that calcites forming in equilibrium with ma-

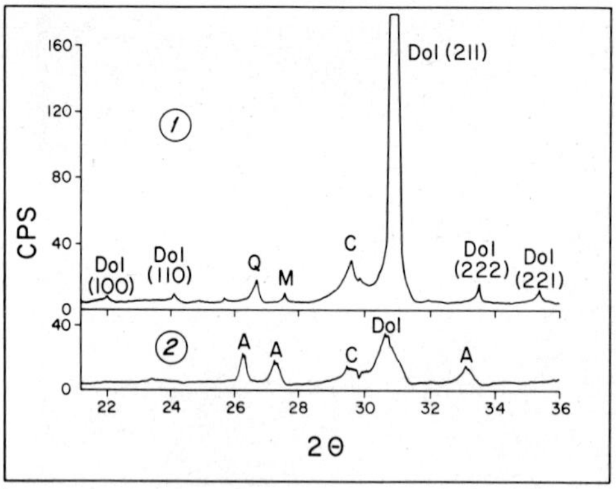

Fig. 7-26. Comparison of X-ray diffractograms of sample of algal stromatolitic dolomite from Cambrian Little Falls Formation (*1*), and one from Recent supratidal dolomitic crust in Bahamas (*2*). Note line broadening and displacement of main dolomite reflection (211) in the protodolomite (*2*). Also note apparent absence or insignificance of ordering reflections (100) and (221). Although not shown here, a longer scan failed to show the presence of the (111) reflection in the Holocene sample. (After Zenger, 1972a, fig. 7, p. 5; courtesy Geological Society of America.)

rine waters should have strontium concentrations of about 1050 p.p.m. Behrens and Land (1972) claimed that dolomites forming in equilibrium from marine reservoirs should have about 600 p.p.m. Sr^{2+}. According to Land and Hoops (1973), recent marine dolomite from Baffin Bay and the Persian Gulf contain Na^{+} over a range of 800 to 2340 p.p.m. They claimed that the low sodium content of most ancient dolomites suggests they are either primary precipitates from solutions containing little sodium or they have recrystallized and re-equilibrated with ground waters low in sodium. Hope Gate dolomites average 1‰ lighter in δC^{13} than Recent Jamaican reef sediments and indicate an addition of C^{12} to the original marine carbon reservoir, probably from Jamaican meteoric water. [1] Hope Gate dolomites are about 5‰ enriched in O^{18} relative to intimately co-existing calcite which, in turn, are 1‰ heavier than Falmouth-age calcite which is presumed to be in equilibrium with Jamaican meteoric water (Land and Epstein, 1970). Land (1973a, pp. 82—83) suggested that the source of additional O^{18} is the large volume of relatively heavy marine carbonates forming the Hope Gate Formation. Land favored at least a 3‰ fractionation between calcite and ordered, stoichiometric dolomite under equilibrium conditions. He did not believe that the +2‰ value for δO^{18} of the

[1] All values for δC^{13} and δO^{18} in this paper are relative to the PDB, or Pedee belemnite standard, University of Chicago.

Hope Gate dolomite indicates hypersaline conditions because δO^{18} of the Holocene marine dolomites is always in excess of +4‰.

Land's model (Fig. 7-27) involves a mixing of phreatic meteoric water and sea water. Dolomitization occurred first as CO_2-rich meteoric waters penetrated seaward-growing reefs. Dolomite precipitated from a fluid the composition of which was controlled by: (1) the composition of the meteoric water, (2) the amount of mixing with sea water (a mixture of CO_2-oversaturated meteoric water and only 3% sea water is oversaturated with regard to dolomite because its Mg^{2+}/Ca^{2+} ratio is above 0.8), and (3) the more significant composition of the primary reef carbonate. Land (1973b), in attempting to strengthen the test of his model for relatively rapid dolomitization by phreatic meteoric processes, also studied Holocene dolomitization in mineralogically unstable Late Pleistocene (Sangamon) limestones which have undergone little vadose diagenesis. There, in the mixing zone, the rocks are more highly altered than those above; dolomite occurs as randomly oriented euhedra 8–25 μm in diameter replacing micrite, as a replacement of red algal allochems, and as a drusy lining of pores.

Land (1973a) believed that this meteoric water—sea water mixing model much more nearly fulfils the petrographic and chemical requirements of the more widespread dolomites of the past than the other models described. Although a marine origin of dolomite could explain the Mg^{2+} ion supply, the

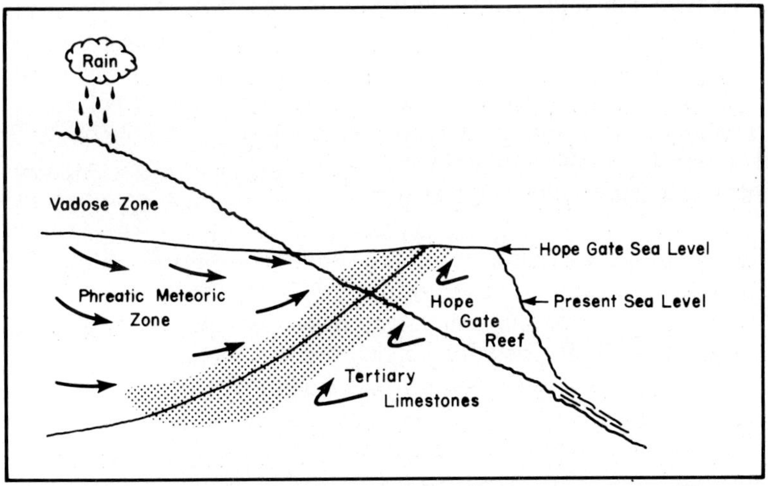

Fig. 7-27. Schematic representation of Hope Gate diagenetic environment and schizohaline dolomitization as proposed by Land (1973a). Diagenesis and dolomitization took place in the complex zone of mixing (shaded), as meteoric waters with a high P_{CO_2} became mixed with sea water. (After Land, 1973a, p. 845; courtesy Bull. Mar. Sci.)

trace-element geochemistry of dolomites argues against their formation from marine solutions. Rather, the mixing model would rely on a magnesium supply form the sediments themselves, especially in view of the abundance of magnesian calcite being formed today, and by the interaction of CO_2-charged meteoric waters with small amounts of sea water in a mixing zone.

Attractive though this model may be, it must be checked in other Pleistocene and Holocene units. There are some uncertainties and problems, such as the difficulty in interpreting the oxygen isotope values in view of the existing controversy over the fractionation between coexisting calcite and dolomite (Land, 1973a, p. 83), as well as that over what the values mean in terms of paleosalinity. Recrystallization is characteristic of most ancient limestones (Kinsman, 1969) and diagenetic effects in dolomites are even more impressive. Consequently, one should expect an equilibration with ground waters and a decrease in Sr^{2+} and Na^{+}, even though higher amounts of these cations may have been present initially. Also, although one must be impressed with the dolomitization in these young Jamaican sediments, does Land's model suffice to account for the extensive sheet-like dolomites of the Early Paleozoic, for example?

Perhaps very significant in this regard is the previously referred to model of Hanshaw et al. (1971) for dolomitization in a ground-water system also involving mixing with saline waters. The model is geochemically similar to Land's, but is on a much larger scale (i.e., the Florida Platform). They described dolomitization of Tertiary carbonates in brackish ground-water zones beneath Florida. The Mg^{2+}/Ca^{2+} ratio in waters from the Tertiary aquifer is as low as 0.05 in the recharge area, where the water is also undersaturated with respect to both calcite and dolomite. Here the model would differ from Land's in the sense that the dolomitizing solution evolves in the subsurface. With time and length of travel in the system, the Mg^{2+}/Ca^{2+} ratio in ground water increases and approaches unity. Downgradient, the solution apparently becomes oversaturated with respect to both calcite and dolomite.

About one-third of the well samples consist of ordered dolomite containing less than 50—58 mole% $CaCO_3$. The dolomites in both the recharge and the downgradient areas are generally massive, homogeneous, sugary, and variable in grain size. Fossils are usually calcitic in the recharge area, but bryozoans and other fossils have been dolomitized downgradient.

It is unlikely that solution of Mg-calcites and dolomite in the potable zone of active circulation is sufficient to supply the necessary Mg^{2+} ions. Rather, the magnesium may be derived from the readily available ocean water in the zone of mixing with the potable water (Fig. 7-28), as the hydrologic regimen is changed because of relative fluctuation of sea level. Thus, a large supply of Mg^{2+} ions rather than a high Mg^{2+}/Ca^{2+} ratio was stressed.

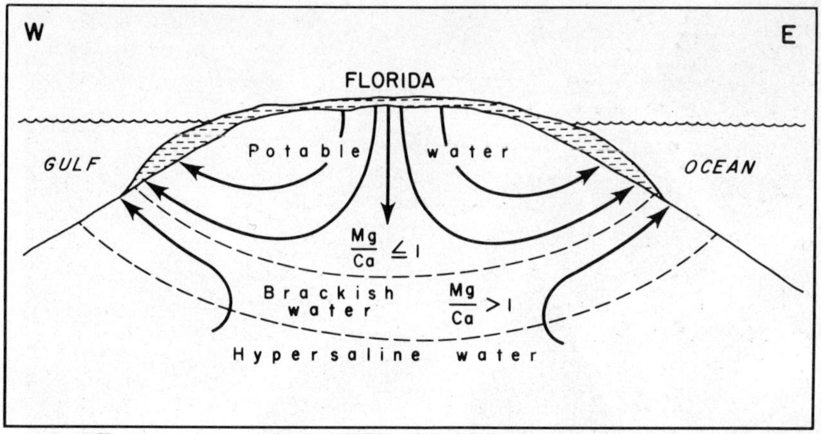

Fig. 7-28. Model for dolomitization by ground water as proposed by Hanshaw et al. (1971, fig. 7, p. 721). (Modified from Vernon, 1969, courtesy Econ. Geol.)

Technically, the dolomitization would be classed as epigenetic because the limestones in the aquifer are lithified.

Other proposed examples of dolomitization by sea water—meteoric water mixtures in the geologic record have been reported. Folk and Siedlecka (1974) designated Late Paleozoic strata on Bear Island, between Spitzbergen and Norway, as the "type locality" of the schizohaline environment, which was formally defined as ". . . one having extreme and rapid changes of salinity from hypersaline to dilute, even fresh. The environment is split ("schizo") between alternating epochs of hyper- and hyposalinity and seldom has normal sea-water salinity. Strictly speaking, the term should be restricted to changes essentially occurring within the time of "early" diagenesis, before deep burial of the rocks" (Folk and Siedlecka, 1974, pp. 1—1). Fabrics and structures in the rocks indicative of hypersalinity are evaporite nodules, length-slow chalcedony, and very finely crystalline, penecontemporaneous dolomite. Indicative of hyposalinity are euhedral limpid dolomite crystals, coarse poikilotopic sparry calcite, and microspar calcite. Zoned and hollow dolomite rhombs (Fig. 7-29) are fabrics of questionable origin. As one possible explanation, Folk and Siedlecka suggested that the hollow centers of the crystals formed under conditions of high salinity, whereas a later overgrowth of limpid dolomite formed as the solutions became more dilute. On weathering, the limpid dolomite becomes more in equilibrium with the meteoric waters, whereas the more soluble nucleus is more easily dissolved.

Badiozamani (1973) applied his dorag dolomitization model to Middle Ordovician carbonates of southwestern Wisconsin and adjacent states. Ap-

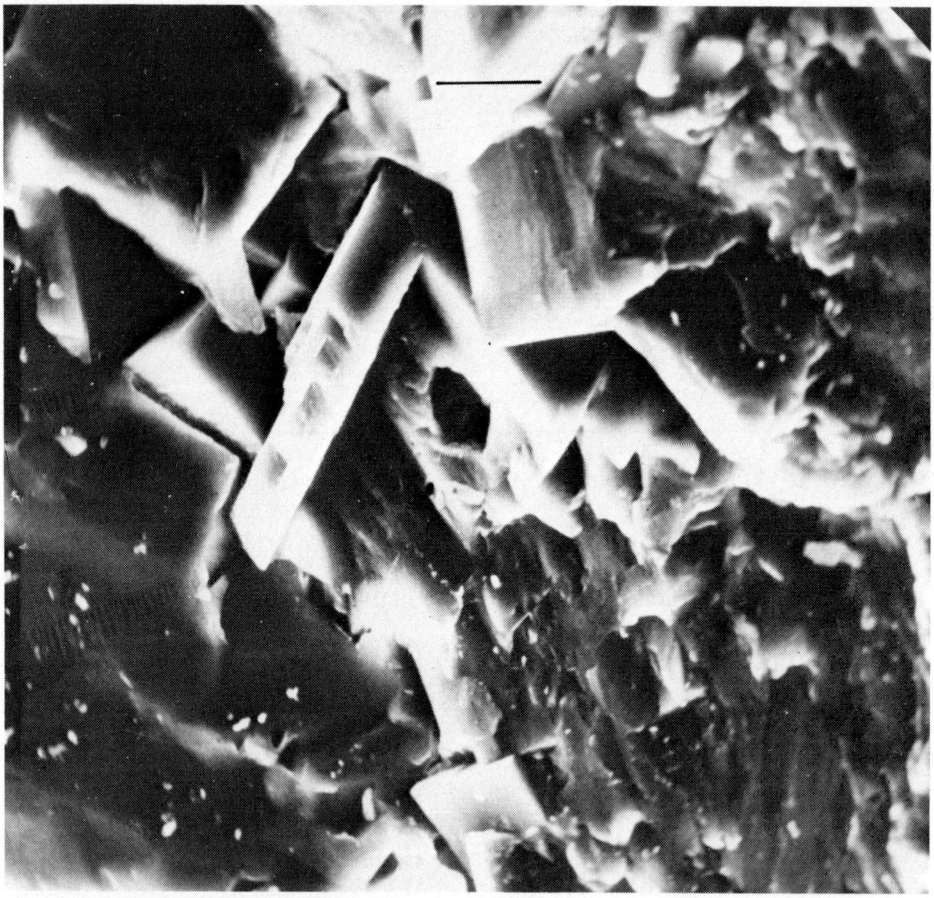

Fig. 7-29. Scanning electron photomicrograph of rhombohedra of limpid dolomite, *Fusulina* Limestone (Upper Carbonaceous), Bear Island. Note large, hollow, box-like crystal. Bar = 5 μm. (After Folk and Siedlecka, 1974, fig. 3; courtesy Sediment. Geol.)

parently the Mifflin Member dolomite represents original carbonate sediments that formed in an open marine environment. Evidence of restricted environments, such as evaporites, algal mats, and mudcracks, is lacking. Compared to modern marine carbonates, the values of δO^{18} and δC^{13} as well as the contents of Sr^{2+} and Na^{+} are low, suggesting an exchange of the Mifflin carbonates with meteoric water. Badiozamini concluded that dolomitization seems to be best explained by mixing of sea water and ground water in the phreatic zone.

Land et al. (1975) described two formations, in addition to the Holocene and Pleistocene dolomite occurrences in Jamaica previously described by Land (1973a, b), that showed evidence of dolomitization resulting from early meteoric influx into marine carbonate deposits: the Eocene El Naqb Formation of Egypt and the Ordovician Allen Bay Formation of Devon Island in the Arctic Islands. In contrast to Holocene "marine" dolomites, these ancient ones consist of crystals rarely less than 10 μm and normally hundreds of μm in size; commonly these are subhedral to euhedral and often limpid. X-ray diffraction patterns of these ancient crystals are sharp and the dolomite is more nearly stoichiometric. Sodium and strontium contents are low, a few hundred p.p.m. at most, and O^{18} is usually depleted. These workers (p. 1622) believed that most carbonate sediments, including these dolomites, are deposited initially from marine-influenced solutions and subsequently become modified by the flushing of "cleaner" continental waters. Their conclusions are supported by microprobe analyses of dolomite crystals from the El Naqb Formation, which were depleted in sodium and strontium in the outer zones relative to the cores, suggesting that the outer parts formed in a more dilute solution. They also suggested that the "hollow rhombs" described by them may have resulted from preferential dissolution of the earlier formed impure cores which were overgrown by the more stable, limpid rims.

The hollow, box-like dolomite crystals just mentioned were also observed by Folk and Siedlecka (1974; see Fig. 7-29) who also theorized that the hollow centers resulted from the solution of a more soluble, impure nucleus. Weaver (1975), in his exquisitely illustrated paper regarding the development of limpid dolomite crystals in Miocene palygorskite deposits showed hollow, poorly formed rhombs of ordered protodolomite, commonly having only three sides (Fig. 7-30). It was Weaver's thought that these hollow crystals, replacing shells in a schizohaline environment, actually *grew* in this manner: "During the early stages of replacement, Mg apparently diffuses into the shell's calcite plates, thus preserving the plate morphology. Once the plates achieve a protodolomite composition, growth starts at the edge of the plates (Fig. 7-31); two- and three-sided and eventually six-sided rhombs are formed, leaving the middle hollow. Growth appears to be layer-by-layer. Folk and Siedlecka (1974) suggested that hollow rhombs are dolomite boxes formed around more soluble nuclei that are later dissolved. This is a possible mechanism in some instances but the platy morphology and arrangement of the plates in the Miocene dolomite is more suggestive of growth than solution." (Weaver, 1975, p. 427).

The mixed sea water—meteoric water models have stirred a new excitement in the study of dolomitization. More work will be presented shortly by Gebelein et al. (pers. communication, 1972, 1973, 1976), who has found pro-

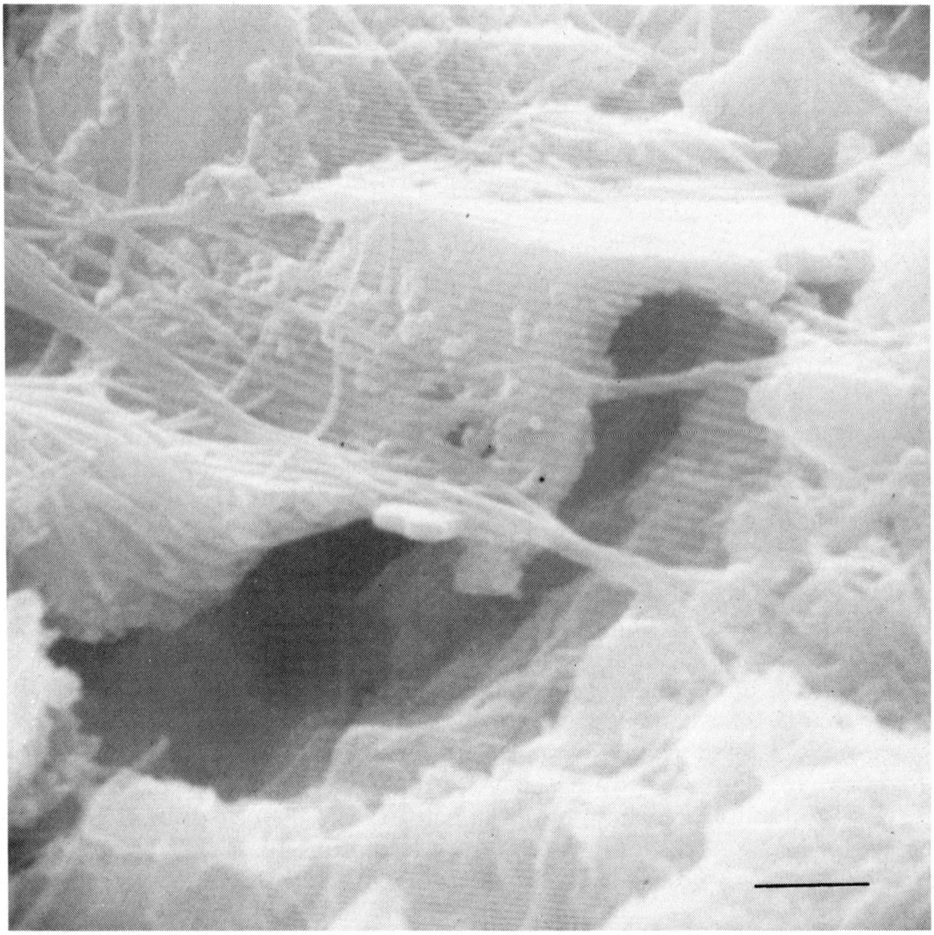

Fig. 7-30. Scanning electron photomicrograph of partial, hollow rhombs of dolomite and palygorskite replacing shell. Bar = 1 μm. (After Weaver, 1975, fig. 1A; reproduced by permission of the Geological Society of America.)

todolomite in association with a zone of mixing between sea-water and fresh-water lenses in tidal sediments on the southwest coast of Andros Island, Bahamas. In the meantime, the adequacy of the models will have to be continually reexamined in light of studies of ancient dolomites. There is already the danger that a "band-wagon effect" has been created.

Ground-water dolomitization. Models and examples discussed here differ from the foregoing in that mixing of meteoric water with brines need not be

Fig. 7-31. Scanning electron photomicrograph of dolomite rhombs in matrix of palygorskite. Note overlapping, flame-shaped growth sheets starting at rhomb edges. Bar = 5 μm. (After Weaver, 1975, fig. 2A; reproduced by permission of the Geological Society of America.)

involved. The depth of dolomitization may be shallow or deep, and dolomitization may be diagenetic or epigenetic, although the emphasis is on the epigenetic, deep-burial type in most instances. The dolomitizing waters may be meteoric or connate.

The first example involves diagenetic dolomitization by meteoric ground waters inland from a zone of mixing of sea water and meteoric water. Work in the Coorong region of South Australia by Von der Borch et al. (1975) has shown that a zone of sea-water and ground-water mixing permits the distinction of three areas of carbonate formation. In a seaward direction is the lagoon and its restricted embayment with the carbonate assemblages dominated by aragonite and Mg-calcite. In the landward direction follows an intermediate zone (probably the zone of mixing) containing ephemeral lakes in which poorly crystallized protodolomite is forming. Next inland are coastal ephemeral lakes which are now remote from the "parent" lagoon and which are fed only by ground waters; they contain an upper unit of dolomite or dolomite plus magnesite. Von der Borch et al. concluded that the bulk of the dolomite in the region is forming from shallow Ca- and Mg-bicarbonate ground waters that actively discharge from carbonate aquifers beneath the coastal plain. There, near the coastline, seaward-flowing ground waters emerge, undergo summer evaporation, and through some as yet undefined processes produce dolomite. They envision the Mg-ions to be derived from either dissolution of Mg-calcite skeletal components in Upper Tertiary limestones and Pleistocene calcareous sands that constitute the shallow aquifer or, more likely, leaching of Tertiary and Quaternary basic volcanic rocks.

An example of epigenetic (postorogenic) dolomitization in Spain was described by Freeman (1972). The situation of a carbonate unit (Muschelkalk, Triassic), with peritidal characteristics and consisting of 75% dolomite, overlain by the evaporitic Keuper deposits seemed to be a potentially ideal one for a study of penecontemporaneous peritidal dolomitization followed by dedolomitization. Instead, Freeman found that the dolomite distribution is joint-controlled and bears no relation to sedimentary facies. Coupled with a practical absence of "dedolomite", the evidence suggested a late but sustained solution with an Mg^{2+}/Ca^{2+} ratio in excess of that expected from an evaporite-bearing assemblage. A study of water-bearing samples from salt-producing evaporite pans in the overlying Keuper deposits indicated that meteoric water descending through them evolves into a brine with Mg^{2+}/Ca^{2+} ratios locally up to 4, despite common occurrences of gypsum and anhydrite. X-ray diffraction analyses of Keuper chlorites show a loss of Mg^{2+} from the lattice. Freeman proposed that this loss of Mg^{2+}, as a result of diagenesis, added to the Keuper meteoric waters and, aided by the geothermal gradient prior to unloading, was responsible for the dolomitization.

Brown (1959) described epigenetic dolomitization in the Maurice Formation (Upper Cambrian) in Montana and Wyoming. Dolomitization followed recrystallization of the oolitic rocks, but preceded fracturing. Zones of scattered oöids are more severely dolomitized than those with closely packed oöids in a finely crystalline calcite matrix. Brown assigned the source of Mg^{2+} to the surrounding argillaceous sediments. Compaction of these relatively high-magnesian sediments produced a connate water with sufficiently high

Mg^{2+}/Ca^{2+} ratios to dolomitize the porous oölitic carbonates along permeability trends oriented roughly parallel to the stratification. Such a mechanism, of compaction and expulsion of fluids from terrigenous magnesium-bearing material, may be significant in some cases of dolomitization. (For details on the role of compaction fluids during diagenesis see Chilingarian and Wolf, 1976).

The effect of subsurface brines may be important in some epigenetic, deep-burial dolomitization. Probably much of this interstitial brine is derived ultimately from sea water (Degens and Chilingar, 1967, p. 480), although modified by salt-filtration and other processes. The chemistry of these connate waters differs from modern ocean water. For example, magnesium is commonly present only in minor amounts in oil field waters, although the opposite is true for calcium. Calcium-chloride waters are not produced in surface environments, but are common in subsurface brines. Degens and Chilingar felt that this aspect could be linked to dolomitization processess. Murata et al. (1972) pointed out that Mg^{2+}/Ca^{2+} ratios within the dolomite stability range are common in oil field brines in California and may explain why dolomite is the predominant diagenetic carbonate in these fields.

Through the mechanism of salt-filtration at shale or compacted clay horizons, dissolved salts could be concentrated on one side of these semipermeable beds as ordinary water moves through them (Berner, 1971, p. 111). These brines might have sufficiently high concentrations of magnesium to dolomitize limestone.

In hydrothermal dolomitization, the solutions are at higher temperatures than that resulting from the normal geothermal gradient. Hewett's classical work on dolomitization and ore deposition (1928, pp. 857—862) includes a consideration of sources of magnesium for hydrothermal dolomitization. He accepted the idea that magnesium could be provided by remote bedded dolomites for the dolomitization of ore-bearing units in southwestern Missouri, but rejected it as an explanation in deformed areas such as Goodsprings, Nevada. As a major source of magnesium, he proposed deeply buried masses of igneous rocks from which magnesium has been driven off during alteration, especially sericitization by hydrothermal solutions. There is serious doubt, however, whether this could have been an adequate supply.

Based on field relations and the probable genetic relations of Mg, Pb, Zn, Fe, and Si, Banks (1967, pp. 65—72) concluded that the most likely source of magnesium for dolomitization of the Leadville Limestone (Mississippian) of Colorado consisted of fluids produced from deep-seated metamorphic reactions such as granitization of a paragneiss. Referring to the widespread hydrothermal dolomitization in central Colorado, Banks stressed that the potential source had to be not only capable of yielding great quantities of magnesium but also be regional in extent.

In his study of hydrothermal dolomitization, Lovering (1969) considered subsurface connate brines as possible agents of dolomitization. If such waters were near intrusive centers, they would become heated and convection would occur in permeable rocks. Such a process would provide large volumes of magnesium which would be available for dolomitization, although these brines might be impoverished in magnesium compared to sea water (see also the reference to Lovering in the section on chemistry of dolomitization). Lovering's heated ground-water model decreases the difficulty of providing an adequate source of magnesium. Further, it "... explains the occurrence of hydrothermal dolomite in areas remote from ore deposits but in which there are deep-reaching fractures or other permeable channels that could conduct hot ground water from depth" (p. 743).

Ilich (1974) proposed a connection between hydrothermal and sedimentary dolomite in the genesis of the dolomite deposits in Miocene fresh-water lakes of the Inner Dinaric Alps, Yugoslavia. Referring to these deposits as "hydrothermal—sedimentary" dolomites, he envisioned hydrothermal solutions, enriched in Mg^{2+} and Ca^{2+} ions derived from neighboring limestones and ultramafics, moving along faults into certain stratigraphic levels in the lacustrine sediments, where dolomite was precipitated. Ilich believed that CO_2-rich hydrothermal waters entering oceans could also give rise to dolomitization without the necessity of an outside source of Mg^{2+} and Ca^{2+}. In addition to the increased P_{co_2} and temperature (conditions he believes to be necessary for dolomitization) provided by the hydrothermal waters, increased salinity of the waters would be practically essential. Although Ilich suggested hydrothermal—sedimentary dolomite as "the missing link" in the dolomite question, it would seem that the special conditions required by his model would account for a relatively small portion of the vast dolomite deposits of the past.

Models stressing "solid" source of magnesium. The theories discussed thus far emphasize the movement of magnesium-bearing waters to the site of dolomitization. The magnesium in these waters is generally supplied by sea water. Of course, several of these workers refer to specific sources for magnesium supply, for example, Freeman (1972), Land (1973a), and Brown (1959), and there is an understandable overlap in the categorization presented by the writers. In many of the examples summarized here, there is emphasis on magnesium supply at the very site of dolomitization.

Recently, Gebelein and Hoffman (1971, 1973) suggested a possible explanation for ancient carbonate rocks consisting of interlaminations of limestone and dolomite on the scale of centimeters or even millimeters. They proposed an analogy with modern algal stromatolites and flat algal laminated sediments, both of which consist of alternating layers of algal mats and particulate carbonate sediment. The particulate layers correspond to the limestone laminae, whereas the algal-mat layer would be the equivalent of the dolomite. Although no thin dolomite layers have been observed in the Holocene analogs, these algal layers contain significant amounts of magnesium. Apparently, the Mg^{2+} ions are fixed organically, as has been shown to be possible by Greenfield (1963), in the algal sheaths. The organic material is stable and does not decompose, and concomitantly releases magnesium, until long after deposition or even lithification. The released magnesium goes to form dolomite more or less in situ, that is, in the "micro"-environment of the algal mat, thus conforming to the bedding. Laboratory experiments were conducted by these workers to test the validity of the hypothesis. No dolomite was formed experimentally, but crystallization of 17—20 mole% magnesian calcite in contact with the sheath material was achieved. On the basis of their calculations (Gebelein and Hoffman, 1973 p. 608), they concluded that enough magnesium is complexed in a 2 mm thick algal-mat layer to produce a dolomite lamination 1 mm thick. They indicated that the actual formation of dolomite may be a long-term process occurring either by the replacement of carbonate material in contact with the sheath and/or by selective replacement of high-magnesian calcite within the algal-mat laminae. Although this model may account for many laminated limestone and dolomite sequences, Gebelein and Hoffman (1973, p. 611) stipulated that the process was not proposed to explain thicker bedded and more massive varieties with a lack of evidence of algal stromatolites. In two of the dolomite units studied in detail by Zenger (Herkimer and Lost Burro Formations), there are no algal stromatolites. These structures are relatively common in the Little Falls Formation (Zenger, 1969; Figs. 7-24 and 7-25), but no limestone is present — all laminae are completely dolomitized. In the Lockport Formation (Zenger, 1965b) there are limestones associated with the dolomite, but not on an interlaminated relationship; where algal stromatolites occur in the unit, they were observed to be all composed of one carbonate or the other.

High-magnesian calcite has commonly been considered a likely source for the magnesium needed for dolomitization. Fairbridge (1957), suggesting that dolomitization generally begins during early diagenesis, believed that the most ideal conditions exist in warm shallow water where there is a relatively high concentration of magnesium in solid solution in calcite. Such magnesium-rich calcite, he hypothesized, could act as nuclei around which more $MgCO_3$ forms, either from sea water or from connate water beneath the sea floor. When the magnesium content reaches a certain level, dolomitization could begin. Largely theoretical, due to the lack of any modern examples, his idea receives some support from the geologic record. Describing the occurrence and diagenesis of Ca—Mg carbonates in the lacustrine environment, a topic not covered in this chapter, Müller et al. (1972, p. 162) explained the transformation of high-Mg calcite into dolomite in two steps: (1) owing to their crystallographic similarity, the surface of the high-Mg calcite acts as a matrix for dolomite nucleation after a sufficiently high Mg^{2+}/Ca^{2+} ratio has been reached, and (2) further dolomite growth involves the dissolution of the high-Mg calcite followed by reprecipitation as dolomite. Schlanger (1957) described and illustrated the localization of dolomite rhombs within segments of high-Mg calcareous algae. He suggested that the dolomite resulted from ordering of the calcium—magnesium solid solution constituting the original algae and that local concentrations of Mg^{2+} ions within the fossil algae formed nuclei for the growth of dolomite rhombs. On the other hand, there are examples of fossils composed of high-magnesium calcite that have resisted dolomitization (Friedman and Sanders, 1967, p. 296). Zenger (1965b) considered high-magnesium calcite as a possible source of magnesium for some dolomitization in the Lockport Formation. Although some parts of the unit are rich in pelmatozoan debris, however, a greater part seems to be devoid of such material. This potential source of magnesium is insufficient to account for all of the dolomitization in the unit. Other workers have also considered this source of magnesium, although not in situ, for dolomitization. Peterson (1962, pp. 26—27) calculated that about half the magnesium in the dolomite of Upper Mississippian carbonate rocks of the Cumberland Plateau could be accounted for by migration of magnesium from the magnesian calcite of crinoidal fragments in these rocks.

Based on a test well drilled on Andros Island, Goodell and Garman (1969) desribed a process termed "solution—cannibalization" to account for subsurface sequences of dolomite beneath the Bahama Platform. They observed four major cycles, each one consisting of deposition, uplift, erosion, and dolomitization. They envisioned that during uplift, solution of the unstable high-magnesium calcite would occur followed by a downward percolation of the dense magnesium-rich brines (seepage refluxion?) to depths that would depend on hydrodynamic and permeability factors and on the volume of brine available. Dolomite growth appears to have begun on submicron-sized abrasion particles of high-magnesium calcite and on high-magnesium calcite crystallites in fossil shells and skeletal fragments. The rate of the process and the proportion of dolomitization is probably a function of brine availability. They believed it unlikely that a protodolomite stage would occur because of the great length of time available for the reaction, i.e., the reaction rate is slow in comparison to that prevailing during the laboratory synthesis of dolomite experiments. Although Goodell and Garman contended that there was enough magnesium to have provided for that dolomitization found in the well, Land (1973a, p. 86) stated that "solution—cannibalization" "... is a workable mechanism but suffers from the classic objection — insufficient magnesium". As with seepage refluxion, the question could be raised whether or not a sufficient head results from the difference in density between the brine and the connate water to permit the downward percolation. And, it does stretch one's imagination to conceive of this model accounting for many of the widespread sheet dolomites of the geologic record.

Kahle (1965), referring to reported positive correlations between dolomite and insoluble residue content, suggested that clay minerals may relate to dolomite formation by providing a source of magnesium, by serving as membranes for ionic migration, or by entering into chemical reactions which produce dolomite. Certain clay minerals, particularly chlorites and vermiculites, contain significant amounts of magnesium and ion exchange could provide Mg^{2+} ions which might contribute to dolomitization. Kahle proposed that this exchange would be most likely during early diagenesis.

In most of the proposed dolomitization models, however, the role of upward moving compaction fluids has not been given proper attention (see Chilingarian and Wolf, 1976).

Indirect organic influence. The effect of organic matter, other than as a source of magnesium, warrants consideration. For example, Murata et al. (1972) pointed out that certain organic acids may decrease the degree of hydration of Mg^{2+} ions and thus promote the entry of magnesium into the calcite lattice. Based on values of δC^{13}, they classed as "light carbon dolomites" (δC^{13} of -31 to $-14‰$) those rich in C^{12}, derived from the oxidation of sedimentary organic matter, and as "heavy carbon dolomites" (δC^{13} of $+5$ to $+21‰$) those rich in heavy carbon produced through isotopic exchange between carbon dioxide and methane according to the following reaction:

$$C^{12}O_2 + C^{13}H_4 \rightarrow C^{13}O_2 + C^{12}H_4 \qquad (7\text{-}15)$$

At Point Fermin, Los Angeles, small dolomite crystals occurring with tar as cement in some Miocene sandstones probably resulted from the alteration of tar or other organic matter (Spotts and Silverman, 1966). Highly negative δC^{13} values for the dolomite indicate that the CO_3^{2-} was partially derived from oxidation of organic matter. The dolomite may have crystallized around oil or tar droplets and carbon dioxide bubbles that formed as discontinuous phases during the oxidation of the organic matter.

Davies and Ferguson (1975) reported the discovery of protodolomite in organic matter above carbonate sediments in a cay in the Great Barrier Reef. They concluded that the dolomite formation is enhanced in a very significant way by the high alkalinity produced by photosynthesis or decaying organic material.

In connection with his emphasis on the significance of alkalinity in dolomitization, Lippmann (1973, pp. 181—186) classed dolomites as "inorganic" and "organic" based on possible sources of CO_3^{2-}: (1) inorganic: alkali carbonate from the terrestrial realm and alteration of siliceous rocks such as basalt, and (2) organic: bacterial reduction of sulfate and the accompanying oxidation of organic matter according to the following equations or combinations thereof:

$$SO_4^{2-} + 2\,C + H_2O \rightarrow CO_3^{2-} + CO_2 + H_2S \qquad (7\text{-}16)$$

$$SO_4^{2-} + 4\,H_2 + CO_2 \rightarrow CO_3^{2-} + 3\,H_2O + H_2S \qquad (7\text{-}17)$$

where C and H_2 represent carbon and hydrogen, respectively, combined in appropriate organic compounds. Sulfate-reducing bacteria can utilize most organic compounds, which are introduced into the sediments. Such bacteria are ubiquitous and become active as soon as they find a suitable environment, i.e., anerobic conditions and the presence of sulfate plus suitable organic matter. Much of the H_2S produced will pass through the sediment—water interface by diffusion and as gas bubbles, but small amounts may be fixed as pyrite or remain in pores and produce fetid dolomite. A relative enrichment in light carbon should characterize dolomite derived from sulfate reduction (Lippmann, 1973, p. 185).

Friedman (1966) described dolomite laminae in lacustrine beds of Quaternary age in Salt Flat Graben, West Texas. The dolomite occurs with H_2S, iron sulfide, and native sulfur. According to Friedman, this dolomite owes its origin to the brines produced by evaporation around the shoreline sinking to the bottom of the lake; the associated gypsum was reduced by bacterial processes and the H_2S reduced iron oxide which blackened the dolomite. Although there is an association between bacterial activity, reducing conditions, and dolomitization, it should be stressed that Friedman did not attribute the dolomite formation to an organic effect. Dolomite formation related to reducing conditions has been reported from some sites of the Deep Sea Drilling Project (Davies and Supko, 1973, p. 386). At Site 138, Leg 14, 1000 km west of Cape Blanc, Africa, at the foot of the continental rise, dolomite occurs in Cretaceous pyrite-bearing silts alternating with sapropelites. Finally, ferroan dolomite is common and according to Peterson (1962, p. 27) the presence of ferrous iron probably signifies a reducing environment.

Detrital-diagenetic recrystallization model. Lindholm's model (1969), based on his studies of the Devonian Onondaga Limestone of New York, attempts to account for those dolomites showing evidence of subtidal accumulation. He suggested a detrital origin for such dolomites, the source being older dolomite or supratidal dolomitic sediments forming nearby (Fig. 7-32). Later diagenetic crystallization on the detrital nuclei and concomitant replacement of the $CaCO_3$ "groundmass" could produce the typical crystalline fabric characteristic of the great bulk of ancient dolomite. The occurrence of detrital dolomite in subtidal sediments is well documented (for example, by Pilkey and Noble, 1966). Evidence suggesting such a model would include correlation between dolomite and insoluble residue content and between the grain size of dolomite and that of quartz. Relict detrital cores of dolomite crystals might be present. Hopefully, one could demonstrate an appropriate paleogeography with an exposed source of dolomite.

"Dolomites and dolomites". There are many other processes and models that have been invoked to explain dolomitization, such as weathering and metamorphism, which are beyond the scope of this paper. In addition to the previously referred to dolomite occurrences in the ephemeral lakes near Coorong Lagoon (Australia) and Deep Spring Lake (California), Ca-Mg carbonates, including dolomite, have been described from lacustrine environments by numerous workers including Eardley (1938), Strakhov (1956), Friedman (1966), Müller et al. (1972), Ilich (1974), Wolfbauer and Surdam (1974), and others. Protodolomite has even been found in caves (Müller and Fischbeck, 1973; Thrailkill, 1968) and in the teeth of living echinoids (Schroeder et al., 1969).

In view of the many models and examples discussed and recalling the section on chemistry, one gets the impression that in a vein similar to Read's (1948) classic paper on granites, it is appropriate to speak of "dolomites and

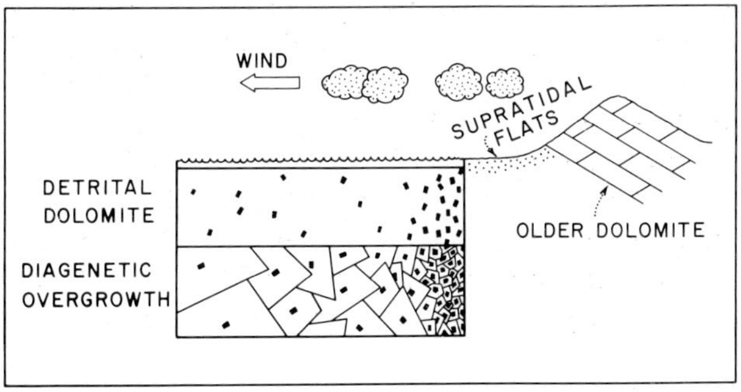

Fig. 7-32. Model for development of dolomite by combination of deposition of detrital dolomite grains followed by diagenetic recrystallization (and concomitant replacement of $CaCO_3$) around the detrital nuclei. (Modified after Lindholm, 1969, p. 1041.)

dolomites". Dolomite may form in a variety of ways in a wide range of environments; there is simply no one tidy model. Some mechanisms, however, are undoubtedly more responsible than others for major dolomitization. In fact, there is considerable disagreement concerning the relative significance of early versus late diagenetic and diagenetic versus epigenetic dolomitization.

Methods of introduction of Mg^{2+} ions into carbonate ooze

The present-day concentrations of Mg^{2+} and Ca^{2+} in the ocean are 1,297 and 408 p.p.m., respectively. Geologists, however, cannot be sure that the Mg^{2+} concentration of ancient seas was not higher at times than it is at present (Chilingar, 1956d, p. 2263).

Influx of Mg^{2+} ions into carbonate muds may occur as a result of reduction of the sulfate ion if Mg^{2+} is combined with SO_4^{2-} (?). According to Strakhov (1956a, p. 19), if the sulfate present in the mud, having a moisture content of 75%, undergoes complete reduction, it will give rise to 0.26% sulfur which can form pyrite in the sediment. Thus, the amount of pyrite in dry sediment should be present in an amount of about 0.5%. Actually, the normal content of pyrite in clayey marine sediments ranges from a small fraction of 1% to 1% (rarely higher). Thus, the sulfate ions used in the formation of pyrite probably come mostly from the initial (connate) water trapped in the sediment, and there seems to be no appreciable diffusion of sulfate ions from the bottom waters into the sediment. But together with SO_4^{2-} anions, Mg^{2+} and Ca^{2+} cations are also trapped in the mud. The amount of magnesium thus can be calculated as follows: at the complete reduction of sulfates in interstitial water, for 0.26% of sulfur there will be 0.13% magnesium (considering that only two-thirds of sulfur is combined with magnesium). This will constitute about one-third of the total magnesium contained in the interstitial water at the time of burial, which is equal to 0.4% of dry sediment.

If it is assumed that all of the magnesium produced as a result of the reduction of sulfates goes to form dolomite, there should be a decrease of magnesium content in the interstitial water in the amount of one-third as compared to its content in the bottom water. Strakhov (1956a, p. 19) suggested that this differential alone could not create a strong influx of magnesium from the bottom water into the mud. Thus, the deposition of 0.13% magnesium would give rise to only 1% of dolomite (i.e., 1% on dry-weight basis). Strakhov also pointed out that the amount of organic matter in carbonate muds he studied is so low, that it is doubtful that any appreciable reduction of sulfates occurred there.

Another reason for the influx of magnesium from bottom waters into sediment would be the decomposition of organic matter by bacteria with generation of CO_2. As a result of this process, first the alkalinity ($[HCO_3^-] + 2[CO_3^{2-}]$) would rapidly increase and then would decrease with loss of CO_2 from interstitial water of the sediment. With increasing alkalinity dolomite could reach a saturation value and precipitate. Removal of Mg^{2+} in this manner could result in its additional influx from bottom waters. This alone, believed Strakhov (1956a, p. 20), could not create any appreciable influx of Mg^{2+} from bottom waters because the amount of organic matter (generating CO_2) is negligible in the carbonate oozes. The process very likely would give rise only to: (1) individual crystals of dolomite, (2) rare and small spots (mottled dolomite), and (3) concretions. Yet, as pointed out by Strakhov, average dolomitization of spotty metasomatic dolomites reaches 30—70% and much more in some instances. It would appear, therefore, that to create such diagenetic dolomitization as is envisioned by geologists, the above process cannot of itself fulfil the requirements.

Possibly, as diagenetic replacement by dolomite progresses, the impoverishment of interstitial waters in magnesium content would enable additional influx of Mg^{2+} ions from bottom waters. Some porosity which results from secondary dolomitization could furnish additional space. This porosity developed during the change of calcite or aragonite to dolomite is possibly sufficient to provide necessary diffusion centers from which Mg^{2+} ions can migrate through the intergranular film (Bissell and Chilingar, 1958).

It is interesting to note here that interstitial water in the dolomite beds of Tertiary carbonates from Experimental Mohole (Guadalupe Island, Mexico) still contains the same Mg^{2+} levels as that of the modern sea (Rittenberg et al., 1963). Degens and Epstein (1964) suggested that possibly this may serve as an indication that sea water entrapped in the sediment is not a significant source for magnesium in the dolomite.

According to Fairbridge (1957), calcareous algae enriched in $MgCO_3$ are geochemically reorganized into low-Mg calcite and dolomite during diagenesis as exemplified by Funafuti dolomite. Possibly the magnesium carbonate present in solid solution in the skeletal and protective structures of organisms (Chilingar, 1962a) comprises one of the sources of Mg necessary for dolomitization. The fact that this $MgCO_3$ is thermodynamically unstable is partly supported by the findings of Chave (1954), Lowenstam (1961), Chave et al. (1962), Chilingar (1962b) and Seibold (1962). Epstein et al. (1964), however, found that the reorganization into calcite and dolomite does not seem to have any significant effect on isotope ratios of Funafuti dolomite; the values of the dolomites are close to those originally fixed in the calcium carbonate.

Jodry (1969, p. 957) pointed out that the process of refluxion has been generally accepted as a mechanism for dolomitization and occlusion of porosity; however, a process opposite to refluxion could result in both dolomitization and compaction of the carbonate sediments. Jodry investigated the stage-by-stage evolution of the growth of Silurian reef complexes in Michigan, as shown in Fig. 7-33 (initial stage) and Fig. 7-34 (final stage), with three intermediate stages presented in his figs. 7, 8, and 9. Compaction of the sediments surrounding the resistant reef framework per se was extensive and resulted in porosity loss, whereas the compaction-resistant framework

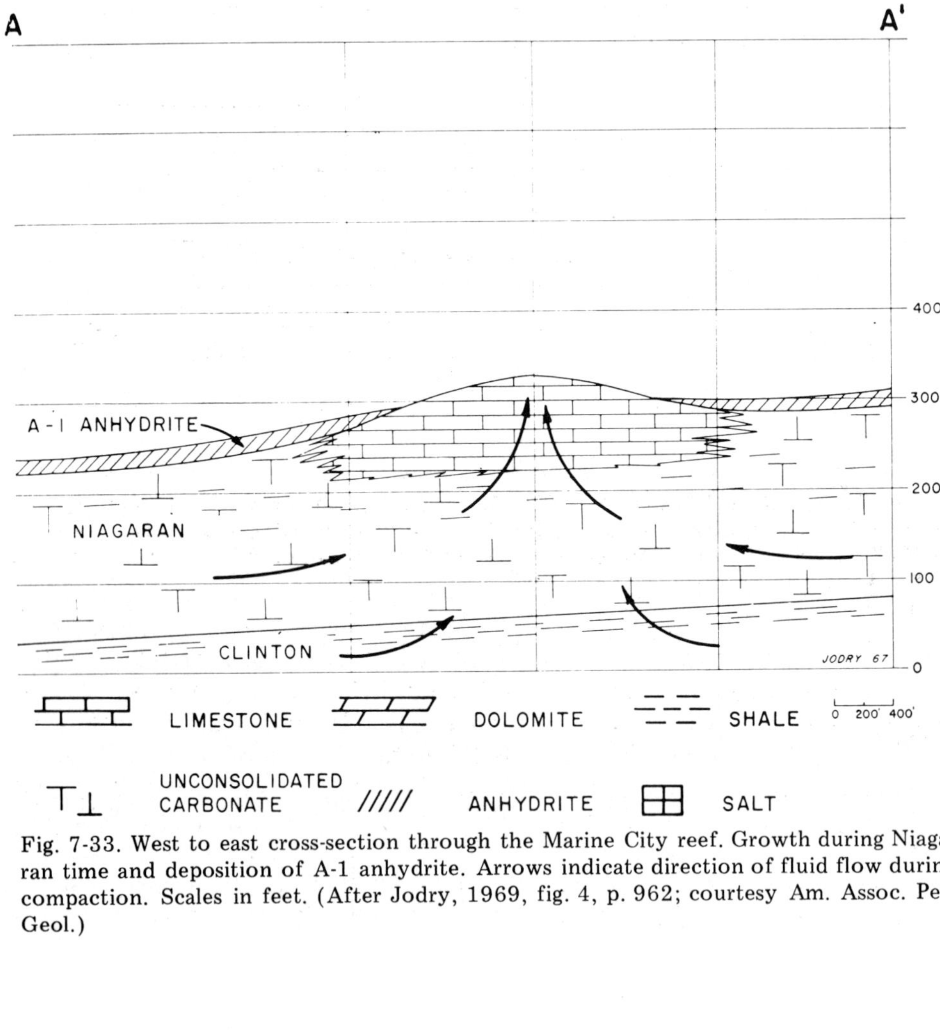

Fig. 7-33. West to east cross-section through the Marine City reef. Growth during Niagaran time and deposition of A-1 anhydrite. Arrows indicate direction of fluid flow during compaction. Scales in feet. (After Jodry, 1969, fig. 4, p. 962; courtesy Am. Assoc. Pet. Geol.)

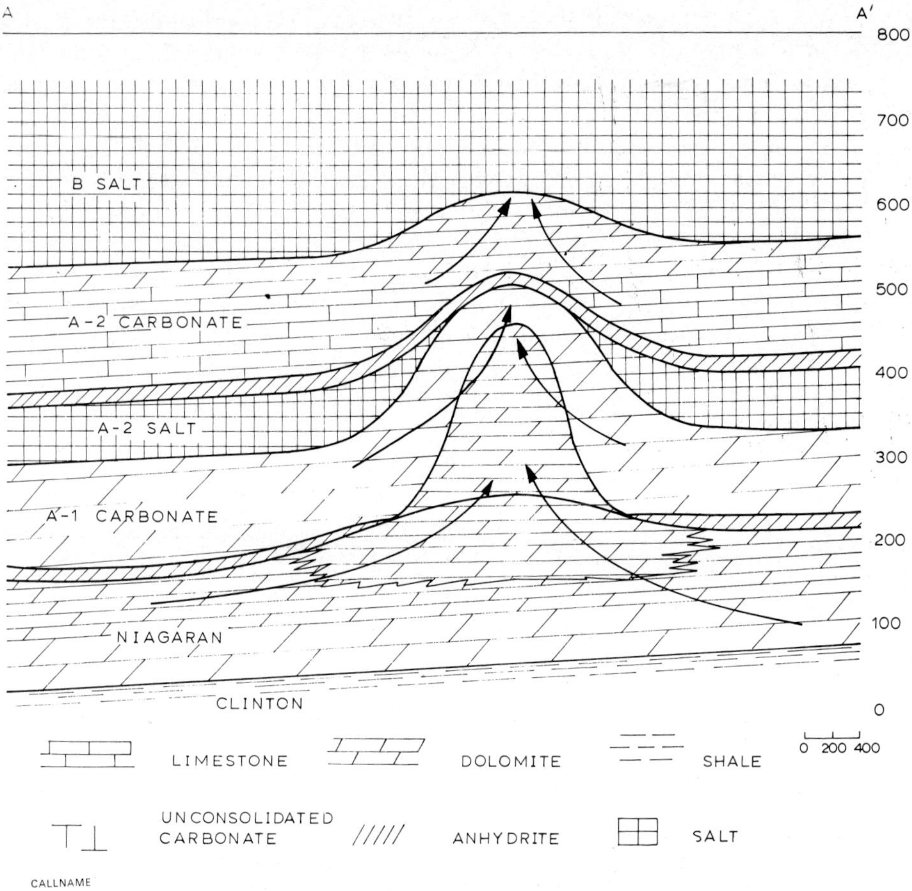

Fig. 7-34. West to east cross-section through Marine City reef. B-salt deposition and final compaction and dolomitization of underlying Silurian carbonates. Arrows indicate direction of fluid flow during compaction. Scales in feet. (After Jodry, 1969, fig. 10, p. 968; courtesy Am. Assoc. Pet. Geol.)

preserved the porosity. The amount of compaction was a function of the relative proportions of the reef framework and surrounding carbonate and other sediments. Both of these lithologies are dolomitized. The pattern of dolomitization (1), the quantity of dolomite in relation to evaporite beds (which supplied the overburden pressure for compaction of the fine-grained sediments) (2), the position of the dolomite in relation to the evaporite beds (3), and the time of dolomitization (4), all indicate that dolomitization was the result of passage of connate solutions (in this case compaction fluids) through the rocks. The refluxion process is not a likely mechanism in this instance because the evaporite sequence overlying the carbonate muds susceptible to compaction prevented vertical escape of compaction fluids. The fluids must have passed laterally to and through the reefs and associated sediments, supplying the Mg^{2+} cation to the relatively small volume of limestone that underwent dolomitization. The outflow of these solutions continued during the cyclic evaporite deposition. This prevented the penetration of the brines into the limestones, which, if it had occurred, would have led to the precipitation of cement and destruction of porosity and permeability.

The compaction actually caused differential movement above and around the reefs (Jodry, 1969, p. 968). Up to 50—75% compaction of the fine-grained basinal (off-reef) sediments took place, as indicated by compaction features around nodules (Jodry, p. 973). Jodry described (pp. 973—974) compaction fluid flow channels, which are clearly preserved in three dimensions and outlined by secondary internal sedimentary accumulations. The dolomitization accomplished by the connate fluids within the reefs increased porosity and the latter has been preserved because of the compaction-resistant nature of the lithified framework. In contrast, the dolomitization of the sediments surrounding the reefs was syn-lithificational in nature with only a minor production of porosity. The compaction of this finer-grained material at the same time caused a reduction in thickness of the beds to form the stratigraphic relationships as seen today.

Lippman (1973) discussed a number of aspects of carbonate diagenesis and dolomitization in which he referred to compaction. For a detailed discussion on the chemistry of compaction fluids and their movement, the reader is referred to the books by Rieke and Chilingarian (1974) and Chilingarian and Wolf (1976).

DIAGENETIC VERSUS EPIGENETIC DOLOMITIZATION

General

Undoubtedly one of the major difficulties in classifying ancient dolomites involves dolomitization that is initiated early but which continues into later stages. It was mentioned that some protodolomites retain their penecontemporaneous earmarks; however, earlier stages of dolomitization no doubt are commonly masked by subsequent diagenesis such as recrystallization. In such instances, one is classifying the latest stage of dolomitization or epigenesis.

Diagenetic dolomitization may include two major stages: (1) early diagenetic and (2) late diagenetic. In the case of early diagenetic dolomitization, due to subsequent compaction the porosity resulting from dolomitization is appreciably lowered (or disappears completely). Teodorovich (1958, p. 306) observed that in marine replacement dolomites, the extremely fine-grained (0.01—0.05 mm) dolomite forming the relics of calcareous skeletons, with original pelitomorphic (<0.005 mm) or micrograined ($\leqslant 0.01$ mm) texture, apparently formed at the same time as the central cores of dolomite grains in the main rock mass. The major part of dolomite grains (that is, crystals) formed during the second phase of dolomitization.

Teodorovich also observed dolomite crystals showing three-stage formation. The peripheral, clearer rims apparently belong to the third stage of dolomitization. In some very fine-grained (0.01—0.1 mm), fine-grained (0.1—0.25 mm), and vary-grained dolomites (0.25—0.5 mm = medium-grained; 0.5—1.0 mm = coarse-grained; >1.0 mm = very coarse-grained) one can observe not only transparent, colorless (in thin-section), rhombohedral grains with sharp contours but also cloudy crystals with irregular contours, often having zonal structure. Possibly, the latter crystals formed earlier in more recent sediments, whereas the former crystals developed later in muds which started to lithify. Dolomite selectively replaces fine-grained $CaCO_3$, but coarse-textured patches of $CaCO_3$ remain relatively unaffected. The order of dolomitization in organisms appears to be as follows: (1) Foraminifera, (2) brachiopods, (3) corals, and (4) crinoids (Chilingar, 1956e). In diagenetic dolomites the c-axes of dolorhombs do not appear to lie within the plane of bedding, and exhibit random orientation (Hohlt, 1948; Chilingar and Terry, 1954).

The epigenetic dolomites result from alteration of completely lithified limestones by downward percolating meteoric solutions or rising hydrothermal solutions (Vishnyakov, 1951). Epigenetic dolomites are cavernous, have obscure stratification, patchy distribution, non-uniform grain size, and relict structure. Idiomorphic rhombohedrons of dolomite with nucleus and zonation are common. The texture is not uniform and the original fauna remains in the form of molds. The Ca/Mg ratio of epigenetic dolomites varies widely over short distances, both vertically and horizontally. It should be remembered, however, that many geologists call epigenetic dolomites "late diagenetic".

According to Strakhov (1956b, p. 199), the apparent increase in degree of dolomitization due to secondary removal of $CaCO_3$ (an epigenetic process) does not seem to be important. For example, in 1 dm^3 of limestone (5% dolomitized; s.g. = 2.7) there is 13.5 g of dolomite. Creation of 10% porosity would necessitate the removal of 27 g of $CaCO_3$ from 270 g of rock; and consequently the percentage dolomitization of the remaining mass would be: $13.5/270-27 \times 100 = 5.5\%$. In other words, it will increase by a negligible amount of only 0.5%. An increase of porosity to 20% would give an apparent increment in degree of dolomitization of 1.2%; to 30% of 2.1%; to 40% of 3.3%; and to 50% of 5%. In the latter case, the rock would have a sponge-like appearance, and yet the apparent degree of dolomitization would increase by 5% only. The above calculations should be kept in mind, because secondary removal of $CaCO_3$ is frequently used to explain high degree of dolomitization.

A photomicrograph of a sample of completely dolomitized Carboniferous limestone, Britain (Fig. 7-35), illustrates a porphyrotopic fabric, the larger crystals showing central zoning and clear rims. Possibly several generations of dolomitization are represented. On the other hand, the successive zones in these larger crystals may have formed within a relatively short period of time.

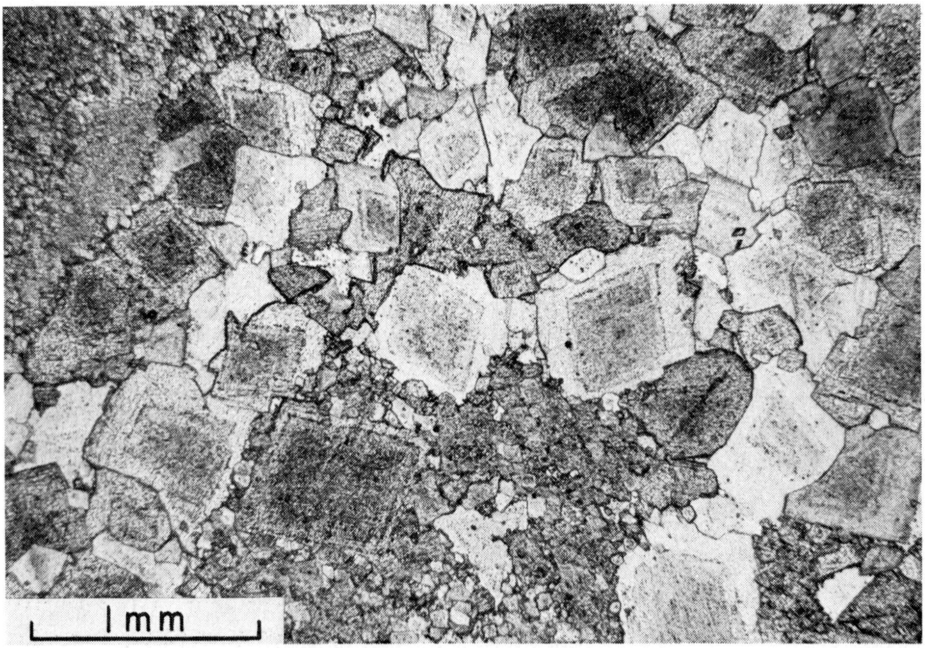

Fig. 7-35. Photomicrograph (plane polarized light) of thin-section of porphyrotopic late diagenetic dolomite containing zoned dolomite crystals; Carboniferous Limestone, Cloud Hill Quarry near Breedon, Leicestershire, England. Inner parts of zoned crystals may be of same age as smaller crystals, whereas clearer rims may be considerably younger. (Photograph by D.H. Zenger.)

Distinguishing characteristics

Field criteria. It has generally been accepted that early diagenetic dolomitization should be quite uniform over a wide area (Parsons, 1918, p. 248), although this "principle" may assume less significance in view of the rapid facies changes in modern carbonate sediments. For example, modern peritidal dolomite shows rapid lateral variations. Fairbridge (1957, p. 158) believed that a persistence of intercalated laminae of limestone and dolomite over hundreds of miles would be characteristic of early diagenetic dolomitization. It seems logical to suspect that such dolomitization would most likely follow primary bedding features rather than replace sediments pervasively. The same might be said for selective dolomitization focused around primary sedimentary structures such as mud cracks (Laporte, 1967; Shinn, 1968) and burrows. Such evidence is suggestive, but not unequivocal. For example, Gebelein and Hoffman (1973, p. 611) recognized that decomposition of algal sheath material and release of Mg^{2+} may occur at an epigenetic stage (postlithification) to produce the interlaminated limestone—dolomite sequences they described. Although epigenetic dolomitization may be less controlled by the primary structures and fabrics, these could act as loci for late-stage replacement.

Fairbridge (1957) took limestones with dolomitic mottling to represent arrested and incomplete dolomitization during diagenesis; however, this relationship is only suggestive. Areas of limestone that differ from the main limestone in terms of porosity, texture, or mineralogy could be dolomitized during epigenesis even though the initial mottling in the limestone may have been diagenetic, possibly penecontemporaneous. Such examples are mentioned and illustrated by Osmond (1956, p. 38).

Cross-cutting relationships have been considered characteristic of epigenetic dolomites. For example, dolomite "fronts" oriented at steep angles relative to the bedding (Figs. 7-36, 7-37a, 7-37b) occur in the Basin and Range Province, U.S.A., where there are common epigenetic dolomites, particularly hydrothermal and postorogenic varieties. In Leicestershire, England, such as at the Cloud Hill quarries, epigenetic dolomitization has occurred at varying distances beneath the angular unconformity with the overlying Triassic (Fig. 7-38), from which dolomitizing waters have percolated into the Carboniferous Limestone (Parsons, 1918, p. 225; King, 1968, p. 129). There is the possibility of developing such a relationship during early diagenesis, however, as a result of seepage refluxion as modeled by Deffeyes et al. (1965). Epigenetic dolomite, particularly the more easily distinguished postorogenic type, is commonly localized along fractures. Hewett (1931) showed several such examples in the Goodsprings, Nevada, area, one being reproduced in Fig. 7-39. At Manystones quarry in Derbyshire, England, some spectacular

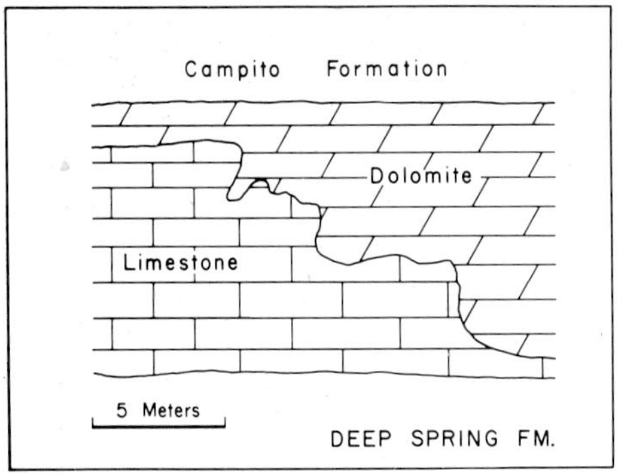

Fig. 7-36. Epigenetic dolomitization along an irregular front in upper part of Deep Spring Formation (Cambrian?), White Mountains, east-central California. (After Bateman, 1965, p. 14.)

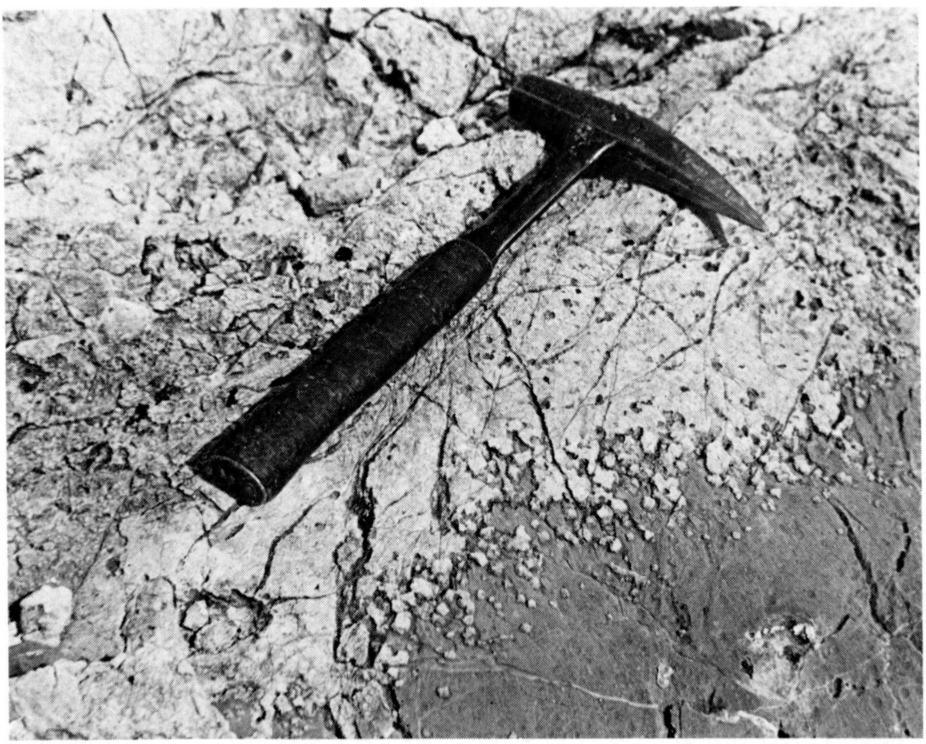

Fig. 7-37a. Dolomite "front" as viewed on bedding plane in Monte Cristo Limestone (Mississippian), Whitney Ridge, Virgin Mountains, Nevada. Note dark-weathering limestone (lower right part) and light dolomite with dolomite blebs weathering out in relief in "advance" of main front. (Photograph, courtesy of John S. Shelton.)

Fig. 7-37b. Dolomitization "front" at high angle to bedding, Lost Burro Formation (Middle and Upper Devonian), Funeral Mountains, east-central California. Abrupt, though irregular, contact separates limestone (dark, left) from epigenetic dolomite (light, right). (Photograph by D.H. Zenger.)

epigenetic dolomitization has occurred in the Carboniferous Limestone. At the top of the quarry, the rocks are largely composed of dolomite, but the dolomite tongues out downward along fractures oriented perpendicular to the bedding (Fig. 7-40). Although erosion has removed any rocks of younger age, T.D. Ford (1968, oral communication) envisioned an original covering of Permian evaporitic sediments from which percolating meteoric waters, descending along fractures, gave rise to the epigenetic dolomitization (Zenger's personal observations). Fracturing, however, is not necessarily a function of time since the formation of sediments that are transected. Marine or peritidal carbonates that are subjected to meteoric processes soon after deposition, owing to uplift or eustatic drop in sea level, are likely to be lithified and fractured relatively early in their history. In such instances, as mentioned previously, diagenesis is of short duration.

Nichols (1974) described secondarily dolomitized carbonates beneath theorized supratidal dolomite of the Panther Canyon Member of the Augusta Mountain Formation (Triassic) in north-central Nevada. Evidence for the

Fig. 7-38 Late diagenetic dolomitization, Cloud Hill Quarry near Breedon, Leicestershire, England. Dolomitization has occurred beneath angular unconformity separating Triassic clastics (dark) from underlying more massive Carboniferous Limestone. (Triassic clastics are 1—2 m in thickness). (Photograph by D.H. Zenger.)

secondary dolomitization consists of a saccharoidal texture and a pattern of dolomitization which cuts across the bedding of unaltered calcareous rocks. She considered it likely that the secondary dolomite is the product of dolo-

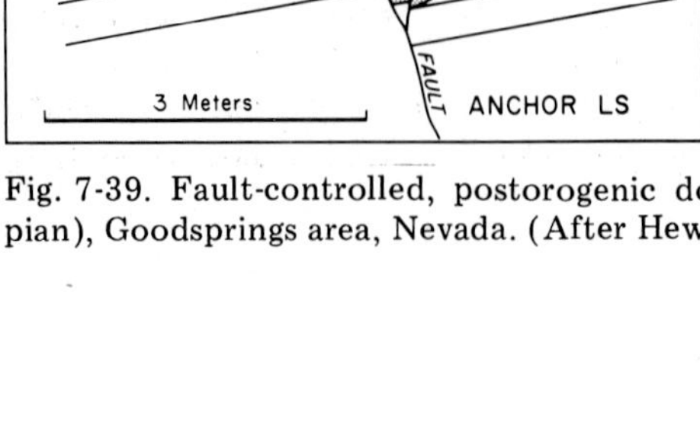

Fig. 7-39. Fault-controlled, postorogenic dolomitization in Anchor Limestone (Missisippian), Goodsprings area, Nevada. (After Hewett, 1931, p. 58.)

Fig. 7-40. Late diagenetic dolomitization along joints in Carboniferous Limestone at Manystones Quarry, Derbyshire, England (see text for explanation). Dolomite weathers darker (one dolomitic zone demarcated by arrows). (Photograph by D.H. Zenger.)

mitization of limestones by solutions associated with the supratidal surface on which penecontemporaneous dolomite of the Panther Canyon was deposited.

On a larger scale, it can be shown that dolomitization can be related to orogenic belts and consequently would be considered as postorogenic. Miller (1966, p. 21f) has stressed the significance of postorogenic dolomite: "... while some of the dolomite related to evaporite and subaerial deposition may be early diagenetic, a great deal formed later in the history of the rock during weathering and structural deformation" and "...I think I can cite a dozen good examples in the Rocky Mountains alone, where secondary dolomitization if mapped carefully will define a structure or reveal the extent of weathering that may be important in creating or enhancing permeability." Miller further stressed the severity of diagenetic alteration, including dolomitization, in the midcontinent area. He showed a relationship between dolomite occurrence and various structural features in the Rocky Mountains, such as the overthrust belt, anticlines, domes, faults, and the Mississippian—Triassic unconformity.

The more obliterative the dolomitization, the greater the chance it has occurred during, or at least continued into, epigenesis. Unfossiliferous dolomite, as contrasted with associated fossiliferous limestone, can frequently be observed in the field. Zenger (personal observation) has noted intercalated limestone and dolomite beds in the Sconondoa Member of the Lockport Formation (Zenger, 1965b) in which the limestones have well preserved specimens of the brachiopod *Howellella* and the ostracode *Herrmannina*, whereas the adjacent dolomites, on first glance, appear unfossiliferous. On close examination, however, the dolomites are observed to have relics of the same fossils. If fossils are completely absent in a dolomite, then one cannot be certain that total obliteration has occurred because an absence of organisms may represent conditions unfavorable to life (e.g., high salinity) at the time of penecontemporaneous or even primary dolomitization.

Microscopic criteria. Optically, Zenger has observed varying degrees of destruction of fossils and structures (oöids and peloids) in completely dolomitized sequences in the Lockport Formation, Herkimer Formation, and the British Carboniferous Limestone, England (Figs. 7-41 to 7-44). The examples of nearly complete obliteration of the ostracode *Eukloedenella* in Fig. 7-41, of the bryozoan in Fig. 7-43, and of the bryozoan(?) in Fig. 7-44 suggest epigenetic dolomitization, at least at the final stage; there is no microscopic evidence indicating when initial dolomitization commenced. Again, it may be emphasized that early lithification may restrict diagenesis to a short geologic time. Zenger (personal observation) has noticed a loss of structure in echinoid grains (composed initially of metastable, high-magnesium calcite) in Bermudan Pleistocene limestones. These are essentially lithified rocks, despite considerable porosity, owing to the diagenesis that has occurred in just a few tens of thousands of years, mainly through vadose and phreatic processes. The better preserved bryozoan from the Herkimer Formation, shown in Fig. 7-42, may well represent diagenetic dolomitization which was not masked completely as a result of later epigenetic processes.

Crystal size can be an aid in the distinction between early and late diagenetic dolomitization and between diagenesis and epigenesis. Generally, the coarser the size, the later the dolomitization, particularly if the sample consists completely of dolomite. In the figures illustrating replacement, the dolomite crystal size is relatively large, mostly >100 μm. Unequivocal epigenetic dolomites are always coarse-grained (Figs. 7-45 and 7-46), up to millimeter-sized. Modern penecontemporaneous dolomites have a grain size on the order of a few microns and crystals in proposed analogs from the geologic record range from that size up to 30 or 40 μm. Although some very fine-grained penecontemporaneous dolomite may remain relatively unaltered, others may be coarsened by subsequent later diagenetic and epigenetic

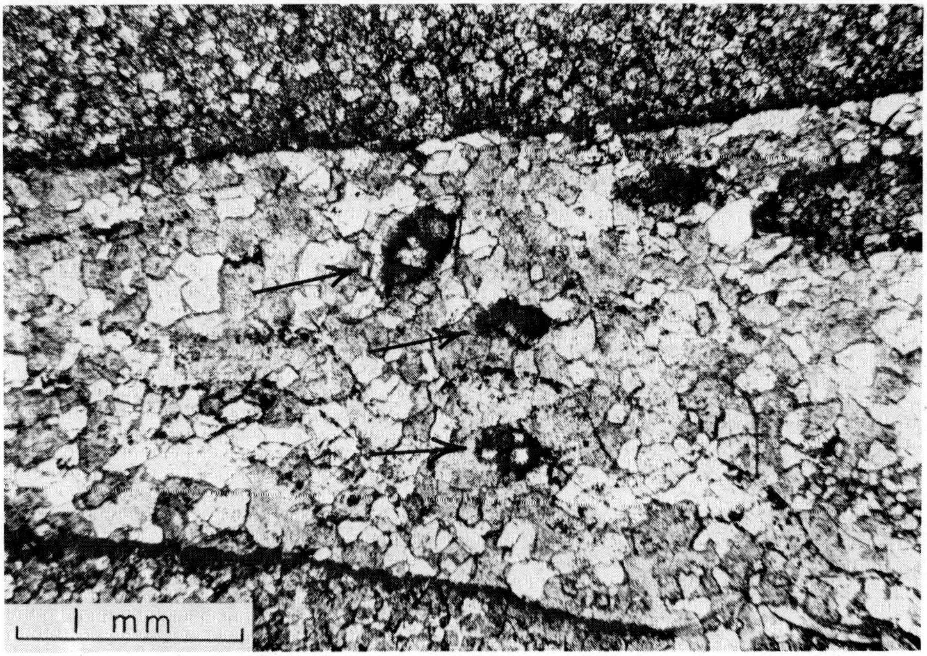

Fig. 7-41. Photomicrograph (plane polarized light) of thin-section of intraclastic dolomite, Lockport Formation (Middle Silurian, New York), showing near obliteration of ostracod tests (*Eukloedenella*, indicated by arrows) by more pervasive, epigenetic dolomitization than that shown in Fig. 7-5. (After Zenger, 1972, fig. 2; courtesy J. Geol. Educ.)

dolomitization and neomorphism. Also, coarse crystal size may be produced relatively early in diagenesis; fine grain size is not without exception in modern carbonates. This is especially true where marine sediments are exposed to meteoric processes very soon after formation. For example, in some Bermudan Pleistocene limestones, calcite spar attains diameters up to a few hundreds of μm (Zenger, personal observations) and the coarser variety of Land's "schizohaline" dolomite in Jamaica consists of "large sparry crystals" (Land, 1973a, pp. 73--74). In both of these cases, however, lithification occurs early and the phrase "early in diagenesis" really is more meaningful with respect to absolute time; such dolomitization may be nearly epigenetic by definition. Deffeyes et al. (1965) reported replacement dolomite in the Plio-Pleistocene of Bonaire of 75 μm diameter; however, these beds are now lithified and despite its "youth", the dolomite is probably epigenetic.

In studying the possibility of preferred orientation of dolomite crystals, Sass (1969) applied an optical procedure making use of the principle that more light is transmitted through the "c" crystallographic axis. He noted a

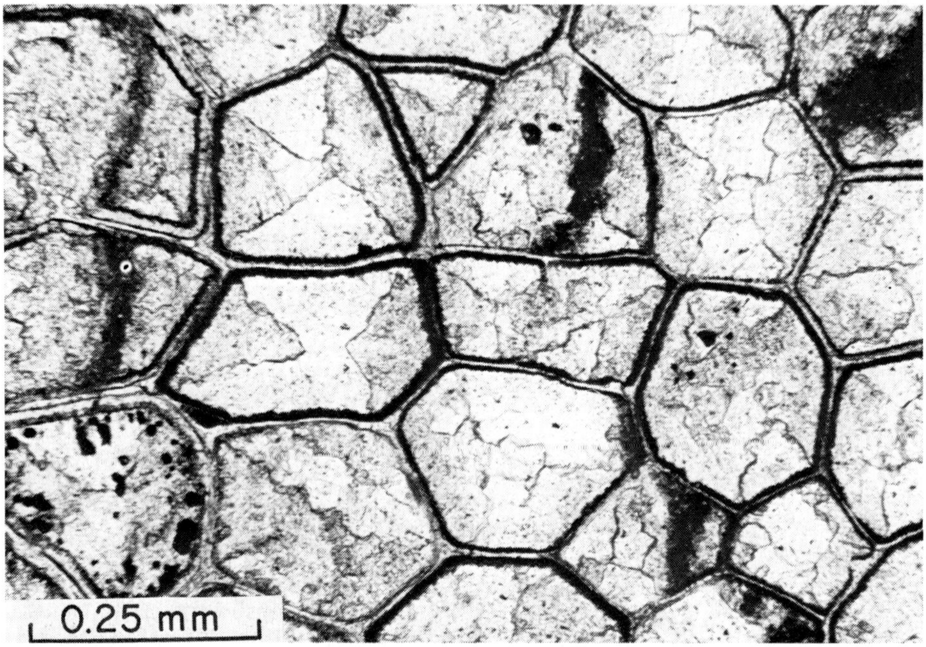

Fig. 7-42. Photomicrograph (ordinary light) of thin-section of well-preserved, though completely dolomitized, bryozoan Joslin Hill Member, Herkimer Formation (Middle Silurian), central New York. Dolomite crystals in walls discrete from coarser replacement(?) dolospar within zooecia. (Photograph by D.H. Zenger.)

preferred orientation of the crystals, the "c" axis being preferentially oriented perpendicular to the bedding. Sass concluded that the dolomite euhedra grew by replacement during very earlier diagenesis in calcareous mud with the stable position in the soft medium being that with the maximum cross section of the crystals oriented horizontally. (It may be worth mentioning that his petrographic study of dolomite crystals was two-dimensional; for unequivocal results it would seem that three-dimensional analysis by universal stage would be necessary.)

Katz (1971, pp. 44—45) believed that zoned dolomite crystals of the Mahmal Formation (Jurassic) of Israel formed by replacement during the early diagenetic stage because of the following evidence:

(1) Dolomite crystals replace skeletal material.

(2) Where dolomite crystals abut against neighboring crystals or grains, the dolomite exhibits anhedral boundaries as though the dolomite was unable to grow against these hard grains. Within the dolomite, however, the euhedral zoning implies growth within a soft matrix which permitted euhe-

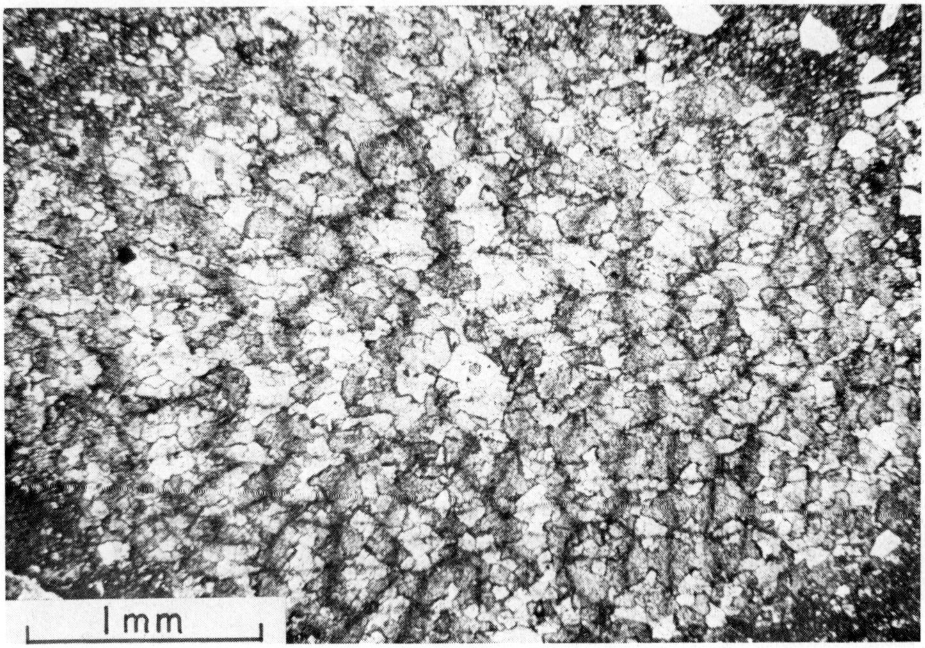

Fig. 7-43. Photomicrograph (plane polarized light) of thin-section of nearly completely obliterated bryozoan zoarium, Joslin Hill Member, Herkimer Formation (Middle Silurian), central New York. Dolomitization interpreted as epigenetic. (Photograph by D.H. Zenger.)

dral development until mutual interference resulted in compromise boundaries.

(3) Larger, more euhedral dolomite crystals protrude from anhedral masses into irregular voids, which were supported by the plastic, slowly lithifying sediment.

Certainly evidences (1) and (3) may be questioned as to their being diagnostic of an early diagenetic origin, and the present writers wish to point out that the occurrence of euhedral dolomite rhombs in calcareous mudstone need not necessarily be indicative of early growth despite many generalizations to this effect. It is possible that dolomite euhedra could replace calcareous rocks as well as sediments providing, of course, there was accessibility for the dolomitizing waters. Although the situation is not entirely similar, porphyroblasts do grow in metamorphic rocks!

Dolomite focused along microfractures and microstylolites, or truncating such structures, implies epigenesis. Figures 7-46 through 7-50 are photomicrographs illustrating such relationships. Freeman (1966) referred to the con-

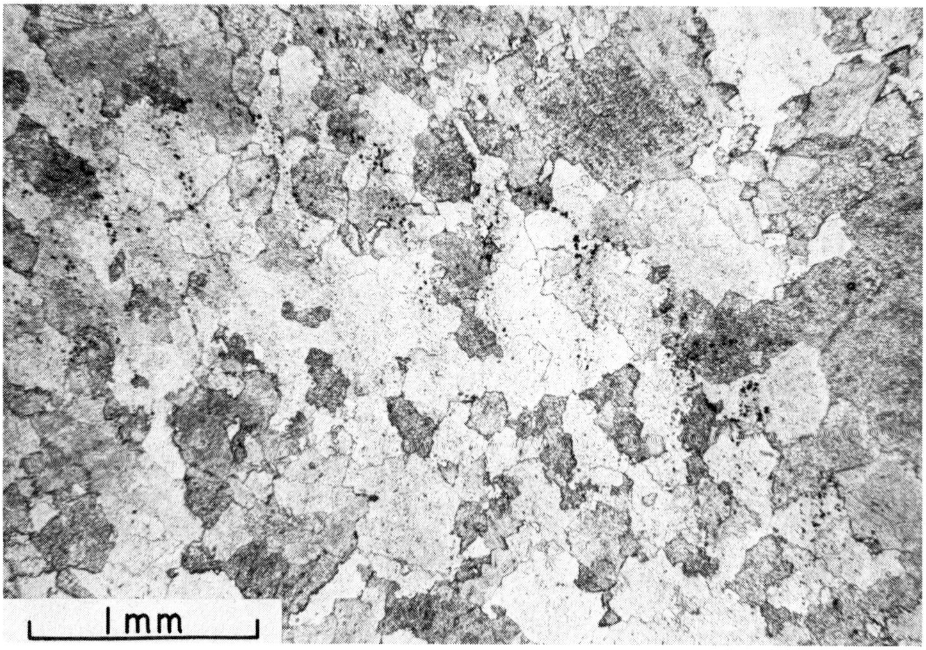

Fig. 7-44. Photomicrograph (plane polarized light) of thin-section of relic bryozoan(?), as outlined by small pyrite crystals (black), in completely dolomitized (epigenetic) Carboniferous Limestone, Cloud Hill Quarry near Breedon, Leicestershire, England. (Photograph by D.H. Zenger.)

centration of dolomite crystals along microstylolites (not as insoluble residue) as partial evidence for epigenetic dolomitization. Microstylolites probably represent deep burial. Microfractures, although indicating rigidity, could form under shallow burial conditions. In a similar way, dolomite crystals concentrated in, or especially floating in, secondary void fillings are suggestive of epigenetic replacement.

Choquette (1971) described late ferroan dolomite cement in the Ste. Genevieve Formation (Mississippian) of the Illinois Basin, and deduced its epigenetic origin using both petrologic and isotopic data (this latter aspect will be discussed below). The ferroan dolomite, occurring throughout whole crystals or as zones, postdated fracturing and brecciation. The zonation resulted from periodic occurrence of iron enrichment as the crystals grew into voids. Crystals of ferroan dolomite are larger than iron-free crystals and possess curved faces and a pronounced undulatory extinction. Choquette (p. 345) suggested that these features be considered as possible general characteristics of epigenetic cement. Beales (1971) described white sparry dolo-

Fig. 7-45. Photomicrograph (plane polarized light) of thin-section of very coarse-grained dolomite, with sutured crystal boundaries, in Wyman Formation (late Precambrian), northeastern Inyo Mountains. Dolomitization interpreted as hydrothermal. (Photograph by D.H. Zenger.)

mite cement of late diagenetic origin in the Middle Devonian Presqu'ile dolomites of the Northwest Territories; some of these crystals attain maximum dimensions of a centimeter. The "zebra" dolomite of the Leadville Limestone (Mississippian) of Colorado and banded, coarse, white dolomite in the Lost Burro Formation in the Nopah Range of east-central California (Zenger, unpublished observations) are both epigenetic, although their exact origin is less obvious than the very definite void-filling spar of Beales. The dolomite spar described by Choquette and Beales resembles the milky colored, saddle-shaped vug linings of epigenetic origin in the Lockport and Little Falls Formation of New York (Zenger, unpublished observations). Also, Zenger (1972c) has recognized ferroan dolomite, as whole crystals and as rims around non-ferroan dolomite crystals, in the Little Falls Formation. The ferroan dolomite formed relatively late in paragenesis and its growth into voids is suggestive of epigenetic development. Undulatory extinction is characteristic of the coarse, void-filling dolospar. Iron content alone, however, is un-

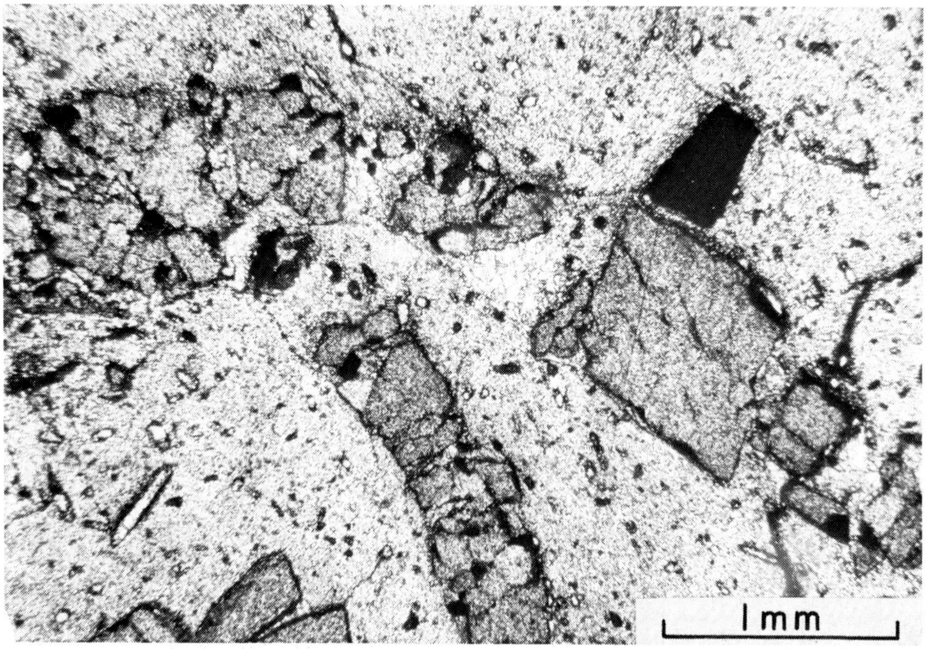

Fig. 7-46. Photomicrograph of acetate peel of epigenetic (postorogenic) dolomitization following secondary fractures in dolomitic limestone sample of Sultan Limestone (Devonian), Mountain Pass, California. Sample collected near fault, toward which dolomitization is more complete. Note coarse crystals of dolomite (darker) contrasted with lighter calcite (specimen provided by S.R. Bereskin; photograph by D.H. Zenger.)

reliable in attempting to establish an epigenetic origin. Very ferroan dolomite of an earlier replacement type is ubiquitous within the Middle Silurian Herkimer Formation, which is associated with the famous Clinton hematitic iron ores. Parsons (1918, p. 249) concluded that hematitic centers and zones in dolomite rhombs in the Carboniferous Limestone of Leicestershire, England, were "subsequent" (i.e., epigenetic). Parsons felt that the hematite was introduced at the time the dolomitization occurred and the iron [as well as the dolomite] was provided by waters percolating through the overlying reddish Triassic. Katz (1971, pp. 48—49) believed that the hematite zones in the Mahmal dolomite crystals resulted from oxidation of the ferroan dolomite zones as the crystal grew, but he stressed that growth occurred during early diagenesis.

$^{18}O/^{16}O$ and $^{13}C/^{12}C$ ratios. It has been noted by Degens and Epstein (in Bissell and Chilingar, 1962), that in the case of co-genetic precipitation, the cal-

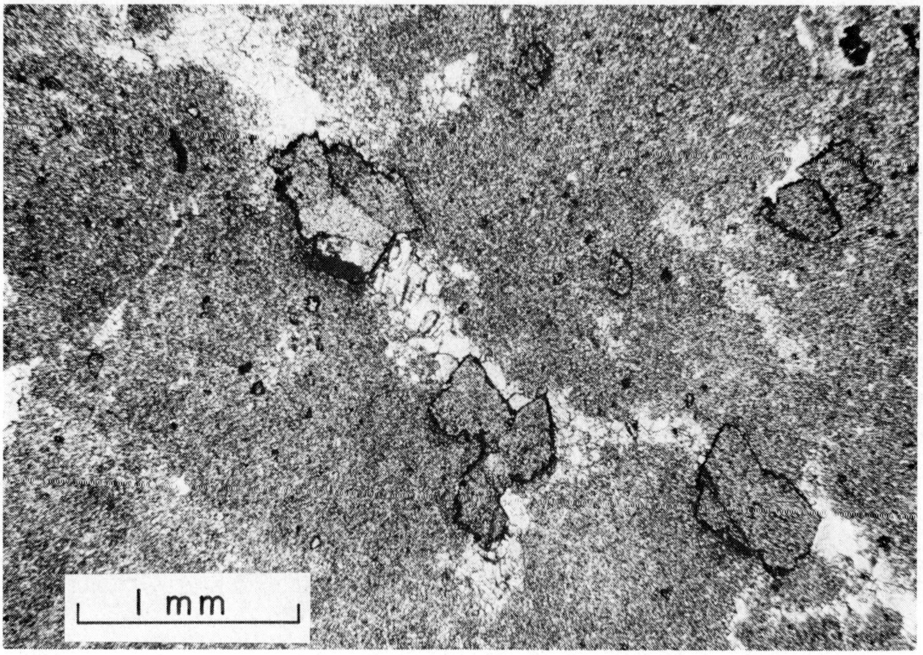

Fig. 7-47. Photomicrograph of acetate peel of epigenetic dolomitization in Lower Ordovician Gailor Dolomite, Saratoga Springs, New York. Dolomite rhombs (dark) centered in fracture-filling spar (light) as contrasted with micrite (dark). (Photograph by D.H. Zenger.)

cites and dolomites should be different by 8‰ in their $^{18}O/^{16}O$ ratio. Thus, if the $\delta^{18}O$ of dolomite is, for example, only 1‰ heavier than that of calcite which occurs together with it (evaporite type dolomite in the salt flats of western Utah; Bissell and Chilingar, 1962, p. 207), then the dolomite is probably of diagenetic origin.

In a comprehensive study of coexisting dolomites and calcites, Degens and Epstein (1964) showed that dolomites from Recent and unconsolidated marine sediments have $\delta^{18}O$ values of —0.8 to +4.9 (+2.1 average) and $\delta^{13}C$ values of —1.2 to +1.4 (—0.2 average). The associated calcites have $\delta^{18}O$ of —0.4 to +4.6 (+1.8 average) and $\delta^{13}C$ of —0.8 to +1.2 (—0.4 average). [1] Thus, the coexisting calcites and dolomites are quite similar isotopically. Consequently, Degens and Epstein concluded that recent dolomites are not precipitated directly out of aqueous solutions, but are formed as a result of meta-

[1] The data are reported as per mil deviation to the PDB Chicago Belemnite Standard (Craig, 1957).

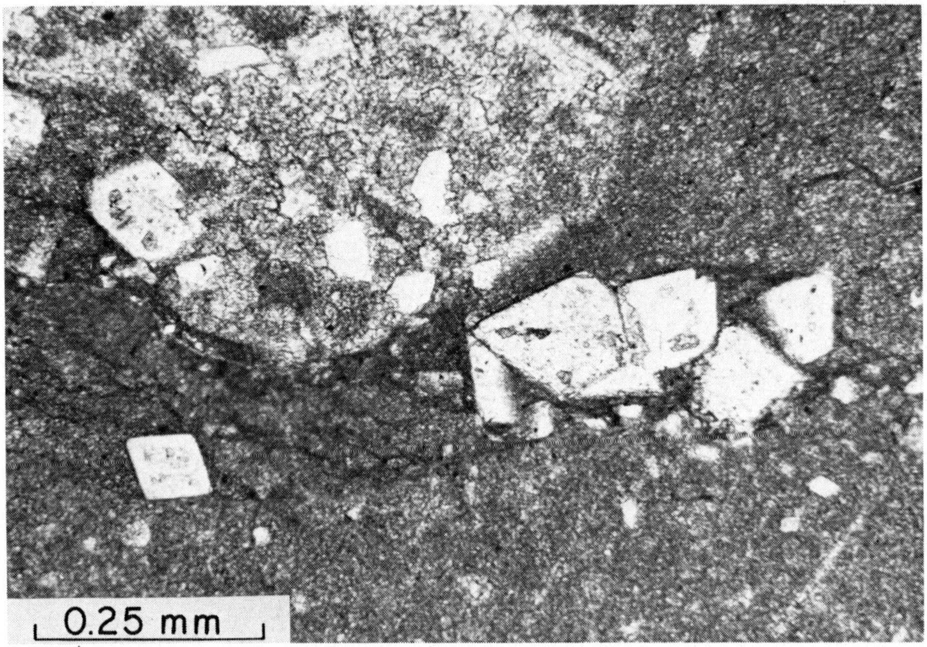

Fig. 7-48. Photomicrograph (plane polarized light) of thin-section of dolomitic, quartzose, intraclastic limestone of Tribes Hill Formation (Lower Ordovician), St. Johnsville, New York, showing relationship of dolomite crystals to microstylolite. Dolomite (light, generally rhombic) apparently transecting microstylolites, or at least focused along them, *not* as insoluble residue; dolomitization interpreted as epigenetic. (Photograph by D.H. Zenger.)

somatism from pre-existing crystalline calcite carbonate. They also stated that geologic and ^{14}C data indicate that dolomitization can start shortly after deposition; and that dolomitization has to proceed under solid-state conditions by a mole for mole exchange of Ca^{2+} for Mg^{2+}, without chemically altering the CO_3^{2-} unit, because no isotope fractionation (either for C or O) occurs during the alteration of the $CaCO_3$ precursor material to dolomite. They ruled out the participation of bicarbonate phase during dolomite formation.

Epstein et al. (1964) demonstrated the reluctance of dolomite to adjust isotopically to changes in temperature and $^{18}O/^{16}O$ ratio of formation waters. This presents a possibility of using dolomites for the evaluation of paleotemperatures of ancient seas. It was also demonstrated that $\delta^{13}C$ apparently is not significantly affected by the dolomitization process (Degens and Epstein, 1964).

There are disagreements, inconsistencies, and uncertainties, however, in

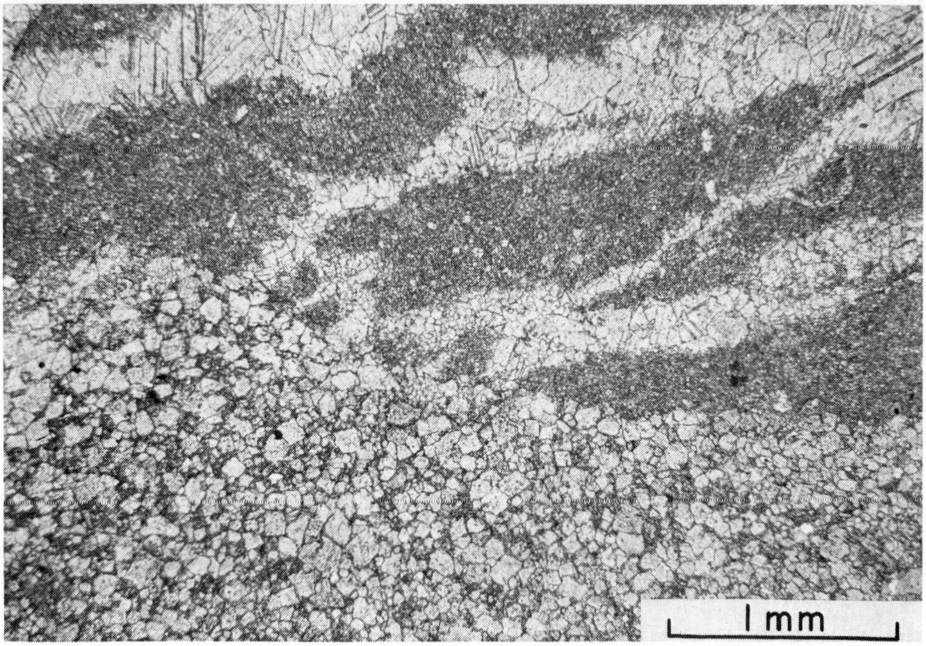

Fig. 7-49. Photomicrograph (ordinary light) of thin-section showing dolomite field truncating fractures with void-filling calcite, in turn transecting micrite. Dolomite (epigenetic) constitutes lower half of photograph. Lost Burro Formation (Middle and Upper Devonian), Funeral Mountains, California. (Photograph by D.H. Zenger.)

the interpretation of δC^{13} and δO^{18} values of dolomite in terms of the source of waters causing dolomitization. Some workers, as mentioned above, favor a lack of fractionation between calcite and the dolomite replacing it (Degens and Epstein, 1964). Land (1973a, p. 83), however, recorded at least a 3‰ fractionation between calcite and ordered, stoichiometric dolomite at equilibrium conditions. Kier (1973) took δO^{18} values of +4.8 to +5.4% for dolomicritic sediments in Baffin Bay, Texas, to indicate precipitation from hypersaline waters. Behrens and Land (1972, p. 160), obtaining nearly the same values as Kier, did *not* assume hypersalinity to account for the relatively high O^{18} content. Rather, they believed that the marine dolomites in Baffin Bay have been relatively enriched by incorporation of O^{18} into their structure other than that which would have been available from some $Ca_xMg_y(CO_3)_z$ precursor. The isotopic composition of ground water is a function of its temperature and isotopic exchange (depending on the degree of equilibration) with minerals encountered en route to the site of dolomitization. In addition to these variables, there is the further possible complica-

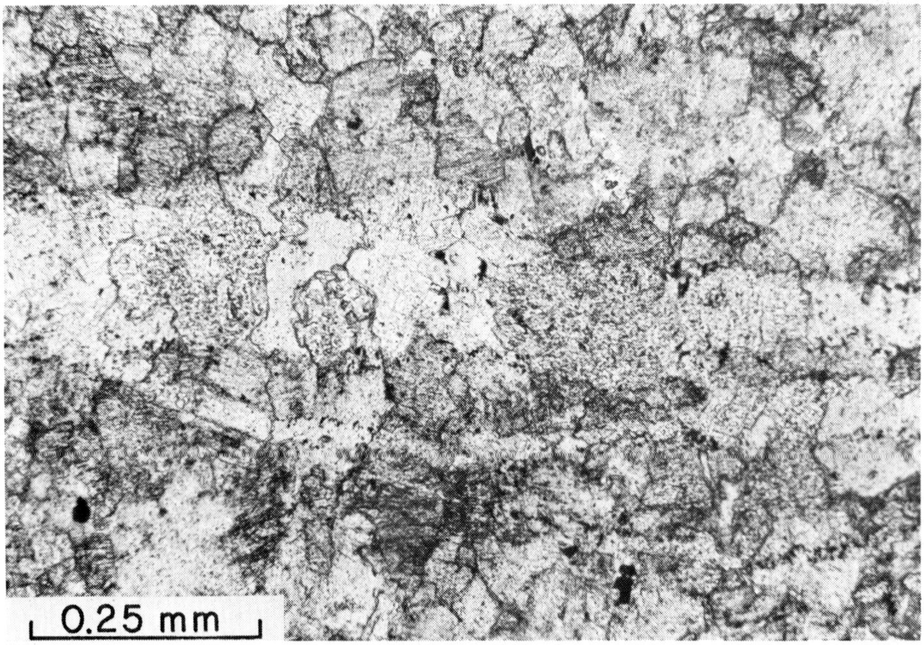

Fig. 7-50. Photomicrograph (plane polarized light) of thin-section showing epigenetic dolomite and "healed" fracture (distinguished by lighter band, and branch, running across center of photograph). Note dolomite crystals transecting the margins of main fracture and branch. Middle Devonian(?), near Torquay, Devonshire, England. (Photograph by D.H. Zenger.)

tion of mixing of deeper waters with lighter surface waters.

The work of Murata et al. (1972) on "diagenetic" dolomites (i.e., mainly epigenetic as defined by the present writers) has been referred to in the section on chemistry. The carbon isotopic composition of these subsurface dolomites is seldom that of marine carbonate (δC^{13} = +4 to −5‰), but is lighter or heavier depending on whether isotopic exchange between CO_2 and methane has been established or not. Light carbon dolomites occur at relatively shallow depths (1000 ft or less) in moderately compacted sediments rich in organic matter, from which the light carbon is derived. At a greater depth, where there is a greater possibility of isotopic exchange between CO_2 and methane, heavy-carbon dolomites (generally low in iron) form.

More work has been done using isotopes to determine the nature of the dolomitizing solutions. In general, the δO^{18} of deep formation brines are low relative to marine water due to the higher temperatures of the former, despite their higher concentration. Meteoric waters are lighter, the O^{16}

content becoming relatively enriched from surface-derived CO_2. Choquette (1968, 1971) recognized two kinds of dolomite in the Ste. Genevieve (Mississippian). He posited that early diagenetic dolomite with an average δO^{18} of +1.9‰ formed in nonequilibrium with marine $CaCO_3$ in intrasediment sea water or related brine, with oxygen derived from both the $CaCO_3$ and the water. The late diagenetic ferroan dolomite with undulatory extinction (referred to previously) ranged in δO^{18} value from −5.0 to −7.3‰. Choquette (1969) realized that such values could be accounted for by light waters close to the earth's surface, perhaps in a vadose zone supplied by rain water. He failed, however, to observe any of the features of this zone, as described by Dunham (1969). Also, a fresh-water origin for the light oxygen would signify that the water, rather than the carbonate host rock, provided the oxygen for the ferroan dolomite. Fracture- and vein-filling calcite and dolomite in limestone and dolomite host rocks, respectively, coupled with similar δC^{13} values between the ferroan dolomite and the "matrix" dolomite, strongly suggest a local source for carbon (and carbonate) for the epigenetic phase. Thus, Choquette concluded that possibly heated waters, enriched in O^{18} relative to sea water (perhaps connate brines or mixtures of connate, magmatic(?) and meteoric waters) were responsible for the origin of the ferroan dolomite.

Two types of dolomite occur in the Leadville Limestone (Mississippian) of Colorado (Banks, 1967, p. xiv). A diagenetic dolomite formed prior to lithification and a hydrothermal dolomite replaced lithified limestone during Laramide times. This hydrothermal dolomite forming the halo around the sulfide ore at Gilman, Colorado, yields δO^{18} values much lower than the diagenetic type (Engel et al., 1958).

Emphasizing mineral and isotopic composition and the sodium trace content, Fritz (1971) studied Middle Devonian carbonates hosting the famous Pine Point lead—zinc deposits in northern Canada. Three groups of dolomite were recognized: (1) supratidal and early diagenetic; (2) epigenetic; and (3) hydrothermal. Because of the close association of these dolomites, but yet distinct geochemical and mineralogical characteristics, Fritz assumed their retention of primary isotopic composition. In terms of isotopes, there is a clear demarcation only between group 1 and the other two groups: (a) the supratidal and early diagenetic (δO^{18} = −1 to −4‰), and (b) the epigenetic (δO^{18} = −7 to −12‰) and hydrothermal (δO^{18} = −6 to −11‰). Zenger (1976) has described dolomite "dikes" (Fig. 7-51) cutting the uppermost, pisolitic limestone part of the latest Precambrian Wyman Formation in the northern Inyo Mountains of California. Relict pisolites and gradational boundaries indicate an in situ replacement origin for these "dikes". Their orientation suggests epigenetic (postorogenic?) origin along older fractures. The average value of δO^{18} for the dolomites is about −13‰. In view of the

Fig. 7-51. Dolomite "dike" (dark weathering) cutting intraclastic, pisolitic limestone of upper Wyman Formation (latest Precambrian), northeastern Inyo Mountains. Dolomite interpreted as hydrothermal, with later generation of fractures. (Photograph by D.H. Zenger.)

local geology, Zenger attributed dolomitization to a mixture of hydrothermal waters and meteoric waters, which descended through the overlying Reed Dolomite.

Zenger (personal observations) analyzed six dolomite samples of varying crystal size from the Little Falls Dolomite (Upper Cambrian, New York) for δO^{18} and δC^{13}. At one end of the spectrum is a very finely crystalline stromatolitic dolomite ($\delta O^{18} = -5.61‰$) and at the other is a milky, secondary dolomite rhomb ($\delta O^{18} = -11.34‰$) from a vug. Although all the values of δO^{18} are negative, there is a general trend toward more negative values with coarser crystallinity, suggesting that the development of these coarser crystals was influenced by heated formation brines or incipient hydrothermal waters.

Fluid inclusion criteria. Analysis of fluid inclusions within dolomite crystals represents a subtle way of distinguishing among dolomite types. The "filling temperature", with certain corrections, represents the temperature at which the inclusions were trapped and the crystals grown. The "freezing temperature", dependent on the concentrations of the original solutions, is useful in determining the source and nature of the dolomitizing fluids. In fluid inclusion studies of the Pine Point ore and of mineral deposits of the southern Appalachian Valley, Roedder (1969, 1971) found that the epigenetic dolomites showed lower freezing temperatures than the calcites, indicating a more saline solution for dolomite formation. The filling temperatures suggested a formation temperature for the dolomite of about 100°C at Pine Point and even higher in the Appalachians. Roedder attributed the dolomitization in both cases primarily to deeply circulating, hot connate brines.

Certain questions have been raised about the reliability of fluid inclusion studies. Most fluid inclusion compositions suggest initially saline water. During recrystallization at depth and in contact with connate brines, could there be an exchange such that fluids of higher salinity would replace original fluids of lower concentration? Also, need the composition of the fluid inclusions be representative of the solutions in which the crystals have formed or could it represent a more extreme residuum? Most investigators, however, agree that such studies are reliable and certainly can be very useful when used in conjunction with other geological evidence.

Significance of diagenetic versus epigenetic dolomitization

In view of the preceding discussion, one can conclude that although individual pieces of evidence may be equivocal, a combination of data from the various lines of study — field, microscopic, and chemical — can often lead to a solid conclusion regarding the origin of a particular dolomite. Still debatable is the subject of which types of dolomitization are the significant ones in accounting for the dolomites of the geologic record.

Primary dolomites (direct precipitates out of sea water) are considered by many investigators to be relatively unimportant. Also, it seems that the reports on modern and ancient penecontemporaneous supratidal dolomites led to an overgeneralization of their significance. Penecontemporaneous dolomitization of tidal flat deposits possibly accounts for a relatively small part of the great bulk of ancient dolomite. It should be mentioned here, however, that in many cases it is extremely difficult to distinguish between the primary and penecontemporaneous dolomites. On the other hand, there are numerous examples of the hydrothermal and postorogenic varieties of epigenetic dolomitization. In fact, Miller (1966) was convinced that most dolomitization occurred late in the history of the rock, that is, during weathering and structural deformation. It has yet to be proven, however, that these epigenetic types of dolomitization can really account for most of the dolomite in the widespread stratiform units such as the Knox, the Ellenburger, the Lockport, the Little Falls—Conococheague, and the Big Horn, to mention some United States examples. It seems that many of these massive dolomites are of late diagenetic origin. Many geologists all over the world, however, believe that many massive Precambrian and Early Paleozoic dolomites are of primary origin.

It appears that the great proportion of dolomitization occurs after the burial of carbonate sediments, before and after lithification but prior to uplift and orogeny, and exclusive of the effect of hydrothermal waters. Thus, the most significant dolomitization begins while the sediment or rock is exposed to surface-generated solutions or compaction fluids, some of which may undergo evolution in the subsurface at temperatures consistent with the geothermal gradient. Such dolomites, then, would range from diagenetic (post-penecontemporaneous) to epigenetic (pre-orogenic). Again one must return to the matter of lithification, which Choquette and Pray (1970, p. 218) have questioned as a definitive element in approaching diagenesis. No doubt the key factor is porosity. Sediments may attain a good deal of rigidity by compaction and some cementation a short distance below the depositional surface (it is realized, however, that some ancient sediments have remained unlithified for long periods of time); porosity may or may not be occluded. Rocks under relatively shallow burial conditions, or in the eogenetic zone, as well as in the upper part of the mesogenetic zone may retain sufficient porosity and permeability to permit fluids to move through them. Some fracturing and vugs may exist and could provide vertical conduits, but movement of waters can be to a large extent lateral. Although fluid movement may be much more restricted with increasing depth of burial, time and higher temperatures are favorable factors and most pore waters are suitable for dolomitization (Usdowski, 1968). Deeper in the mesogenetic zones porosity occlusion is dominant and dolomitization is greatly impeded; when orogeny or, possibly, epeirogeny occurs, fractures are created that permit infiltration of hydrothermal solutions and later, during uplift, meteoric water.

As Wilson (1975) has pointed out, in many instances petrographic evidence on the timing of dolomitization is equivocal and contradictory. In sediments exhibiting progressive and partial replacement, the process must have occurred before complete calcitic cementation. On the other hand, many dolomite beds show pervasive replacement of fragments and matrix alike; inasmuch as fabric permeability exercised no control on the degree of dolomitization, this dolomitization must have occurred after thorough cementation. It should be pointed out, however, that in such late epigenetic dolomite, the presence of any original early diagenetic dolomite would be camouflaged by the later, more pervasive type.

Under favorable geologic and hydrologic conditions, solution cavities, vugs, and channels form during the epigenetic stage. As a result, the vertical permeability often exceeds the horizontal one manyfold. (See Sarkisyan et al., 1973.) On the other hand, in the case of diagenetic dolomitization the horizontal permeability is usually higher than the vertical one, as in the case of sandstones.

The major dolomitization of the extensive sheet dolomites seems to occur during post-penecontemporaneous diagenesis and during pre-orogenic epigenesis. This is essentially the "diagenetic" dolomitization of Friedman and Sanders (1967) except for the penecontemporaneous replacement dolomites that those workers would include.

DEDOLOMITIZATION

Only a brief mention of dedolomitization is made here, or as Smit and Swett (1969) strongly recommended, calcitization. Some statement is justified because this process is related to the subject at hand. Although petrographic evidence for the replacement of dolomite by calcite shows that it is of fairly widespread occurrence, there is little documentation for the production of thick and extensive masses of limestone from dolomite precursors. Lippmann (1973, p. 166) believed it doubtful that the generally used equation for dedolomitization:

$$CaMg(CO_3)_2 + Ca^{2+} \rightarrow 2\ CaCO_3 + Mg^{2+} \qquad (7\text{-}18)$$

is of much consequence in nature because of the relatively high solubility of calcite under those conditions proposed for the replacement.

The papers by Shearman et al. (1961) and by Evamy (1967) described many of the petrographic fabrics associated with dedolomitization. Yanat'eva (1955; see also Chilingar, 1956b) and De Groot (1967) contributed substantially to the knowledge of the chemistry of dedolomitization. Tatarskiy (1949) reviewed Soviet work on the subject.

On the basis of such work, geologists accept such textural evidence as the polycrystalline calcite rhomb (Fig. 7-52). Dedolomitization must be a near-surface process where P_{CO_2} is much less than atmospheric and temperature is below 50°C. Further, a solution with a high Ca^{2+}/Mg^{2+} ratio, for example, gypsiferous waters, and a rapid flow rate seem to be necessary for effective dedolomitization. Under such conditions, dolomite will dissolve incongruently permitting reaction 7-18 to occur. [1] Field and laboratory studies

[1] Despite the generalization that calcite is more soluble than dolomite, Langmuir (1971, pp. 1035, 1038) has shown that dolomite solubility is greater than that of calcite in pure water at low temperature. Under the same conditions, however, the *rate* of solubility of calcite is generally greater than that of dolomite (Graf and Lamar, 1955, p. 663), following the relationship that increased solubility rate parallels increased ionic radius of the cation, which probably indicates weaker bonds.

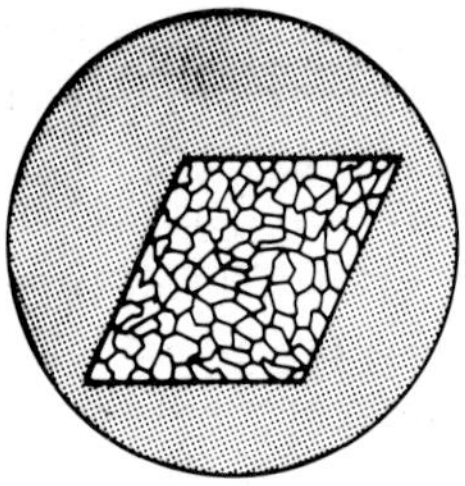

Fig. 7-52. The polycrystalline composite calcite rhombohedron, exhibiting a common dedolomitization fabric. (Modified after Evamy, 1967, p. 1208.)

have documented a common association of dedolomitization and dissolution of gypsum (Zenger, 1973). It has been assumed that dedolomitization occurs beneath erosion surfaces, which in the geologic record would be represented by unconformities.

Evamy, (1963), Folkman (1969), and Braun and Friedman (1969) observed an association between hematite and dedolomitization. They concluded that sulfate ions were formed during the oxidation of pyrite. These sulfate ions, probably in combination with Ca^{2+} and Fe^{2+} as calcium sulfate and ferrous sulfate (Evamy, 1963, p. 168), then promoted dedolomitization. Langmuir (1974, written communication), however, has questioned this explanation claiming that there is very little "combining" between Fe^{2+} and SO_4^{2-} as molecular $FeSO_4^0$; most of Fe^{2+} and SO_4^{-2} ions are free and uncomplexed and react in that way, not in combination. He has further suggested (1973, written communication) the following possible reactions for producing conditions suitable for dedolomitization, begining with initial oxidation of pyrite:

$$FeS_2 + 7/2\ O_2 + H_2O \rightarrow Fe^{2+} + 2\ SO_4^{2-} + 2\ H^+ \qquad (7\text{-}19)$$

and followed by further oxidation:

$$Fe^{2+} + 1/4\ O_2 + 5/2\ H_2O \rightarrow Fe(OH)_3 + 2\ H^+ \qquad (7\text{-}20)$$

The protons can attack dolomite or calcite and dedolomitization is possible under these conditions according to the following reaction:

$$CaMg(CO_3)_2 + 2\ H^+ \rightarrow CaCO_3 + Mg^{2+} + CO_2 + H_2O \qquad (7\text{-}21)$$

if, for example, CO_2 can be exsolved or Mg^{2+} ion removed, possibly by cation exchange equilibria. The replacement process also depends on the Mg^{2+}/Ca^{2+} activity ratio as evident from eq. 7-18.

Based on his observation of the ubiquitous occurrence of hematite (ferric hydroxide?) pseudomorphic after pyrite, Zenger (1973) considered the oxidation of pyrite as a likely mechanism for the production of some probable dedolomitization fabrics in the Little Falls Formation (Upper Cambrian) of east-central New York. One particular fabric consists of epitaxial "calcite rims" on dolomite crystals. The rims are even and uniform in thickness, ranging from 5 to 25 μm (Figs. 7-53 and 7-54). Analysis of the fabric geometry and mineral paragenesis strongly suggests that these rims have formed by marginal, or centripetal, dedolomitization rather than by passive precipitation (Figs. 7-55 and 7-56). Somewhat similar rims have been described by Braun and Friedman (1969, pp. 120—121) in the Tribes Hill Formation (Lower Ordovician) of east-central New York and by Fritz (1967, p. 1251) in Malm (Jurassic) carbonates of southern Germany, and were also attributed to centripetal dedolomitization.

A fabric probably indicating centrifugal dedolomitization is that of optically continuous calcite centers of dolomite crystals, such as those observed by Zenger (1973) in the Little Falls Formation and in the overlying Tribes

Fig. 7-53. Photomicrograph (plane polarized light) of stained thin-section of epitaxial calcite rim on dolomite crystal, just left of center. Stained calcite rim (darker) is distinguishable from dolomite crystal (*D*). Little Falls Formation (Upper Cambrian), Middleville, New York. (After Zenger, 1973, fig. 2A; courtesy J. Sediment. Petrol.)

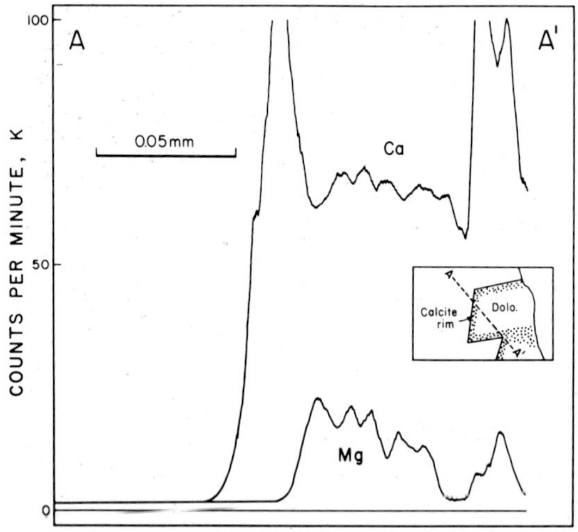

Fig. 7-54. Electron Probe Microanalyzer stage traverse for CaK∝ and MgK∝ across calcite rims along line *A-A'* as shown on sketch in insert. (Modified after Zenger, 1973, fig. 6; courtesy J. Sediment. Petrol.)

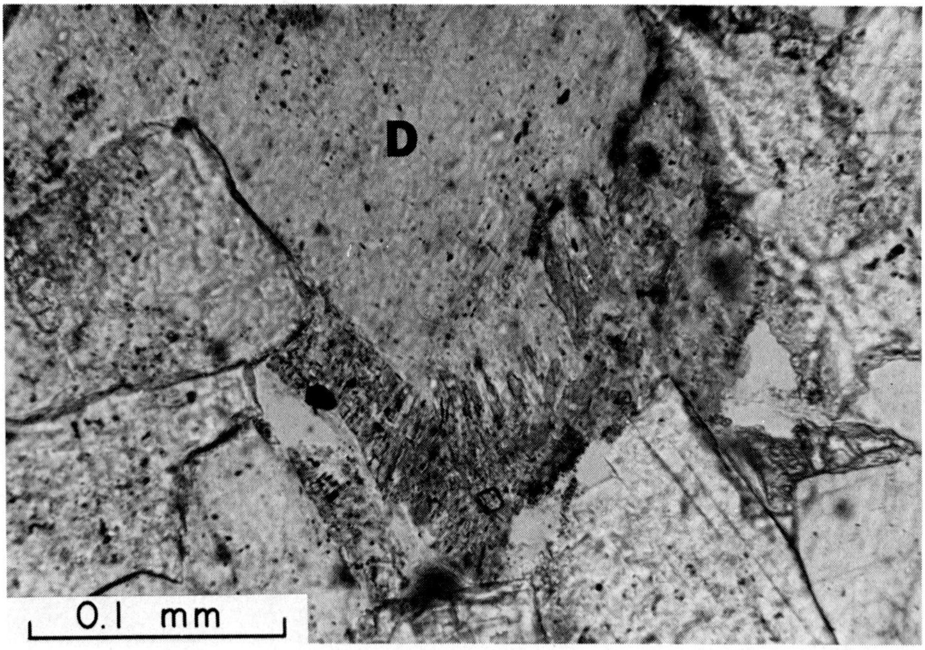

Fig. 7-55. Photomicrograph (plane polarized light) of stained thin-section showing jagged inner edges of calcite rim (darker) along lower margin of dolomite crystal (*D*). Irregular inner edge suggests dedolomitization. Little Falls Formation (Upper Cambrian), Little Falls, New York. (After Zenger, 1973, fig. 8; courtesy J. Sediment. Petrol.)

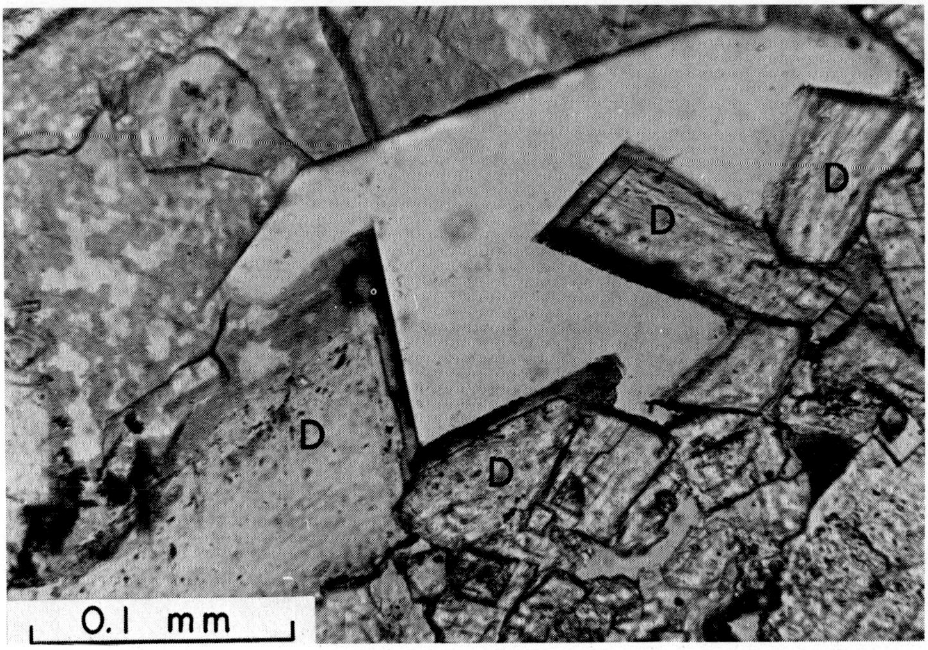

Fig. 7-56. Photomicrograph (plane polarized light) of stained thin-section illustrating relations between calcite rims, void-filling calcite, and authigenic quartz. Along upper part of photo are enfacial junctions of void-filling calcite (darker due to staining) against authigenic quartz crystal (white). Extending into quartz from lower side are several dolomite crystals (*D*) with darker calcite rims. Calcite is of later origin than dolomite and secondary quartz. Little Falls Formation (Upper Cambrian), Middleville, New York. (After Zenger, 1973, fig. 7; courtesy J. Sediment. Petrol.)

Hill Formation (Fig. 7-57). Calcite centers of dolomite crystals have been reported by other workers (Fritz, 1967, p. 1251; Koch, 1970, pp. 45, 47; Bourrouilh-LeJan, 1973) although only those of Bourrouilh-LeJan are optically continuous with the surrounding dolomite. The dedolomitized centers reported by Fritz and Koch consist of polycrystalline calcite.

Evamy's analysis (1967) of rhombohedral pores was that they represented solution of a dedolomitization product of metastable carbonate, perhaps magnesium calcite, instead of less soluble calcite. Al-Hashimi and Hemingway (1973) desribed the development of magnesian calcite by Recent and sub-Recent dedolomitization of ferroan dolomite in "rusty" weathering crusts of the Carboniferous Limestone of northeastern England. The distinctive coloration of these crusts, they believed, is due to the formation of iron

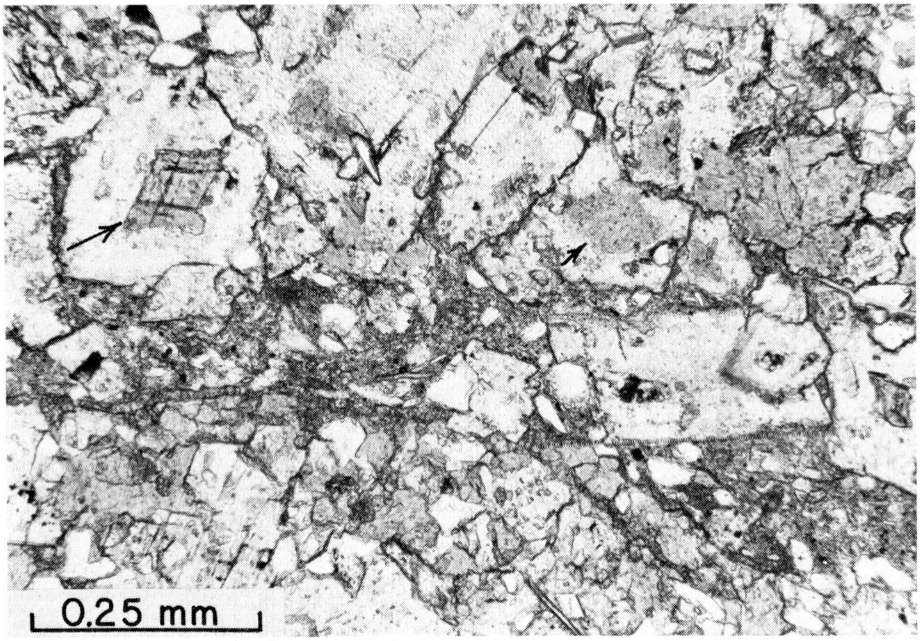

Fig. 7-57. Photomicrograph (ordinary light) of stained thin-section showing optically continuous calcite centers (slightly darker due to staining, see arrows) of dolomite crystals (lighter). Interpreted as dedolomitization fabric. Tribes Hill Formation (Lower Ordovician), St Johnsville, New York. (Photograph by D.H. Zenger.)

hydroxide from the ferroan dolomite according to the following equations:

$$4[Ca(MgFe)(CO_3)_2] + 2\ H_2O + 1/2\ O_2 \rightarrow 4\ CaCO_3 + Fe_2O_3 + 2\ Mg^{2+} + 2\ CO_3^{2-} + 2\ H_2CO_3 \quad (7\text{-}22)$$

and:

$$Fe_2O_3 + H_2O \rightarrow 2(FeOOH) \quad (7\text{-}23)$$

The iron hydroxides, present as a coating on all calcite rhombs, but absent internally, may either have been replaced or expelled to the outside of the rhomb by the growing calcite. On the other hand, Langmuir (1971a) has pointed out that whereas goethite (FeOOH) may dehydrate to give hematite, rehydration of hematite is kinetically unfeasible.

In a paper referred to earlier, Katz (1971) concluded that the zonation in the Mahmal (Jurassic of Israel) dolomite crystals was produced as the crystals grew early in diagenesis in unlithified sediments, and that the growth stages of the crystal can be related to compositional changes of the dolomitizing solutions. One type of zone consists of polycrystalline calcite that Katz attributed to the replacement of a calcian dolomite precursor. He proposed that the dedolomitization occurred concomitantly with a dilution of the brines leading to sufficient reductions of the Mg^{2+}/Ca^{2+} ratios of the solutions to bring them into the stability field of calcite. Thus, Katz was describing a process of dedolomitization that occurred at intervals during crystal growth. He could see no necessity for recourse to late-stage surface phenomena because the depositional environment of the Mahmal sediments could have provided the essential conditions of low P_{CO_2}, temperature < 50°C, and high Ca^{2+}/Mg^{2+} ratios. Katz found no relation between the dedolomite and gypsum and/or pyrite. Wood and Armstrong (1975; G.V. Wood, 1974, written communication) have also described such internal calcite zones in the Lisburne Limestone (Carboniferous) of Alaska, which they feel tend to support the argument of Katz. The dedolomitization in this instance is judged to have occurred prior to the pore-filling stage of calcite cementation. This work demonstrates that many investigators have been premature in ascribing all dedolomitization to sites beneath modern or ancient weathering surfaces (i.e., unconformities). Perhaps dedolomitization can be eogenetic as well as telogenetic.

It is interesting to note here that Mattaveli et al. (1969), who studied the Taormina Formation of the Gela oil field in Sicily, Italy, observed that porosity generally increases with increasing $CaCO_3$ content. The higher the amount of calcite, the more pronounced the dissolution phenomena, and the higher the secondary porosity. A complex sequence of diagenetic processes occurred in the Taormina Formation: (1) syndiagenetic dolomitization which, because of the lack of solid framework, commonly does not give rise to porosity; (2) diagenetic dolomitization, which creates porosity; (3) cementation, which destroys porosity; (4) fracturing; (5) dedolomitization, which destroys porosity; and (6) secondary solution of calcite after dedolomitization, which results in porosity increase.

GLOSSARY

Calcitization. The process of calcite replacing dolomite as used by Smit and Swett (1969). Dedolomitization, a synonym for this restricted meaning of calcitization, is preferred here because of its wider usage and also because calcitization has been used more generally for the replacement by calcite of various precursors.

Capillary concentration. Capillary movement of sea water, of normal or near-normal salinity, upward through sea-marginal sediments and subsequent concentration by evaporation at the sediment—air interface (see Friedman and Sanders, 1967, pp. 280—281).

Congruent dissolution. Dissolution of a mineral such that equal molar quantities of the components are taken into solution. The term is contrasted with *Incongruent dissolution.*

Dedolomitization. The process of calcite replacing dolomite. Synonym: calcitization, as used by Smit and Swett (1969).

Dehydration barrier. The energy (enthalpy) that is required to break the cation—H_2O bond in a solution in order to permit the cation to enter an anhydrous crystal structure (see Lippmann, 1973, pp. 76—91).

Desolvation. Process of removal of solvent (i.e., water) molecules from a cation, in particular the magnesium ion (see Gaines, 1974, and Gaines and Heffner, 1973).

Detrital—diagenetic recrystallization model. A model proposed by Lindholm (1969) to account for subtidal formation of dolomitic sediments by transportation and deposition of detrital dolomite grains in carbonate sediments, followed by replacement and recrystallization around these clastic nuclei.

Diagenesis. It includes all physicochemical, biochemical and physical processes modifying sediments between deposition and lithification, or cementation, at low temperatures and pressures characteristic of surface and near-surface environments. In general, diagenesis is divisible into pre-, syn-, and post-cementation or lithification processes. Diagenesis, as defined in this chapter, takes an intermediate position between syngenesis and epigenesis, the former grading into diagenesis by syndiagenesis, and the latter grading into metamorphism. Under unusual conditions, however, diagenesis as defined here may grade directly into metamorphism (see *Epigenesis*). Because reef limestones, and other limestones which are constructed in situ by organic frame-builders, are largely in and of themselves lithified to a degree, the definition must be expanded for this particular group of limestones to include the interactions between sediments and the fluids contained within them below the temperature and pressure levels of metamorphism sensu stricto, and in a similar sense between fluids and framework, infilled detritus framework, and combination thereof.

Disordered dolomite. Dolomite in which the Ca^{2+} and Mg^{2+} ions do not occupy their correct sites, as contrasted with ordered dolomite. Although ordered and disordered dolomites have certain *hkl* reflections in common, the ordered form provides an additional set of reflections termed superstructure or superlattice reflections. This material is typical of most modern, natural dolomite and of that produced synthetically in the more successful laboratory experiments run at conditions similar to those on the earth's surface.

Dololutite to dolorudite. A range of terms applied to sedimentary dolomites composed of constituents ranging in size from clay to gravel, similar to those in limestones, as follows:
dololutite = clay-sized dolomite particles
dolosiltite = silt-sized dolomite particles
dolarenite = sand-sized dolomite particles
dolorudite = gravel-sized dolomite particles.

Dolomite. (1) A mineral, $CaMg(CO_3)_2$, hexagonal rhombohedral. (2) A carbonate rock composed predominantly of the mineral dolomite; in normal routine petrographic work, dolomite (or dolostone of some geologists) is a carbonate rock composed of more than 50% by weight of the mineral dolomite. More practically, areal percentages are used instead of weight percentages.

Dolomite "dikes". Tabular bodies of dolomite generally at high angles to the bedding in the Precambrian Wyman Formation, east-central California. The word "dike" is descriptive only, with no genetic implication intended.

Dolomite front. A relatively sharp contact between laterally adjacent limestone and dolomite.

Dolomitic. Where used in a rock name, "dolomitic" refers to those rocks that contain 5—50% of the mineral dolomite, as cement and/or grains or crystals. Dolomitic can be applied to the large spectrum of sedimentary rocks that are dolomite-bearing, and also to limestones which have been dolomitized to a degree but not completely.

Dolomitic mottling. Incipient or arrested dolomitization, or arrested (or incomplete) dedolomitization. Common to limestones that have large particulate skeletal or non-skeletal material embedded in fine-textured matrix. Under the effects of dolomitization there is a preferential replacement or alteration of the matrix but not of the large particles. Also common to more or less homogeneous textured limestones that have been incompletely dolomitized, leaving patches, blotches, laminae, or other structures unaffected.

Dolomitized. Refers to rocks or portions of rocks in which limestone textures are discernible, but which have been changed to dolomite.

Dorag dolomitization. A model proposed by Badiozamani (1973) for dolomitization by a mixture of meteoric ground water and sea water. ("Dorag" is a Persian word for "mixed blood".)

Eogenetic stage. The time interval between final deposition and burial of the newly deposited sediment or rock below the depth of significant influences by processes that either operate from the surface or depend for their effectiveness on proximity to the surface (Choquette and Pray, 1970, p. 269).

Epigenesis. As used here, it includes all processes at low temperature and pressure that affect sedimentary rocks after diagenesis and up to metamorphism. Epigenesis has been subdivided into juxta- and apo-epigenesis (Wolf, 1963b, 1965c). It is possible that under unusual conditions syngenesis and diagenesis grade directly into metamorphism. For example, unconsolidated sediments may be exposed to volcanic high temperatures and metasomatic material and undergo metamorphism before diagenesis. Also, sediments partly undergoing cementation may be metamorphosed by shallow intrusions causing an increase of temperature and possibly pressure before epigenesis could occur:

syngenesis }
diagenesis } → metamorphism
epigenesis }
↓
metamorphism

It appears, however, that the term has been used by different workers in three ways. It has been used, for example, by many Soviet workers to embrace the modifications and transformations occurring within rocks, prior to weathering and metamorphism. Friedman and Sanders (1967, p. 308) used the term "epigenetic dolomitization" for that which was structurally controlled. Additionally, Freeman (1972, pp. 445—446) used "epigenetic" to refer to dolomitization that occurred as a result of dolomitizing waters that evolved mainly in the subsurface. Owing largely to these different usages, the term "catagenesis" is commonly used instead of "epigenesis" in the U.S.S.R.

Epitaxial calcite rim. Refers to thin, even borders of calcite around dolomite crystals, the calcite being in optical continuity with the dolomite. Originally defined as syntaxial "calcite rim" (Zenger, 1973, p. 118).

Evaporative pumping. Fluid movement through porous media induced by an upward decrease in hydrodynamic potential resulting from surficial evaporative loss. Experiments showed that saline interstitial waters underneath an arid coastal plain could move in response to such a gradient, even through relatively impermeable sediments, and cause dolomitization (see Hsü and Siegenthaler, 1969).

Ground-water dolomitization. An artificial category used herein to include the various models for dolomitization by cold or hot meteoric or connate waters in shallow or deep situations, but excluding sea water—meteoric ground-water mixing (i.e., "schizohaline" and "dorag" models).

Hydrothermal. "Of or pertaining to heated water, to the action of heated water, or to the products of the action of heated water, such as a mineral deposit precipitated from a hot aqueous solution, with or without demonstrable association with igneous processes . . ." (*AGI Glossary of Geology, 1972*). Lovering (1969, p. 743) has set off from hydrothermal solutions waters of "low temperature", which he defined as those characterized by ". . . temperatures below those that will allow organic life to exist, or below the boiling point of water at geologically low pressures, approximating a maximum temperature of 100°C". At the other limit, hydrothermal alteration grades into hydrothermal metamorphism and may be essentially indistinguishable, or indeed even inseparable, from it.

Incongruent dissolution. Dissolution of a salt such that equal molar quantities are not taken into solution; that is, the more soluble part of the mineral is dissolved to a greater degree. Opposite of *Congruent dissolution.*

Kinetics (chemical). That branch of physical chemistry that deals with the speed of chemical reactions.

Limpid dolomite. Dolomite crystals, characteristically clear and near-perfect, which are a product of dilute waters, such as are found in schizohaline environments (see Folk and Land, 1975, pp. 65—66).

Lithification. "The conversion of a newly deposited, unconsolidated sediment into a coherent and solid rock, involving processes such as cementation, compaction, desiccation, crystallization, recrystallization, and compression. It may occur concurrent with, or shortly or long after, deposition . . ." (*AGI Glossary of Geology, 1972*).

Magnesium-rich calcite. Calcite which has a relatively high mole % of $MgCO_3$, generally greater than 11%. Low-magnesium calcite commonly contains less than 5 mole % $MgCO_3$, there being a sparsity of known samples in the range of 6—10 mole % $MgCO_3$ (see Bathurst, 1971, p. 235).

Mesogenetic stage. The time interval in which the sediments or rocks are buried at depth below the major influence of processes directly operating from or closely related to the surface (Choquette and Pray, 1970, p. 220).

Metasomatism. A synonym of *Replacement.* Implied is a process of practically simultaneous capillary solution and deposition through which an old mineral is transformed into a new one (see *AGI Glossary of Geology, 1972*).

Neomorphism. A term introduced by Folk (1965, pp. 20—21) to include ". . . all transformations between one mineral and itself or a polymorph — whether inversion or recrystallization, whether the new crystals are larger or smaller or simply differ in shape from the previous ones".

Ordered dolomite. Dolomite in which the Ca^{2+} and Mg^{2+} ions occupy their correct sites and from which superstructure X-ray reflections are obtained.

Peloid. A grain composed of microcrystalline or cryptocrystalline calcite, regardless of origin (see McKee and Gutschick, 1969, p. 101). As pointed out by Bathurst (1971, p. 547) the term is useful in that it permits reference to grains consisting of micrite or microspar without a genetic implication as, for example, the term "pellet" would have.

Peritidal. A term to include the whole of the tidal flat, that is, both intertidal and supratidal zones.

Phreatic. The zone of saturation, that is, beneath the ground-water table which separates it from the overlying zone of aeration or vadose zone; water occurring in that zone.

Poikilotopic. A fabric of inequigranular carbonate rocks which have been precipitated or recrystallized, in which the crystals are of more than one size, and in which larger crystals enclose smaller crystals of another mineral (see Friedman, 1965, p. 651).

Polycrystalline calcite rhomb. A fabric, taken to be diagnostic of dedolomitization, consisting of a rhomb-shaped aggregate of calcite crystals.

Porphyrotopic. A fabric of inequigranular sedimentary rocks which have been precipitated or recrystallized, in which the crystals are of more than one size, and in which the larger crystals (porphyrotopes) are enclosed in a finer-grained matrix (see Friedman, 1965, p. 649).

Postorogenic dolomite. Dolomite that is localized along tectonic structures such as folds and fractures. Equivalent of "epigenetic dolomite" as used here.

Primary dolomite. Dolomite formed by direct precipitation from solution during the depositional phase.

Protodolomite. Material consisting of ". . . single-phase rhombohedral carbonates which deviate from the composition of the dolomite that is stable in a given environment, or are imperfectly ordered, or both . . ." (Graf and Goldsmith, 1956, p. 184). Included in the original definition was the point that protodolomite would transform to dolomite if equilibrium were established. No doubt most ancient "protodolomite" transformed to dolomite. Owing to the persistence of some "protodolomite" in the geologic record, however, it seems most practical to maintain the term in a descriptive way only.

Recrystallization. A kind of neomorphism that involves a textural or fabric alteration unaccompanied by a mineralogical change; the mineral remains identical through the process (see Folk, 1965, p. 23). "Recrystallization embraces any change in the fabric of a mineral or a monomineralic sediment" (Bathurst, 1971, p. 476).

Replacement. The transformation of one mineral into another of differing composition. Synonym of *Metasomatism*, according to Folk (1965, p. 21).

Sabkha. A supratidal zone consisting of a band of salt flats parallel to the coast and inland from intertidal algal flats in the Persian Gulf region. The term has also been applied to supposed ancient analogs.

Schizohaline dolomitization. The replacement of carbonate sediments or rocks by brackish waters in an environment alternating between hypersaline and near-surface fresh water conditions such as in a floodable sabkha or a phreatic mixing zone (see Folk and Land, 1972).

Salt-filtration. A process of filtering of dissolved salts from connate brines by compacted clay or shale that exhibit Donnan behavior (i.e., exclusion of anions, and also cations in order to maintain electroneutrality, as a result of double layer effects of colloidal particles). As a consequence, dissolved salts build up in a water at the point where it enters such a compacted clay or shale (Berner, 1971, pp. 36—39, 108—111).

Sea water—meteoric water mixing. An artificial category used herein to include the various models for dolomitization by a mixture of sea water and fresh water; includes the "schizohaline" dolomitization of Folk and Land (1972), the "dorag" dolomitization of Badiozamani (1973), and the dolomitization in the Tertiary limestone aquifer beneath Florida, as described by Hanshaw et al. (1971).

Seepage refluxion. A process, popularized by Adams and Rhodes (1960), by which hypersaline brines, refluxing from evaporite lagoons, seep into underlying porous carbonate sediments or rocks, displacing connate waters and acting as a source of magnesium to promote dolomitization and as a vehicle for removing displaced calcium. The concept has been somewhat broadened to include brines derived from arid supratidal flats.

Solubility product. The equilibrium constant for a saturated solution of an ionic solid, i.e., the product of the concentration of the ions at a particular temperature.

Solution—cannibalization. A model for dolomitization proposed by Goodell and Garman (1969), which involves solution of mixed carbonate phases at sea level to furnish the magnesium for dolomitization of material below. They applied their model to subsurface rocks beneath Andros Island, Bahamas.

Sucrosic: contraction of saccharoidal, thus meaning "sugary" texture.

Supratidal. That zone, commonly the landward part of the tidal flat, above and/or beyond the normal high-tide mark, which is reached by sea water during excessively high tides or floods, or simply spray, as well as by capillary concentration or evaporative pumping in an arid climate.

Telogenetic stage. The time interval during which long-buried carbonate rocks are influenced significantly by processes associated with the formation of an unconformity (see Choquette and Pray, 1970, p. 220).

Thermodynamics. The mathematical treatment of the relation of heat to mechanical and other forms of energy (*AGI Glossary of Geology, 1972*). Modern thermodynamics deal with the properties of systems for the description of which temperature is a necessary coordinate (*Random House Dictionary of the English Language, 1966*).

Vadose. The zone of aeration, above the ground-water table; water in that zone.

REFERENCES

Adams, J.E. and Rhodes, M.L., 1960. Dolomitization by seepage refluxion. *Bull. Am. Assoc. Pet. Geol.*, 44: 1912—1921.

Alderman, A.R. and Skinner, H.C.W., 1957. Dolomite sedimentation in the southeast of South Australia. *Am. J. Sci.*, 255: 561—567.

Al-Hashimi, W.S. and Hemingway, J.E., 1973. Recent dedolomitization and the origin of the rusty crusts of Northumberland. *J. Sediment. Petrol.*, 43: 82—91.

Asquith, G.B., 1967. The marine dolomitization of the Mifflin Member Platteville Limestone in southwest Wisconsin. *J. Sediment. Petrol.*, 37: 311—326.

Athy, L.F., 1930. Density, porosity and compaction of sedimentary rocks. *Bull. Am. Assoc. Pet. Geol.*, 14: 1—24.

Atwood, D.K. and Bubb, J.N., 1970. Distribution of dolomite in a tidal flat environment, Sugarloaf Key, Florida. *J. Geol.*, 78: 499—505.

Badiozamani, K., 1973. The Dorag dolomitization model — application to the Middle Ordovician of Wisconsin. *J. Sediment. Petrol.*, 43: 965—984.

Badiozamani, K., 1976. Reply to discussion of the Dorag dolomitization model, by Alden B. Carpenter. *J. Sediment. Petrol.*, 46: 256—258.

Banks, N.G., 1967. *Geology and Geochemistry of the Leadville Limestone (Mississippian, Colorado) and its Diagenetic, Supergene, Hydrothermal and Metamorphic Derivatives.* Thesis, University of California, San Diego, 298 pp. (unpublished).

Barnes, I. and Back, W., 1964. Dolomite solubility in ground water. In: *Short Papers in Geology and Hydrology—U.S. Geol. Surv. Prof. Pap.*, 475-D: 179—180.

Barnes, I. and O'Neil, J.R., 1969. The nature of secondary dolomite formation (abstract). *Am. Geophys. Union Trans.*, 50: 347.

Barnes, I. and O'Neil, J.R., 1971. Calcium-magnesium carbonate solid solutions from Holocene conglomerate cements and travertines in the Coast Range of California. *Geochim. Cosmochim. Acta*, 35: 699—718.

Baron, G., 1960. Sur la synthèse de la dolomie; application au phénomène de dolomitisation. *Rev. Inst. Fr. Pétrol. Ann. Combust. Liquides*, 15: 3—68.

Baron, G. and Favre, J., 1959. Recherches expérimentales sur le rôle des facteurs physicochimique dans la synthèse de la dolomie. *World Pet. Congr., Proc., 5th, N.Y., 1959*, 1 (3): 19—25.

Bateman, P., 1965. Geology and tungsten mineralization of the Bishop District, California. *U.S. Geol. Surv. Prof. Pap.*, 470: 208 pp.

Bathurst, R.G.C., 1971. *Carbonate Sediments and Their Diagenesis*. Elsevier, Amsterdam, 620 pp.

Beales, F.W., 1971. Cementation by white sparry dolomite. In: O.P. Bricker (Editor), *Carbonate Cements*. The Johns Hopkins Press, Baltimore, pp. 330—338.

Behrens, E.W. and Land, L.S., 1972. Subtidal Holocene dolomite, Baffin Bay, Texas. *J. Sediment. Petrol.*, 42: 155—161.

Beloborodova, G.V., Davydov, Yu.V., Kazanskiy, Yu.P., Kataeva, V.N. and Mandrikova, N.T., 1972. Possibility of chemical precipitation of dolomite from sea waters of Riphaean Era. *Izv. Akad. Nauk. S.S.S.R., Ser. Geol.*, 4: 152—155.

Berner, R.A., 1965. Dolomitization of the mid-Pacific atolls. *Science*, 147: 1297—1299.

Berner, R.A., 1971. *Principles of Chemical Sedimentology*. McGraw-Hill, New York N.Y., 240 pp.

Berry, W.B.N. and Boucot, A.J., 1971. Correlation of the North American Silurian rocks. *Geol. Soc. Am., Spec. Pap.*, 102: 189 pp.

Bissell, H.J., 1959. Silica in sediments of the Upper Paleozoic of the Cordilleran area. In: H.A. Ireland (Editor), Silica in Sediments. *Soc. Econ. Paleontol. Mineral., Spec. Publ.*, 7: 150—185.

Bissell, H.J. and Chilingar, G.V., 1958. Notes on diagenetic dolomitization. *J. Sediment. Petrol.*, 28: 490—497.

Bissell, H.J. and Chilingar, G.V., 1962. Evaporite type dolomite in salt flats of Western Utah. *Sedimentology*, 1: 200—210.

Bissell, H.J. and Chilingar, G.V., 1967. Classification of sedimentary carbonate rocks. In: G.V. Chilingar, H.J. Bissell and R.W. Fairbridge (Editors), *Carbonate Rocks. Origin, Occurrence and Classification*. Elsevier, Amsterdam, pp. 87—168.

Blatt, H., Middleton, G.V. and Murray, R.C., 1972. *Origin of Sedimentary Rocks*. Prentice-Hall Inc., Englewood Cliffs, N.J., 634 pp.

Bourrouilh-Le Jan, F., 1973. Les dolomies et leurs genèses. *Bull. Cent. Rech., Pau, S.N.P.A.*, 7: 111—135.

Braun, M. and Friedman, G.M., 1969. Carbonate lithofacies and environments of the Tribes Hill Formation (Lower Ordovician) of the Mohawk Valley, New York. *J. Sediment. Petrol.*, 39: 113—135.

Brown, C.W., 1959. Diagenesis of Late Cambrian oölitic limestone, Maurice Formation, Montana and Wyoming. *J. Sediment. Petrol.*, 29: 260—266.

Brown, P.R., 1969. Compaction of fine-grained terrigenous and carbonate sediments. A review. *Bull. Can. Pet. Geol.*, 17: 486—495.

Carpenter, A.B., 1976. Discussion: Dorag dolomitization model: by K. Badiozamani. *J. Sediment. Petrol.*, 46: 254—256.

Chave, K.E., 1954. Aspects of the biogeochemistry of magnesium. 2. Calcareous sediments and rocks. *J. Geol.*, 62: 587—599.

Chazen, P.O. and Ehrlich, T., 1973. Low-temperature synthesis of dolomite from aragonite. *Bull. Geol. Soc.*, 84: 3627—3634.

Chilingar, G.V., 1953. Use of Ca/Mg ratio in limestones as a geologic tool. *Compass*, 30: 29—34.

Chilingar, G.V., 1955a. Physical-chemical characteristics of basins and sediments of Taman peninsula. *Bull. Am. Assoc. Pet. Geol.*, 39: 1668—1669.

Chilingar, G.V., 1955b. Review of Soviet literature on petroleum source rocks. *Bull. Am. Assoc. Pet. Geol.*, 39: 764—768.

Chilingar, G.V., 1956a. Relationship between Ca/Mg ratio and geologic age. *Bull. Am. Assoc. Pet. Geol.*, 40: 2256—2266.

Chilingar, G.V., 1956b. Dedolomitization: a review. *Bull. Am. Assoc. Pet. Geol.*, 40: 762—764.

Chilingar, G.V., 1956c. Note on direct precipitation of dolomite out of sea water. *Compass*, 34: 29—34.

Chilingar, G.V., 1956d. Relationship between Ca/Mg ratio and geologic age. *Bull. Am. Assoc. Pet. Geol.*, 40: 2256—2266.

Chilingar, G.V., 1956e. Use of Ca/Mg ratio in porosity studies. *Bull. Am. Assoc. Pet. Geol.*, 40: 2489—2493.

Chilingar, G.V., 1956f. Joint occurrence of glauconite and chlorite in sedimentary rocks. A review. *Bull. Am. Assoc. Pet. Geol.*, 40: 394—398.

Chilingar, G.V., 1956g. Solubility of calcite, dolomite, and magnesite, and mixtures of these carbonates. *Bull. Am. Assoc. Pet. Geol.*, 40: 2770—2773.

Chilingar, G.V., 1957. Classification of limestones and dolomites on basis of Ca/Mg ratio. *J. Sediment. Petrol.*, 27: 187—189.

Chilingar, G.V., 1958a. A short note on authigenic minerals. *Compass*, 35: 294—297.

Chilingar, G.V., 1958b. Some data on diagenesis obtained from Soviet literature. *Geochim. Cosmochim. Acta*, 13: 213—217.

Chilingar, G.V., 1959a. The Caspian Sea and its sediments. A review. *Int. Geol. Rev.*, 1 (1): 105—111.

Chilingar, 1959b. Formation of sediments in recent basins. A review. *Int. Geol. Rev.*, 1 (3): 74—81.

Chilingar, G.V., 1960. Notes on classification of carbonate rocks on basis of chemical composition. *J. Sediment. Petrol.*, 30: 157—158.

Chilingar, G.V., 1962a. Dependence on temperature of Ca/Mg ratio of skeletal structures of organisms and direct precipitates out of sea water. *South. Calif. Acad. Sci. Bull.*, 61 (1): 45—60.

Chilingar, G.V., 1962b. Possible loss of magnesium from fossils to the surrounding environment. *J. Sediment. Petrol.*, 32: 136—139.

Chilingar, G.V. and Bissell, H.J., 1961. Dolomitization by seepage refluxion. *Bull. Am. Assoc. Pet. Geol.*, 45: 679—681.

Chilingar, G.V. and Bissell, H.J., 1963a. Is dolomite formation favored by high or low pH? *Sedimentology*, 2: 171—172.

Chilingar, G.V. and Bissell, H.J., 1963b. Formation of dolomite in sulfate and chloride solutions. *J. Sediment. Petrol.*, 33: 801—803.

Chilingar, G.V. and Terry, R.D., 1954. Relationship between porosity and chemical composition of carbonate rocks. *Pet. Eng.*, 26: 341—342.

Chilingarian, G.V. and Wolf, K.H., 1976. *Compaction of Coarse-Grained Sediments, II.* Elsevier, Amsterdam, 808 pp.

Chilingar, G.V., Bissell, H.J. and Wolf, K.H., 1967. Diagenesis of carbonate rocks. In: G. Larsen and G.V. Chilingar (Editors), *Diagenesis in Sediments.* Elsevier, Amsterdam, pp. 179—322.

Chilingar, G.V., Mannon, R.W. and Rieke III, H.H., 1972. *Oil and Gas Production from Carbonate Rocks.* Am. Elsevier, New York, 408 pp.

Choquette, P.W., 1968. Marine diagenesis of shallow marine lime-mud sediments: insights from δO^{18} and δC^{13} data. *Science*, 161: 1130—1132.

Choquette, P.W., 1971. Late ferroan dolomite cement, Mississippian carbonates, Illinois Basin, U.S.A. In: O.P. Bricker (Editor), *Carbonate Cements.* The Johns Hopkins Press, Baltimore, pp. 330—338.

Choquette, P.W. and Pray, L.C., 1970. Geologic nomenclature and classification of porosity in sedimentary carbonates. *Bull. Am. Assoc. Pet. Geol.*, 54: 207—250.

Clayton, R.N., Jones, B.F. and Berner, R.A., 1968. Isotope studies of dolomite formation under sedimentary conditions. *Geochim. Cosmochim. Acta*, 32: 415—432.
Coogan, A.H., 1970. Measurements of compaction in oölitic grainstone. *J. Sediment. Petrol.*, 40: 921—929.
Curtis, R., Evans, G., Kinsman, D.J.J. and Shearman, D.J., 1963. Association of dolomite and anhydrite in the Recent sediments of the Persian Gulf. *Nature*, 197: 679—680.
Davies, G.R., 1970. Algal-laminated sediments, Gladstone Embayment, Shark Bay, western Australia. In: B.W. Logan, G.R. Davies, J.F. Read and D.E. Cebulski (Editors), *Carbonate Sedimentation and Environments, Shark Bay, Western Australia — Am. Assoc. Pet. Geol. Mem.*, 13: 169—205.
Davies, P.J. and Ferguson, J., 1975. Dolomite and organic material. *Nature*, 255: 472—473.
Davies, T.A. and Supko, P.R., 1973. Oceanic sediments and their diagenesis: some examples from deep-sea drilling. *J. Sediment. Petrol.*, 43: 381—390.
Deelman, J.C., 1975. Dolomite synthesis and crystal growth. *Geology*, 3: 471—472.
Deffeyes, K.S., Lucia, F.J. and Weyl, P.K., 1965. Dolomitization of Recent and Plio-Pleistocene sediments by marine evaporite waters on Bonaire, Netherlands Antilles. In: L.C. Pray and R.C. Murray (Editors), *Dolomitization and Limestone Diagenesis — Soc. Econ. Paleontol. Mineral., Spec. Publ.*, 13: 71—87.
Degens, E.T. and Chilingar, G.V., 1967. Diagenesis of subsurface waters. In: G. Larsen and G.V. Chilingar (Editors), *Diagenesis in Sediments*. Elsevier, Amsterdam, pp. 477—502.
Degens, E.T. and Epstein, S., 1964. Oxygen and carbon isotope ratios in coexisting calcites and dolomites from Recent and ancient sediments. *Geochim. Cosmochim. Acta*, 28: 23—44.
De Groot, K., 1967. Experimental dedolomitization. *J. Sediment. Petrol.*, 37: 1216—1220.
Dolomieu, D. de, 1791. Sur un genre de pierres calcaires très-peu effervescentes avec les acides, et phosphorescentes par la collision. *J. Phys.*, 39: 3—10.
Dunbar, C.O. and Rodgers, J., 1957. *Principles of Stratigraphy*. Wiley, New York, N.Y., 356 pp.
Dunham, R.J., 1969. Early vadose silt in Townsend Mound (reef), New Mexico. In: G.M. Friedman (Editor), *Depositional Environments in Carbonate Rocks — Soc. Econ. Paleontol. Mineral., Spec. Publ.*, 14: 139—181.
Eardley, A.J., 1938. Sediments of Great Salt Lake, Utah. *Bull. Am. Assoc. Pet. Geol.*, 22: 1305—1411.
Engel, A.E.J., Clayton, R.N. and Epstein, S., 1958. Variations in isotopic composition of oxygen and carbon in Leadville Limestone (Mississippian, Colorado) and in its hydrothermal and metamorphic phases. *J. Geol.*, 66: 374—393.
Epstein, S., Graf, D.L. and Degens, E.T., 1964. Oxygen isotope studies on the origin of dolomites. In: H. Craig, S.L. Miller and G.J. Wasserburg (Editors), *Isotopic and Cosmic Chemistry* (dedicated to H.C. Urey on the occasion of his 70th birthday). North Holland, Amsterdam, pp. 169—180.
Evamy, B.D., 1963. The application of a chemical staining technique to a study of dedolomitization. *Sedimentology*, 2: 164—170.
Evamy, B.D., 1967. Dedolomitization and the development of rhombohedral pores in limestones. *J. Sediment. Petrol.*, 37: 1204—1215.
Fadeyev, F.I., 1962. Upper Carboniferous dolomite in the Kuybyshev area of the Volga. *Dokl. Acad. Nauk S.S.S.R.*, 147: 180—182.
Fagerstrom, J.A. and Burchett, R.R., 1972. Upper Pennsylvanian shoreline deposits from

Iowa and Nebraska — their recognition, variation, and significance. *Bull. Geol. Soc. Am.*, 83: 367—387.

Fairbridge, R.W., 1957. The dolomite question. In: R.J. LeBlanc and J.G. Breeding (Editors), *Regional Aspects of Carbonate Deposition — Soc. Econ. Paleontol. Mineral., Spec. Publ.*, 5: 125—178.

Fairbridge, R.W., 1967. Phases of diagenesis and authigenesis. In: G. Larsen and G.V. Chilingar (Editors), *Diagenesis in Sediments.* Elsevier, Amsterdam, pp. 19—89.

Folk, R.L., 1959. Practical petrographic classification of limestones. *Bull. Am. Assoc. Pet. Geol.*, 43: 1—38.

Folk, R.L., 1965. Recrystallization in ancient limestones. In: L.C. Pray and R.C. Murray (Editors), *Dolomitization and Limestone Diagenesis — Soc. Econ. Paleontol. Mineral., Spec. Publ.*, 13: 14—48.

Folk, R.L., 1970. Evidence of peritidal origin for part of the Caballos Novaculite (Devonian radiolarian chert), Marathon Basin, Texas (abstract). *Abstr. Prog. Geol. Soc. Am., Annu. Meet.*, 2: 552—553.

Folk, R.L. and Land, L.S., 1975. Mg/Ca ratio and salinity: two controls over crystallization of dolomite. *Bull. Am. Assoc. Pet. Geol.*, 59: 60—68.

Folk, R.L. and Siedlecka, A., 1974. The "schizohaline" environment: its sedimentary and diagenetic fabrics as exemplified by Late Paleozoic rocks of Bear Island, Svalbard. *Sediment. Geol.*, 11: 1—15.

Folkman, Y., 1969. Diagenetic dedolomitization in the Albian-Cenomanian Yagur Dolomite on Mount Carmel (northern Israel). *J. Sediment. Petrol.*, 39: 380—385.

Freeman, T., 1966. Postlithification dolomite in the Joachim and Plattin formations (Ordovician), northern Arkansas (abstract). *Geol. Soc. Am., Spec. Pap.*, 87: 59.

Freeman, T., 1972. Sedimentology and dolomitization of Muschelkalk carbonates (Triassic), Iberian Range, Spain. *Bull. Am. Assoc. Pet. Geol.*, 56: 434—453.

Friedman, G.M., 1965. Terminology of crystallization textures and fabrics in sedimentary rocks. *J. Sediment. Petrol.*, 35: 643—655.

Friedman, G.M., 1966. Occurrence and origin of Quaternary dolomite of Salt Flat, West Texas. *J. Sediment. Petrol.*, 36: 263—267.

Friedman, G.M. and Sanders, J.E., 1967. Origin and occurrence of dolostones. In: G.V. Chilingar, H.J. Bissell and R.W. Fairbridge (Editors), *Carbonate Rocks. Origin, Occurrence, and Classification.* Elsevier, Amsterdam, pp. 267—348.

Fritz, P., 1967. Oxygen and carbon isotopic composition of carbonates from the Jura of southern Germany. *Can. J. Earth Sci.*, 4: 1247—1267.

Fritz, P., 1971. Geochemical characteristics of dolomites and the δO^{18} content of Middle Devonian oceans. *Earth Planet. Sci. Lett.*, 11: 277—282.

Fritz, P. and Smith, D.G.W., 1970. The isotopic composition of secondary dolomites. *Geochim. Cosmochim. Acta*, 34: 1161—1173.

Fruth Jr., L.S., Orme, G.R. and Donath, F.A., 1966. Experimental compaction effects in carbonate sediments. *J. Sediment. Petrol.*, 36: 747—754.

Füchtbauer, H. and Goldschmidt, H., 1965. Beziehungen zwischen Calciumgehalt und Bildungsbedingungen der Dolomite. *Geol. Rundsch.*, 55: 29—40 (not seen).

Gaines, A.M., 1974. Protodolomite synthesis at 100°C and atmospheric pressure. *Science*, 183: 518—520.

Gaines, A.M. and Heffner, J.D., 1973. Dolomitization: a new experimental approach. *Abst. Prog. Geol. Soc. Am., Annu. Meet.*, 5: 628.

Garrels, R.M. and Thompson, M.E., 1962. A chemical model for sea water at 25°C and one atmosphere total pressure. *Am. J. Sci.*, 260: 57—66.

Garrett, P., 1970. Phanerozoic stromatolites: noncompetitive ecologic restriction by grazing and burrowing animals. *Science*, 169: 171—173.

Gebelein, C.D., 1969. Distribution, morphology, and accretion rate of Recent subtidal algal stromatolites, Bermuda. *J. Sediment. Petrol.*, 39: 46—69.
Gebelein, C.D. and Hoffman, P., 1971. Algal origin of dolomite in interlaminated limestone-dolomite sedimentary rocks. In: O.P. Bricker (Editor), *Carbonate Cements.* The Johns Hopkins Press, Baltimore, pp. 319—326.
Gebelein, C.D. and Hoffman, P., 1973. Algal origin of dolomite laminations in stromatolitic limestone. *J. Sediment. Petrol.*, 43: 603—613.
Ginsburg, R.N., 1957. Early diagenesis and lithification of shallow-water carbonate sediments in South Florida. In: R.J. LeBlanc and J.G. Breeding (Editors), *Regional Aspects of Carbonate Deposition — Soc. Econ. Paleontol. Mineral., Spec. Publ.*, 5: 80—100.
Glover, E.D. and Sippel, R.F., 1967. Synthesis of magnesium calcites. *Geochim. Cosmochim. Acta,* 31: 603—613.
Goldsmith, J.R. and Graf, D.L., 1958. Structural and compositional variations in some natural dolomites. *J. Geol.*, 66: 678—693.
Goldsmith, J.R., Graf, D.L. and Heard, H.C., 1961. Lattice constants of the calcium-magnesium carbonates. *Am. Mineral.*, 46: 453—457.
Goodell, H.G. and Garman, R.K., 1969. Carbonate geochemistry of Superior Deep Test Well, Andros Island, Bahamas. *Bull. Am. Assoc. Pet. Geol.*, 53: 513—536.
Graf, D.L. and Goldsmith, J.R., 1955. Dolomite-magnesian calcite relations at elevated temperatures and CO_2 pressures. *Geochim. Cosmochim. Acta,* 7: 109—128.
Graf, D.L. and Goldsmith, J.R., 1956. Some hydrothermal syntheses of dolomite and protodolomite. *J. Geol.*, 64: 173—186.
Graf, D.L. and Lamar, J.E., 1955. Properties of calcium and magnesium carbonates and their bearing on some uses of carbonate rocks. *Econ. Geol.*, (Fiftieth Anniversary Vol.), pp. 639—713.
Greenfield, L.J., 1963. Metabolism and concentration of calcium and magnesium and precipitation of calcium carbonate by a marine bacterium. *N.Y. Acad. Sci. Ann.*, 109: 23—45.
Halla, F., Chilingar, G.V. and Bissell, H.J., 1962. Thermodynamic studies on dolomite formation and their geologic implications: an interim report. *Sedimentology,* 1: 296—303.
Hanshaw, B.B., Back, W. and Deike, R.G., 1971. A geochemical hypothesis for dolomitization by ground water. *Econ. Geology,* 66: 710—724.
Hatfield, C.B. and Rohrbacher, T.J., 1966. Dolomite-insoluble residue relationships in the Ten Mile Creek Dolomite (Middle Devonian) near Toledo, Ohio. *J. Sediment. Petrol.*, 36: 828—831.
Hewett, D.F., 1931. Geology and ore deposits of the Goodsprings Quadrangle, Nevada. *U.S. Geol. Surv. Prof. Pap.*, 162: 172 pp.
Hewett, D.G., 1928. Dolomitization and ore deposition. *Econ. Geol.*, 23: 821—863.
Hohlt, R.B., 1948. The nature and origin of limestone porosity. *Q. Colo. Sch. Mines,* 43 (4): 51 pp.
Holland, H.D., Kirsipu, T.V., Huebner, J.C. and Oxborgh, U.M., 1964. On some aspects of the chemical evolution of cave waters. *J. Geol.*, 72: 36—37.
Hsü, K.J., 1967. Chemistry of dolomite formation. In: G.V. Chilingar, H.J. Bissell and R.W. Fairbridge (Editors), *Carbonate Rocks. Origin, Occurrence, and Classification.* Elsevier, Amsterdam, pp. 169—191.
Hsü, K.J. and Siegenthaler, C., 1969. Preliminary experiments on hydrodynamic movement induced by evaporation and their bearing on the dolomite problem. *Sedimentology,* 12: 11—25.

Hudson, J.D., 1970. Algal limestones with pseudomorphs after gypsum from the Middle Jurassic of Scotland. *Lethaia*, 3: 11—40.

Ilich, M., 1974. Hydrothermal-sedimentary dolomite: the missing link? *Bull. Am. Assoc. Pet. Geol.*, 58: 1331—1347.

Illing, L.V., Wells, A.J. and Taylor, J.C.M., 1965. Penecontemporaneous dolomite in the Persian Gulf. In: L.C. Pray and R.C. Murray (Editors), *Dolomitization and Limestone Diagenesis — Soc. Econ. Paleontol. Mineral., Spec. Publ.*, 13: 89—111.

Ingerson, E., 1962. Problems of the geochemistry of sedimentary carbonate rocks. *Geochim. Cosmochim. Acta*, 26: 815—847.

Jones, B.F., 1961. Zoning of saline minerals at Deep Spring Lake. In: *Geological Survey Research, 1961: Short Papers in the Geologic and Hydrologic Sciences. — U.S. Geol. Surv., Prof. Pap.*, 424-B: 199—202.

Kahle, C.F., 1965. Possible roles of clay minerals in the formation of dolomite. *J. Sediment. Petrol.*, 35: 448—453.

Katz, A., 1971. Zoned dolomite crystals. *J. Geol.*, 79: 38—51.

Khvorova, I.V., 1958. *Atlas of Carbonate Rocks of the Middle and Upper Carboniferous of the Russian Platform.* Akad. Nauk. S.S.S.R., Moscow, 170 pp.

Kier, J.S., 1973. Primary subtidal dolomicrite from Baffin Bay, Texas (abstract). *Bull. Am. Assoc. Petr. Geol.*, 57: 788.

King, R.J., 1968. Permo-Triassic mineralization. In P.C. Sylvester-Bradley and T.D. Ford (Editors), *The Geology of the East Midlands.* Leicester University Press, Leicester, pp. 123—132.

Kinsman, D.J.J., 1969. Interpretation of Sr^{2+} concentrations in carbonate minerals and rocks. *J. Sediment. Petrol.*, 39: 486—508.

Kinsman, D.J.J., 1973. Dolomitization process in sabkha environment (abstract). *Bull. Am. Assoc. Pet. Geol.*, 57: 788—789.

Koch, D.L., 1970. Stratigraphy of the Upper Devonian Shell Rock Formation of North Central Iowa. *Iowa Geol. Surv. Rep. Invest.*, 10, 123 pp.

Krauskopf, K.B., 1967. *Introduction to Geochemistry.* McGraw-Hill, New York, N.Y., 721 pp.

Land, L.S., 1967. Diagenesis of skeletal carbonates. *J. Sediment. Petrol.*, 37: 914—930.

Land, L.S., 1973a. Contemporaneous dolomitization of Middle Pleistocene reefs by meteoric water, North Jamaica. *Bull Mar. Sci.*, 23: 64—92.

Land, L.S., 1973b. Holocene meteoric dolomitization of Pleistocene limestones, North Jamaica. *Sedimentology*, 20: 411—424.

Land, L.S. and Epstein, S., 1970. Late Pleistocene diagenesis and dolomitization, north Jamaica. *Sedimentology*, 14: 187—200.

Land, L.S. and Hoops, G.K., 1973. Sodium in carbonate sediments and rocks: a possible index to the salinity of diagenetic solutions. *J. Sediment. Petrol.*, 43: 614—617.

Land, L.S., Salem, M.R.I. and Morrow, D.W., 1975. Paleohydrology of ancient dolomites: geochemical evidence. *Bull. Am. Assoc. Pet. Geol.*, 59: 1602—1625.

Langmuir, D., 1971a. Particle size effect on the reaction goethite = hematite + water. *Am. J. Sci.*, 271: 147—156.

Langmuir, D., 1971b. The geochemistry of some carbonate ground waters in central Pennsylvania. *Geochim. Cosmochim. Acta*, 35: 1023—1045.

Laporte, L.F., 1967. Carbonate deposition near mean sea-level and resultant facies mosaic, Manlius Formation (Lower Devonian) of New York State. *Bull. Am. Assoc. Pet. Geol.*, 51: 73—101.

Leighton, M.W. and Pendexter, C., 1962. Carbonate rock types. In: W.E. Ham (Editor), *Classification of Carbonate Rocks — Am. Assoc. Pet. Geol., Mem.*, 1: 33—61.

Liebermann, O., 1967. Synthesis of dolomite. *Nature*, 213: 241—245.
Lindholm, R.C., 1969. Detrital dolomite in Onondaga Limestone (Middle Devonian) of New York: its implications to the "dolomite question". *Bull. Am. Assoc. Pet. Geol.*, 53: 1035—1042.
Lippmann, F., 1968. Synthesis of $BaMg(CO_3)_2$ (norsethite) at 20°C and the formation of dolomite in sediments. In: G. Müller and G.M. Friedman (Editors), *Recent Developments in Carbonate Sedimentology in Central Europe.* Springer, Heidelberg, pp. 33—37.
Lippmann, F., 1973. *Sedimentary Carbonate Minerals.* Springer, Berlin, 228 pp.
Logan, B.W., Rezak, R. and Ginsburg, R.N., 1964. Classification and environmental significance of algal stromatolites. *J. Geol.*, 72: 68—83.
Lovering, T.S., 1969. The origin of hydrothermal and low-temperature dolomite. *Econ. Geol.*, 64: 743—754.
Lowenstam, H.A., 1961. Mineralogy, $^{18}O/^{16}O$ ratios and magnesium contents of recent and fossil brachiopods and their bearing on the history of the oceans. *J. Geol.*, 69: 241—260.
Lucia, F.J., 1968. Recent sediments and diagenesis of South Bonaire, Netherlands Antilles. *J. Sediment. Petrol.*, 38: 845—858.
Lucia, F.J., 1972. Recognition of evaporite-carbonate shoreline sedimentation. In: J.K. Rigby and W.K. Hamblin (Editors), *Recognition of Ancient Sedimentary Environments — Soc. Econ. Paleontol. Mineral., Spec. Publ.*, 16: 160—191.
Lumsden, D.N., 1973. Dolomite, limestone facies, and insoluble residue — a relationship? (Abstract). *Bull. Am. Assoc. Pet. Geol.*, 57: 791.
Lumsden, D.N., 1974. Relationships among insoluble residue, dolostone, and limestone facies. *J. Sediment. Petrol.*, 44: 450—455.
Mattavelli, L., Chilingarian, G.V. and Storer, D., 1969. Petrography and diagenesis of the Taormina Formation, Gela Oil Field, Sicily, Italy. *Sediment. Geol.*, 3: 59—86.
Matter, A., 1967. Tidal flat deposits in the Ordovician of western Maryland. *J. Sediment. Petrol.*, 37: 601—609.
McCauley, J.W. and Roy, R., 1966. Evidence for epitaxial control of $CaCO_3$ phase formation as the mechanism of the influence of impurity ions (abstract). *Am. Geophys. Union Trans.*, 47: 202—203.
McKee, E.O. and Gutschick, R.C., 1969. History of Redwall Limestone of northern Arizona. *Geol. Soc. Am. Mem.*, 114: 1—726.
Miller Jr., D.N., 1961. Early diagenetic dolomite associated with salt extraction process, Inagua, Bahamas. *J. Sediment. Petrol.*, 31: 473—476.
Miller Jr., D.N., 1966. Diagenesis of sedimentary rocks. In: *Symposium on Recently Developed Geologic Principles and Sedimentation of the Permo-Pennsylvanian of the Rocky Mountains — Wyo. Geol. Assoc., 12th Annu. Conf.*, pp. 17—21k.
Mossler, J.H., 1971. Diagenesis and dolomitization of Swope Limestone (Upper Pennsylvanian) southeast Kansas. *J. Sediment. Petrol.*, 41: 961—970.
Mottl, M.J. and Holland, H.D., 1975. Basalt—sea water interaction, sea-floor spreading, and the dolomite problem (abstract). *Am. Geophys. Union Trans.*, 56: 1074.
Mueller, G., 1960. The theory of formation of north Chilean nitrate deposits through "capillary concentration". *Int. Geol. Congr., 21st, Copenhagen, Rep. Sess., Norden*, 1: 79—86.
Muffler, L.J.P. and White, D.E., 1969. Active metamorphism of Upper Cenozoic sediments in the Salton Sea geothermal field and the Salton Trough, southeastern California. *Bull. Geol. Soc. Am.*, 80: 157—182.
Müller, G. and Fischbeck, R., 1973. Possible natural mechanism for protodolomite formation. *Nat. Phys. Sci.*, 242: 139—141.

Müller, G. and Tietz, G., 1966. Recent dolomitization of Quaternary biocalcarenites from Fuerteventura (Canary Islands). *Contrib. Mineral. Petrol.*, 13: 89—96.

Müller, G., Irion, G. and Förstner, V., 1972. Formation and diagenesis of inorganic Ca—Mg carbonates in the lacustrine environment. *Naturwissenschaften*, 59: 158—164.

Murata, K.J., Friedman, I. and Cremer, M., 1972. Geochemistry of diagenetic dolomites in Miocene marine formations of California and Oregon. *U.S. Geol. Surv., Prof. Pap.*, 724—C: 12 pp.

Murray, R.C., 1969. Hydrology of south Bonaire, N.A. A rock selective dolomitization model. *J. Sediment. Petrol.*, 39: 1007—1013.

Murray, R.C. and Lucia, F.J., 1967. Cause and control of dolomite distribution by rock selectivity. *Geol. Soc. Am. Bull.*, 78: 21—36.

Newell, N.D., Rigby, S.K., Fischer, A.G., Whiteman, A.J., Hickox, J.E. and Bradley, J.S., 1953. *The Permian Reef Complex of the Guadalupe Mountains Region, Texas and New Mexico.* Freeman, San Francisco, Calif., 236 pp.

Nichols, K.M., 1974. Coextensive supratidal dolomite and underlying secondary dolomite in the Triassic of north-central Nevada. *J. Sediment. Petrol.*, 44: 783—789.

Osmond, J.C., 1956. Mottled carbonate rocks in the Middle Devonian of eastern Nevada. *J. Sediment. Petrol.*, 26: 32—41.

Parsons, L.M., 1918. Dolomitization and the Leicestershire dolomites. *Geol. Mag.*, 5: 246—258.

Peterson, M.N.A., 1962. The mineralogy and petrology of Upper Mississippian carbonate rocks of the Cumberland Plateau in Tennessee. *J. Geol.*, 70: 1—31.

Peterson, M.N.A., Von der Borch, C.C. and Bien, G.S., 1966. Growth of dolomite crystals. *Am. J. Sci.*, 264: 257—272.

Pilkey, O.H. and Noble, D., 1966. Carbonate and clay mineralogy of the Persian Gulf. *Deep-sea Res.*, 13: 1—16.

Powers, R.W., 1962. Arabian Upper Jurassic carbonate reservoir rocks. In: W.E. Ham (Editor), *Classification of Carbonate Rocks — Am. Assoc. Pet. Geol., Mem.*, 1: 122—192.

Pray, L.C., 1960. Compaction in calcilutites (abstract). *Bull. Geol. Soc. Am.*, 71: 1946.

Read, H.H., 1948. Granites and granites. In: J. Gilluly (Editor), *Origin of Granite—Geol. Soc. Am., Mem.*, 28: 1—19.

Rieke III, H.H. and Chilingarian, G.V., 1974. *Compaction of Argillaceous Sediments.* Elsevier, Amsterdam, 424 pp.

Rittenberg, S.C., Emery, K.O., Hülsemann, J., Degens, E.T., Fay, R.C., Reuter, J.H. Grady, J.R., Richardson, S.H. and Bray, E.E., 1963. Biogeochemistry of sediments in Experimental Mohole. *J. Sediment. Petrol.*, 33: 140—172.

Roedder, E., 1968. Temperature, salinity, and origin of the ore-forming fluids at Pine Point, Northwest Territories, Canada, from fluid inclusion studies. *Econ. Geol.*, 63: 439—450.

Roedder, E., 1971. Fluid-inclusion evidence on the environment of formation of mineral deposits of the southern Appalachian Valley. *Econ. Geol.*, 66: 777—791.

Roehl, P.W., 1967. Stony Mountain (Ordovician) and Interlake (Silurian) facies analogs of Recent low-energy marine and subaerial carbonates, Bahamas. *Bull. Am. Assoc. Pet. Geol.*, 51: 1979—2032.

Sarkisyan, S.G., Politykina, M.A. and Chilingarian, G.V., 1973. Effect of postsedimentation processes on carbonate reservoir rocks in Volga-Urals Region, U.S.S.R. *Bull. Am. Assoc. Pet. Geol.*, 57 (7): 1305—1313.

Sass, E., 1965. Dolomite—calcite relationships in sea water: theoretical considerations and preliminary experimental results. *J. Sediment. Petrol.*, 35: 339—347.

Sass, E., 1969. Microphotometric determination of preferred orientation in undeformed dolomites. *Science*, 165: 802—803.
Schenk, P.E., 1967. The Macumber Formation of the Maritime Provinces, Canada. A Mississippian analogue to Recent strand-line carbonates of the Persian Gulf. *J. Sediment. Petrol.*, 37: 365—376.
Schlanger, S.O., 1957. Dolomite growth in coralline algae. *J. Sediment. Petrol.*, 27: 181—186.
Schmidt, V., 1965. Facies, diagenesis, and related reservoir properties in the Gigas Beds (Upper Jurassic), northwestern Germany. In: L.C. Pray and R.C. Murray (Editors), *Dolomitization and Limestone Diagenesis — Soc. Econ. Paleontol. Mineral., Spec. Publ.*, 13: 124—168.
Schroeder, J.H., Dwornik, E.J. and Papike, J.J., 1969. Primary protodolomite in echinoid skeletons. *Bull. Geol. Soc. Am.*, 80: 1613—1616.
Seibold, E., 1962. Kalk-Konkretionen und karbonatisch gebundenes Magnesium. *Geochim. Cosmochim. Acta*, 26: 899—909.
Shearman, D.J., Khouri, J. and Taha, S., 1961. On the replacement of dolomite by calcite in some Mesozoic limestones from the French Jura. *Geol. Assoc. London, Proc.*, 72: 1—12.
Shinn, E.A., 1968. Selective dolomitization of some Recent sedimentary structures. *J. Sediment. Petrol.*, 38: 612—616.
Shinn, E.A. and Ginsburg, R.N., 1964. Formation of Recent dolomite in Florida and the Bahamas (abstract). *Bull. Am. Assoc. Pet. Geol.*, 48: 547.
Shinn, E.A., Ginsburg, R.N. and Lloyd, R.M., 1965. Recent supratidal dolomite from Andros Island, Bahamas. In: L.C. Pray and R.C. Murray (Editors), *Dolomitization and Limestone Diagenesis — Soc. Econ. Paleontol. Mineral., Spec. Publ.*, 13: 112—123.
Shinn, E.A., Lloyd, R.M. and Ginsburg, R.N., 1969. Anatomy of a modern carbonate tidal flat, Andros Island, Bahamas. *J. Sediment. Petrol.*, 39: 1201—1228.
Shinn, E.A., Halley, R.B., Hudson, J.H. and Lidz, B.H., 1977. Limestone compaction: an enigma. *Geology*, 5: 21—24.
Siegel, F.R., 1961. Factors influencing the precipitation of dolomite carbonates. *Kans. Geol. Surv., Bull.*, 152: 127—158.
Skinner, H.C.W., 1963. Precipitation of calcian dolomites and magnesian calcites in the southeast of South Australia. *Am. J. Sci.*, 261: 449—472.
Smit, D.E. and Swett, K., 1969. Devaluation of "dedolomitization". *J. Sediment. Petrol.*, 39: 379—380.
Sonnenfeld, P., 1964. Dolomites and dolomitization: A review. *Can. Pet. Geol. Bull.*, 12: 101—132.
Spotts, J.H. and Silverman, S.R., 1966. Organic dolomite from Point Fermin, California. *Am. Mineral.*, 51: 1144—1155.
Steidtmann, E., 1911. Evolution of limestone and dolomite. *J. Geol.*, 19: 323—345; 393—428.
Strakhov, N.M., 1953. Diagenesis of sediments and its significance for sedimentary ore formation. *Izv. Akad. Nauk S.S.S.R., Ser. Geol.*, 5: 12—49.
Strakhov, N.M., 1956a. Types and genesis of dolomite rocks. In: *Types of Dolomite Rocks and their Genesis. Akad. Nauk S.S.S.R.*, Moscow, pp. 5—27.
Strakhov, N.M., 1956b. Distribution and genesis of Upper Carboniferous dolomite rocks of Samarskaya Luka. In: *Types of Dolomite Rocks and their Genesis.* Akad. Nauk S.S.S.R., Moscow, pp. 185—208.
Strakhov, N.M., 1958. *Méthodes d'Etude des Roches sédimentaires.* Bur. Rech. Géol. Géophys. Minières, Paris, pp. 535—542.

Strakhov, N.M., 1959 (Editor). *Toward Knowledge of Diagenesis of Sediments.* Akad. Nauk S.S.S.R., Moscow, 295 pp.
Strakhov, N.M., 1960. *Principles of Theory of Lithogenesis, 1. Types of Lithogenesis and their Distribution on the Earth's Surface.* Akad. Nauk S.S.S.R., Moscow, 212 pp.
Strakhov, N.M., Brodskaya, N.G., Knyazeva, L.M., Razzhivina, A.N., Rateev, M.A., Sapozhnikov, D.G. and Shishova, E.S., 1954. *Formation of Sediments in Recent Basins.* Izd. Akad. Nauk S.S.S.R., Moscow, 791 pp.
Tatarskiy, V.B., 1949. About the occurrence of rocks in which dolomite is replaced by calcite. *Dokl. Akad. Nauk S.S.S.R.*, 49: 849—851.
Teodorovich, G.I., 1950. *Lithology of Carbonate Rocks. Paleozoic of Ural-Volga Region.* Akad. Nauk S.S.S.R., Moscow, 213 pp.
Teodorovich, G.I., 1955. Towards the question of formation of sedimentary calcareous-dolomite rocks. *Tr. Inst. Nefti, Akad. Nauk S.S.S.R.*, 5: 75—107.
Teodorovich, G.I., 1958. *Study of Sedimentary Rocks.* Gostoptekhizdat, Leningrad, 572 pp.
Teodorovich, G.I., 1961. *Authigenic Minerals in Sedimentary Rocks.* Consultants Bureau, New York, N.Y., 120 pp.
Textoris, D.A., 1969. Supratidal origin of Appalachian Basin dolostone (abstract). *Geol. Soc. Am., Spec. Pap.*, 121: 470—471.
Thomas, G.E., 1962. Grouping of carbonate rocks into textural and porosity units for mapping purposes. In: W.E. Ham (Editor), *Classification of Carbonate Rocks — Am. Assoc. Pet. Geol., Mem.*, 1: 193—223.
Thompson, A.M., 1970. Tidal-flat deposition and early dolomitization in Upper Ordovician rocks of southern Appalachian Valley and Ridge. *J. Sediment. Petrol.*, 40: 1271—1286.
Thraikill, J., 1968. Dolomite cave deposits from Carlsbad Caverns. *J. Sediment. Petrol.*, 38: 141—145.
Usdowski, H.-E., 1968. The formation of dolomite in sediments. In: G. Müller and G.M. Friedman (Editors), *Recent Developments in Carbonate Sedimentology in Central Europe.* Springer, Heidelberg, pp. 21—32.
Valyashko, M.G., 1962. *Geochemical Regularities in the Formation of Potassium Salt Deposits.* Akad. Nauk, S.S.S.R., Moscow, 397 pp.
Van Tuyl, F.M., 1916. The origin of dolomite. *Iowa Geol. Surv. Annu. Rep.*, 25: 251—422.
Vernon, R.O., 1969. Geology and hydrology associated with a zone of high permeability (Boulder zone) in Florida. *Soc. Min. Eng.*, Preprint 69-AG-12: 24 pp.
Vishnyakov, S.G., 1951. Genetic types of dolomite rocks. *Dokl. Akad. Nauk S.S.S.R.*, 76 (1): 112—113.
Von der Borch, C.C., Lock, D.E. and Schwebel, D., 1975. Ground-water formation of dolomite in the Coorong region of South Australia. *Geology*, 3: 283—285.
Weaver, C.E., 1975. Construction of limpid dolomite. *Geology*, 3: 425—428.
Weber, J.N., 1964. Trace-element composition of dolostones and dolomites and its bearing on the dolomite problem. *Geochim. Cosmochim. Acta*, 28: 1817—1868.
West, I.M., Brandon, A. and Smith, M., 1968. A tidal-flat evaporitic facies in the Visean of Ireland. *J. Sediment. Petrol.*, 38: 1079—1093.
Wilson, J.L., 1975. *Carbonate Facies in Geologic History.* Springer, Berlin, 471 pp.
Winland, H.D., 1969. Stability of calcium carbonate polymorphs in warm, shallow seawater. *J. Sediment. Petrol.*, 39: 1579—1587.
Wolfbauer, C.A. and Surdam, R.C., 1974. Origin of nonmarine dolomite in Eocene Lake Gosiute, Green River Basin, Wyoming. *Bull. Geol. Soc. Am.*, 85: 1733—1740.

Wood, G.V. and Armstrong, A.K., 1975. Diagenesis and stratigraphy of the Lisburne Group Limestones of the Sadlerochit Mountains and adjacent areas of northeastern Alaska. *U.S. Geol. Surv. Prof. Pap.*, 857: 47 pp.

Yanat'eva, O.K., 1950. The solubility of dolomite in aqueous salt solutions. *Izv. Sekt. Fiz. Khim. Anal., Inst. Obsh. Neorgan. Khim., Akad. Nauk S.S.S.R.*, 20: 252--268.

Yanat'eva, O.K., 1954. About physical-chemical characteristics of some carbonate rocks. *Dokl. Akad. Nauk S.S.S.R.*, 96: (4): 777—779.

Yanat'eva, O.K., 1955. Effect of aqueous solutions of gypsum on dolomite in the presence of carbon dioxide. *Dokl. Akad. Nauk S.S.S.R.*, 101: 911—912.

Yanat'eva, O.K. 1957. On the solubility polytherm of the system ($CaCO_3 + MgSO_4 = CaSO_4 + MgCO_3$) — H_2O. *Proc. Acad. Sci. U.S.S.R.* (Chem. Sect., English transl.), 1957: 155—157.

Zenger, D.H., 1965a. Calcite-dolomite ratios vs. insoluble content in the Lockport Formation (Niagaran) in New York State. *J. Sediment. Petrol.*, 35: 262—265.

Zenger, D.H., 1965b. Stratigraphy of the Lockport Formation (Middle Silurian) in New York State. *N.Y. State Mus. Sci. Serv. Bull.*, 404: 210 pp.

Zenger, D.H., 1969. Stratigraphy of the Little Falls Formation (Upper Cambrian) in the type area, east-central New York (abstract). *Abstr. Prog. Geol. Soc. Am., Northeast. Sect.*, 1: 67—68.

Zenger, D.H., 1971. Uppermost Clinton (Middle Silurian) stratigraphy and petrology, East-Central New York. *N.Y. State Mus. Sci. Serv. Bull.*, 417: 58 pp.

Zenger, D.H., 1972a. Significance of supratidal dolomitization in the geologic record. *Bull. Geol. Soc. Am.*, 83: 1—12.

Zenger, D.H., 1972b. Dolomitization and uniformitarianism. *J. Geol. Educ.*, 20: 107—124.

Zenger, D.H., 1972c. Diagenesis of peritidal carbonates: Little Falls Formation (Upper Cambrian), east-central New York (abstract). *Abstr. Programs Geol. Soc. Am., Northeast. Sect.*, 4: 55.

Zenger, D.H., 1973. Syntaxial calcite borders on dolomite crystals, Little Falls Formation (Upper Cambrian), New York. *J. Sediment. Petrol.*, 43: 118—124.

Zenger, D.H., 1976. Dolomitization and dolomite "dikes" in the Wyman Formation (Precambrian), northeastern Inyo Mountains, California. *J. Sediment. Petrol.*, 46: 457—462.

Zenger, D.H. and Pearson, E.F., 1969. Stratigraphy and petrology of Lost Burro Formation, Panamint Range, California. In: *Short Contributions to California Geology — Calif. Div. Mines Geol., Spec. Rept.*, 100: 45—65.

ZoBell, C.E., 1942. Changes produced by micro-organisms in sediments after deposition. *J. Sediment. Petrol.*, 12: 127—136.

ZoBell, C.E., 1946. Studies on redox potential of marine sediments. *Bull. Am. Assoc. Pet. Geol.*, 40: 477—513.

References added in proof

Craig, H., 1957. Isotopic standards for carbon and oxygen and correction factors for mass-spectrometric analysis of carbon dioxide. *Geochim. Cosmochim. Acta*, 12: 133—149.

Harbaugh, J.W., 1967. Carbonate oil reservoir rocks. In: G.V. Chilingar, H.J. Bissell and R.W. Fairbridge (Editors), *Carbonate Rocks: Origin, Occurrence, and Classification. Develop. Sedimentol.*, 9A — Elsevier, Amsterdam, pp. 349—398.

Jodry, R.L., 1969. Growth and dolomitization of Silurian reefs, St. Clair County, Michigan. *Bull. Am. Assoc. Pet. Geol.*, 53: 957—981.

Süjkowski, Zb.L., 1958. Diagenesis. *Bull. Am. Assoc. Pet. Geol.*, 42: 2692—2717.

AUTHOR INDEX

Numbers in italics refer to references lists

SUBJECT INDEX *

---

* Prepared by Mr. Nai-Hwei Ying, Miss Christine Estrada, and Dr. George V. Chilingar.